Contributions to Statistics

V. Fedorov / W. G. Müller / I. N. Vuchkov (Eds.) Model-Oriented Data Analysis, XII/248 pages. 1992

J. Antoch (Ed.) Computational Aspects of Model Choice, VII / 285 pages, 1993

W. G. Müller / H. P. Wynn / A. A. Zhigljavsky (Eds.) Model-Oriented Data Analysis XIII/287 pages, 1993

Petr Mandl
Marie Hušková (Eds.)

Asymptotic Statistics

Proceedings of the Fifth Prague Symposium,
held from September 4-9, 1993

With 25 Figures

Springer-Verlag Berlin Heidelberg GmbH

Series Editors
Werner A. Müller
Peter Schuster

Editors
Professor Petr Mandl
Professor Marie Hušková
Department of Probability and
Mathematical Statistics
Charles University
Sokolovská 83
CZ-186 00 Praha 8
Czech Republic

ISBN 978-3-7908-0770-7

CIP-Titelaufnahme der Deutschen Bibliothek
Asymptotic statistics : proceedings of the fifth Prague
symposium, held from September 4 - 9, 1993 / [Prague
Symposia on Asymptotic Statistics]. Petr Mandl ; Marie
Hušková (ed.). - Heidelberg : Physica-Verl., 1994
(Contributions to statistics)
ISBN 978-3-7908-0770-7 ISBN 978-3-642-57984-4 (eBook)
DOI 10.1007/978-3-642-57984-4
NE: Mandl, Petr [Hrsg.]; Symposium on Asymptotic Statistics <5, 1993,
Praha>

Originally published by Physica-Verlag Heidelberg in 1994

2201/2202-543210 - Printed on acid-free paper

Preface

The Prague Symposia on Asymptotic Statistics represent a twenty years' tradition of contacts between Czech mathematical statisticians and the conference participants. Both, as the organizers hope, return from the Symposia to their work with fresh ideas and new information.

The Fifth Prague Symposium was held from September 4 to September 9, 1993 at the Faculty of Mathematics and Physics, Charles University. It was sponsored by the Bernoulli Society for Mathematical Statistics and Probability, the Czech Statistical Society, the Czech Society of Actuaries, Česká Pojišťovna–Insurance and Reinsurance Corporation, and the IFIP WG 7.7. Asymptotic Statistics, a prolific source of methodological concepts, dominated the program of the Symposium. Special sessions were devoted to Mathematics of Insurance and Finance and to Stochastic Programming.

The papers presented at the Symposium are published in two parts. Part 1 is this volume. Part 2 is Number 3, Volume 30 (1994) of the journal Kybernetika, comprising the papers of the authors who were not able to meet the early editorial deadline.

The editors of the Proceedings would like to express their sincere thanks to the authors for valuable contributions, to the reviewers for prompt and careful reading the papers, to J. Antoch for his advice with technical part of the Proceedings. Finally they also express their appreciation of the kind cooperation with the Publishing House Physica - Verlag and the journal Kybernetika in bringing out the volumes.

Part of the Proceedings was typeset by $\mathcal{AMS}$-TEX, the macrosystem of the American Mathematical Society.

Praha, January 1994

Marie Hušková, Petr Mandl

Contents

Invited papers

Contributed papers

Contributions in a special issue of the journal Kybernetika

Beirlant J. and Győrfi L., *Pitman efficiencies of L_p-goodness-of-fit tests*

Teräsvirta T., *Testing linearity and modelling nonlinear time series*

Aguirre N. and Nikulin M., *Chi-squared goodness-of-fit test for the family of logistic distribution*

Browne J., *A comparison of algorithms to filter noisy observations of a linear differential system driven by Brownian motion and a simple Markov switching process*

Chadoeuf J. and Beneš V., *On some estimation variances in spatial statistics*

Hála M., *Method of Ritz for random eigenvalue problems*

Hanousek J., *Bayesian-type estimators. Robust and sensitivity analyses*

Hennecart F., *Stirling distributions and Stirling numbers of the second kind. Computational problems in statistics*

Mizera I., *On consistent M-estimators: tuning constants, unimodality and breakdown*

Nikulin M. and Voinov V., *On power series, Bell polynomials, Hardy-Rademacher problem and its statistical applications*

Schwabe R., *On Bather's stochastic approximation algorithm*

Štěpán J., *A general bounded continuous moment problem and its set of uniqness*

Vajda I. and Österreicher F., *Statistical analysis and applications of log-optimal investments*

Zwanzig S., *On adaptive estimation in nonlinear regression*

Invited Papers

PROCEDURES FOR THE DETECTION OF MULTIPLE CHANGES IN SERIES OF INDEPENDENT OBSERVATIONS

Jaromír ANTOCH, Marie HUŠKOVÁ
Charles University, Prague

ABSTRACT. The paper concerns the problem of detection and identification of one or more change points in a series of independent observations. Procedures based on M-estimators and M-residuals are introduced, their limit properties studied and typical simulation results presented.

1990 MATHEMATICS SUBJECT CLASSIFICATION: 62F35, 62G10, 62F03

KEY WORDS AND PHRASES: Change point problem, M-estimators, M-residuals, abrupt changes, continuous change, MOSUM.

1. INTRODUCTION

Let $X_1, \ldots, X_n$ be independent random variables with distribution functions $F(x;\theta_0 + \delta_n r_{n,1}), \ldots, F(x;\theta_0 + \delta_n r_{n,n})$, where θ_0 and δ_n are unknown parameters. As to the constants $r_{n,1}, \ldots, r_{n,n}$, they are also unknown, however, fulfil

$$\int_0^1 \left(r_{n,\lfloor un \rfloor} - r(u)\right)^2 du \to 0 \qquad \text{as} \qquad n \to \infty, \tag{1.1}$$

where $r(u)$ is a function on $(0,1)$ belonging to a certain class of functions, e. g. class of functions continuous on $(0,1)$ or class of jump functions etc., and $\lfloor \cdot \rfloor$ denotes the function integer part.

If $r_{n,i} = 0$, $i = 1, \ldots, n$ (hence $r(u) = 0$, $u \in (0,1)$), then $X_1, \ldots, X_n$ are independent identically distributed (i. i. d.) random variables and there is no change point. On the other hand, if $r(u)$ is continuous with $\int_0^1 r^2(u)\,du > 0$, we speak of a continuous (or gradual) change in the model. If $r(u)$ has some jumps, we speak on abrupt changes in the model and points with jumps call change points.

Great attention was paid to the case with the detection of at most one abrupt change. More information can be found in one of several survey papers on this topic, see Shaban (1980), Zacks (1983, 1991), Broemling and Tsurumi (1986), Krishnaiah and Miao (1988), Csörgő and Horváth (1988), Hušková and Sen (1989) or Antoch and Hušková (1992) among others.

The problem with more than one change was studied less intensively. Wolfe and Schechtman (1984) developed procedures based on ranks. Krishnaiah and Miao (1988) mentioned some procedures to estimate multiple change points. Lombard (1987) proposed a class of Bayesian type rank statistics for detection both abrupt and continuous changes. Koziol (1987), Darkhovskii and Brodskii (1987) and Barry and Hartigan (1993) constructed a Bayesian type test procedures to detect multiple abrupt changes.

In this paper:

- three different classes of procedures based on M-residuals studied by Hušková (1993) for the case of at most one change are examined to find how they reflect continuous and abrupt changes;
- a new procedure based on differences of moving sums is introduced and its properties studied.

The paper is organized as follows. At the end of this section the assumptions we work with are formulated. The test and estimation procedures are described and studied in Section 2. Section 3. discusses in more details how different test statistics reflect various types of alternatives. Properties of the estimators discussed in the paper are summarized in Section 4. Typical simulation results are surveyed in Section 5. Technical theorems and sketches of their proofs are postponed to Section 6. General conclusions can be found throughout the paper.

The assumptions we use throughout the paper read as follows:

A.1. Random variables $X_1, \dots, X_n$ are independent and X_i, $i = 1, \dots, n$, has distribution function $F(x; \theta_0 + \delta_n r_{n,i})$, $x \in R_1$, $\theta_0 \in \Theta \subseteq R_1$, Θ open, $\delta_n \in R_1$ and $r_{n,1}, \dots, r_{n,n}$ fulfil (1.1).

A.2. The score function $\psi(x; \theta)$, $x \in R_1$, $\theta \in \Theta$, is nondecreasing in the second argument,

$$\int \psi(x; \theta_0)\, dF(x; \theta_0) = 0$$

and

$$0 < \int \psi^2(x; \theta_0)\, dF(x; \theta_0) < \infty.$$

A.3. There exist positive constants D_1, D_2 and β such that

$$\int \Big(\psi(x; \theta_0 + t_1) - \psi(x; \theta_0 + t_2)\Big)^2 dF(x; \theta_0 + t_3) \le D_1 |t_1 - t_2|^\beta,$$

$$|\, t_i \,| \le D_2, \; i = 1, 2, 3.$$

A.4. There exist positive constants λ, D_3, D_4 and $0 < \eta \le 1$ such that

$$\Big| \int \Big(\psi(x; \theta_0 + t_1) - \psi(x; \theta_0 + t_2)\Big)\, dF(x; \theta_0 + t_3) - \lambda(t_1 - t_2) \Big| \le$$

$$\le D_3 |t_1 - t_2|^{1+\eta}, \qquad |\, t_i \,| \le D_4, \; i = 1, 2, 3.$$

A.5. There exist positive constants D_5 and D_6 such that (for η from A.4.)

$$\Big| \int \psi(x; \theta_0)\, d\Big(F(x; \theta_0 + t) - F(x; \theta_0)\Big) + \lambda t \Big| \le D_5 |t|^{1+\eta}, \qquad |\, t \,| \le D_6.$$

A.6. There exist positive constants κ, D_7 and D_8 such that

$$\int \Big|\psi(x; \theta_0 + t)\Big|^{2+\kappa} dF(x; \theta_0) \le D_7, \qquad |\, t \,| \le D_8.$$

A.7. $G/n \to 0$ and $G^{-1} n^{2/(4+\kappa)} \log n \to 0$ as $n \to \infty$ (for κ from A.6.).

2. TEST PROCEDURES

Considering the test problem of at most one change, i. e. test of the hypothesis

$$H_0 : r_{n,i} = 0, \qquad i = 1, \dots, n,$$

against

$$H_1 : \text{ there exists } 1 \le m < n \text{ such that}$$

$$(0 =) r_{n,1} = \dots = r_{n,m} \ne r_{n,m+1} = \dots = r_{n,n},$$

Hušková (1993) studied the test procedures based on the following statistics:

$$T_{n,1}(\psi) = \max_{1 \le k < n} \left\{ \frac{1}{\sqrt{n}} \left| S_{n,k}(\psi) \right| \right\}, \tag{2.1}$$

$$T_{n,2}(\psi) = \max_{1 \le k < n} \left\{ \sqrt{\frac{n}{k(n-k)}} \left| S_{n,k}(\psi) \right| \right\}, \tag{2.2}$$

$$\begin{aligned} T_{n,3}(\psi) &= \max_{G < k \le n} \left\{ \frac{1}{\sqrt{G}} \left| S_{n,k}(\psi) - S_{n,k-G}(\psi) \right| \right\} = \\ &= \max_{G < k \le n} \left\{ \frac{1}{\sqrt{G}} \left| \sum_{i=k-G+1}^{k} \psi(X_i; \theta_n) \right| \right\}, \end{aligned} \tag{2.3}$$

where G satisfies A.7.,

$$S_{n,k}(\psi) = \sum_{i=1}^{k} \psi\Big(X_i; \widehat{\theta}_n(\psi)\Big), \qquad k = 1, \dots, n, \tag{2.4}$$

ψ is score function fulfilling assumptions formulated above and $\widehat{\theta}_n(\psi)$ is the M-estimator generated by the function ψ and defined as any solution of the equation

$$\sum_{i=1}^{n} \psi(X_i; \theta) = 0.$$

Throughout $\psi\Big(X_i; \widehat{\theta}_n(\psi)\Big)$, $i = 1, \dots, n$, are called M-residuals. Statistics (2.1) and (2.2) are of the maximum likelihood type, while (2.3) is based on the maximum of moving partial sums. Aside them, we shall consider a new class of test statistics based on the maximum of differences of moving sums

$$\begin{aligned} T_{n,4}(\psi) &= \max_{2G < k \le n} \left\{ \frac{1}{\sqrt{2G}} \left| S_{n,k}(\psi) - 2S_{n,k-G}(\psi) + S_{n,k-2G}(\psi) \right| \right\} = \\ &= \max_{2G < k \le n} \left\{ \left| \frac{1}{\sqrt{2G}} \sum_{i=k-2G+1}^{k} \psi(X_i; \widehat{\theta}_n) - \frac{1}{\sqrt{2G}} \sum_{i=k-2G+1}^{k} \psi(X_i; \widehat{\theta}_n) \right| \right\}. \end{aligned} \tag{2.5}$$

First we establish in Theorem 2.1. the representations for M-estimator $\widehat{\theta}_n(\psi)$ and test statistics $T_{n,1}(\psi)$ and $T_{n,2}(\psi)$. The analogous results for test statistics $T_{n,3}(\psi)$ and $T_{n,4}(\psi)$ are given in Theorem 2.2.

Theorem 2.1. *Let Assumptions A.1.–A.5. be satisfied. Moreover, let*

$$\delta_n \to 0 \qquad \text{as} \qquad n \to \infty$$

and

$$\int_0^1 r^2(u)\,du < \infty.$$

Then, as $n \to \infty$,

$$\widehat{\theta}_n(\psi) - \theta_0 = \delta_n \bar{r}_n - \frac{Z_{n,n}(\psi)}{\lambda n} + O_p\left(\frac{B_n}{\sqrt{n}}\right), \tag{2.6}$$

$$\max_{1\le k<n}\left\{\left|\frac{1}{\sqrt{n}}\Big(S_{n,k}(\psi) - Z_{n,k}(\psi)\Big) - \lambda\delta_n\sqrt{n}\,q_{n,1}(k)\right|\right\} = O_p(B_n) \tag{2.7}$$

and

$$\max_{1\le k<n}\left\{\left|\sqrt{\frac{n}{k(n-k)}}\Big(S_{n,k}(\psi) - Z_{n,k}(\psi)\Big) - \lambda\delta_n\sqrt{n}\,q_{n,2}(k)\right|\right\} = \tag{2.8}$$

$$= O_p\left(B_n + |\delta_n|^{1+\kappa}\max_{1\le k<n}\frac{1}{\sqrt{k}}\sum_{i=1}^{k}|r_{n,i} - \bar{r}_n|^{1+\kappa}\right),$$

where

$$\bar{r}_n = \frac{1}{n}\sum_{i=1}^{n} r_{n,i}\,,$$

$$Z_{n,k}(\psi) = \sum_{i=1}^{k}\Big(\psi(X_i;\theta_0) - E\,\psi(X_i;\theta_0)\Big), \qquad k = 1,\dots,n, \tag{2.9}$$

$$B_n = \Big(\max(|\delta_n|, n^{-1/2})\Big)^{\beta/2}\log\log n +$$

$$+ \sqrt{n}\Big(\max(|\delta_n|, n^{-1/2})\Big)^{1+\eta} + \frac{\sqrt{n}}{\log n}\max(|\delta_n|, n^{-1/2}),$$

$$q_{n,1}(k) = \frac{1}{n}\sum_{i=1}^{k}\Big(r_{n,i} - \bar{r}_n\Big), \qquad k = 1,\dots,n-1,$$

and

$$q_{n,2}(k) = \frac{1}{\sqrt{k(n-k)}}\sum_{i=1}^{k}\Big(r_{n,i} - \bar{r}_n\Big) = \frac{n}{\sqrt{k(n-k)}}\,q_{n,1}(k), \qquad k = 1,\dots,n-1.$$

Proof. To show (2.6) we proceed similarly as in the proof of Theorem 2.1 in Hušková (1993) using Theorem 6.1 (6.1–6.3) of the present paper instead of Theorem 5.2 in Hušková (1993).

By (2.6) and assumptions A.4. and A.5. we have

$$(2.10)\qquad \frac{1}{\sqrt{k}}\sum_{i=1}^{k}\int \psi\left(x;\widehat{\theta}_n(\psi)\right)\, dF(x;\theta_0+\delta_n r_{n,i}) =$$

$$= \frac{\lambda}{\sqrt{k}}\sum_{i=1}^{k}\left(\widehat{\theta}_n(\psi)-\theta_0-\delta_n r_{n,i}\right)+O_p\left(\frac{1}{\sqrt{k}}\sum_{i=1}^{k}\left(|\widehat{\theta}_n(\psi)-\theta_0|^{1+\kappa}+|\delta_n r_{n,i}|^{1+\kappa}\right)\right)=$$

$$= \frac{1}{\sqrt{k}}\left\{-\frac{k}{n}Z_{n,n}(\psi)-\delta_n\lambda\sum_{i=1}^{k}(r_{n,i}-\overline{r}_n)\right\}+O_p\left(B_n+\frac{|\delta_n|^{1+\kappa}}{\sqrt{k}}\sum_{i=1}^{k}|r_{n,i}-\overline{r}_n|^{1+\kappa}\right)$$

uniformly in k.

Combining (2.6) together with Theorem 6.1. we receive, as $n\to\infty$,

$$(2.11)\qquad \max_{1\le k<n}\left\{\frac{1}{\sqrt{k}}\left|\sum_{i=1}^{k}\left(\psi(X_i;\widehat{\theta}_n(\psi))-\psi(X_i;\theta_0)\right)-\right.\right.$$

$$\left.\left.-\int\left(\psi(x;\widehat{\theta}_n(\psi))-\psi(x;\theta_0)\right)dF(x;\theta_0+\delta_n r_{n,i})\right|\right\}=$$

$$=O_p\left(\left(\max(|\delta_n|,n^{-1/2})\right)^{\beta/2}\log\log n+\frac{\sqrt{n}}{\log n}\max(|\delta_n|,n^{-1/2})\right).$$

Assertion (2.7) now follows immediately from (2.10)–(2.11).
Since

$$(2.12)\qquad \sum_{i=1}^{k}\psi\left(X_i;\widehat{\theta}_n(\psi)\right)=-\sum_{i=k+1}^{n}\psi\left(X_i;\widehat{\theta}_n(\psi)\right),\qquad k=1,\dots,n-1,$$

and since assertions (2.10) and (2.11) remain true if summation for $1\le i\le k$ is replaced by summation for $k+1\le i\le n$, we conclude that also (2.8) holds true.
□

Theorem 2.2. *Let the assumptions of Theorem 2.1. and Assumptions A.6. and A.7. be satisfied. Then, as* $n\to\infty$,

$$(2.13)\qquad \max_{G<k\le n}\left\{\left|\frac{1}{\sqrt{G}}\left(\left[S_{n,k}(\psi)-S_{n,k-G}(\psi)\right]-\left[Z_{n,k}(\psi)-Z_{n,k-G}(\psi)\right]\right)-\right.\right.$$

$$\left.\left.-\lambda\delta_n\sqrt{G}q_{n,3}(k,G)\right|\right\}=O_p\left(B_n(G)+\frac{|\delta_n|^{1+\kappa}}{\sqrt{G}}\max_{G<k\le n}\sum_{i=k-G+1}^{k}|r_{n,i}-\overline{r}_n|^{1+\kappa}\right)$$

and

$$(2.14)\qquad \max_{2G<k\le n}\left\{\left|\frac{1}{\sqrt{2G}}\left(\left[S_{n,k}(\psi)-2S_{n,k-G}(\psi)+S_{n,k-2G}(\psi)\right]-\right.\right.\right.$$

$$-\Big[Z_{n,k}(\psi) - 2Z_{n,k-G}(\psi) + Z_{n,k-2G}(\psi)\Big]\Big) - \lambda\delta_n\sqrt{G}\,q_{n,4}(k,G)\Big|\Big\} =$$

$$= O_p\Big(B_n(G) + \frac{|\delta_n|^{1+\kappa}}{\sqrt{G}} \max_{G<k\le n} \sum_{i=k-G+1}^{k} |r_{n,i} - \overline{r}_n|^{1+\kappa}\Big),$$

where $Z_{n,k}(\psi)$ is given by (2.9),

$$B_n(G) = \Big(\max(|\delta_n|, n^{-1/2})\Big)^{\beta/(4+\kappa)} \log\frac{G}{n} + \sqrt{G}\,\Big(\max(|\delta_n|, n^{-1/2})\Big)^{1+\eta} +$$

$$+ \sqrt{\frac{G}{n}} + \frac{\log n}{\sqrt{G}} \max(|\delta_n|, n^{-1/2}),$$

(2.15) $$q_{n,3}\,(k,G) = \frac{1}{G} \sum_{i=k-G+1}^{k} \Big(r_{n,i} - \overline{r}_n\Big), \qquad k = G,\dots,n,$$

and

(2.16) $$q_{n,4}\,(k,G) = \frac{1}{G}\left(\sum_{i=k-G+1}^{k} r_{n,i} - \sum_{i=k-2G+1}^{k-G} r_{n,i}\right), \qquad k = 2G,\dots,n.$$

Proof. Assertion (2.13) follows in a standard way from (6.5) and the following relation:

$$\frac{1}{\sqrt{G}} \sum_{i=k-G+1}^{k} \int \psi\left(x; \widehat{\theta}_n(\psi)\right) dF(x; \theta_0 + \delta_n r_{n,i}) =$$

$$= \frac{1}{\sqrt{G}} \sum_{i=k-G+1}^{k} \Big\{ \int \Big(\psi(x;\widehat{\theta}_n(\psi)) - \psi(x;\theta_0)\Big)\, dF(x;\theta_0 + \delta_n r_{n,i}) +$$

$$+ \int \psi(x;\theta_0)\, d\Big(F(x;\theta_0 + \delta_n r_{n,i}) - F(x;\theta_0)\Big)\Big\} =$$

$$= \frac{1}{\sqrt{G}} \sum_{i=k-G+1}^{k} \Big\{ \Big(\widehat{\theta}_n(\psi) - \theta_0\Big)\,\lambda + O_p\Big(|\widehat{\theta}_n(\psi) - \theta_0|^{1+\kappa}\Big) -$$

$$- \delta_n r_{n,i} + O_p\Big(|\delta_n|^{1+\kappa} \cdot |r_{n,i}|^{1+\kappa}\Big)\Big\} =$$

$$= -\frac{1}{\sqrt{G}} \sum_{i=k-G+1}^{k} (r_{n,i} - \overline{r}_n) + O_p\left(\sqrt{\frac{G}{n}} + \frac{|\delta_n|^{1+\kappa}}{\sqrt{G}} \sum_{i=k-G+1}^{k} |r_{n,i} - \overline{r}_n|^{1+\kappa}\right).$$

□

From theorems 2.1, 2.2 and 6.2 we obtain the limit distribution of $T_{n,j}(\psi)$, $j = 1,\dots,4$, both under the null and local alternatives. Particularly, for the test problem

$$H_0 : r_{n,i} = 0, \qquad i = 1,\dots,n,$$

against

$$H_{n,1} : \text{ at least some } r_{n,i} \text{ are nonzero,}$$

we receive the following critical regions with (asymptotic) level α:

$$T_{n,1}(\psi) \geq \widehat{\sigma}_n(\psi) \cdot b_\alpha, \tag{2.17}$$

$$T_{n,2}(\psi) \geq \widehat{\sigma}_n(\psi) \cdot b\left(-\log\log\frac{1}{\sqrt{1-\alpha}}, \log n\right), \tag{2.18}$$

$$T_{n,3}(\psi) \geq \widehat{\sigma}_n(\psi) \cdot b\left(-\log\log\frac{1}{\sqrt{1-\alpha}}, \frac{n}{G}\right), \tag{2.19}$$

$$T_{n,4}(\psi) \geq \widehat{\sigma}_n(\psi) \cdot b\left(-\log\log\frac{1}{\sqrt{1-\alpha}}, \frac{3n}{2G} - 3\right), \tag{2.20}$$

b_α fulfils

$$P\Big(\max_{t\in(0,1)} |B(t)| > b_\alpha\Big) = \alpha$$

with $\{B(t),\, t \in (0,1)\}$ being the Brownian bridge, $\widehat{\sigma}_n^2(\psi)$ is an estimator of

$$\sigma^2(\psi) = \int \psi^2(x;\theta_0)\, dF(x;\theta_0) \tag{2.21}$$

and

(2.22)

$$b(z,y) = \frac{1}{\sqrt{2\log z}}\left(y + 2\log z + \frac{1}{2}\log\log z - \frac{1}{2}\log \pi\right), \qquad y \in R_1,\ z > 1.$$

The usual choice of the estimator of $\sigma^2(\psi)$ is

$$\widehat{\sigma}_n^2(\psi) = \frac{1}{n}\sum_{i=1}^{n}\left(\psi\Big(X_i;\widehat{\theta}_n(\psi)\Big) - \frac{1}{n}\sum_{j=1}^{n}\psi\Big(X_j;\widehat{\theta}_n(\psi)\Big)\right)^2. \tag{2.23}$$

3. CONTINUOUS AND ABRUPT CHANGES

We shall examine how the test statistics $T_{n,j}(\psi)$, $j = 1,\dots,4$, reflect various types of alternatives. It reduces to the study of the drift functions $q_{n,1}(k)$, $q_{n,2}(k)$, $q_{n,3}(k,G)$ and $q_{n,4}(k,G)$.

a) Consider the continuous change of the form

$$r_{n,i} = r\left(\frac{i}{n+1}\right), \qquad i = 1,\dots,n, \tag{3.1}$$

where $r(u)$ is a square-integrable, nonconstant and continuous function on (0,1). Then we have, as $n \to \infty$,

$$q_{n,1}(k) = \int_0^{k/(n+1)} \left(r(u) - \overline{r}\right) du + o\left(\frac{\sqrt{k(n-k)}}{n}\right), \tag{3.2}$$

$$q_{n,2}(k) = \frac{n}{\sqrt{k(n-k)}} \int_0^{k/(n+1)} \left(r(u) - \overline{r}\right) du + o(1), \tag{3.3}$$

$$q_{n,3}(k,G) = r\left(\frac{k}{n+1}\right) - \overline{r} + o(1), \tag{3.4}$$

$$q_{n,4}(k,G) = o(1), \tag{3.5}$$

uniformly in k, where $\overline{r} = \int_0^1 r(u)\, du$.

b) Consider the case of abrupt changes of the form

$$r_{n,i} = a_j, \qquad b_{j-1} < \frac{i}{n} \leq b_j, \qquad j = 1, \dots, L, \tag{3.6}$$

where $b_0 = 0 < b_1 < \cdots < b_{L-1} < b_L = 1$, $a_{j-1} \neq a_j$, $j = 1, \dots, L$, i.e. $\lfloor nb_1 \rfloor, \dots, \lfloor nb_{L-1} \rfloor$ are change points. Then, as $n \to \infty$,

$$q_{n,1}(k) = q_1^*\left(\frac{k}{n}\right) - \overline{q}^* + O\left(\frac{1}{n}\right), \tag{3.7}$$

$$q_{n,2}(k) = \frac{n}{\sqrt{k(n-k)}} \left(q_1^*\left(\frac{k}{n}\right) - \overline{q}^* + O\left(\frac{1}{n}\right)\right), \tag{3.8}$$

$$q_{n,3}(k,G) = q_{n,3}^*\left(\frac{k}{n}\right) - \overline{q}^* + O\left(\frac{1}{G}\right), \tag{3.9}$$

$$q_{n,4}(k,G) = q_{n,4}^*\left(\frac{k}{n}\right) + O\left(\frac{1}{G}\right), \tag{3.10}$$

uniformly in k, where

$$\begin{aligned}
q_1^*(u) &= \sum_{v=1}^{j-1} a_v (b_v - b_{v-1}) + a_j (u - b_{j-1}), && b_{j-1} < u \leq b_j,\ 1 \leq j \leq L,\\
q_{n,3}^*(u) &= a_j, && b_{j-1} < u \leq b_j,\ 1 \leq j \leq L,\\
&= a_j + (a_{j+1} - a_j)(u - b_j)\frac{n}{G} && b_j < u \leq b_j + \frac{G}{n},\ 1 \leq j \leq L-1,\\
q_{n,4}^*(u) &= 0, && b_{j-1} + \frac{G}{2} < u \leq b_j,\ 1 \leq j \leq L,\\
&= a_j + (a_{j+1} - a_j)(u - b_j)\frac{n}{G} && b_j < u \leq b_j + \frac{G}{n},\ 1 \leq j \leq L-1,\\
&= (a_{j+1} - a_j)\left(b_j + \frac{2G}{n} - u\right)\frac{n}{G}, && b_j + \frac{G}{n} < u \leq b_j + \frac{2G}{n},\ 1 \leq j \leq L-1.
\end{aligned}$$

Relations (3.2)–(3.5) and (3.7)–(3.10) together with Theorems 2.1, 2.2, and (2.17)–(2.20) lead to the following conclusions:

- The tests based on $T_{n,1}(\psi)$, $T_{n,2}(\psi)$ and $T_{n,3}(\psi)$ are sensitive to continuous as well as to abrupt changes. On the other hand $T_{n,4}(\psi)$ is sensitive to abrupt changes but not to the continuous ones.
- Asymptotically, $T_{n,1}(\psi)$ should distinguish alternatives with $\delta_n = O(n^{-1/2})$ and $T_{n,2}(\psi)$ with $\delta_n = O\left(n^{-1/2}\sqrt{\log\log n}\right)$. On the other hand, $T_{n,3}(\psi)$ and $T_{n,4}(\psi)$ should distinguish alternatives with $\delta_n = O\left(G^{-1/2}\sqrt{\log(n/G)}\right)$.
- $T_{n,1}(\psi)$ is more sensitive to changes *in the middle*, while $T_{n,2}(\psi)$ is more sensitive to changes at the *beginning* and *at the end* of the observational period.
- The drift function q_1^* is a piece wise linear, $q_{n,3}^*$ is constant on the intervals $(b_{j-1} + G/n, b_j]$ and linear with the slope $(a_{j+1} - a_j)n/G$ on interval $(b_{j-1}, b_{j-1} + G/n]$, $j = 1, \dots, L$. The drift function $q_{n,4}^*$ is zero on the intervals $(b_{j-1} + 2G/n, b_j]$ and linear (nonconstant) on the intervals $(b_j, b_j + G/n]$, $(b_j + G/n + 2G/n]$, $j = 1, \dots, L-1$.
- Statistics

$$\frac{1}{\sqrt{2G}}\left|S_{n,k}(\psi) - 2S_{n,k-G}(\psi) + S_{n,k-2G}(\psi)\right|, \qquad k = 2G, \dots, n,$$

reflect in a transparent way abrupt changes.

Here is one of the possible recommendations how to proceed:

- If the test procedure based on $T_{n,1}(\psi)$ or $T_{n,2}(\psi)$ lead to the rejection of H_0 but not that based on $T_{n,4}(\psi)$, we conclude that there are only continuous (gradual) changes but not abrupt changes.
- If the test procedure based on $T_{n,4}(\psi)$ rejects H_0, we conclude that there is at least one abrupt change.
- To decide whether there is one or more changes, we calculate local maxima of

$$\frac{1}{\sqrt{2G}}\left|S_{n,k}(\psi) - 2S_{n,k-G}(\psi) + S_{n,k-2G}(\psi)\right|, \qquad k = 2G, \dots, n,$$

and compare them with the critical value $b\left(-\log\log(1/\sqrt{1-\alpha}), (3n/2G) - 3\right)$. If only the global maximum exceeds the critical value, we make conclusion that there is only one abrupt change, otherwise we decide for more changes.

4. ESTIMATORS

Here we focus on the problem to estimate the unknown parameters in case of abrupt changes described by (3.6).

We start with estimators of change points $\lfloor nb_1 \rfloor, \dots, \lfloor nb_{L-1} \rfloor$ and proceed as follows:

- Find

$$\widehat{m}_1(\psi, G) = \underset{k \in \{G, \dots, n\}}{\arg\max}\left\{\left|S_{n,k}(\psi) - 2S_{n,k-G}(\psi) + S_{n,k-2G}(\psi)\right|\right\} - G.$$

- Generally, if we have $\widehat{m}_j(\psi, G)$, $j = 1, \ldots, v$, and find that

$$\max_{k \notin \cup_{j=1}^{v}(\widehat{m}_j(\psi)-G,\, \widehat{m}_j(\psi)+G)} \Big\{ \Big| S_{n,k}(\psi) - 2S_{n,k-G}(\psi) + S_{n,k-2G}(\psi) \Big| \Big\} \le$$

$$\le \sqrt{2G}\, \widehat{\sigma}_n(\psi)\, b\Big(-\log\log \frac{1}{\sqrt{1-\alpha}}, \frac{3n}{2G} - 3 \Big),$$

we stop the procedure and $m_j(\psi, G)$, $j = 1, \ldots, v$, are the estimators of the change points. In the opposite case, we find

$$\widehat{m}_{v+1}(\psi) = \arg\max_{k \notin \cup_{j=1}^{v}(\widehat{m}_j(\psi)-G,\, \widehat{m}_j(\psi)+G)} \Big\{ |S_{n,k}(\psi) - 2S_{n,k-G}(\psi) + S_{n,k-2G}(\psi)| \Big\}$$

and repeat the procedure with v replaced by $v+1$.

Denote by J_n the total number of detected change points. The properties of these estimators are the following:

Theorem 4.1. *Let Assumptions A.1. – A.7. be satisfied and let, as* $n \to \infty$, $\delta_n \log n \to 0$, $G|\delta_n|^{\beta/(4+\kappa)} \log n \to \infty$ *and let* $r_{n,i}$*'s are defined by (3.6). Then*

$$P(J_n = L-1) \to 1 \qquad as \qquad n \to \infty.$$

If $J_n = L-1$, *then there exists a permutation* $\{s_1, \ldots, s_{L-1}\}$ *of numbers* $\{1, \ldots, L-1\}$ *such that*

$$\delta_n^2 \Big(\widehat{m}_j(\psi, G) - nb_{s_j} \Big) = O_p(1) \qquad as \qquad n \to \infty, \tag{4.1}$$

and

$$\Big(\frac{\lambda \delta_n}{\sigma(\psi)} \Big)^2 \Big(\widehat{m}_j(\psi, G) - nb_{s_j} \Big) \xrightarrow{\mathcal{D}} Q_j \qquad as \qquad n \to \infty,\ j = 1, \ldots, L-1,$$

where $\widehat{m}_1, \ldots \widehat{m}_{L-1}$ *are estimators introduced above with* $\alpha = \alpha_n$, $\alpha_n \to 0$, $n\alpha_n \to \infty$, $\sigma^2(\psi)$ *and* λ *are defined by (2.23) and A.4., respectively.* $Q_1, \ldots, Q_{L-1}$ *are i. i. d. random variables defined by*

$$Q_j = \arg\max_{z \in R_1} \Big\{ W(6z) - |z| \Big\},$$

where

$$W(z) = W_1(-z), \qquad z \le 0,$$
$$= W_2(z), \qquad z > 0,$$

with W_1 *and* W_2 *being independent standard Wiener processes on* $(0, \infty)$.

Proof. It is omitted since it is quite close to that of Theorem 3.2. in Hušková (1993). □

Now we turn to the problem to estimate the amount of changes and $\sigma^2(\psi)$. For simplicity we shall assume that $\widehat{m}_1(\psi,G) < \widehat{m}_2(\psi,G) < \cdots < \widehat{m}_{L-1}(\psi,G)$. Denote by $\widehat{\theta}_{j-1,j}(\psi)$ any solution of the equation

$$\sum_{i=\widehat{m}_{j-1}(\psi,G)+1}^{\widehat{m}_j(\psi,G)} \psi(X_i,t) = 0, \qquad j = 1,\dots,L,$$

where $\widehat{m}_0(\psi,G) = 0$, $\widehat{m}_L(\psi,G) = n$, i.e. $\widehat{\theta}_{j-1,j}(\psi)$ is the M-estimator with the score function ψ based on $X_{\widehat{m}_{j-1}(\psi,G)+1},\dots,X_{\widehat{m}_j(\psi,G)}$. We introduce the estimator $\widehat{D}_j$, $j = 1,\dots,L-1$, of the amount of changes $D_j = (a_j - a_{j-1})\delta_n$, $j = 1,\dots,L-1$, as

$$\widehat{D}_j = \widehat{\theta}_{j,j+1}(\psi) - \widehat{\theta}_{j-1,j}(\psi), \qquad j = 1,\dots,L-1.$$

Using standard tools it can be shown that these estimators are $\sqrt{n}$-consistent and asymptotically normally distributed.

At last we discuss possible estimators of $\sigma^2(\psi)$ defined by (2.21). Under both the null and contiguous alternatives the estimator

$$\widetilde{\sigma}_n^2(\psi) = \frac{1}{n}\sum_{i=1}^{n} \psi^2\Big(X_i; \widehat{\theta}_n(\psi)\Big)$$

is consistent. If

$$\left|\int \psi^2(x;\theta_0)\, d\Big(F(x;\theta_0) - F(x;\theta_0 + \delta_n \max_{1\le i\le n} r_{n,i})\Big)\right| = O\left(|\delta_n|^{\beta/2}\right),$$

where $\beta > 0$ is from A.3., then $\widetilde{\sigma}_n^2(\psi)$ is $|\delta_n|^{\beta/2}$-consistent also for local alternatives, i.e. $\|\delta_n\|^2 \to 0$. However, the simulation study suggests to apply rather

$$\overline{\sigma}_n^2(\psi) = \frac{1}{n-L+1}\sum_{j=1}^{L-1}\sum_{i=\widehat{m}_{j-1}(\psi,G)+1}^{\widehat{m}_j(\psi,G)} \left(\psi(x_i;\widehat{\theta}_{j-1,j}(\psi)) - \overline{\psi}_j\right)^2,$$

where

$$\overline{\psi}_j = \frac{1}{\widehat{m}_j(\psi;G) - \widehat{m}_{j-1}(\psi;G)} \sum_{\widehat{m}_{j-1}(\psi;G)+1}^{\widehat{m}_j(\psi;G)} \psi\Big(X_i;\widehat{\theta}_{j-1,j}(\psi)\Big)$$

see Hawkins (1989), e.g., who has the same consistency property as $\widetilde{\sigma}_n(\psi)$.

5. SIMULATION EXPERIMENT

To illustrate the behavior of the procedures described above, we prepared a small Monte Carlo simulation experiment. Typical simulation results are presented in the sequel.

The basic data model consists of $n = 100$ observations with two change points, the first one being between the 35th and 36th observation and the second one between the 60th and 61st observation. Three pairs of changes are considered:

- jumps of different size in opposite directions, we put $\delta_1 = 3.5$ and $\delta_2 = -1.5$;
- jumps of the same size in the same direction, we put $\delta_1 = 3$ and $\delta_2 = 3$;
- jumps of different size in the same direction, we put $\delta_1 = 2$ and $\delta_2 = 4$.

The errors follow standard normal distribution $N(0,1)$. Respective data sets, denoted as Data I. – Data III., are graphically summarized in Figures 1. – 3. Standard Huber's ψ–function with the constant $k = 1.7$ was used.

Instead of the presentation of many tables full of numerical results we have decided to present several typical results in graphical form and to add concise comments on them.

In figures 4. – 6. and 7. – 9. the statistics

$$V\_k,1 = \frac{1}{\sqrt{n}} \left| S_{n,k}(\psi) \right|, \qquad k = 1,\dots,n,$$

and

$$V\_k,2 = \sqrt{\frac{n}{k(n-k)}} \left| S_{n,k}(\psi) \right|, \qquad k = 1,\dots,n-1,$$

respectively, are plotted for all three data sets (they are related to $T_{n,1}(\psi)$ and $T_{n,2}(\psi)$, respectively). As for the figures related to $T_{n,3}(\psi, G)$ and $T_{n,4}(\psi, G)$, respectively, we plot

$$V\_k,3 = \frac{1}{\sqrt{G}} \left| S_{n,k}(\psi) - S_{n,k-G}(\psi) \right|, \qquad k = G+1,\dots,n,$$

$$V^*\_k,3 = \frac{1}{\sqrt{G}} \left(S_{n,k}(\psi) - S_{n,k-G}(\psi) \right), \qquad k = G+1,\dots,n,$$

$$V\_k,4 = \frac{1}{\sqrt{2G}} \left| S_{n,k}(\psi) - 2S_{n,k-G}(\psi) + S_{n,k-2G}(\psi) \right|, \qquad k = 2G+1,\dots,n,$$

$$V^*\_k,4 = \frac{1}{\sqrt{2G}} \left(S_{n,k}(\psi) - 2S_{n,k-G}(\psi) + S_{n,k-2G}(\psi) \right), \qquad k = 2G+1,\dots,n,$$

in Figures 10. – 19. for all three data sets. Notice that $V^*\_k,3$ and $V^*\_k,4$ reflect changes in the model in a more transparent way than $V\_k,3$ and $V\_k,4$; compare the shape of Figures 11. and 13. or 16. and 17., respectively.

Figures 10. – 12. show the influence of the increase of the window length G on statistics $V^*\_k,3$. As expected, the increase of the width of the MOSUM window G results in "smoothing" of the values of statistics $V^*\_k,3$ and points out the problem which has not been still solved satisfactorily, i. e. the problem of the optimal choice of G. The effect on the statistics $V^*\_k,4$ is of the same type.

Generally said, all statistics under the consideration, and especially $V^*\_k,3$ and $V^*\_k,4$, reflect the changes in the model according to our expectation.

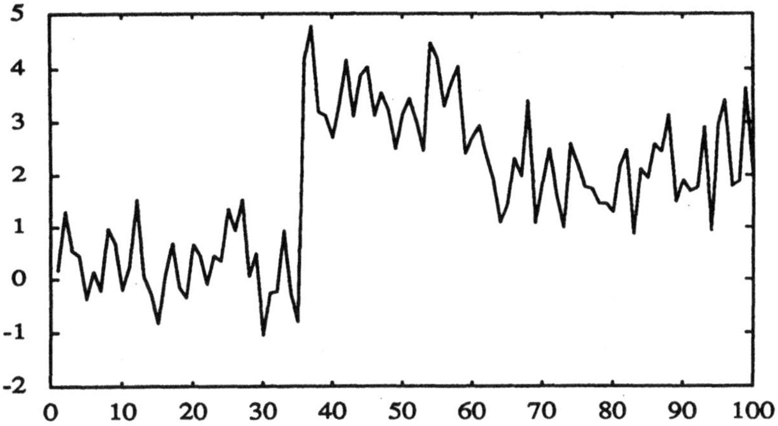

Figure 1. Data I.

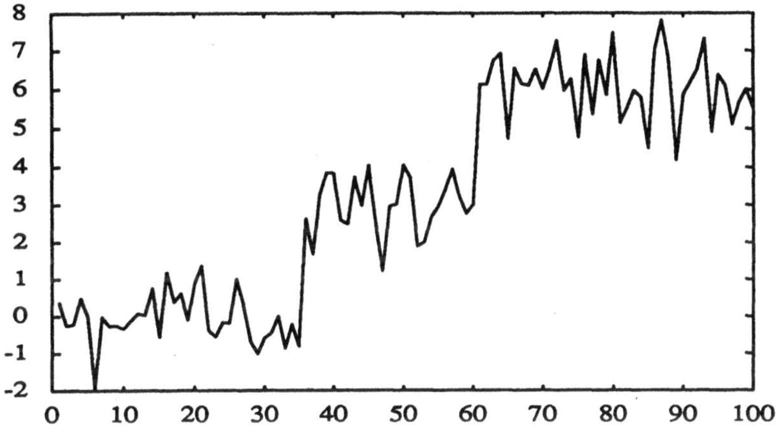

Figure 2. Data II.

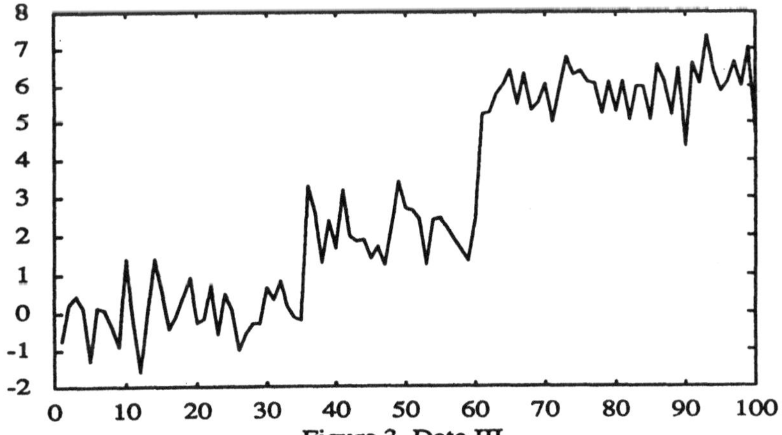

Figure 3. Data III.

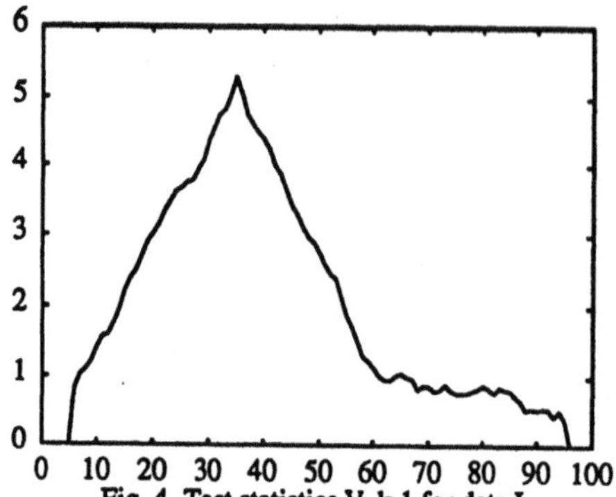

Fig. 4. Test statistics V_k,1 for data I.

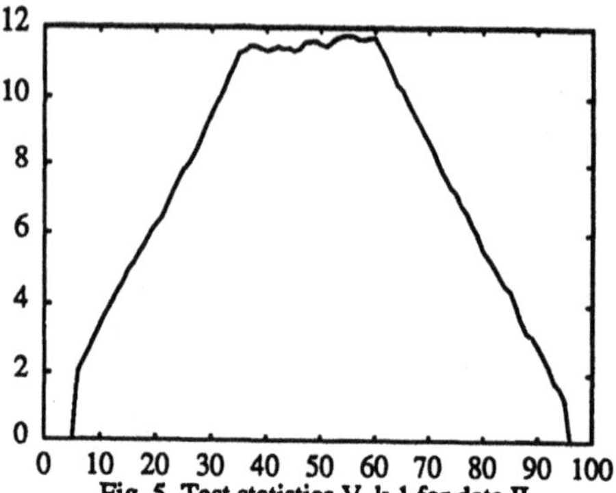

Fig. 5. Test statistics V_k,1 for data II.

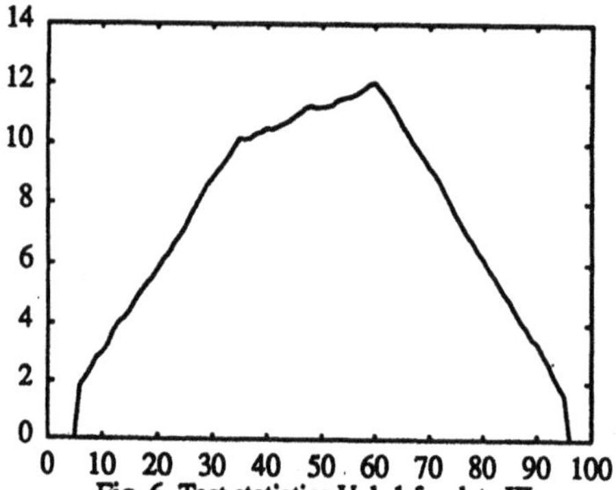

Fig. 6. Test statistics V_k,1 for data III.

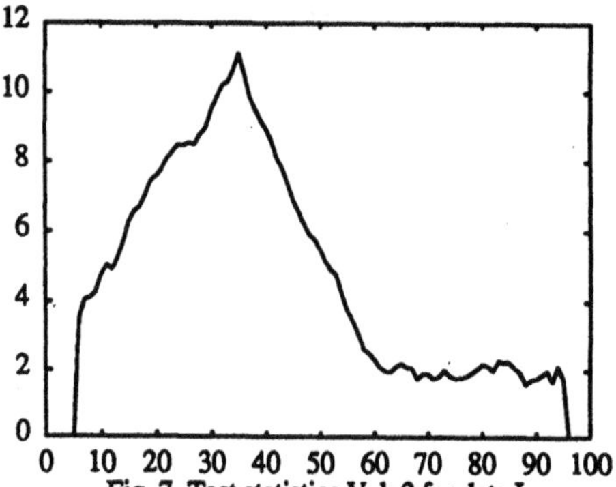

Fig. 7. Test statistics V_k,2 for data I.

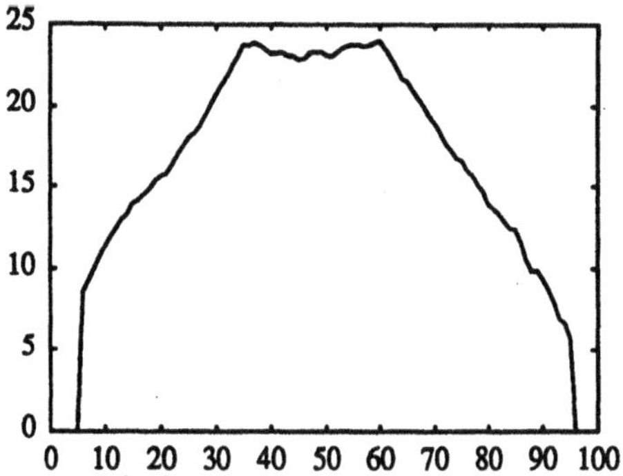

Fig. 8. Test statistics V_k,2 for data II.

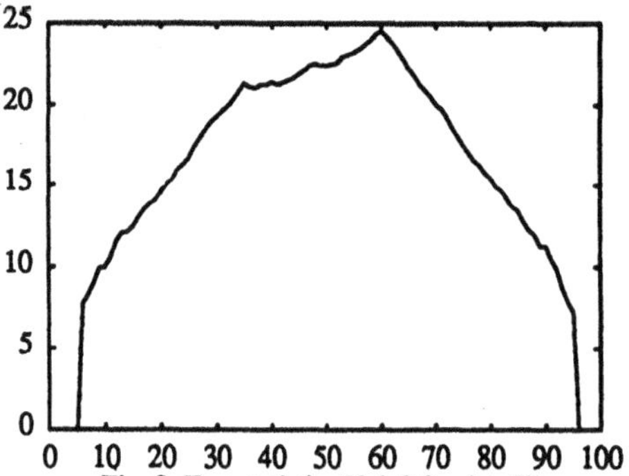

Fig. 9. Test statistics V_k,2 for data III.

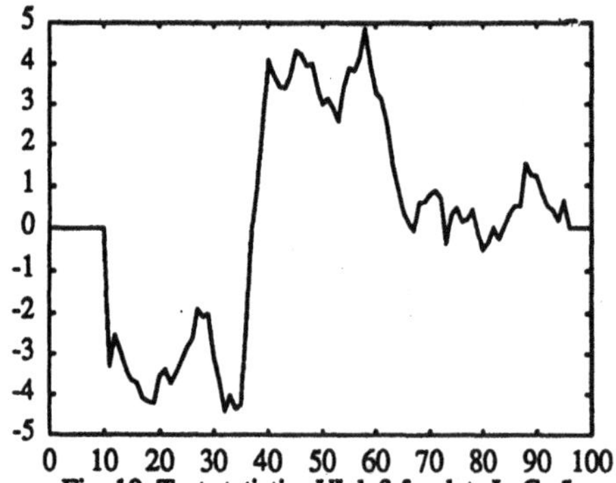

Fig. 10. Test statistics V"_k,3 for data I., G=5.

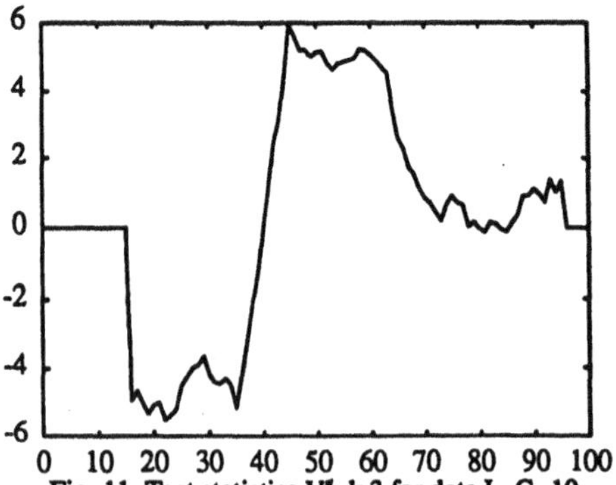

Fig. 11. Test statistics V"_k,3 for data I., G=10.

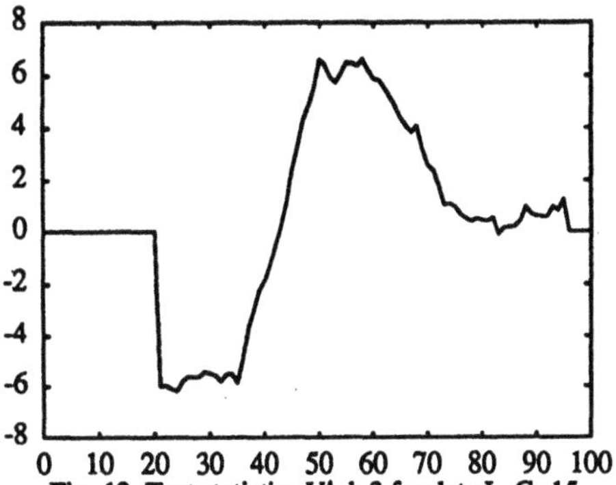

Fig. 12. Test statistics V^{A}_k,3 for data I., G=15.

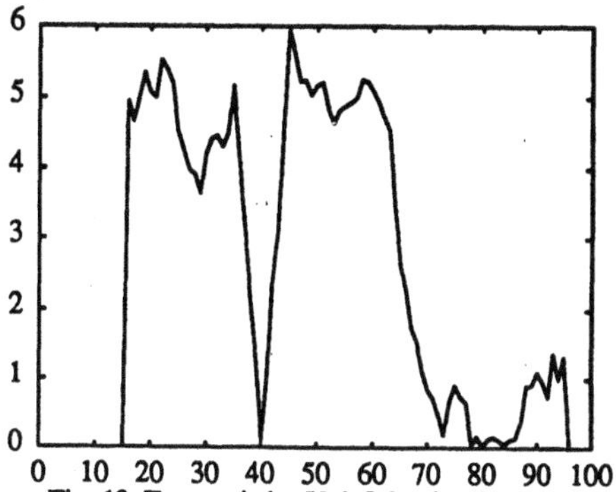

Fig. 13. Test statistics V_k,3 for data I., G=10.

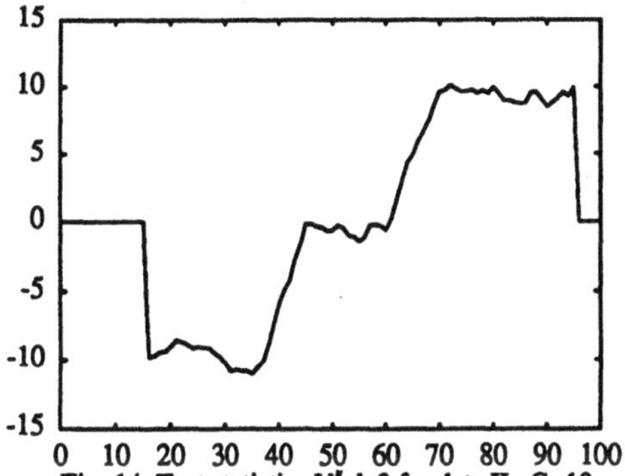

Fig. 14. Test statistics V^{A}_k,3 for data II., G=10.

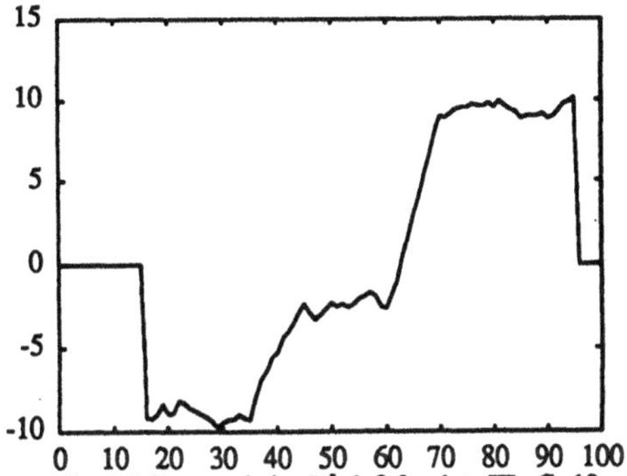

Fig. 15. Test statistics V^{A}_k,3 for data III., G=10.

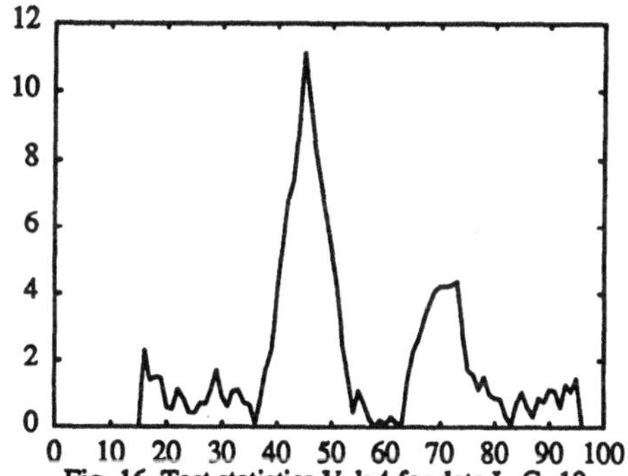

Fig. 16. Test statistics V_k,4 for data I., G=10.

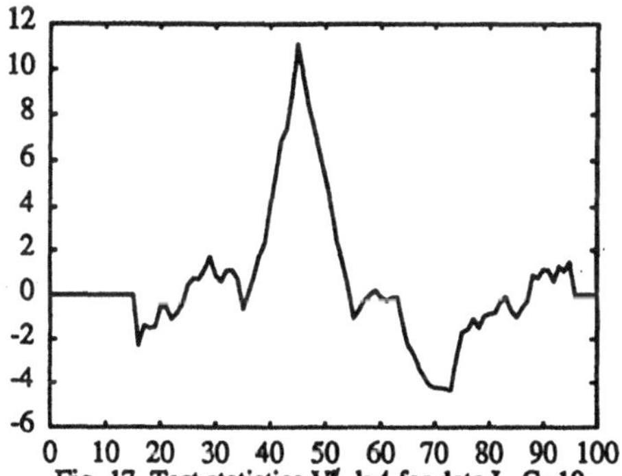

Fig. 17. Test statistics V^{w}_k,4 for data I., G=10.

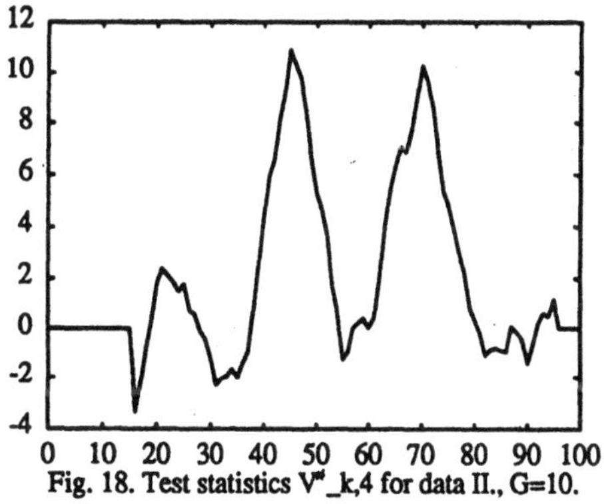

Fig. 18. Test statistics V^{w}_k,4 for data II., G=10.

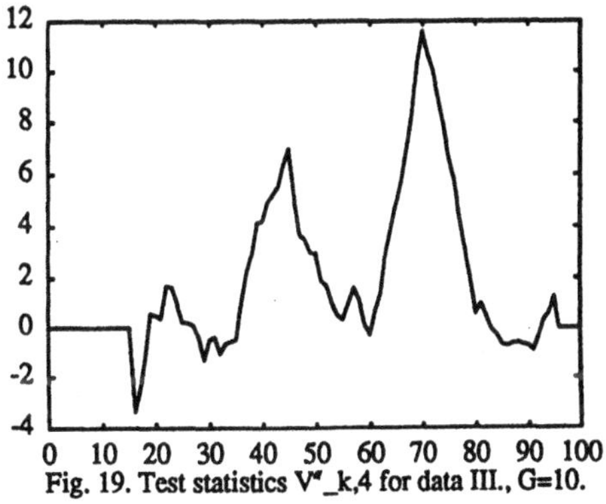

Fig. 19. Test statistics V^{w}_k,4 for data III., G=10.

6. TECHNICAL THEOREMS

Here we present theorems proved elsewhere which are crucial to obtain the desired limit properties.

Theorem 6.1. *Let Assumptions A.1. – A.5. be satisfied and let $a_n \to 0$ as $n \to \infty$. Then, as $n \to \infty$,*

$$\max_{1\le k\le n}\left\{\left|\sum_{i=1}^{k}\left[\psi(X_i;\theta_0+a_n)-\psi(X_i;\theta_0)-E\Big(\psi(X_i;\theta_0+a_n)-\psi(X_i;\theta_0)\Big)\right]\right|\right\}$$

$$= O_p\left(|a_n|^{\beta/2}\sqrt{n}\right), \tag{6.1}$$

$$\max_{1\le k\le n}\left\{\frac{1}{\sqrt{k}}\left|\sum_{i=1}^{k}\left[\psi(X_i;\theta_0+a_n)-\psi(X_i;\theta_0)-E\Big(\psi(X_i;\theta_0+a_n)-\psi(X_i;\theta_0)\Big)\right]\right|\right\}$$

$$= O_p\left(|a_n|^{\beta/2}\log\log n\right) \tag{6.2}$$

and

$$\max_{1\le k\le n}\sup_{|t|\le 1}\left\{\frac{1}{\sqrt{k}}\left|\sum_{i=1}^{k}\left[\psi(X_i;\theta_0+a_nt)-\psi(X_i;\theta_0)- \right.\right.\right.$$

$$\left.\left.\left.-E\Big(\psi(X_i;\theta_0+a_nt)-\psi(X_i;\theta_0)\Big)\right]\right|\right\}$$

$$= O_p\left(|a_n|^{\beta/2}\log\log n+\frac{|a_n|\sqrt{n}}{\log n}\right). \tag{6.3}$$

If, moreover, Assumptions A.6. and A.7. are fulfilled, then, as $n \to \infty$,

$$\max_{G\le k\le n}\left\{\frac{1}{\sqrt{G}}\left|\sum_{i=k-G+1}^{k}\left[\psi(X_i;\theta_0+a_n)-\psi(X_i;\theta_0)-E\Big(\psi(X_i;\theta_0+a_n)-\psi(X_i;\theta_0)\Big)\right]\right|\right\}$$

$$= O_p\left(|a_n|n^{\beta/(4+\kappa)}\log(n/G)\right) \tag{6.4}$$

and

$$\max_{G\le k\le n}\sup_{|t|\le 1}\left\{\frac{1}{\sqrt{G}}\left|\sum_{i=k-G+1}^{k}\left[\psi(X_i;\theta_0+a_nt)-\psi(X_i;\theta_0)- \right.\right.\right.$$

$$\left.\left.\left.-E\Big(\psi(X_i;\theta_0+a_nt)-\psi(X_i;\theta_0)\Big)\right]\right|\right\}$$

$$= O_p\left(|a_n|^{\beta/(4+\kappa)}\log\left(\frac{n}{G}\right)+\frac{\sqrt{G}\,|a_n|}{\log n}\right). \tag{6.5}$$

Proof. Checking the proof of Theorem 5.2. in Hušková(1993), the difference is that in the setup considered here X_i has d.f. $F(x;\theta_0+\delta_n r_{n,i})$, $i=1,\ldots,n$, we find that (6.1) – (6.5) holds true. □

Theorem 6.2. *Let $Y_1, \dots, Y_n$ be i.i.d. random variables with zero mean, unit variance and $E|Y_i|^{2+\eta} < \infty$ for some $\eta > 0$. Then:*

a) Darling and Erdős (1956). For any $y \in R_1$

$$P\left(\max_{1\le k\le n}\left\{\frac{1}{\sqrt{k}}\left|\sum_{i=1}^{k} Y_i\right|\right\} \le b(\log n, y)\right) \to \exp\{-2e^{-y}\} \quad \text{as } n \quad \to \infty$$

and

$$P\left(\max_{1\le k<n}\left\{\sqrt{\frac{n}{k(n-k)}}\left|\sum_{i=1}^{k}\left(Y_i - \overline{Y}_n\right)\right|\right\} \le b(\log n, y)\right) \to \exp\{-2e^{-y}\},$$

as $n \to \infty$, where $b(y,z)$ is defined by (2.22) and $\overline{Y}_n = n^{-1}\sum_{i=1}^{n} Y_i$.

b) Deheuvels and Revész (1987). For any $y \in R_1$

$$P\left(\max_{G\le k\le n}\left\{\frac{1}{\sqrt{G}}\left|\sum_{i=k-G+1}^{k} Y_i\right|\right\} \le b(n/G, y)\right) \to \exp\{-2e^{-y}\} \quad \text{as} \quad n \to \infty.$$

c) Chen Xiru (1984). If, moreover, Assumption A.6. is satisfied, then for any $y \in R_1$

$$P\left(\max_{2G\le k\le n}\left\{\frac{1}{\sqrt{2G}}\left|\sum_{i=k-G+1}^{k} Y_i - \sum_{i=k-2G+1}^{k-G} Y_i\right|\right\} \le b\left(\frac{3n}{2G} - 3, y\right)\right)$$

$$\to \exp\{-2e^{-y}\}$$

as $n \to \infty$.

□

REFERENCES

[1] Antoch J. and Hušková M., *Some M-tests for detection of a change in linear models*, Proceedings of the 4th Prague Symposium on Asymptotic Statistics (Hušková M. and Mandl P., eds.), Charles University Press, Praha, 1989, pp. 123-136.

[2] Antoch J. and Hušková M., *Change point problem*, Computational Aspects of Model Choice (Antoch J., ed.), Physica Verlag, Heidelberg, 1992.

[3] Barry D. and Hartigan J. A., *A Bayesian analysis for change point problems*, Journal of the American Statistical Association **88** (1993), 309-319.

[4] Broemling L. D. and Tsurumi H., *Econometrics and Structural Change*, Series Statistics, vol. 74, M. Dekker, New York, 1986.

[5] Chen X., *Inference in a simple change-point problem*, Scientia Sinica **A 31** (1988), 654-667.

[6] Csörgő M. and Horváth L., *Nonparametric methods for the change point problem*, Handbook of Statistics, vol. 7 (Krishnaiah P. R. and Rao C. R., eds.), J. Wiley, New York, 1988, pp. 403-425.

[7] Darkhovskii B. S., Brodskii B. E., *A nonparametric method for fastest detection of a change in the mean of a random sequence*, Theory of Probability and its Applications **32** (1987), 640-648.

[8] Darling and Erdös P., *A limit theorem for the maximum of normalized sums of independent random variables*, Duke Mathematical Journal **23** (1956), 143 – 155.
[9] Deheuvels P. and Révész P., *Weak laws for the increments of Wiener processes, Brownian bridges, empirical processes and partial sums of i.i.d. r.v.'s.*, Mathematical Statistics and Probability Theory, vol. A (Puri M. L. et al., eds.), D. Reidel, Berlin, 1987, pp. 69 – 88.
[10] Hawkins D. L., *A U – I approach to retrospective testing for shifting parameters in a linear model*, Communications in Statistics – Theory Methodology **18** (1989), 3117 – 3134.
[11] Hušková M., *Recursive M-tests for detection of change*, Sequential Analysis **7** (1988 a), 75 – 90.
[12] Hušková M., *Tests and estimators for the change point problem based on M – statistics.* (submitted)
[13] Hušková M. and Sen. P. K., *Nonparametric tests for shift and change in regression at an unknown time point*, Statistical Analysis and Forecasting of Economic Structural Change (Hackl P., ed.), Springer – Verlag, New York, 1989, pp. 71 – 85.
[14] Koziol J. A., *A note on non-parametric tests for the change – point problem*, Biometrical. Journal **29** (1987), 323 – 330.
[15] Krishnaiah P. R. and Miao B. Q., *Review about estimation of change points*, Handbook of Statistics, vol. 7 (Krishnaiah P. R. and Rao R. C., eds.), Elsevier, Amsterdam, 1988, pp. 375 – 402.
[16] Lombard E., *Rank tests for change point problems*, Biometrika **74** (1987), 615 – 624.
[17] Shaban S. A., *Change point problem and two phase regression: An annotated bibliography*, International Statistical Review **48** (1980), 83 – 93.
[18] Steinebach J., *Change point and jump estimates in an AMOC renewal model.* (this volume)
[19] Wolfe D. A. and Schechtman E., *Nonparametric statistical procedures for the change point problem*, Journal of Statistical Planning and Inference **9** (1984), 389 – 396.
[20] Zacks S., *Survey of classical and Bayesian approaches to the change point problem: Fixed sample and sequential procedures of testing and estimation*, Recent Advances in Statistics. Papers in Honor of Herman Chernoff's Sixtieth Birthday (Rizvi M. H., ed.), Academic Press, New York, 1983, pp. 245 – 269.
[21] Zacks S., *Detection and change – point problem*, Handbook of Sequential Analysis (Ghosh B. K. and Sen P. K., eds.), Series Statistics, vol. 118, M. Dekker, New York, 1991.

Charles University, Department of Probability and Mathematical Statistics, Sokolovská 83, CZ – 186 00 Praha 8 – Karlín, Czech Republic; tel: (+ 42 2) 231 78 62; fax: (+ 42 2) 664 150 95; telex: 121 673 mff

E-mail: antoch@cspguk11.bitnet, huskova@cspguk11.bitnet

PROBING FOR INFORMATION IN TWO-STAGE STOCHASTIC PROGRAMMING AND THE ASSOCIATED CONSISTENCY

BY ZVI ARTSTEIN

The Weizmann Institute of Science

Abstract. Information structure is introduced as a decision variable in two-stage stochastic programming. To this end the notion of sensors is employed. The outcome resembles a three-stage stochastic program, and hence can be analyzed with standard tools. This is demonstrated by establishing a strong law of large numbers for the two-stage problem with the information variable.

Key Words: stochastic programming, sensors, SLLN.

AMS Subject Classification: 90C15

1. INTRODUCTION

In this paper, and following Artstein and Wets (1993), we offer a model where the information structure is in itself a decision variable in a two-stage stochastic program. Consider first the standard two-stage stochastic programming problem given by

$$\text{(1.1)} \qquad \underset{x \in X}{\text{minimize}} \left(c(x) + E^{P(d\xi)} \left(\min_{y \in Y} h(x, \xi, y) \right) \right) .$$

Here x is the decision variable of the first stage, with $c(x)$ the associated cost. The random variable ξ is governed by the probability distribution $P(d\xi)$, and is realized and becomes available to the programmer before the decision y of the second stage takes place. Hence the cost $h(x, \xi, y)$ of the second stage depends on both x and ξ, as well as on y. The minimization in the first stage is with respect to the expected cost, denoted $E^{P(d\xi)}$. (We list the technical conditions on the data in (1.1) in the next section.)

Concrete examples of both abstract problems and such that arise in applications of two-stage stochastic programs of the form (1.1), are widely available.

Note that the underlying assumption in (1.1) is that the programmer is informed of the realization of the random variable ξ before the decision y is taken. If, for instance, this information is not available, the problem becomes a one-stage program, namely,

$$\text{(1.2)} \qquad \underset{x \in X}{\text{minimize}} \left(c(x) + \min_{y \in Y} E^{P(d\xi)}(h(x,\xi,y)) \right) .$$

In many cases neither of the information structures of (1.1) and (1.2) prevail, but rather partial information about the realization of ξ is available. Furthermore, the information revealed at the second stage may itself be affected by the decisions at the first stage, and then the programmer should weigh the benefit from having the information at the second stage against the cost of obtaining it. How to incorporate such a decision variable into the two-stage stochastic program is the subject of this paper.

As a framework for the analysis of information we use the notion of sensors, as developed by Artstein and Wets (1993); that paper elaborates on the structure of sensors, with concrete analytic and numerical examples. Here we show how sensors can be incorporated into the two-stage problems. The model is displayed in the next section.

In Section 3 we identify the decision structure associated with the model. The structure resembles a three-stage stochastic program, and thus can be analyzed with standard tools; some possibilities, and some difficulties, are listed. One property is elaborated upon in Section 4, namely, the strong law of large numbers, or the consistency, of the stochastic procedure.

2. PROBING FOR INFORMATION

We start with the technical framework, then recall the sensor concept and show how it can be incorporated into the two-stage problem. Some remarks close the section.

(i) The spaces X and Y are complete separable metric spaces.

(ii) The function $c(x) : X \to (-\infty, \infty]$ is lower semicontinuous.

(iii) The space Ξ is a complete separable metric space endowed with the Borel σ-field. The random element ξ is an element in Ξ.

(iv) The function $h(x,\xi,y) : X \times \Xi \times Y \to (-\infty,\infty]$ is lower semicontinuous in the pair (x,y) and measurable in (x,ξ,y).

The conditions above concern the data appearing in, e.g., (1.1) and (1.2). We describe now the motivation behind the model of probing for information.

In the model we offer, part of the decision in the first stage concerns getting information about the random variable before the second stage materializes. Information about ξ is a probability distribution $\mu(d\xi)$ that governs the occurrence of ξ, and thus the decision x will lead to revealing a probability $\mu(d\xi)$. But at the time x is chosen, the specific probability μ is not known; rather, x determines a distribution, say S_x, on the family of probability distributions μ that may occur in the second stage. We refer to such a probability measure S as a sensor. The specific μ is then a random variable itself, governed by the sensor S_x, which we can denote by $S_x(d\mu)$. The remarks in the sequel elaborate on the model. We place now the technical conditions and display the model.

(v) Let $\mathcal{M}$ be the space of probability measures on Ξ. We consider $\mathcal{M}$ with the Prohorov metric of weak convergence of measures; then $\mathcal{M}$ is a complete separable metric space, see Billingsley (1968). Denote by $\mathcal{S}$ the space of probability measures on $\mathcal{M}$, also considered with the Prohorov metric of weak convergence.

(vi) Let $x \to S_x$ be a mapping from X into $\mathcal{S}$; we assume that this mapping is continuous.

With the data and conditions as in (i)–(vi), the two-stage stochastic programming model with the information as a decision variable is as follows.

$$(*) \qquad \underset{x\in X}{\text{minimize}} \left(c(x) + E^{S_x(d\mu)} \left(\min_{y\in Y} E^{\mu(d\xi)}(h(x,\xi,y)) \right) \right) .$$

We recall the interpretation: In the first stage a decision x which costs $c(x)$ is taken. It determines both the cost functional $h(x,\cdot,\cdot)$ of the second stage, and the distribution $S_x(d\mu)$ from which the information $\mu(d\xi)$ available in the second stage is drawn. At the second stage $\mu(d\xi)$ is known to the programmer, who minimizes expected cost in both stages.

Remark 2.1. Note that the model $(*)$ contains in particular the stochastic programs (1.1) and (1.2). In both, of course, x does not affect the information, hence S is independent of x. To get (1.1) choose a sensor S in $(*)$ which is supported on Dirac measures δ_ξ, and with $S(\{\delta_\xi : \xi \in A\}) = P(A)$. To get (1.2) choose S which is an atom, namely supported on $\{P(d\xi)\}$. Consult with Artstein and Wets (1993) for other models of information, e.g., the σ-field or the signalling models, that are particular cases of our model.

Remark 2.2. The model (*) is motivated to a large extent by the Bayesian sampling approach. Consider the following example. The decision x involves both, the level of advertisement for a product, and the extent of a market research to assess the success of this advertisement. The second stage determines the level of production, and this is done when the outcome $\mu(d\xi)$ of the market research is known. A typical market research results in a normal distribution $N(m, s^2)$, depicting the predicted number of customers. The variance s^2 is a direct consequence of the decision, namely the extent of the market research. The mean m, however, is unknown at the decision time, only its distribution can be calculated if a prior is available. The actual distribution can then be calculated using the Bayesian approach, as in optimal statistical decisions, see DeGroot (1970). In our case the decision x would determine s^2, and the resulting sensor would then be supported on the family $\{N(m, s^2) : m \in A\}$ for some A. Note that unlike the cases studied in optimal statistics, the priors which determine the sensors in the stochastic programming case may be affected by the decisions, say by the advertisement in the first stage.

Remark 2.3. The technical conditions (i)–(iv) may not be necessary for the formulation of the problem (1.1), but are considered as minimal conditions. Additional assumptions should then be added to derive properties of the model. In the same spirit we demand the continuity of $x \to S_x$ in condition (vi).

3. THE DECISION PROCESS

Note the resemblance of the two-stage problem with information as given in (*), to a three-stage stochastic program. In a typical three-stage problem there are two random events separating three decision stages. The problem (*) has a similar structure with a degenerate third decision. Hence standard tools of multi-stage stochastic programming can, in principle, be utilized for the analysis of (*). Yet, the model with information as a decision variable has some special features which make it worthwhile to spell out the characteristics of the decision process.

Definition 3.1. Denote by $\mathcal{Y}$ the space of measurable mappings $y(\mu) : \mathcal{M} \to Y$. The decision space associated with (*) is the space $\mathcal{D} = X \times \mathcal{Y}$. The cost associated with $(x, y(\cdot)) \in \mathcal{D}$ is given by

$$\text{cost}(x, y(\cdot)) = c(x) + E^{S_x(d\mu)}(E^{\mu(d\xi)}(h(x, \xi, y(\mu)))) \tag{3.1}$$

where we allow $+\infty$ and $-\infty$ as values, and set the cost as $+\infty$ if the extended integration is not defined.

The motivation for the preceding definition is clear; indeed, the decision y is a function of $\mu \in \mathcal{M}$, and measurability seems to be a minimal requirement. The same is true indeed for the standard stochastic program (1.1), but here $y(\xi) : \Xi \to Y$ is the appropriate second stage decision. We denote the respective cost by

$$\text{cost}_{(1.1)}(x, y(\cdot)) = c(x) + E^{P(d\xi)}(h(x, \xi, y(\xi))) \tag{3.2}$$

and again allow $\pm\infty$ as values.

Note that the decision space $\mathcal{D}$ (and the analogous space for (1.1)) are not endowed with a topological structure. Pointwise convergence seems to be a natural convergence structure. In the case of (1.1), and since $P(d\xi)$ is fixed, convergence in measure with respect to P (which we refer to as convergence in P-measure) is also natural, and it is metrizable. The following result points toward a difference between the two cases.

Proposition 3.2. Suppose $h(x, \xi, y)$ is bounded. Under conditions (i)–(iv) the functional $\text{cost}_{(1.1)}$ is lower semicontinuous with respect to convergence on X and pointwise convergence of $y(\cdot)$, namely $x_k \to x_0$ and $y_k(\cdot)$ converges pointwise to $y_0(\cdot)$ then

$$\liminf \text{cost}_{(1.1)}(x_k, y_k(\cdot)) \geq \text{cost}_{(1.1)}(x_0, y_0(\cdot)) .$$

(In particular the lower semicontinuity holds for convergence in P-measure on $y(\cdot)$.) The analogous lower semicontinuity property does not hold for the functional $\text{cost}(x, y(\cdot))$ on $\mathcal{D}$.

Proof. Had $h(x, \xi, y)$ been continuous in (x, y), rather than only lower semicontinuous, the Lebesgue dominated convergence theorem would imply that the integral

$$E^{P(d\xi)}(h(x, \xi, y(\xi))) \tag{3.3}$$

is continuous in $(x, y(\cdot))$. Indeed, $x_k \to x_0$ and $y_k(\cdot) \to y_0(\cdot)$ pointwise, would then imply pointwise convergence of $h(x_k, \xi, y_k(\xi))$ to $h(x_0, \xi, y_0(\xi))$, and the boundedness allows us to use dominated convergence. The extension to lower semicontinuity is done as follows. The function $h(x, \xi, y)$ can be approximated from above by, say, $h_j(x, \xi, y)$ such that $h_j(x, \xi, y)$ decrease to $h(x, \xi, y)$ pointwise, each h_j is continuous in (x, y) and measurable in (x, ξ, y). For a fixed ξ the approximation is standard, and the measurability in ξ can be achieved by using a constructive algorithm, e.g., the one proposed by Beer (1987). This algorithm defines $h_j(x, \xi, y) = h_{j-1}(x, \xi, y) -$

$\varphi(h_{j-1}(x,\xi,y),h(\cdot,\xi,\cdot))$, with φ being the distance of the point $h_j(x,\xi,y)$ from the graph of the function $h(\cdot,\xi,\cdot)$. The function h_0 can be chosen as a constant dominating h. Since φ is clearly measurable in ξ (with respect to the P-completion of the Borel field), it follows that each h_j is measurable. For each j the expectation

$$E^{P(d\xi)}(h_j(x,\xi,y(\xi))$$

is then continuous in $(x,y(\cdot))$ by the previous arguments. In the limit, and since $h_j(x,\xi,y) \geq h(x,\xi,y)$, we get that (3.3) is lower semicontinuous in $(x,y(\cdot))$. Since by (ii) the function $c(x)$ is lower semicontinuous, the lower semicontinuity of $\text{cost}_{(1.1)}$ follows.

To get a counterexample for the case (3.1), it is enough to consider a case with no second decision, but with h discontinuous in ξ, or alternatively a fixed $y(\cdot)$ discontinuous in μ. Indeed, it is well known that $E^P(h(\cdot))$ is neither continuous nor semicontinuous in the variable P if $h(\cdot)$ is discontinuous. For instance, take $h(\xi) = -1$ for $-1 \leq \xi < 0$, $h(0) = 0$ and $h(\xi) = 1$ for $0 < \xi \leq 1$. Let P_k be concentrated on $\{\frac{1}{k}\}$ and P_j concentrated on $\{\frac{-1}{j}\}$. Then both P_k and P_j converge to P_0, which concentrate on $\{0\}$, yet $E^{P_k(d\xi)}(h(\xi)) = 1$, $E^{P_j(d\xi)}(h(\xi)) = -1$ while $E^{P_0(d\xi)}(h(\xi)) = 0$. Thus, for the fixed discontinuous integrand $h(\xi)$ in the inner integration of (3.1), it is possible to arrange that $x \to S_x$ (even $x \to P_x(d\xi)$) be such that the cost is not semicontinuous, for instance $x \in [-1,1]$ and $x = k^{-1}$ is mapped into P_k. A similar example can be worked out with $y(\cdot)$ dicontinuous. This completes the proof.

To get a positive result for the problem (∗) we have to require more from the decision functions, or from the cost function h. The following is such a result, where more continuity properties are assumed. We note that continuity seems a natural requirement in many applications. To emphasize our point, we impose also compactness conditions which allow a simple proof. These can be relaxed and replaced by equi-continuity conditions in line with Section 8 of Artstein and Wets (1993). Note that compactness also arises naturally in many applications.

The following result uses the property that $h(x,\xi,y)$ is lower semicontinuous in x uniformly in (ξ,y). By that we mean that for every $\varepsilon > 0$ and every x_0 a $\delta > 0$ exists such that $h(x,\xi,y) - h(x_0,\xi,y) \geq -\varepsilon$ whenever $d(x,x_0) < \delta$ and for all (ξ,y). A similar condition was used in Artstein and Wets (1993) when examining the continuity of the value of a sensor.

Proposition 3.3. In addition to (i)–(vi), assume that $h(x,\xi,y)$ is bounded, continuous in the pair (ξ,y), lower semicontinuous in x uniformly in (ξ,y), and also assume that Y and Ξ are compact. Consider the subset $\mathcal{D}_c$ of $\mathcal{D}$ which consists of continuous functions $h : \mathcal{M} \to Y$, and on these functions consider the uniform convergence. Then $\text{cost}(x, y(\cdot))$ is lower semicontinuous on $\mathcal{D}_c$. If in addition $c(x)$ is continuous and $h(x, \xi, y)$ is continuous, then $\text{cost}(x, y(\cdot))$ is continuous.

Proof. Consider first the mapping

$$(3.4) \qquad (\mu, y) \longrightarrow E^{\mu(d\xi)}(h(x, \xi, y))$$

from $\mathcal{M} \times Y$ into the reals, with x as a parameter. It is clear that for a fixed x this mapping is continuous (with respect to weak convergence in $\mathcal{M}$ and metric convergence in Y), and by the compactness of Ξ, which implies compactness of $\mathcal{M}$, it follows that the mapping (3.4) is uniformly continuous in (ξ, y) for x fixed.

Consider next the mapping

$$(3.5) \qquad (S, y(\cdot)) \longrightarrow E^{S(d\mu)} E^{\mu(d\xi)}(h(x, \xi, y(\mu)))$$

from $\mathcal{S} \times \mathcal{D}_c$, but x fixed. The uniform continuity of (3.4) makes the integrand (for fixed x) of the outer expectation, continuous, and uniformly so when uniform continuity is imposed on $y(\cdot)$. Hence the properties of weak convergence on measures imply that (3.5) is continuous for fixed x.

Let now $x_k \to x_0$ and $y_k(\cdot) \to y_0(\cdot)$, the latter in the sup norm. Then

$$(3.6) \qquad E^{S_{x_k}(d\mu)} E^{\mu(d\xi)}(h(x_k, \xi, y_k(\xi)) - h(x_0, \xi, y_k(\xi))) \geq -\varepsilon$$

for k large enough, this by the uniform lower semicontinuity. On the other hand

$$(3.7) \qquad E^{S_{x_k}(d\mu)} E^{\mu(d\xi)}(h(x_0, \xi, y_k(\xi))) - E^{S_{x_0}(d\mu)} E^{\mu(d\xi)}(h(x_0, \xi, y_0(\xi))) \longrightarrow 0$$

as $k \to 0$; this by the continuity established in (3.5) for x_0 fixed, indeed $x_k \to x_0$ implies by (vi) that $S_{x_k} \to S_{x_0}$. Combining (3.6) and (3.7) implies that the integral part in (3.1) is indeed lower semicontinuous in $(x, y(\cdot))$ on D_c, and since $c(x)$ was assumed in (ii) to be lower semicontinuous, it follows that $\text{cost}(x, y(\cdot))$ is indeed lower semicontinuous.

The dual argument would imply upper semicontinuity of the cost under upper semicontinuity of the functionals, hence continuity of $c(x)$ and continuity of $h(x, \xi, y)$ (which is actually uniformly continuous by compactness), imply the continuity of $\text{cost}(x, y(\cdot))$ on D_c. This completes the proof.

Remark 3.4. Note that lower semicontinuity implies that if $(x_k, y_k(\cdot))$ is a minimizing sequence which converges, say to $(x_0, y_0(\cdot))$, then the latter is an optimal strategy. In this connection, the drawback of the previous result is that the norm on D_c is not compact, hence a minimizing sequence may not have a converging subsequence.

Remark 3.5. The continuity condition imposed on $y(\cdot)$ seems to be severe, and in particular may outrule optimal decisions. Indeed, simple examples show that optimal second stage decisions may not be continuous in the random element. Yet, this continuity (or alternative assumptions) are significant in stability analysis – consult e.g., the analysis in Robinson and Wets (1987), Dupačhová (1991), Römisch and Schultz (1991), Schultz (1992, 1993) and Artstein and Wets $(1993)_a$ (the last three papers replace continuity of the integrand by additional conditions on the probability measures). The techniques offered in these references can be modified such that the model (*) is included (see also Artstein and Wets (1993)). We leave out the details. But note that at any rate, even if the optimal $(x, y(\cdot))$ is discontinuous, the function $y(\cdot)$ can be approximated by a continuous one with only a small deviation of the cost. This follows from standard measure theory arguments.

4. CONSISTENCY

Laws of large numbers for stochastic programs are of interest, among other reasons, as they establish the consistency of the sampling procedure with respect to the expectation operator. We follow in this section the paradigm set by King and Wets (1991), and Attouch and Wets (1991), and establish a strong law of large numbers for the model (*). However, as an additional random event is involved, namely the one governed by the sensor, we have first to set the framework and identify the random process with respect to which the limit law is verified.

Definition 4.1. Consider $\mathcal{M} \times \Xi$ as a measurable space induced by the Borel σ-field of the product metric. With a given x we associate a measure ν_x on $\mathcal{M} \times \Xi$ obtained by integrating $\mu(d\xi)$ with respect to $S_x(d\mu)$; namely, for $M \times A$ a rectangle in $\mathcal{M} \times \Xi$ we have

$$\nu_x(M \times A) = \int_M \mu(A) S_x(d\mu) ,$$

see e.g., Bourbaki (1959).

Remark 4.2. It is clear that once x is chosen, the random process of sampling μ first according to S_x, and then sampling ξ according to μ, is stochastically equivalent to sampling (μ, ξ) according to ν_x.

The desired consistency is that for a sequence of random realizations (μ_i, ξ_i) the averaged actual cost converges almost surely to the expected cost. We now set the notations which make this statement precise.

For a decision $(x, y(\cdot))$ in $\mathcal{D}$, and a realization (μ, ξ) of the random variable, we denote the actual cost by

$$\text{(4.1)} \qquad \operatorname{cost}(x, y(\cdot), \mu, \xi) = c(x) + h(x, \xi, y(\mu)) .$$

For a given sequence (μ_i, ξ_i), $i = 1, 2, \ldots$, of realizations, and a given element in $\mathcal{D}$, we denote the sequence of averaged actual costs by

$$\text{(4.2)} \qquad g_N(x, y(\cdot)) = \frac{1}{N} \sum_{i=1}^{N} \operatorname{cost}(x, y(\cdot), \mu_i, \xi_i) .$$

The standard strong law of large numbers provides quite general conditions implying that for a fixed $(x, y(\cdot))$ the sequence $g_N(x, y(\cdot))$ converges almost surely to $\operatorname{cost}(x, y(\cdot))$ given by (3.1). Here is such a result.

Proposition 4.3. Suppose $h(x, \cdot, y(\cdot))$ is ν_x-integrable on $\mathcal{M} \times \Xi$. If (μ_i, ξ_i) is a sequence of independent and identically ν_x-distributed random variables, then $g_N(x, y(\cdot))$ converges almost surely to $\operatorname{cost}(x, y(\cdot))$.

Proof. Follows from the standard strong law of large numbers.

But consistency means more than the almost sure convergence for any fixed decision. One would like to be sure that the functionals $g_N(\cdot, \cdot)$ on $\mathcal{D}$ (or a subset of $\mathcal{D}$) converge almost surely to the functional $\operatorname{cost}(\cdot, \cdot)$. See King and Wets (1991) and Attouch and Wets (1991). In our framework we encounter a problem not addressed by these references, namely that the probability space governing the random event is affected by the decision (this is not particular only to the information model). We address this problem later and now present a partial result, namely that for a fixed x, the consistency holds with respect to $y(\cdot)$.

To this end we should agree upon a convergence notion for functionals, with respect to which the convergence of $g_N(x, \cdot)$ to $\operatorname{cost}(x, \cdot)$ is checked. We use here the notion of epi-convergence; it proved to be adequate and useful in stochastic analysis in general, and variational convergence in particular; see Rockafellar and Wets (1984).

Definition 4.4. (See e.g., Attouch and Wets (1991).) Let Z be a metric space and $f_N : Z \to (-\infty, \infty]$ a sequence of lower semicontinuous functions, each not identically $+\infty$. Then f_N epi-converge to f_0 if

$$\liminf \; f_N(x_N) \geq f_0(x_0)$$

whenever $x_N \to x_0$, and equality holds for at least one sequence $x_N \to x_0$.

Remark 4.5. Note that the definition applies also when Z is endowed with only a convergence structure, as our space $\mathcal{Y}$. On the other hand the epi-convergence is, in many cases, itself metrizable as it can be expressed as set convergence of the epigraphs $\{(x,r) : r \geq f_N(x)\}$ of the functions in the sequence; see Attouch and Wets (1991) and Rockafellar and Wets (1984).

The paper by Attouch and Wets (1991) provides a very general theory of consistency, or strong laws of large numbers, for functionals with respect to epi-convergence. One could find the conditions on the data in (*) (and possibly on a subset of $\mathcal{D}$, as our functionals on $\mathcal{D}$ are not lower semicontinuous, see Proposition 3.2), that fit the framework set by Attouch and Wets. We proceed in a different way. We set, may be, stronger conditions which allow in turn a simple and transparent proof. It is based on the following abstract result.

Let Z be a metric space and let (H,ν) be a probability measure space. Let $f(z,\eta) : Z \times H \to (-\infty, \infty]$ be lower semicontinuous in z and measurable in (z,η).

Proposition 4.6. Suppose that the family $f(\cdot,\eta)$, $\eta \in H$, of functions is contained in a set which is compact and metrizable with respect to epi-convergence. Let $\eta_1, \eta_2, \ldots$ be a sequence of independent and identically ν-distributed random variables. Then $\frac{1}{N}\sum_{i=1}^{N} f(\cdot,\eta_i)$ epi-converge almost surely to $E^{\nu(d\eta)}(f(\cdot,\eta))$.

Proof. This is a standard law of large numbers result once it is observed that the addition

$$(f_1 + f_2)(z,\eta) = f_1(z,\eta) + f_2(z,\eta)$$

is bi-continuous with respect to epi-convergence (which is trivial). Indeed, for a function $f(\cdot,\eta)$ which takes only a finite number of values, the result follows from the weak law of large numbers. The compact set of values can be uniformly approximated by a finitely valued mapping and a standard approximation argument completes the proof.

Corollary 4.7. In addition to (i)–(vi) assume that $h(x,\xi,y)$ is bounded, continuous in the pair (ξ,y), and that Y and Ξ are compact. Consider the family of functionals $\mathrm{cost}(x,\cdot,\mu,\xi)$ on $\mathcal{D}_c$, with $(\mu,\xi) \in \mathcal{M} \times \Xi$ (and x fixed). If this family is contained in a set which is compact and metrizable with respect to epi-convergence, then for an identically and ν_x-distributed sequence (μ_i,ξ_i), the functionals $g_N(x,\cdot)$ epi-converge almost surely to $\mathrm{cost}(x,\cdot)$.

Proof. In view of Proposition 3.3 with x fixed, all the requirements of Proposition 4.6 hold. (Note that the restriction to continuous functions $y \in \mathcal{Y}$ is needed in view of Proposition 3.2.)

We note that it is easy to come up with conditions on the data in (*), that would imply the compactness demanded in the preceding result.

To get consistency for the entire decision space $\mathcal{D}_C$, namely with varying x as well, one encounters the problem that the random event (μ, ξ) depends on x, as is reflected in Definition 4.1. This leads to some technical problems. In the rest of the section we set the framework of the desired result, and display one simple example.

Definition 4.8. Let Π be the infinite product of $\mathcal{M} \times \Xi$ by itself, indexed by X, namely

$$\Pi = (\mathcal{M} \times \Xi)^X .$$

An element in Π is denoted by $(\boldsymbol{\mu}, \boldsymbol{\xi})$ and its x-coordinate by $(\boldsymbol{\mu}^x, \boldsymbol{\xi}^x)$. Let $\boldsymbol{\nu}$ be the probability measure on Π obtained as the product of the ν_x defined on coordinates of Π (see e.g. Neveu (1965)). Consider the extension of the cost functional given in (4.1) from $\mathcal{D} \times (\mathcal{M} \times \Xi)$ to $\mathcal{D} \times \Pi$ in the natural way, namely

$$\text{cost}(x, y(\cdot), \boldsymbol{\mu}, \boldsymbol{\xi}) = \text{cost}(x, y(\cdot), \boldsymbol{\mu}^x, \boldsymbol{\xi}^x) \tag{4.3}$$

and similarly, $g_N(x, y(\cdot))$ is defined by (4.2) for a sequence $(\boldsymbol{\mu}_i, \boldsymbol{\xi}_i)$ of elements in Π.

With the framework described in the previous definition, the desired strong law of large numbers would state that for a sequence $(\boldsymbol{\mu}_i, \boldsymbol{\xi}_i)$ of independent and identically $\boldsymbol{\nu}$-distributed drawings, the functionals $g_N(\cdot, \cdot)$ epi-converge almost surely to $\text{cost}(\cdot, \cdot)$. As mentioned, such a result concerns technical difficulties which are beyond the scope of the present paper. We therefore conclude with a result that can be derived from our previous analysis in a simple way.

Proposition 4.9. Suppose that the conditions of Corollary 4.7 hold, with the boundedness and modulus of continuity being uniform in the x variable. Suppose also that $h(x, \xi, y)$ is lower semicontinuous in x, uniformly in (ξ, y). Suppose that X is denumerable. Let $(\boldsymbol{\mu}_i, \boldsymbol{\xi}_i)$ be a sequence of independent and $\boldsymbol{\nu}$-identically distributed drawings from Π. Then $g_N(\cdot, \cdot)$ epi-converge almost surely on $\mathcal{D}_C$ to $\text{cost}(\cdot, \cdot)$.

Proof. The countability of X enables us to deduce from the almost sure convergence for every fixed x given in Corollary 4.7, the almost sure convergence uniform in x. It remains to show that the epi-convergence holds also with respect to the x variable. This follows easily from the lower semicontinuity in x established in Proposition 3.3.

Acknowledgement. The author is the incumbent of the Hettie H. Heineman Professorial Chair in Mathematics. The research was supported by a grant from the United States–Israel Binational Science Foundation.

REFERENCES

Artstein, Z. and Wets, R. J-B (1993). Sensors and information in optimization under stochastic uncertainty. *Math. Oper. Res.* 18, 523–547.

Artstein, Z. and Wets, R. J-B $(1993)_a$. Stability results for stochastic programs and sensors, allowing for discontinuous objective functions. Preprint, to appear in *SIAM J. Optimization.*

Attouch, H. and Wets, R. J-B (1991). Epigraphical processes: laws of large numbers for random lsc functions. Preprint, University of California, Davis.

Beer, G. (1987). A geometric algorithm for approximating semicontinuous functions. *J. Approx. Theory* 49, 31–40.

Billingsley, P. (1968). *Convergence of Probability Measures.* John Wiley, New York.

Bourbaki, N. (1959). *Integration, Chapitre 6.* Herman, Paris.

DeGroot, M.H. (1970). *Optimal Statistical Decisions.* McGraw-Hill, New York.

Dupačhová, J. (1991). On statistical sensitivity analysis in stochastic programming. *Ann. of Oper. Res.* 30, 199–214.

King, A.J. and Wets, R. J-B (1991). Epi-consistency of convex stochastic programs. *Stochastics and Stochastics Reports* 34, 83–92.

Neveu, J. (1965). *Mathematical Foundation of the Calculus of Probability.* Holden Day, San Francisco.

Robinson, S.M. and Wets, R. J-B (1987). Stability in two stage stochastic programming. *SIAM J. Control and Opt.* 25, 1409–1416.

Rockafellar, R.T. and Wets, R. J-B (1984). Variational analysis: an introduction. In *Multifunctions and Integrands, Stochastic Analysis and Optimization,* G. Salinetti, ed., Lecture Notes in Mathematics 1091, Springer-Verlag, Berlin, pp. 1–54.

Römisch, W. and Schultz, R. (1991). Stability analysis for stochastic programs. *Ann. of Oper. Res.* 30, 241–266.

Schultz, R. (1992). Continuity and stabilization in two-stage stochastic integer programming. In *Stochastic Optimization, Numerical Methods and Technical Applications*, K. Marti, ed., Lecture Notes in Economics and Mathematical Systems 379, Springer-Verlag, Berlin, pp. 81-92.

Schultz, R. (1992). Continuity properties of the expectation functions in stochastic integer programming. *Math. of Oper. Res.* 18, 578–589.

Department of Theoretical Mathematics
The Weizmann Institute of Science
Rehovot 76100, Israel

Outliers and Switches in Time Series

A. C. Atkinson*, Siem Jan Koopman* and Neil Shephard†
* Department of Statistical and Mathematical Sciences
London School of Economics, London WC2A 2AE, UK and
†Nuffield College, Oxford, OX1 1NF, UK

Abstract

For unobserved components or structural time series models, the effect of elaborations of the model on inferences can be investigated by the use of interventions involving a single parameter, such as trend or level changes. The effect of the intervention is measured by the change in the estimates of the individual variances. We examine the effect on these estimated parameters of moving various kinds of intervention along the series. The horrendous computational problems involved are overcome by the use of score statistics combined with recent developments in filtering and smoothing. Interpretation of the resulting time series plots of diagnostics is aided by simulation envelopes using an approximation to the score statistic. The procedure is illustrated with an example in which the intervention is in the form of a 'switch'.

Keywords: added variable; deletion methods; diagnostics; dynamic linear model; EM-algorithm; Kalman filter; Monte–Carlo test; outlier; score test; shock; simulation envelope; structural change; structural time series model; unobserved components model.

1 Purpose

The purpose of this paper is to outline and exemplify powerful methods for detecting specific inadequacies in time series models. We use the unobserved component or structural approach to time series modelling, that is dynamic linear models, to provide procedures which have much in common with the well-known diagnostics for multiple regression models.

A single outlier in a regression model can be detected by the effect of its deletion on the residual sum of squares. An equivalent procedure is the simple intervention in which an extra parameter is added for the mean of the observation in question. The effect on the residual sum of squares is identical in the two cases. Similarly, for unobserved component time series models, the effect of changes in the data on the model can be investigated by the use of interventions corresponding to physically identifiable features involving a single parameter, such as trend or level changes. Because these time series models include more than one variance, the effect of the intervention is measured by the effect on the individual variances, which we call the parameters of the model. Atkinson, Koopman, and Shephard (1993) describe in detail the working out of this simple idea. Here we summarize the main results and give a new example in which a 'switch' is include in the model to allow, for example, for stockpiling of goods before an anticipated strike.

The large and growing literature on regression diagnostics, is summarized by Atkinson et al. (1993). Central to this literature is the distinction between outlier detection and the influence of observations on specific inferences. There is likewise a large literature on the detection of outliers in time series, although there has been appreciably less work on influential observations than on outlier detection. Our approach is based on inspecting plausibly interesting and interpretable departures, some of which are related to those examined by the more conventional diagnostics. These departures are produced by a generalized form of intervention analysis. In standard intervention analysis interest centres on the effect of a known exogenous change on the model, the specific times of intervention or of shock to the system being known. Our purpose is the more general diagnostic one of determining whether inferences from the model, specifically parameter estimates and forecasts, are fragile: that is, do they respond significantly to changes in the fitted model? To achieve this we use an intervention for each unobserved component and examine the effect as the intervention is moved along the series. Thus for the irregular component of the model we look at the effect of modelling a single additive outlier at each time point.

Plausible departures from the model are given in §2 and illustrated in Fig.1. In §3 we discuss how to assess the effect of the intervention or shock. Possible measures include the change in loglikelihood or the change in the estimated parameter. For inferential purposes we work with a marginal likelihood function. Since we look at five or six forms of intervention for each time point it is clearly not possible to re-estimate the parameters for each intervention. We therefore use a score statistic which avoids such re-estimation. From this we derive a one-step estimate of the effect of the proposed change in the model on the parameter. The statistical significance of these estimates is assessed by plots incorporating simulation envelopes. The derivation of the score function and one-step estimates is given in §4.

In standard regression models the effect of the addition of an extra variable can be assessed by the use of an added-variable plot in which the residuals of the response are plotted against the residuals of the extra variable. Both sets of residuals are from regression on all variables already in the model. Regression of one set of residuals on the other yields the coefficient for the new variable in the multiple regression model containing all variables, together with its associated

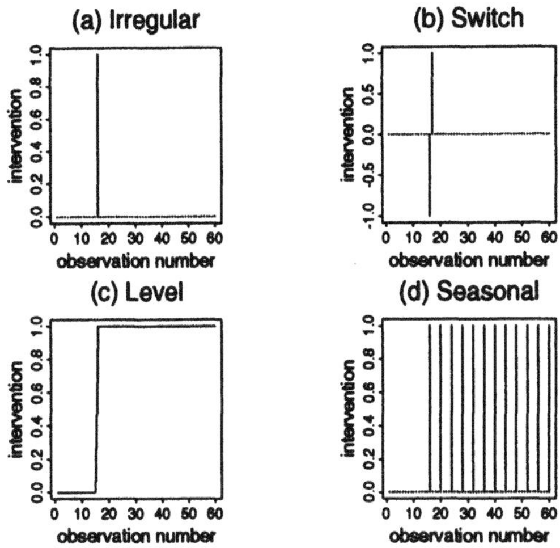

Figure 1: Four kinds of added variables for interventions which are assessed by being moved along the series

t test. Related, although more complicated, results for unobserved component models are given in §4.1. The combination of these results for added variables with the results on score tests provides elegant and speedy methods for the calculation of simulation envelopes and stresses the relationship with diagnostics for regression models. These results are illustrated in §6 with an analysis of data on UK coal consumption.

2 Inadequate Models and Interventions

The unobserved components or Gaussian state space model, often called the dynamic linear model, is described in many books, for example Harvey (1989), especially Chapter 3, and West and Harrison (1989). We give general expressions for the measurement and transition equations in §4. The model for the observations y_t may contain regression variables and, indeed, this is how the shocks are introduced in §4. In all our examples y_t contains an irregular component. In addition there may be level, slope, seasonal or other components such as cycles. To test how fragile the estimated model is we modify one of the components, or introduce a new one, and see whether there are changes in important features of the fitted model. In particular, tests for stability to shocks in those components which are included, are determined from the effect of the addition of regression variables of the form shown in Fig.1. We calculate the effect of moving each regression variable along the series, so adding one extra parameter to the model at each time point in turn, and see how the fit of the model changes as a result of this extra parameter. A large decrease in the estimate of one of the variances associated with the measurement or transition equations indicates sensitivity of the model to the intervention. Suppose, for example, there were an unusually large level break in the series at $t = 16$. The fitted model would then have an inflated value of the variance for the stochastic level. The addition of the level break variable of Fig.1c at $t = 16$ would dramatically reduce the variance, which would also be somewhat reduced by adding the variable at $t = 15$ or 17. As a result, inference about the exact change point might be 'smeered' over several neighbouring observations. However introducing the variable at appreciably smaller or larger values of t would have no, or a negligible, effect. The results of Atkinson et al. (1993) show that this smeering effect can be ignored for the outlier intervention and does not seem to cause problems of interpretation for the other interventions such as level or slope breaks.

The result of our analysis for any particular shock is a time-series plot accompanied by a simulation envelope. In line with the approach of regression diagnostics, we focus on statistics associated with individual observations and specific features of the data or model.

3 Measuring the Effect of Shocks

Let the parameter of interest be θ, which is a vector function of the variances in the measurement and transition equations, and let $\hat{\theta}$ be its maximum likelihood estimator. It is the value of θ which controls the rate of discounting of past observations and so determines the time series properties of the model. We want to measure the effect on the estimation of θ of the intervention modelled by inclusion of a new regression variable which has coefficient β. One possibility, in line with the approach to added and constructed variables in regression, is to calculate the t statistic for β. For the remainder of this section we however focus on estimation of θ. The close relationship between the t statistic and changes in the estimate of θ due to the intervention is described in §4.1.

To avoid biases in our inferences about the time series parameter θ in the presence of the nuisance parameter β we work with a marginal likelihood (that is, modified profile likelihood) function, equivalent, for Gaussian time-series models, to a Bayesian procedure with a weak prior (Shephard (1993), Tunnicliffe Wilson (1989)).

Let the variable applied at time t be of the jth type, perhaps one of the four of Fig.1. To stress the connection with deletion diagnostics in regression we call the estimate of the parameter from the marginal likelihood $\hat{\theta}_{(t,j)}$. If the maximized log likelihood for the unperturbed model is $L(\hat{\theta})$, the difference in the loglikelihood due to addition of the explanatory variable is

$$L_{(t,j)}(\hat{\theta}_{(t,j)}) - L(\hat{\theta}). \tag{1}$$

Because $L_{(t,j)}(.)$ is a marginal loglikelihood this difference can be positive or negative: the standard asymptotic chi-squared distribution is inappropriate. Calculation of (1) requires estimation of the parameter $\hat{\theta}_{(t,j)}$. Unfortunately the estimation of the parameters in an unobserved components model is the computationally most intensive step, involving many passes of the Kalman filter. We therefore work with a score version of the difference in loglikelihoods which avoids re-estimation of the parameters for each combination of t and j.

Let the score for the marginal loglikelihood be

$$s_{(t,j)}(\theta) = \frac{\partial L_{(t,j)}(\theta)}{\partial \theta}$$

with the corresponding definition for $s(\theta)$ as a derivative of $L(\theta)$. Then the change or difference in scores due to the intervention evaluated at some θ is

$$c_{(t,j)}(\theta) = s_{(t,j)}(\theta) - s(\theta). \tag{2}$$

The second term on the right hand side of (2) is identically zero when $\theta = \hat{\theta}$. Although these scores are of interest, they are relatively highly correlated across the various elements of θ. A more nearly orthogonal set of variables is obtained from the one-step estimates of the parameters

$$\tilde{\theta}_{(t,j)} = \hat{\theta} + \jmath^{-1}(\hat{\theta})s_{(t,j)}(\hat{\theta}) = \hat{\theta} + \jmath^{-1}(\hat{\theta})c_{(t,j)}(\hat{\theta}). \tag{3}$$

In (3) $\jmath(\hat{\theta})$ is the observed information found from the numerical matrix of second derivatives from the likelihood fitting. Explicit expressions for the score and one-step estimates for vector θ are given in the next section. For plotting we standardize the one-step estimates to give Wald-like t statistics for the elements of $\tilde{\theta}_{(t,j)}$.

The empirical distribution of $\tilde{\theta}_{(t,j)}$ can be found by simulation using $\hat{\theta}$ as the parameter value to generate new series $y^{(k)}$ for k from 1 to K. The maximum likelihood estimate of θ for this simulation is $\hat{\theta}^{(k)}$ and we could find the simulated one-step estimate $\tilde{\theta}^{(k)}_{(t,j)}$ from $c^{(k)}_{(t,j)}\{\hat{\theta}^{(k)}\}$ given by (2) when the data are $y^{(k)}$ and $\theta = \hat{\theta}^{(k)}$. However we avoid calculation of $\hat{\theta}^{(k)}$ by noting that $c_{(t,j)}(\theta)$ is the difference of two scores. If the loglikelihood is approximately quadratic, as it must be for asymptotic theory to be useful, the two scores will be approximately linear in θ so that, on taking differences, the precise value of θ used to evaluate the scores is not important. So, from (3), we take the empirical distribution of

$$\breve{\theta}^{(k)}_{(t,j)} = \hat{\theta} + \jmath^{-1}(\hat{\theta})c^{(k)}_{(t,j)}(\hat{\theta}). \tag{4}$$

where

$$c^{(k)}_{(t,j)}(\hat{\theta}) = s^{(k)}_{(t,j)}(\hat{\theta}) - s^{(k)}(\hat{\theta}), \tag{5}$$

but now $s^{(k)}(\hat{\theta})$ is, in general, not equal to zero. It is hard to overstress the importance of this simple approximation which makes feasible the calculation of simulation envelopes for the computationally intensive methods of this paper.

4 Evaluating the Score

The models we analyse in this paper can be placed into the Gaussian state space form

$$y_t = Z_t\alpha_t + \epsilon_t,\ \alpha_t = T_t\alpha_{t-1} + \eta_t,\ t = 1, ..., n,$$
$$\alpha_0|Y_0 \sim N(a_0, P_0),\ \epsilon_t \sim N(0, H_t),\ \eta_t \sim N(0, Q_t), \tag{6}$$

where Y_t denotes the information available at time t. Here $\alpha_0|Y_0$, (ϵ_t) and (η_s) are independent of one another for all t and s and the initial conditions a_0 and P_0 are known. We will assume, for simplicity of exposition, that all the parameters which index this model are in the H_t and Q_t matrices. Then the score can be derived in two stages, following Koopman and Shephard (1992) who exploit an argument similar to that leading to the EM algorithm. First a Kalman filter is run through the data

$$a_{t+1|t} = T_{t+1}a_{t|t-1} + K_t v_t,\ P_{t+1|t} = T_{t+1}P_{t|t-1}L_t' + Q_{t+1},$$
$$v_t = y_t - Z_t a_{t|t-1},\ F_t = Z_t P_{t|t-1} Z_t' + H_t,$$
$$K_t = T_{t+1}P_{t|t-1}Z_t'F_t^{-1},\ L_t = T_{t+1} - K_t Z_t,\ t = 1, ..., n. \tag{7}$$

storing only v_t, F_t^{-1} and K_t. This is followed by the disturbance smoother of Koopman (1993) which is passed through the data, starting with $r_n = 0$ and $N_n = 0$ and letting

$$e_t = F_t^{-1}v_t - K_t'r_t,\ r_{t-1} = Z_t'F_t^{-1}v_t + L_t'r_t,$$
$$D_t = F_t^{-1} + K_t'N_tK_t,\ N_{t-1} = Z_t'F_t^{-1}Z_t + L_t'N_tL_t, t = n, ..., 1. \tag{8}$$

In (8) e_t, r_{t-1}, D_t and N_{t-1} have a simple intuitive interpretation: H_te_t is the smoothed estimate of ϵ_t and Q_tr_{t-1} the corresponding estimate for η_t. The mean squared errors are, respectively, $H_t - H_tD_tH_t'$ and $Q_t - Q_tN_{t-1}Q_t'$. These two recursive calculations yield the score function which has the form

$$\frac{\partial \log L(\theta)}{\partial \theta_i} = \frac{1}{2}\sum_{t=1}^{n} tr\left((e_te_t' - D_t)\frac{\partial H_t}{\partial \theta_i}\right) + \frac{1}{2}\sum_{t=1}^{n} tr\left((r_{t-1}r_{t-1}' - N_{t-1})\frac{\partial Q_t}{\partial \theta_i}\right), \tag{9}$$

which is one of the Koopman and Shephard results.

As Koopman's disturbance smoother only involves the multiplication of typically sparse vectors and matrices it tends to be much more rapid than the conventional smoothing algorithms. Usually it takes about the same time as does the Kalman filter for a forward pass.

4.1 Algebra of Added Variables

Suppose we now change the model to

$$y_t = Z_t\alpha_t + X_t\beta + \epsilon_t,\ \alpha_t = T_t\alpha_{t-1} + \eta_t,\ t = 1, ..., n,$$

$$\begin{pmatrix} \beta \\ \alpha_0 \end{pmatrix} | Y_0 \sim N\left(\begin{pmatrix} S_0^{-1}s_0 \\ a_0 \end{pmatrix}, \begin{pmatrix} S_0^{-1} & 0 \\ 0 & P_0 \end{pmatrix},\right),\ \epsilon_t \sim N(0, H_t),\ \eta_t \sim N(0, Q_t), \quad (10)$$

where the prior is now independent of ϵ_t and η_s for all t and s. Usually S_0 and s_0 may be taken to be a matrix and vector of zeros, respectively. Here X_t allows the introduction of explanatory variables, and hence interventions, into the model.

At first sight it seems that if we wish to manipulate the filtered and smoothed quantities, or the corresponding score, for this new model we have to rerun both (7) and (8), followed by recalculation of (9). However, use of the auxiliary recursions of de Jong (1989) circumvents this necessity. His recursions have a number of parts. The first part, the auxiliary filter, corrects the old run of the Kalman filter for the presence of the new explanatory variables. It does this by first computing

$$E\beta|Y_t = b_t = S_t^{-1}s_t, \text{ and } E(\beta - b_t)(\beta - b_t)'|Y_t = S_t^{-1} \quad (11)$$

where S_t and s_t are found recursively through

$$s_t = s_{t-1} + V_t'F_t^{-1}v_t,\ S_t = S_{t-1} + V_t'F_t^{-1}V_t,$$
$$V_t = -X_t - Z_tA_{t|t-1},\ A_{t+1|t} = T_tA_{t|t-1} + K_tV_t, t = 1, ..., n.$$

starting with $A_{1|0} = T_1a_0$.

The auxiliary filter is the generalization to time series models of the calculations leading to the added variable plot for regression models mentioned in §1. There the effect of a new regression variable X_t was found by first calculating its residual after regression on the other variables. Regression of the residuals of y_t on these new residuals provided quantities for the adjustment of the fitted model for inclusion of the extra variable. The same intuition applies to the auxiliary filter. We rerun the Kalman filter not on y_t, but instead on $-X_t$. The model is now

$$-X_{tj} = Z_t\alpha_t^* + \epsilon_t,\ \alpha_t^* = T_t\alpha_{t-1}^* + \eta_t,\ t = 1, ..., n\ ,\ j = 1, ..., p, \quad (12)$$

that is each vector of the matrix X_t is fitted by the old model. This generates the sequence of innovations which are stacked as V_t, each with mean square error F_t and Kalman gain K_t. They are unchanged from the old Kalman filter because the dynamics of the model are not altered and so do not have to be recomputed. Having run both y_t and $-X_t$ through the filter, the estimate of β is then b_t. This follows from the standard theory of generalized least squares as v_t and V_t are the recursive residuals from the two fits. It is useful to note that if y_t and β are univariate then all of these recursions, except for $A_{t+1|t}$, will be scalar. As $A_{t+1|t}$ is a vector and is formed by multiplying sparse matrices and vectors it will not take long to run: running through this filter is computationally trivial. Here only V_t will need to be stored for later use by the auxiliary smoother.

This result allows the efficient computation of the t statistic for β, for testing whether the intervention is significant. The estimates from the old Kalman filter $a_{t+1|t}$ and $P_{t+1|t}$ can be corrected for the effect of introducing the regression variables, but instead we turn our attention to correcting the disturbance smoother.

The second part of de Jong's recursions, the auxiliary smoother, has the form

$$E_t = F_t^{-1}V_t - K_t'R_t,\ R_{t-1} = Z_t'E_t + T_t'R_t, t = n, ..., 1$$

starting with $R_n = 0$. Again E_t and R_{t-1} have a simple intuitive interpretation. H_tE_t and Q_tR_{t-1} are the smoothed estimates of ϵ_t and η_{t-1} using $-X_t$ as the observations.

If, in an obvious notation, we write the new smoothing quantities for the model which includes the explanatory variables as e_t^*, R_t^*, D_t^* and N_t^* then

$$e_t^* = e_t + E_tb_n,\ r_{t-1}^* = r_{t-1} + R_{t-1}b_n$$
$$D_t^* = D_t - E_tS_n^{-1}E_t',\ N_{t-1}^* = N_{t-1} - R_{t-1}S_n^{-1}R_{t-1}'.$$

These simple expressions mean that the score for the marginal loglikelihood when X_t is the jth intervention at time t is

$$\frac{\partial \log L_{(t,j)}(\theta)}{\partial \theta_i} = \frac{1}{2}\sum_{t=1}^{n} tr\left((e_t^*e_t^{*\prime} - D_t^*)\frac{\partial H_t}{\partial \theta_i}\right) + \frac{1}{2}\sum_{t=1}^{n} tr\left((r_{t-1}^*r_{t-1}^{*\prime} - N_{t-1}^*)\frac{\partial Q_t}{\partial \theta_i}\right),$$

For the special case of y_t univariate and β scalar, which obtains in the example of §6 and those of Atkinson et al. (1993),

$$\frac{\partial \log L_{(t,j)}(\theta)}{\partial \theta_i} - \frac{\partial \log L(\theta)}{\partial \theta_i} = \frac{b_n^2 - S_n^{-1}}{2}\sum_{t=1}^{n}\left(E_t^2\frac{\partial H_t}{\partial \theta_i} + tr\left(R_{t-1}R_{t-1}'\frac{\partial Q_t}{\partial \theta_i}\right)\right)$$
$$+\ b_n\sum_{t=1}^{n}\left(e_tE_t\frac{\partial H_t}{\partial \theta_i} + tr\left(r_{t-1}R_{t-1}'\frac{\partial Q_t}{\partial \theta_i}\right)\right). \quad (13)$$

A full discussion of the implications of this expression is given by Atkinson et al. (1993), including the relationship with the t statistic for β.

4.2 One Step Re-estimation

The central theme of this paper is the assessment of the sensitivity or lack of robustness, of the inferences from the data analysis to small changes in the specification of the model. In time series modelling these inferences will be mainly determined by the value of the estimated parameters H_t and Q_t, for these control the rate of discounting and so influence forecasting, seasonal adjustment, detrending, etc. To focus attention on these parameters we argued in §3 that we transform from the scores into one-step scoring estimates of θ. With $\jmath(\theta)$ as the observed information matrix at θ, we are interested in the distance

$$\tilde{\theta}_{(t,j)} - \hat{\theta} = \jmath^{-1}(\hat{\theta}) \, \frac{\partial \, \log L_{(t,j)}(\hat{\theta})}{\partial \, \theta}. \tag{14}$$

or more precisely its scaled analog

$$d^i_{(t,j)} = \frac{\tilde{\theta}^i_{(t,j)} - \hat{\theta}^i}{\jmath^{-\frac{1}{2}ii}(\hat{\theta})} \tag{15}$$

where the superscripts denote elements of vectors and matrices. To detect sensitivity to departures we graph the elements of $d_{(t,j)}$ against t, for each value of j which indexes the type of intervention.

5 Simulation of Intervention Measures

Although the plots of $\tilde{\theta}_{(t,j)} - \hat{\theta}$ are scaled by an estimate of the standard error of $\hat{\theta}^i$, the small-sample distribution of $d_{(t,j)}$ is unknown. In regression diagnostics a similar type of problem exists when Cook's distance is used (Atkinson (1985), p.25). There simulation envelopes are often employed to give some impression of likely scatter. Here we follow a similar path.

As outlined in §3, we simulate a new time series $y^{(k)} = (y_1^{(k)}, ..., y_n^{(k)})'$ using $\hat{\theta}$ as the true value and record the value of $s(\hat{\theta}; y^{(k)})$. The distance this new score moves when we change the model is

$$\frac{\partial \, \log L_{(t,j)}(\hat{\theta}; y^{(k)})}{\partial \, \theta} - s(\hat{\theta}; y^{(k)})$$

which, as $n \to \infty$, will be an observation from the distribution we require. Hence we can produce a Monte Carlo sample of

$$d^{i(k)}_{t,j} = \jmath^{\frac{1}{2}ii}(\hat{\theta})\jmath^{-1}(\hat{\theta}) \left(\frac{\partial \, \log L_{(t,j)}(\hat{\theta}; y^{(k)})}{\partial \, \theta} - s(\hat{\theta}; y^{(k)}) \right), \tag{16}$$

where k=1,...,K, and so find asymptotically appropriate pointwise bounds for the plotted distance measures.

At first sight the computations involved in producing these replications look horrendous. However, as the parameters are always fixed at $\hat{\theta}$, most of the Kalman filter and smoothing algorithms have only to be run once since $P_{t+1|t}$, F_t, K_t, L_t, D_t and N_t do not depend on the data. As these are the most cumbersome parts of the computations these replications can be computed very quickly.

We use the K replications to construct one-sided pointwise confidence intervals; one-sided as we introduce new features into the model in order to reduce the variation in, for example, the irregular or slope components. Typically we will want to display 95% and 99% intervals. The results of Atkinson et al. (1993) indicate that taking $K = 399$ provides a good compromise between computation time and smoothness, so that the 4th and 20th largest simulated values provide the desired envelopes. These are smoothed using the S-Plus function 'lowess'.

6 Example: Coal Consumption

A series on the quarterly consumption of coal in the UK from 60Q1 to 83Q4 is analysed on pp.2 and 95 of Harvey (1989). The coal users are classified as 'other final users', a category including public administration, commerce and agriculture. The series shows a continual fall as the UK diversifies its energy sources and a strong seasonal pattern, which seems quite steady. After taking a logarithmic transformation we fitted a model with a stochastic trend, trigonometric seasonal and an additive irregular component. The fitted model has fixed, that is non-stochastic, slope and seasonal terms. The presence of fixed terms is not unusual in applied work.

Some summary statistics for the fit are given in Table 1. With two random components in the model there are two variances to be estimated: σ_ϵ^2 for the irregular and σ_η^2 for the level components. Of the skewness and kurtosis statistics S and K, the latter indicates some departure from normality of the innovations, which is confirmed by the value of 10.63 for the Bowman and Shenton normality test N, to be compared with χ_2^2. It is however much more illuminating to move from these aggregate statistics to the index plots of our new statistic. Fig.2 shows the effect of moving the intervention for an additive outlier along the series. The plot provides strong evidence for outliers at observations 37 and 39 and also at observations 59 and 61. When the model is refitted taking out these four additive outliers the value of N drops to 2.37, so that the evidence of non-normality disappears.

Although an improved model is obtained by allowing for outliers, the movements around the 60th observation, the 74Q4 value, are tricky to analyse. Atkinson et al. (1993) present evidence for a level break at this time — not only by use of a level intervention of the form of Fig.1c but also from visual inspection of the raw data. However, there may also be some additive outliers around this point, which corresponds to a major strike in the British coal mining industry and the

Statistic	Model 1	Model 2	Model 3
$\log_e \hat{\sigma}_\epsilon$	-2.22	-2.43	-2.43
$\log_e \hat{\sigma}_\eta$	-3.36	-3.37	-3.37
S	-0.58	-0.32	-0.33
K	4.18	2.54	2.48
N	10.63	2.37	2.71

Table 1: Parameter estimates and normality tests for the coal data: Model 1, original data; Model 2, outlier interventions at observations 37, 39, 59 and 61; Model 3, spaced switch interventions at observations 37/39 and 59/61. Variances: σ^2_ϵ, irregular; σ^2_η, level. S, K and N are the skewness, kurtosis and Bowman-Shenton normality test.

resultant turbulence of the restocking period. This turbulence is just the kind of departure which we would expect the switching intervention of Fig.1b to explain. Fig.3 shows the statistic for this intervention. The results are surprisingly like those for the additive outlier. To obtain Fig.3 a variable of the form ...0, 0, -1, 1, 0, 0, ... was used (the sign clearly being irrelevant). But inspection of Fig.2 led to identification of two pairs of outliers, each separated by one observation. Accordingly we try the spaced switch intervention of the form ...0, 0, -1, 0, 1, 0, 0, ... As Fig.4 shows the results confirm this choice (in comparing the three figures note that the third has been rescaled). There are two clear spaced switch interventions at observations 37/39 and 59/61 with perhaps another at observation 49/51. The two major switches are appreciably more significant, when judged against the simulation envelope, than the level break diagnostic given by Atkinson et al. (1993). The results of Table 1 show that the variances and statistics for the model with two switches are indistinguishable from those accomodating four outliers.

A final comment is on the form of intervention we have found useful. The switch variables were suggested by the interaction of economic knowledge and inspection of plots such as those of Fig.2. The spacing in the switch variable could be caused by the effect of quarterly data collection on a series in which consumption declines and gradually recovers after a strike or other upset. We do however find surprising the difference in the effect of the two kinds of intervention shown in Fig.3 and 4.

7 Discussion

In this paper we have exploited a new type of intervention analysis to suggest interesting directions for investigation and used score statistics and simulation

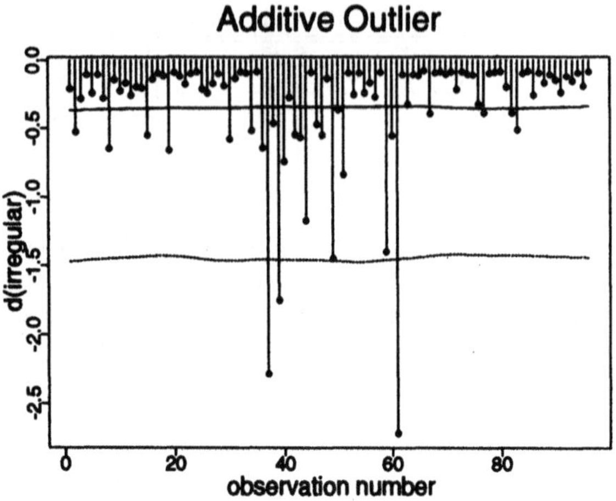

Figure 2: Coal Data: effect of an additive outlier.

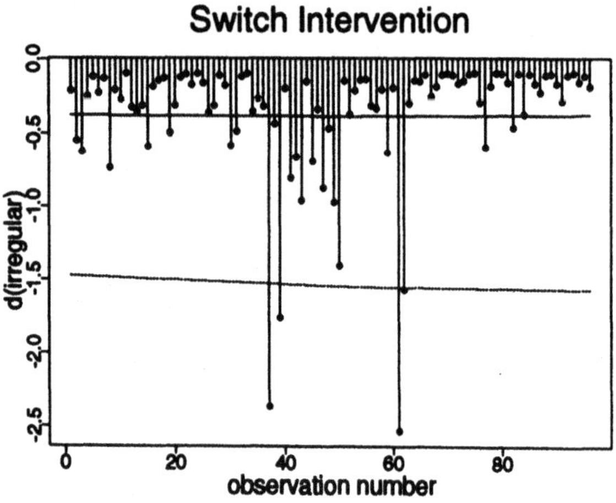

Figure 3: Coal Data: effect of a switch.

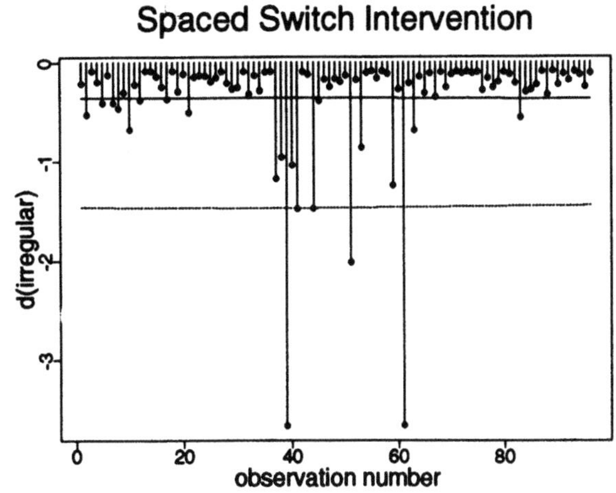

Figure 4: Coal Data: effect of the spaced switch.

envelopes to form a scaled distance to assess the importance of the interventions. The use of simulation envelopes depends crucially on the approximation to the score statistic introduced in §3. We also require a smoothing algorithm if the amount of computing is to be kept within reasonable limits. Our choice was the S-plus function 'lowess' (Becker, Chambers, and Wilks (1988), p.497) with the parameter f which controls the fraction of data used in smoothing equal to 0.35. Although we experimented with values of f for our examples, we did not experiment with other smoothers. The widespread application of our method would seem to require a robust smoothing algorithm which can carry out this part of the analysis in an automatic and satisfactory way – in particular in a way that automatically selects the bandwidth.

The example in the previous section raises questions about alternative forms of modelling and, for example, the meaning of a level shift in a model which already includes a stochastic level. A discussion of these and other topics in modelling time series is given by Atkinson et al. (1993).

References

Atkinson, A. C. (1985). *Plots, Transformations, and Regression.* Oxford: Clarendon Press.

Atkinson, A.C., S. J. Koopman, and N. Shephard (1993). Detecting shocks:

Outliers and breaks in time series. Technical report, Tinbergen Institute, Rotterdam, NL.

Becker, R. A., C. Chambers, and A. R. Wilks (1988). *The New S Language.* Pacific Grove, California: Wadsworth and Brooks/Cole.

de Jong, P. (1989). Smoothing and interpolation with the state space model. *Journal of the American Statistical Association 84*, 1085–88.

Harvey, A. C. (1989). *Forecasting, Structural Time Series Models and the Kalman Filter.* Cambridge: Cambridge University Press.

Koopman, S. J. (1993). Disturbance smoother for state space models. *Biometrika 80*, 117–126.

Koopman, S. J. and N. Shephard (1992). The exact score for time series models in state space form. *Biometrika 79*, 823–826.

Shephard, N. (1993). Maximum likelihood estimation of regression models with stochastic trend components. *Journal of the American Statistical Association 88*, 590–595.

Tunnicliffe Wilson, G. (1989). On the use of marginal likelihood in time series model estimation. *Journal of the Royal Statistical Society, Series B 51*, 15–27.

West, M. and P. J. Harrison (1989). *Bayesian Forecasting and Dynamic Models.* New York: Springer–Verlag.

FRÉCHET DIFFERENTIABILITY AND ROBUST ESTIMATION

Tadeusz Bednarski
Polish Academy of Sciences, Wrocław

Abstract

Differentiability of statistical functionals is naturally linked to robustness. A robust estimator and a smooth one have frequently similar formal sense. Some notions of differentiability used in statistics are discussed in this context and it is concluded that Fréchet's notion, for the supremum norm, gives a reasonable alternative. It is shown, under very mild assumptions, that estimators regular in a small model's vicinity are asymptotically equivalent to M-estimators resulting from Fréchet differentiable functionals. A general result concerning questions of existence of the differentiable functionals for one-dimensional parametric models is also presented.

Key words: robust estimation, von Mises functionals, M-estimation

AMS Subject Classification: 62G35

1 Introduction

The basic idea behind the calculation of von Mises statistical functionals (von Mises (1947)) is the approximation of a given statistic with a simpler one, facilitating asymptotic evaluations. Von Mises original ideas were not very much used by statisticians till late 60ties. The renaissance came with the birth of robust statistics when a number of new complex procedures appeared. Reeds (1976) presented the merits of von Mises method in connection with compact differentiability, the notion he skillfully developed and applied. One of the advantages of the calculus stressed by Reeds was a convenient separation of the analytic part (differentiability) and the probabilistic part consisting in approximation of the estimator by sums of independent random variables. The compact differentiability is now well known in the statistical literature (compare for instance Fernholz (1983), Gill (1989) and Rieder (1991)), also as a tool in optimal estimation of statistical functionals (van der Vaart (1991)).

Smoothness is naturally linked to robustness since it is an intuitive requirement there. When estimators result from statistical functionals then smoothness of the functionals becomes a natural requirement. Differentiability of the functionals would seem then to be the right formal concept to encompass at least some of the well established characteristics of robust estimators. In fact, because of the variety of different differentiability notions the situation complicates at this point. To give an example for the best known notion of differentiability: a functional compactly differentiable need not be even continuous in the supremum norm for the cumulative distribution functions. Compactly differentiable functionals need not satisfy such elementary robustness conditions as bounded influence function or non-zero breakdown point (compare again van der Vaart (1991)). As a matter of fact much stronger objections can be formulated with respect to Gateaux differentiability.

The Fréchet differentiability which appeared very suitable there, was at least for some time considered a too rare event in statistics. Fortunately Boos (1979) and Clarke (1983, 1986) have shown that some well known robust estimators result from Fréchet differentiable functionals (see also Serfling (1980)).

It seems that Fréchet differentiable functionals may still find their place in the statistical inference. A reasonable estimation procedure should be nearly optimal at the conceived model and it should be stable under small deviations from the model (we say small to indicate that the statistical model should, at least in an approximative sense, reflect the law of the studied random mechanism). It should also be effectively computable and finally we should be able to assess its variability in a credible way. In Bednarski(1991) and Bednarski and Zontek(1993) it is shown that Fréchet differentiable functionals satisfying the above qualitative requirements can be constructed for the important estimation problems in some basic statistical models: estimation of the regression parameter in the Cox model and estimation of variance components in the normal model.

Some important implications of Fréchet differentiability are immediate. The differentiability implies continuity in the norm thus leading to nonzero breakdown point (the breakdown point is here understood according to the original Hampel's definition). It also implies estimator's regularity in a shrinking nonparametric neighborhood of the model; the estimator has an expansion which holds under product measures with marginals from the shrinking neighborhoods. This strong expansion in turn assures the possibility of effective assessment of estimator's variability near the model; a property which must be appreciated in any estimation problem. The regularity of the estimator (functional) in a small vicinity of the model yields also boundedness of the influence function. Having finally concluded the bounded gross error sensitivity we can see that the essential robustness objectives are fulfilled for estimators resulting from Fréchet differentiable functionals.

In the following section basic definitions of differentiability are recalled and a discussion of their relevant statistical properties is given. Section 3 presents a refinement of a result of Bednarski, Clarke and Kolkiewicz(1991) concerning relations between strong expansions and Fréchet differentiability of statistical functionals. In Section 4 a general question of existence and constructability of Fréchet differentiable functionals for regular models is discussed.

2 Differentiability Concepts and Infinitesimal Supermodels

We begin with formal definitions of the most common concepts of differentiability used in statistics. This will constitute a ground for our further discussion. Take $\mathcal{L}$ to be a linear space equipped with a norm "$\| \ \|$" and suppose $\mathcal{G}$ is a family of subsets of $\mathcal{L}$. A functional $T : \mathcal{L} \mapsto R^k$ is differentiable at $F \in \mathcal{L}$ with respect to $\mathcal{G}$ if there is a linear functional $TL : \mathcal{L} \mapsto R^k$ such that for all $A \in \mathcal{G}$

$$\sup_{x \in \mathcal{A}} | T(F + hx) - T(F) - hTL(x) | = o(h).$$

If $\mathcal{G}$ is composed of singletons from $\mathcal{L}$, then we call the functional Gateaux differentiable. When $\mathcal{G}$ is the family of compact sets then it is said to be compactly (Hadamard) differentiable. The Fréchet differentiability (bounded) holds when $\mathcal{G}$ is the family of bounded sets. Gateaux differentiability is usually used as a heuristic tool which yields the influence function for the functional T. Hadamard and Fréchet differentials are used to assess the asymptotic behavior of the estimator $T(F_n)$ and they give a more precise meaning to the influence function (Hampel(1974)).

The formal definition of differentiability given above has elements that are not essential in the statistical context (asymptotic derivations). For instance x is usually an element of a convex cone spanned by differences of cumulative distribution functions. The linear functional TL is always a result of construction of a functional and it is useful when represented as an integral. The choice for norms is also limited for they should induce neighborhoods containing empirical distribution functions. P-variation norms were used by Dudley (1992) to demonstrate that the inverse and composition operators are Fréchet differentiable. Dudley's differentiability was applied by Florczak (1993) to handle the asymptotic behavior of some generalized functionals which include measures of spread and dispersion, generalized L-functionals and V-statistics. The above cases were not manageable by the supremum norm, which leads to a stronger differentiability notion. In fact we shall be more interested in the stronger differentiability notion having in mind rather a perspective of construction than of studying the asymptotic behavior of known statistical procedures. There are also other reasons favoring this simple norm. First of all neighborhoods induced by this norm are large, thus giving high smoothness to the functional. Secondly, by virtue of the Kolmogorov-Smirnov theorem for comparing two distributions, this norm appears naturally in statistics. Finally we are going to use the notion of infinitesimal neighborhood $\{G : \| G - F \| \leq \epsilon/\sqrt{n}\}$ which complies then with the ability of testing goodness of fit. More precisely if F is a member of a parametric family and Kolmogorov type statistic is used to verify goodness of fit, then the hypothesis can be accepted with asymptotically positive probability for distributions from these infinitesimal neighborhoods only.

If a functional T is Fréchet differentiable at a distribution F and F_n is the empirical distribution function based on a sample of size n from F then usually for some function ψ

$$T(F_n) - T(F) = \int \psi dF_n + o_P(\| F_n - F \|).$$

Therefore by the inequality of Kiefer (1961) (see also Dvoretzky, Kiefer and Wolfowitz (1956) and Serfling (1980)), the asymptotic normality of $\sqrt{n}[T(F_n)-T(F)]$ under $F^{\otimes n}$ (product distribution) follows directly. Let us notice that if in fact F_n is the empirical distribution based on n-sample from the distribution G_n for which $\sqrt{n} \parallel G_n - F \parallel$ is bounded, then the stochastic expansion implied by the above formula holds also under $G_n^{\otimes n}$. Therefore the asymptotic behavior of $T(F_n)$ can be established for the infinitesimal neighborhoods of the form $\{G :\parallel G - F \parallel\leq \epsilon/\sqrt{n}\}$ directly from the expansion ! This is a very important property which is not shared by other concepts of differentiability.

Any Fréchet differentiable functional gives an estimator whose distribution can be approximated in the whole nonparametric shrinking neighborhood. This forms a natural link between the differentiation and consideration of shrinking supermodels. The idea of this kind of supermodels raises in fact quite frequently objections of the following nature: how can we suppose that marginal distributions "move" with the sample size, or in our case, how can the size of contamination change with sample size. In a real experiment we are "probably" not far from the truth imagining that we have a fixed marginal.

From the theoretical point of view this local asymptotic approach leads to formulations of reasonable optimality criterions as it is done in the purely parametric statistics (Bickel(1981)). From the practical perspective we should understand firstly the approximative value of the approach. For a nearly true parametric model with a reasonable sample size the theory should give good approximation to distributions of statistics of interest. We should have

$$T(F_n) - \theta = \int \psi(x,\theta)dF_n + o_{G^{\otimes n}}(1/\sqrt{n})$$

for G near F_θ. The philosophy is exactly the same as the one which let us believe that the Poisson distribution approximates well the binomial when the probability of success converges to zero at the rate $1/n$.

3 Expansions and Fréchet Differentiability

This section shows that an estimator $T(F_n)$ which has the strong expansion is in fact, under very natural additional conditions, an M-estimator which induces a Fréchet differentiable functional.

Let $\{F_\theta\}_{\theta\in\Theta}$ be a family of probability measures defined on the Borel σ-field of R^m, where Θ is an open subset of R^k. Denote by $d(\cdot,\cdot)$ a metric on the space $\mathcal{G}$ of probability measures on R^m and fix $\theta \in \Theta$. We shall further consider sequences of parameters approaching θ and sequences of probability measures approaching F_θ, all with the rate $1/\sqrt{n}$:

$$J_n(b) = \{\tau : \sqrt{n}|\tau - \theta| \leq b\},$$

$$\mathcal{F}_{n,\tau}(\eta) = \{Q \in \mathcal{G} : \sqrt{n}d(Q, F_\tau) < \eta\},$$

where $|\cdot|$ stands for the usual Euclidean norm and $\tau \in U$ a neighborhood of θ. We shall write J_n instead of $J_n(b)$ and $\mathcal{F}_n$ instead of $\mathcal{F}_{n,\theta}(\eta)$ when it will not be essential what b and η are.

The sets $\mathcal{F}_{n,\tau}$ form the shrinking supermodel under which we shall express assumptions concerning the expansions. It will be supposed in the sequel that the model satisfies the following smoothness condition:
for some positive constants $k_1 < k_2$ we have

$$k_1|\theta - \tau| \leq d(F_\theta, F_\tau) \leq k_2|\theta - \tau| \tag{1}$$

for all $\tau \in U$.

Though we shall formulate the basic conditions in terms of a general metric d one can actually think of the supremum norm for the distributions. The norm generates neighborhoods large enough to imply robustness of estimators there.

Conditions B

B1 There is a measurable function $\psi : R^m \times \Theta \to R^k$ such that

$$sup_{\tau \in J_n} \sqrt{n}|T(F_n) - \tau - \int \psi(x,\tau)dF_n|$$

converges to zero in $G_n^{\otimes n}$, where F_n is the empirical distribution function based on n-sample and $G_n \in \mathcal{F}_{n,\theta}$.

B2 The random variables $\sqrt{n}(T(F_n) - \theta)$ are relatively compact under $G_n^{\otimes n}$, for all $G_n \in \mathcal{F}_n$

B3 The functions $K_G(\cdot) = \int \psi(x,\cdot)dG(x)$ are continuous in U for all $G \in \mathcal{F}_n$ if n is sufficiently large.

The above conditions are less restrictive than those given in Bednarski, Clarke and Kolkiewicz (1991) and they also imply the Fréchet differentiability of any "reasonable" functional T^* defined to be a solution of the equation $K_{F_n}(\tau) = 0$. The conditions simply say that for a smooth model we are given a robust, in a sense, estimator (B2), which has a strong expansion (B1). In fact (B1) means that the asymptotic variance of the estimator can be assessed via the influence function also at a small vicinity of the model. The assumption (B3) is included to make sure that we can define T^* which will be shown equivalent asymptotically to T and Fréchet differentiable. The choice of T^* may in fact be a serious problem and not only theoretical if we try to define usable M-estimators. Since here we are concerned only with an existence problem of a Fréchet differentiable functional we can define T^* with use of a minimum distance functional as in Clarke (1986).

Theorem 3.1. Expansions and Differentiability. Under the conditions B1-B3 the functional T^* is locally uniformly Fréchet differentiable with respect to d, that is

$$|T^*[G] - \tau - K_G(\tau)| = o(d(G, F_\tau) + d(F_\tau, F_\theta)).$$

The theorem follows from the following two lemmas and arguments given in Bednarski, Clarke and Kolkiewicz (1991).

Lemma 3.1. There is a neighborhood U_0 of θ such that ψ is bounded in $R^m \times U_0$.

Proof: The condition B3 implies that $K_{F_\theta}(\tau)$ is bounded in U_0. Suppose then by contradiction that there is a sequence $\{(x_n, \tau_n)\}$ for which $\psi(x_n, \tau_n) \to \infty$ (or $-\infty$). Take $G_n = (1 - \eta/\sqrt{n})F_\theta + \eta/\sqrt{n} \times \delta_{x_n}$ and note that the law of ψ under G_n is the same as that of

$$Y = I\psi(X, \tau_n) + (1 - I)\psi(x_n, \tau_n)$$

with X distributed according to F_θ and I independent indicator variable with $P(I = 1) = 1 - \eta/\sqrt{n}$.

Therefore $K_{F_n}(\tau_n)$ under G_n has the same law as the sum of n-independent copies of Y. Since by B1 $\sqrt{n}K_{F_n}(\tau_n)$ is bounded in probability, the same is true for $(1/\sqrt{n})\sum_{i=1}^n I_i\psi(X_i, \tau_n)$. On the other hand $(1/\sqrt{n})\sum_{i=1}^n(1 - I_i)$ have variances approaching zero and moreover a constant and positive expectation. Thus

$$(1/\sqrt{n})\psi(x_n, \tau_n)\sum_{i=1}^{n}(1 - I_i)$$

is not bounded in probability and in effect $(1/\sqrt{n})\sum Y_i$ cannot be bounded.

Lemma 3.2. The function $K_{F_\theta}(\tau)$ is differentiable at $\tau = \theta$ and the derivative is the identity matrix.

Proof: It follows from B1 that under sequences $\{G_n\}$, $G_n \in \mathcal{F}_n$, we have

$$\sqrt{n}[\theta - \tau_n + \int(\psi(x, \theta) - \psi(x, \tau_n))dF_n] \tag{2}$$

converges to zero in probability. It will be sufficient then to show that the expected values of (2) under F_θ tend to zero.

Let us suppose by contradiction that the expected values do not converge to zero. Then, there must be a sequence of x_n and τ_n so that $a_n = \sqrt{n}[\psi(x_n, \theta) - \psi(x_n, \tau_n)]$ tend to ∞ (or $-\infty$). We can take again as in the previous proof the measures $G_n = (1 - \eta/\sqrt{n})F_\theta + \eta/\sqrt{n}\delta_{x_n}$ and argue similarly to conclude that (2) cannot hold under G_n. Therefore we obtain

$$lim_n sup_{\tau \in J_n}\sqrt{n}[\theta - \tau_n + K_{F_\theta}(\theta) - K_{F_\theta}(\tau_n)] = 0.$$

The main message of the theorem is that strongly expendable estimators are equivalent in the asymptotic sense to Hubers's M-estimators (Huber(1981)), when they result from Fréchet differentiable functionals. It is well known however that M-functionals need not be in general associated with Fréchet differentiable functionals. Another point one can make here is connected with results of Maronna (1976) and Huber (1977) who prove that affine-equivariant M-estimates of location and scatter have breakdown points bounded above by $1/k$ in dimension k. Donoho and Gasko (1992) comment it that "...in high dimensions, such "robust" estimators can be upset by a relatively small fraction of outliers strategically placed." In view of Theorem 3.1 we can suspect that in the mentioned case the estimators with high breakdown point will not possess the strong expansion and in consequence we may expect that their variance will be "difficult" to assess.

4 Existence of Fréchet differentiable functionals for smooth parametric families

Minimax asymptotic testing problems for supermodels formed by shrinking capacities were studied by Huber-Carol(1970), Rieder(1978) and Bednarski(1985,1986). Resulting facts and techniques were then used to answer certain general questions concerning robust estimation for smooth one-dimensional parametric models (see Rieder(1980) and Bednarski(1985a)). It turns out that the methodology based on minimax testing between capacities (cf. Huber and Strassen (1972) and Bednarski (1981),(1982)) largely developed in the above quoted papers can be used to show that Fréchet differentiable functionals always exist for the smooth models. In the argumentation below we shall employ Rieder's(1978, 1980) results to outline the functional's existence proof for a particular type of shrinking neighborhoods given by the total variation distance. In fact we shall also argue an optimality of the differentiable functional.

Suppose that F_θ is a one-dimensional model which satisfies the regularity conditions stated in Rieder(1980). Define a shrinking supermodel $\mathcal{F}_{\theta,n}$ with the use of the total variation distance

$$\mathcal{F}_{n,\theta}(\delta) = \{Q : \|Q - F_\theta\|_{t.v.} \leq \delta/\sqrt{n}\} .$$

Fix θ and consider local asymptotic testing problems

$$\theta_{1n} = \theta_0 - \tau\sqrt{n} \quad against \quad \theta_{2n} = \theta_0 + \tau\sqrt{n}$$

($\tau > 0$) for n samples. The N P lemma provides then sequences of test statistics based on the log-likelihood ratios

$$\sum_{i=1}^{n} log\{dF_{\theta_{2n}}/dF_{\theta_{1n}}\}(X_i)$$

which are asymptotically equivalent to tests based on $\sum \Lambda_\theta(X_i)$; the sum of scores. A robustification of these testing problem (cp. Huber and Strassen (1972) and Bednarski(1992)) consists in considering asymptotically minimax tests for $\mathcal{F}_{\theta-\tau/\sqrt{n},n}$ against $\mathcal{F}_{\theta+\tau/\sqrt{n},n}$. When the supermodel is defined by Choquet capacities then, as in the model case, a simple form of test statistic is obtainable; for the total variation neighborhoods it is simply $\sum \widetilde{\Lambda}_\theta(X_i)$ where $\widetilde{\Lambda}_\theta = d'_\theta \vee \Lambda_\theta \wedge d''_\theta$ with functions d'_θ and d''_θ given uniquely by the equations

$$\int (d'_\theta - \Lambda_\theta)^+ = \int (\Lambda_\theta - d''_\theta)^+ = \delta/\tau .$$

In Rieder(1980) this robust score function is a basic component of the construction of asymptotically minimax estimators for the shrinking supermodels. Rieder's optimality criterion is, roughly speaking, the probability of covering the true parameter value by confidence intervals. The main result specifying the form of the optimal robust estimator is not given for the entire supermodel; it is restricted to its contiguous part. Rieder however briefly outlines reasoning which validates his results for the full shrinking supermodel (see the section on "Complementary remarks" of his paper). We shall take it for granted further.

The optimal robust estimator T_n satisfies then in particular the following conditions:

(4.1) $\sqrt{n}|T_n - \theta - \int IC_\theta dF_n|$ converges to zero in probability under product measures with marginals from $\mathcal{F}_{n,\theta}(\delta)$, where $IC_\theta = \tilde{\Lambda}_\theta / \dot{\lambda}(\theta)$ and $\dot{\lambda}(\theta) = -\int \tilde{\Lambda}_\theta \Lambda_\theta dF_\theta$.

(4.2) $\lambda_\theta(\xi) = \int \tilde{\Lambda}_\xi dF_\theta$ is differentiable at $\xi = \theta$ with the derivative equal to $\dot{\lambda}(\theta)$

The conditions (4.1) and (4.2) easily imply our condition B1, while B2 is a consequence of the Lindeberg-Feller theorem and the fact that $(1/\sqrt{n}) \sum IC_\theta(X_i)$ are relatively compact under product measures with marginals from the shrinking neighborhoods. Since the condition B3 is not restrictive for smooth statistical models we conclude the Fréchet differentiability of a functional T^* defined implicitly by $\int \tilde{\Lambda}_\theta dG = 0$.

Argumentation given above is in fact very complicated in details. Firstly, it is based on a very mathematical theory of minimax testing for capacities. Secondly it relies on a complicated and technical approach to the asymptotic minimax testing for shrinking capacities. This all certainly puts limits for patience of possibly interested researchers. There is moreover another restriction: the methods and results used above are limited to the one-dimensional parametric models. This seems fortunately of importance only when we think of optimal functionals, otherwise it should not cause difficulties in transferring results to the multidimensional case. The most constraining aspect is however that we have proved the Fréchet differentiability for the total variation norm and from that point there is still a long way to the asymptotic derivations. It would certainly be more practical to know that T^* can be defined over the neighborhoods of F_θ generated by the supremum norm. Below we phrase regularity conditions which by Bednarski, Clarke and Kolkiewicz (1991) p. 94 lead in a straightforward way to the desired differentiability.

Take $\tilde{\Lambda}_\theta$ to be Rieder's score function and let $\dot{\lambda}(\theta) = -\int \tilde{\Lambda}_\theta \Lambda_\theta dF_\theta$.

C1 $\lambda_\theta(\xi) = \int \tilde{\Lambda}_\xi dF_\theta$ is differentiable at $\xi = \theta$ with the derivative equal to $\dot{\lambda}(\theta)$

C2 There is a neighborhood of θ such that for all G in a sup-norm neighborhood of F_θ the integrals $\int \tilde{\Lambda}_\tau dG$ are continuous.

C3 $\int \tilde{\Lambda}_\theta d(G - F_\theta) = O(\| G - F_\theta \|_{sup})$

C4 $\int (\tilde{\Lambda}_\theta - \tilde{\Lambda}_\tau) d(G - F_\theta) = o(|\theta - \tau| + \| G - F_\theta \|_{sup})$

The first two conditions were discussed already. The last two follow if $\widetilde{\Lambda}_\theta$ is smooth enough. Notice however that the usual regularity assumptions require Λ_θ to be smooth, while $\widetilde{\Lambda}_\theta$ is just its smooth and bounded truncation.

In any case it appears promising to learn that Fréchet differentiability for the supremum norm is at least potentially as common in statistical inference as common are regular models.

REFERENCES

Bednarski, T. (1981): On solutions of minimax test problems for special capacities. Z. Wachrsheinlichkeitstheorie und Verw. Gebiete. 58, 397-405.

Bednarski, T. (1982): Binary experiments, minimax tests and 2-alternating capacities. Ann.Statist. Vol. 10, 226-232.

Bednarski, T. (1985a): A robust asymptotic testing model for special capacities. Statistics 16, 507-519.

Bednarski, T. (1985b): On minimum bias and variance estimation for parametric models with shrinking contamination. Probability and Mathematical Statistics, 6, 121-129.

Bednarski, T., Clarke, B.R., and Kolkiewicz, W. (1991): Statistical expansions and locally uniform Fréchet differentiability. J. Australian Math. Soc., Ser. A, 50, 88-97.

Bednarski, T. (1991): Robust estimation in Cox's regression model. Preprint of the Institute of Mathematics of the Polish Ac. of Sciences. To appear in the Scand. J. Statist.

Bednarski and Zontek, S. (1993): On robust estimation of variance components via von Mises functionals. Institute of Mathematics, Polish Academy of Sciences. Preprint 504.

Bickel, P.J. (1981): Quelques aspects de la statistique robuste. Lecture Notes in Math. 876, 1-72.

Boos, D.D.(1979): A differential for L-statistics. Ann. Statist., 7, 955-959.

Clarke, B.R.(1983): Uniqueness and Fréchet differentiability of functional solutions to maximum likelihood type equations. Ann. Statist., 11, 1196-1206.

Clarke, B.R.(1986): Nonsmooth analysis and Fréchet differentiability of M-functionals. Probab. Th. Rel. Fields 73, 197-209.

Donoho, D.L. and Gasko, M. (1992): Breakdown properties of location estimates based on halfspace depth and projected outlyingness. Ann. Statist., Vol. 20, 1803-1827.

Dudley, R.M. (1989): Nonlinear functionals of empirical measures and the bootstrap. Unpublished manuscript.

Dvoretzky, A., Kiefer, J. and Wolfowitz, J. (1956): Asymptotic minimax character of the sample distribution function and of the classical multinomial estimator. Ann. Math. Statist., 27, 642-669.

Fernholz, L.T.(1983): Von Mises calculus for statistical functionals. Lecture Notes in Statistics. Vol. 19. Springer Verlag, New York.

Florczak, W. (1993): Asymptotic representation of generalized functionals. Unpublished manuscript.

Gill, R.D. (1989): Non- and semi-parametric maximum likelihood estimators and the von Mises method. Part 1. Scand. J. Statist. 16, 97-128.

Hampel, F.R. (1974): The influence function and its role in robust estimation. J. Amer. Math. Assoc. 62, 1179-1186.

Huber, P.J. (1977): Robust covariances. In Statistical Decision Theory and Related Topics II (S.S. Gupta and D.S. Moore eds.) 165-191. Academic, New York.

Huber, P.J. (1981): Robust Statistics. Wiley, New York.

Huber-Carol, C. (1970): Etude asymptotique des tests robustes. PhD thesis ETH, Zurich.

Kiefer, J. (1961): On large deviations of the empiric D.F. of vector chance variables and a law of iterated logarithm. Pacific J. Math., 11, 649-660.

Maronna, R.A. (1976): Robust M-estimates of multivariate location and scatter. Ann. Statist. 4, 51-67.

Reeds, J.A. (1976): On the definition of von Mises functions. Ph.D. thesis. Dept. of Statist. Harvard Univ., Cambridge.

Rieder, H. (1978): A robust asymptotic testing model. Ann. Statist. Vol. 6, 610-618.

Rieder, H. (1980): Estimates derived from robust tests. Annals of Statist., 8, 106-115.

Rieder, H. (1991): Robust Statistics I. Asymptotic Statistics. University of Bayreuth. Preliminary version of a book.

Serfling, R.J. (1980): Approximation Theorems of Mathematical Statistics. Wiley, New York.

Van der Vaart, A.(1991): Efficiency and Hadamard differentiability. Scand. J. Statist. 18, 63-75.

Von Mises, R.(1947): On the asymptotic distribution of differentiable statistical functionals. Ann. Math. Statist. 18, 309-348.

Institute of Mathematics, Polish Academy of Sciences
ul.Kopernika 18, 51-617 Wroclaw, Poland

STEIN PREDICTORS AND PREDICTION REGIONS

Rudolf Beran
University of California, Berkeley

Consider the problem of predicting a q-random vector Y on the basis of observing a q-random vector X. Here X and Y are independent random vectors, whose distributions are $N(\xi, \sigma_X^2 I)$ and $N(\xi, \sigma_Y^2 I)$ respectively. The two variances σ_X^2 and σ_Y^2 are assumed known, while the mean vector ξ is unknown. This paper studies the Stein predictor $\hat{Y}_S = [1-(q-2)\sigma_X^2/|X|^2]X$ for Y and associated Stein prediction regions. A Stein prediction region is a sphere centered at $\hat{Y}_S$ whose radius $\hat{d}$ is chosen so that coverage probability for Y converges to α as q increases. Our main results are:

- The Stein predictor $\hat{Y}_S$ dominates the classical predictor X under quadratic loss and achieves an asymptotic minimax bound on large compact balls centered at $\xi = 0$.
- Stein prediction regions dominate the classical prediction sphere for Y, as q tends to ∞, under a geometrical risk criterion.
- Valid bootstrap critical values for a Stein prediction region require resampling X from a $N(\hat{\xi}, \sigma_X^2 I)$ distribution and resampling Y from a $N(\hat{\xi}, \sigma_Y^2 I)$ distribution, where $|\hat{\xi}|$ estimates $|\xi|$ well. Neither $\hat{\xi} = X$ nor $\hat{\xi} = \hat{Y}_S$ will do.
- Proper choice of the radius $\hat{d}$ reduces the coverage probability error of a Stein prediction region from $O(q^{-1/2})$ to $O(q^{-1})$.

Key words. Stein estimator, coverage probability, geometrical risk, bootstrap.
AMS 1991 subject classifications. Primary 62M20; secondary 62G07.

1. Introduction

Consider the problem of predicting a discretized random curve, measured at q time points, after observing a learning sample of n such discretized curves. The work on gait analysis by Olshen et al. (1989) exhibits how this question may arise in practice. If the random curves are taken to be Gaussian, a canonical formulation of the prediction problem is as follows. Let X and Y be independent q-dimensional random vectors, whose distributions are $N(\xi, \sigma_X^2 I)$ and $N(\xi, \sigma_Y^2 I)$ respectively. The mean vector ξ is unknown, while the two variances σ_X^2 and σ_Y^2 are assumed known. Having observed the learning variable X, we seek to predict the future observable Y. For the example of discretized Gaussian random curves, the curve X would be the average of the n curves in the learning sample.

The quality of a predictor $\hat{Y} = \hat{Y}(X)$ for Y is measured by the quadratic risk

$$R_q(\hat{Y}, \xi) = q^{-1} E_\xi |\hat{Y} - Y|^2, \tag{1.1}$$

where $|\cdot|$ is Euclidean norm and E_ξ denotes expectation with respect to the normal distribution P_ξ of (X, Y). This model for (X, Y) is completely general as regards the unknown mean ξ, which can be any vector in R^q. The normality assumption is used fully only in the asymptotic minimax results of Sections 2.1 and 3.1. More restrictive is the assumption that the variances σ_X^2 and σ_Y^2 are known. In principle, the assumption on variances can be relaxed if σ_Y^2 has a known relationship to σ_X^2 and σ_X^2 is estimable from the learning sample. We will not pursue such extensions in this paper.

Because

$$R_q(\hat{Y}, \xi) = q^{-1} E_\xi |\hat{Y} - \xi|^2 + \sigma_Y^2, \tag{1.2}$$

the classical predictor $\hat{Y}_C = X$ has risk equal to $\sigma_X^2 + \sigma_Y^2$. When q equals 1 or 2, $\hat{Y}_C$ is admissible in that no other predictor has risk smaller or equal to the risk of $\hat{Y}_C$. However, when $q \geq 3$, the Stein predictor

$$\hat{Y}_S = [1 - (q-2)\sigma_X^2 / |X|^2] X \tag{1.3}$$

has risk

$$R_q(\hat{Y}_S, \xi) = \sigma_X^2 \{1 - q^{-1} E_\xi [(q-2)^2 \sigma_X^2 / |X|^2]\} + \sigma_Y^2 < \sigma_X^2 + \sigma_Y^2, \tag{1.4}$$

which is strictly less than the risk of $\hat{Y}_C$ at every ξ and reaches a minimum value of $2\sigma_X^2/q + \sigma_Y^2$ at $\xi = 0$ (cf. James and Stein, 1961). The extent to which $\hat{Y}_S$ improves on $\hat{Y}_C$ in risk can thus be considerable when ξ is near 0.

Though $\hat{Y}_S$ is not quite admissible itself, it is very nearly a minimax predictor, when q is large, on large compact balls centered at $\xi = 0$. The classical predictor $\hat{Y}_C$ lacks this approximate minimax property. Section 2 develops this topic by drawing on analogous asymptotic minimax bounds for estimators of ξ when dimension q is large. The latter bounds follow from ideas in Stein (1956), as shown in Beran (1992), or from a result in Pinsker (1980), as described by Donoho and Johnstone (1989).

Constructing good prediction regions for Y on the basis of X is the second major theme of the paper. Section 3 formulates the problem of finding a prediction sphere for Y,

$$D(\hat{Y}, \hat{d}) = \{t \in R^q : |\hat{Y} - t| \leq \hat{d}\}, \tag{1.5}$$

that is centered at $\hat{Y} = \hat{Y}(X)$ and has radius $\hat{d} = \hat{d}(X)$. The geometrical risk of prediction region D is defined to be

$$G_q(D,\xi) = q^{-1/2} E_\xi \sup_{t \in D} |t - Y| = q^{-1/2} E_\xi |\hat{Y} - Y| + q^{-1/2} E_\xi \hat{d}. \tag{1.6}$$

The aim is to construct D so as to minimize geometrical risk subject to the requirement that coverage probability $P_\xi[Y \in D(\hat{Y}, \hat{d})]$ be at least α. It is shown that Stein prediction regions of the form

$$D_S = D(\hat{Y}_S, \hat{d}) \tag{1.7}$$

have smaller asymptotic geometrical risk than do classical prediction regions of the same asymptotic coverage probability that are centered at $\hat{Y}_C = X$.

The critical value $\hat{d}$ in (1.6) may be obtained directly from an asymptotic distribution or indirectly from bootstrap approaches. Straightforward asymptotic or bootstrap constructions of $\hat{d}$ result in a coverage probability error of $O(q^{-1/2})$. A variance stabilizing transformation before bootstrapping reduces coverage probability error to $O(q^{-1})$, as shown in Section 3. It proves crucial, in the bootstrap approaches, to resample X and Y from the $N(\hat{\xi}, \sigma_X^2 I)$ and $N(\hat{\xi}, \sigma_Y^2 I)$ distributions respectively, where $|\hat{\xi}|$ estimates $|\xi|$ well. Taking $\hat{\xi}$ to be X or the Stein estimator $\hat{Y}_S$ does not work.

The Stein predictor and prediction regions studied in this paper shrink X towards the origin in R^q. Of course the origin can be replaced by any other constant in R^q. More generally, related asymptotics can be used to design and analyze prediction procedures that shrink X towards a data-dependent point in R^q, such as a fitted linear model for ξ. Such shrinkage procedures robustify prediction based on the fitted linear model against departures from that linear model assumption on ξ. These points will be developed elsewhere.

2. Stein Predictors

The risk decomposition (1.2) shows that a good estimator of ξ, based on X, is also a good predictor of Y under quadratic loss. This section maps asymptotic minimax results for Stein estimation into their counterparts for prediction. Both analytical and geometrical arguments are developed.

2.1. Asymptotic minimax bounds

Fundamental for point prediction of Y under quadratic loss is the asymptotic minimax bound stated below. This bound is attained asymptotically by the Stein predictor $\hat{Y}_S$ defined in (1.3) and by alternative predictors to be described in Section 2.3. All proofs in this paper are deferred to Section 4.

THEOREM 2.1. *For every finite* $c > 0$,

$$\liminf_{q\to\infty} \inf_{\hat{Y}} \sup_{|\xi|^2 \le qc} R_q(\hat{Y}, \xi) \ge r(c/\sigma_X^2) \tag{2.1}$$

where the infimum is taken over all predictors $\hat{Y} = \hat{Y}(X)$ *and*

$$r(t) = [t/(1+t)]\sigma_X^2 + \sigma_Y^2. \tag{2.2}$$

The Stein predictor $\hat{Y}_S$ *attains the bound (2.1) in that*

$$\lim_{q\to\infty} \sup_{|\xi|^2 \le qc} R_q(\hat{Y}_S, \xi) = r(c/\sigma_X^2) \tag{2.3}$$

for every finite $c > 0$.

The classical point predictor $\hat{Y}_C = X$ is suboptimal as dimension q increases because

$$\lim_{q\to\infty} \sup_{|\xi| \le qc} R_q(Y_C, \xi) = \sigma_X^2 + \sigma_Y^2 \tag{2.4}$$

for every finite $c > 0$. The right side of (2.4) exceeds the limit in (2.3).

Two extreme cases illustrate the scope of Theorem 2.1. When $\sigma_Y^2 = 0$, then the future observable Y equals ξ with probability one and the prediction problem coincides with the problem of estimating ξ. The asymptotic minimax bound of Theorem 2.1 now reduces to a known bound for estimators of ξ, proved by Pinsker (1980) and derived from ideas in Stein (1956) by Beran (1992). On the other hand, when $\sigma_X^2 = 0$, then X equals ξ with probability one and observation of X is equivalent to knowledge of ξ. Now the best predictor of Y under quadratic loss is ξ itself. Both $\hat{Y}_S$ and $\hat{Y}_C$ reduce to this ideal predictor when $\sigma_X^2 = 0$.

While normality of (X, Y) is important for deriving asymptotic minimax bound (2.1), much weaker conditions suffice for the convergence of maximum risk in (2.3) and (2.4). For instance, suppose that

$$X = \xi + V_X, \qquad Y = \xi + V_Y, \tag{2.5}$$

where the components of V_X are i.i.d. with mean 0 and variance σ_X^2, the components of V_Y are i.i.d. with mean 0 and variance σ_Y^2, and V_X and V_Y are independent. Let Q_ξ denote the joint distribution of (X, Y) in this weaker model. Suppose $\{\xi_q \in R^q\}$ is any sequence such that $|\xi_q|^2/q \to a\sigma_X^2 < \infty$ as q increases. Then, by the weak law of large numbers,

$$q^{-1}|\hat{Y}_S - Y|^2 \to r(a), \qquad q^{-1}|\hat{Y}_C - Y|^2 \to \sigma_X^2 + \sigma_Y^2 \tag{2.6}$$

in Q_{ξ_q}-probability. Under further assumptions that ensure uniform integrability, these two convergences in probability imply the convergences (2.3) and (2.4). Thus, the Stein predictor can improve substantially upon the classical predictor outside the normal model.

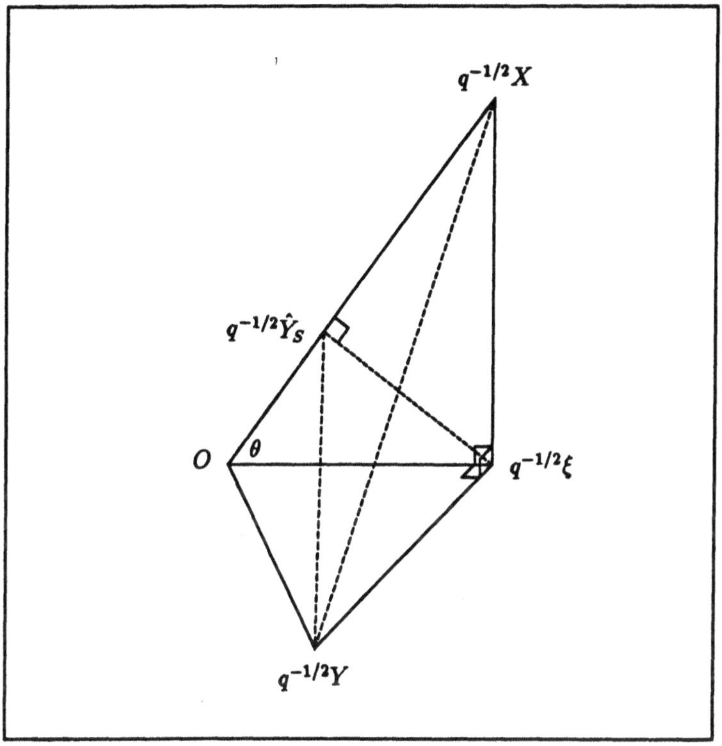

FIGURE 1. *Asymptotic geometry, when dimension q is large, of the Stein predictor $\hat{Y}_S$ compared with the classical predictor X for the future observable Y. The two right-angled triangles with solid sides and with vector $q^{-1/2}\xi$ as common base lie in two orthogonal planes. The distance from $q^{-1/2}\hat{Y}_S$ to $q^{-1/2}Y$ is less than the distance from $q^{-1/2}X$ to $q^{-1/2}Y$. Thus the loss, and so the risk, of the Stein predictor is smaller than that of the classical predictor.*

2.2. Asymptotic geometry

A three-dimensional object, shown in Figure 1, helps to explain Theorem 2.1 and the form of the Stein predictor $\hat{Y}_S$. Underlying the geometry of the prediction problem when dimension q is very large are the following algebraic approximations. Let $\{\xi_q \in R^q\}$ be any sequence such that $|\xi_q|^2/q = a\sigma_X^2$. From the normal model P_{ξ_q} and the weak law of large numbers, the following relations are true with high P_{ξ_q}-probability when q is large:

$$(2.7)\quad |q^{-1/2}\xi_q|^2 = a\sigma_X^2, \qquad |q^{-1/2}X - q^{-1/2}\xi_q|^2 \approx \sigma_X^2, \qquad |q^{-1/2}X|^2 \approx (1+a)\sigma_X^2$$

and

$$(2.8)\qquad |q^{-1/2}Y - q^{-1/2}\xi_q|^2 \approx \sigma_Y^2, \qquad |q^{-1/2}Y|^2 \approx a\sigma_X^2 + \sigma_Y^2$$

and

$$(2.9)\qquad (q^{-1/2}X - q^{-1/2}\xi_q)'(q^{-1/2}Y - q^{-1/2}\xi_q) \approx 0.$$

Consequently, the two triangles in Figure 1 with common base vector $q^{-1/2}\xi_q$ and with hypotenuses $q^{-1/2}X$ and $q^{-1/2}Y$ respectively are very nearly right-angled for large q. The angle θ between $q^{-1/2}X$ and $q^{-1/2}\xi_q$ is determined approximately by

$$(2.10)\qquad \cos^2(\theta) \approx a/(1+a).$$

Moreover, the vectors $q^{-1/2}X - q^{-1/2}\xi_q$ and $q^{-1/2}Y - q^{-1/2}\xi_q$ are very nearly orthogonal.

In seeking a minimax predictor of Y, it follows from (1.2) that we are seeking a minimax estimator of ξ. By the Hunt-Stein theorem and the compactness of the orthogonal group, we need consider only estimators that are orthogonally equivariant. Any orthogonally equivariant estimator has the form

$$(2.11)\qquad \hat{Y}(X) = h(|X|)X$$

for some real-valued measurable function h (Stein 1956) and so lies along the vector $q^{-1/2}X$ in Figure 1.

The scaled orthogonally equivariant estimator $q^{-1/2}\hat{Y}$ that asymptotically minimizes the estimation loss $|q^{-1/2}\hat{Y} - q^{-1/2}\xi_q|^2$, and so minimizes the prediction loss $|q^{-1/2}\hat{Y} - q^{-1/2}Y|^2$, is the orthogonal projection of $q^{-1/2}\xi_q$ onto $q^{-1/2}X$. For large q, this best $\hat{Y}$ satisfies

$$(2.12)\qquad q^{-1/2}\hat{Y} \approx |q^{-1/2}\xi_q|\cos(\theta)X/|X| \approx [a/(1+a)]q^{-1/2}X \approx q^{-1/2}\hat{Y}_S$$

with high probability. This geometrical argument explains both the structure and asymptotic optimality of the predictor $\hat{Y}_S$. Moreover, by the geometry of the projection in Figure 1 and (2.7), (2.8), (2.10), the minimized loss—and so minimized risk—is asymptotically

$$(2.13)\qquad \begin{aligned} q^{-1}|\hat{Y}_S - Y|^2 &\approx q^{-1}|\hat{Y}_S - \xi_q|^2 + q^{-1}|Y - \xi_q|^2 \\ &\approx a\sigma_X^2 \sin^2(\theta) + \sigma_Y^2 \approx [a/(1+a)]\sigma_X^2 + \sigma_Y^2, \end{aligned}$$

in agreement with (2.3).

The reasoning in the last two paragraphs shows heuristically that, when q is large and the parameter is restricted to the sphere $|\xi|^2 = qa\sigma_X^2$, the estimator $\hat{Y}_S$ is approximately minimax for ξ under quadratic loss. The approximate minimax risk is $[a/(1+a)]\sigma_X^2$, which is monotone increasing in a. Consequently, $\hat{Y}_S$ must be asymptotically minimax over the ball $|\xi|^2/q \leq c$, as stated in the paragraph that follows Theorem 2.1. This is the basic geometrical rationale for Stein estimation, implicit in Stein's (1956) paper.

2.3. Other good predictors

In the normal submodel where $|\xi|$ is assumed known, the best orthogonally equivariant predictor of Y exists and is minimax admissible. It takes the form

$$\hat{Y}_E(|\xi|) = |\xi| A_q(|\xi||X|/\sigma_X^2)X/|X|, \tag{2.14}$$

where

$$A_q(z) = I_{q/2}(z)/I_{q/2-1}(z), \qquad z \geq 0 \tag{2.15}$$

and $I_\nu(z)$ is the modified Bessel function of the first kind and order ν. This result follows from (1.2) and Theorem 2.1 in Beran (1992). The function A_q satisfies the recursion

$$A_q(z) = 1/A_{q-2}(z) - (q-2)/z, \qquad q \geq 3 \tag{2.16}$$

by Schou (1978, Appendix A). In particular,

$$\begin{aligned} A_1(z) &= \tanh(z) \\ A_3(z) &= \coth(z) - 1/z \end{aligned} \tag{2.17}$$

and so forth for odd orders q.

In the full model, where $|\xi|$ is unknown, the uniformly minimum variance unbiased estimator of $|\xi|^2$ is $|X|^2 - q\sigma_X^2$. This fact, together with the form of the Stein estimator, suggests estimating $|\xi|$ by $|\hat{\xi}_{CL}|$, where

$$\hat{\xi}_{CL} = [1 - (q-2)\sigma_X^2/|X|^2]_+^{1/2} X. \tag{2.18}$$

Note the square root in (2.18), in contrast to the definition (1.3) of $\hat{Y}_S$. The estimator $|\hat{\xi}_{CL}|$ is consistent and locally asymptotically minimax for $|\xi|$, as will be discussed in Section 3.1. For these asymptotics in q, the constant 2 in (2.18) can be replaced by any constant.

In the full model, the predictor $\hat{Y}_E(|\hat{\xi}_{CL}|)$ satisfies

$$\lim_{q\to\infty} \sup_{|\xi|^2 \leq qc} R_q(\hat{Y}_E(|\hat{\xi}_{CL}|), \xi) = r(c/\sigma_X^2) \tag{2.19}$$

for every finite positive c, just like the Stein predictor $\hat{Y}_S$. This assertion follows from (1.2) and Theorem 3.4 of Beran (1992). By Theorem 3.3 of that same paper, the positive-part Stein predictor

$$\hat{Y}_{PS} = [1 - (q-2)\sigma_X^2/|X|^2]_+ X \tag{2.20}$$

approximates $\hat{Y}_E(|\hat{\xi}_{CL}|)$ in that

$$\lim_{q\to\infty} \sup_{|\xi|^2 \le qc} q^{-1} E_\xi |\hat{Y}_{PS} - \hat{Y}_E(|\hat{\xi}_{CL}|)|^2 = 0 \tag{2.21}$$

for every finite $c > 0$. Moreover, $\hat{Y}_{PS}$ also attains the asymptotic minimax bound of Theorem 2.1 and dominates $\hat{Y}_S$ for every finite q.

3. Stein Prediction Regions

The classical prediction sphere for Y is $D_C = D(X, \hat{d}_C)$, in the notation (1.5), where

$$\hat{d}_C = (\sigma_X^2 + \sigma_Y^2)^{1/2} \chi_q^{-1}(\alpha) \tag{3.1}$$

and the square of $\chi_q^{-1}(\alpha)$ is the αth quantile of the chi-squared distribution with q degrees of freedom. Clearly, $P_\xi(Y \in D_C)$ is exactly α, for every ξ. Moreover, by the triangular array central limit theorem, as explained in Section 4,

$$\lim_{q\to\infty} \sup_{|\xi|^2 \le qc} |G_q(D_C, \xi) - 2(\sigma_X^2 + \sigma_Y^2)^{1/2}| = 0 \tag{3.2}$$

for every finite $c > 0$.

This section treats the contrasting behavior of Stein prediction regions $D_S = D(\hat{Y}_S, \hat{d})$. The key findings are that the radius $\hat{d}$ can be chosen to make coverage probability converge quickly to α as q increases and that the geometrical risk $G_q(D_S, \xi)$ dominates $G_q(D_C, \xi)$ asymptotically on large compact balls centered at $\xi = 0$.

3.1. Basic critical values

To construct a suitable critical value $\hat{d}$ for D_S, consider the quantity

$$T_q(X, Y, \xi) = q^{-1/2}\{|\hat{Y}_S - Y|^2 - [q\sigma_X^2 - (q-2)^2\sigma_X^4/|X|^2 + q\sigma_Y^2]\}, \tag{3.3}$$

which compares the loss of $\hat{Y}_S$ with an unbiased estimator of its risk (cf. the final paragraphs of Stein 1981). Thus, the expectation of $T_q(X, Y, \xi)$ is zero. By orthogonal invariance, the distribution of $T_q(X, Y, \xi)$ depends on ξ only through $|\xi|$, and so can be written as $H_q(|\xi|^2)$. The following weak convergence result is basic to the construction of Stein prediction regions.

THEOREM 3.1. *Suppose that* $\{\xi_q \in R^q\}$ *is any sequence such that* $|\xi_q|^2/q \to a\sigma_X^2 < \infty$ *as* q *increases. Then*

$$H_q(|\xi_q|^2) \Rightarrow N(0, \tau^2(a)), \tag{3.4}$$

where

$$\tau^2(a) = 2\{(1+a)^{-2}\sigma_X^4 + [a(1+a)^{-1}\sigma_X^2 + \sigma_Y^2]^2\}. \tag{3.5}$$

Let

$$\hat{d}(|\xi|^2) = [q\sigma_X^2 - (q-2)^2\sigma_X^4/|X|^2 + q\sigma_Y^2 + q^{1/2}\tau[|\xi|^2/(q\sigma_X^2)]\Phi^{-1}(\alpha)]_+^{1/2}, \tag{3.6}$$

where Φ^{-1} denotes the standard normal quantile function. If $|\xi|$ were known, then the prediction region $D(\hat{Y}_S, \hat{d}(|\xi|^2))$ would have asymptotic coverage probability α for Y, under the conditions of Theorem 3.1. The estimator $|\hat{\xi}_{CL}|$, where

$$\hat{\xi}_{CL} = [1 - (q-2)\sigma_X^2/|X|^2]_+^{1/2} X \tag{3.7}$$

as in Section 2.2, is obviously consistent in the following sense: if $\{\xi_q \in R^q\}$ is any sequence such that $|\xi_q|^2/q \to a\sigma_X^2 < \infty$, then $|\hat{\xi}_{CL}|^2/q \to a\sigma_X^2$ in P_{ξ_q}-probability. The implied consistent estimator of $\tau[|\xi|^2/(q\sigma_X^2)]$ is then

$$\hat{\tau}_A = \tau[|\hat{\xi}_{CL}|^2/(q\sigma_X^2)], \tag{3.8}$$

the function τ being defined by (3.5).

The estimator $\hat{\tau}_A$ is, in addition, asymptotically minimax for $\tau[|\xi|^2/(q\sigma_X^2)]$. More generally, suppose that $f: R^+ \to R$ is differentiable on its domain, with derivative f', and that u is a non-negative monotone increasing function defined on R^+, with $u(0) = 0$. Let

$$\rho_u(t) = Eu[(2+4t)^{1/2}|f'(t)Z|], \tag{3.9}$$

where Z is a standard normal random variable. For every finite $c > 0$,

$$\liminf_{q\to\infty} \inf_{\hat{f}} \sup_{|\xi|^2 \le qc} E_\xi u(q^{1/2}|\hat{f} - f[|\xi|^2/(q\sigma_X^2)]|) \ge \rho_u(c/\sigma_X^2), \tag{3.10}$$

the infimum being taken over all estimators $\hat{f}$. When u is bounded, the estimator $\hat{f}_A = f[|\hat{\xi}_{CL}|^2/(q\sigma_X^2)]$ attains lower bound (3.10) asymptotically in that

$$\lim_{q\to\infty} \sup_{|\xi|^2 \le qc} E_\xi u(q^{1/2}|\hat{f}_A - f[|\xi|^2/(q\sigma_X^2)]|) = \rho_u(c/\sigma_X^2) \tag{3.11}$$

for every finite $c > 0$.

The lower bound (3.10) is implied by a local asymptotic minimax bound for estimators of $f[|\xi|^2/(q\sigma_X^2)]$ that is given in Beran (1993) for the case $\sigma_X^2 = 1$. Closely related is an even more local LAM bound, Theorem 2 in Hasminski and Nussbaum (1984). On the other hand, (3.11) is a consequence of Lemma 4.2 in Section 4.

An alternative to direct asymptotic approximation of $H_q(|\xi|^2)$ by the $N(0, \hat{\tau}_A^2)$ distribution is bootstrapping. Suppose that $\{\xi_q \in R^q\}$ is any sequence such that $|\xi_q|^2/q \to a\sigma_X^2 < \infty$ as q increases. Then, by Theorem 3.1 and the weak law of large numbers,

$$H_q(|\hat{\xi}_{CL}|^2) \Rightarrow N(0, \tau^2(a)) \tag{3.12}$$

while

$$H_q(|X|^2) \Rightarrow N(0, \tau^2(1+a)) \tag{3.13}$$

and

$$H_q(|\hat{Y}_S|^2) \Rightarrow N(0, \tau^2[a^2/(1+a)]), \tag{3.14}$$

all three convergences being in P_{ξ_q}-probability. Thus, the bootstrap estimator

$$\hat{H}_B = H_q(|\hat{\xi}_{CL}|^2) \tag{3.15}$$

is consistent for $H_q(|\xi|^2)$ while its plausible competitors in (3.13) and (3.14) are not.

Equation (3.15) defines the bootstrap distribution as a random probability measure. Let X^* and Y^* be random vectors that are conditionally independent, given X, and whose respective conditional distributions, given X, are $N(\hat{\xi}_{CL}, \sigma_X^2 I)$ and $N(\hat{\xi}_{CL}, \sigma_Y^2 I)$. Then

$$\hat{H}_B = \mathcal{L}[T_q(X^*, Y^*, \hat{\xi}_{CL})|X]. \tag{3.16}$$

This representation as a conditional distribution leads readily to Monte Carlo algorithms for approximating $\hat{H}_B$.

The *asymptotic Stein prediction region* for Y is defined to be $D_{SA} = D(\hat{Y}_S, \hat{d}_A)$, where

$$\hat{d}_A = \hat{d}(|\hat{\xi}_{CL}|^2) \tag{3.17}$$

for $\hat{d}(\cdot)$ given by (3.6). The analogous *bootstrap Stein prediction region* is then $D_{SB} = D(\hat{Y}_S, \hat{d}_B)$, where

$$\hat{d}_B = [q\sigma_X^2 - (q-2)^2\sigma_X^4/|X|^2 + q\sigma_Y^2 + q^{1/2}\hat{H}_B^{-1}(\alpha)]_+^{1/2} \tag{3.18}$$

and $\hat{H}_B^{-1}$ is the quantile function of $\hat{H}_B$. The first-order asymptotic behavior of these two prediction regions for Y is described in the following theorem.

THEOREM 3.2. *For every finite positive c,*

$$\text{(3.19)} \qquad \lim_{q\to\infty} \sup_{|\xi|^2 \le qc} |P_\xi(Y \in D_{SA}) - \alpha| = 0$$

and

$$\text{(3.20)} \qquad \lim_{q\to\infty} \sup_{|\xi|^2 \le qc} |G_q(D_{SA}, \xi) - 2r^{1/2}[|\xi|^2/(q\sigma_X^2)]| = 0,$$

where $r(t)$ is given by (2.2). Both assertions hold as well for prediction region D_{SB}.

Thus, the geometrical risk (3.20) of the Stein prediction regions D_{SA} and D_{SB} is asymptotically smaller, on large compact balls about the origin, than the geometrical risk (3.2) of the classical prediction region D_C. In Figure 1, the scaled prediction regions $q^{-1/2}D_{SA}$ and $q^{-1/2}D_{SB}$ both converge, as q increases, to the ball centered at $q^{-1/2}\hat{Y}_S$ whose surface passes through $q^{-1/2}Y$. The asymptotic geometrical risk in (3.20) is just twice the radius of this limit ball. Similarly, the scaled prediction region $q^{-1/2}D_C$ converges to the ball centered at $q^{-1/2}X$ whose surface passes through $q^{-1/2}Y$; and twice the radius of this ball is the limiting geometrical risk in (3.2).

3.2. Better bootstrap critical values

The cdf $H_q(\cdot, |\xi|^2)$ of $H_q(|\xi|^2)$ has an Edgeworth expansion whose first term is $\Phi\{\cdot/\tau[|\xi|^2/(q\sigma_X^2)]\}$ and whose second term is a skewness correction of order $O(q^{-1/2})$. By Lemma 4.2 and (3.8), the estimator $\hat{\tau}_A$ makes an error of $O_p(q^{-1/2})$. Consequently, approximating $H(\cdot, |\xi|^2)$ by $\Phi(\cdot/\hat{\tau}_A)$ commits an error of $O_p(q^{-1/2})$ from two sources, by estimating $\tau[|\xi|^2/(q\sigma_X^2)]$ in the first term of the Edgeworth expansion and by ignoring the skewness term. This consideration suggests that the coverage probability error of D_{SA} in (3.19) is $O(q^{-1/2})$.

On the other hand, approximating $H(\cdot, |\xi|^2)$ by the bootstrap cdf $H(\cdot, |\hat{\xi}_{CL}|^2)$ commits an error of $O_p(q^{-1/2})$ from only one source, by estimating $\tau[|\xi|^2/(q\sigma_X^2)]$ in the first term of the Edgeworth expansion. The error in estimating the skewness term is now of order $O_p(q^{-1})$. These considerations suggest that the coverage probability error of D_{SB} is also $O(q^{-1/2})$, but may be smaller than the coverage probability error of D_{SA} when the skewness term is non-negligible.

To improve the rate of convergence of coverage probability, it is desirable to bootstrap not $T_q(X, Y, \xi)$ but a related quantity whose limiting distribution does not depend on the unknown $|\xi|$. One possibility—bootstrapping $T_q(X, Y, \xi)/\hat{\tau}_A$—did not give accurate coverage probability for small values of q in exploratory simulations. More promising is the following approach, which relies on a variance stabilizing transformation.

Consider the strictly monotone increasing function

$$f(t) = 2^{-1}\log[2t - \sigma^2 + 2^{1/2}(2t^2 - 2\sigma^2 t + \sigma^4)^{1/2}], \tag{3.21}$$

where $\sigma^2 = \sigma_X^2 + \sigma_Y^2$. The first derivative of f is

$$f'(t) = 2^{-1/2}(2t^2 - 2\sigma^2 t + \sigma^4)^{-1/2} > 0. \tag{3.22}$$

Let

$$S_q(X, Y, \xi) = q^{1/2}\{f[|\hat{Y}_S - Y|^2/q] - f[\sigma_X^2 - (q-2)^2\sigma_X^4/(q|X|^2) + \sigma_Y^2]\} \tag{3.23}$$

and let $J_q(|\xi|^2)$ denote the distribution of $S_q(X, Y, \xi)$ under P_ξ. Under the assumptions of Theorem 3.1, it follows from (3.22) and (3.4) that

$$J_q(|\xi_q|^2) \Rightarrow N(0,1). \tag{3.24}$$

Consequently, as in (3.12), the bootstrap distribution

$$\hat{J}_B = J_q(|\hat{\xi}_{CL}|^2) \Rightarrow N(0,1) \tag{3.25}$$

in P_{ξ_q}-probability.

The *better bootstrap Stein prediction region* for Y is now defined to be $D_{SBB} = D(\hat{Y}_S, \hat{d}_{BB})$, where

$$\hat{d}_{BB} = [qf^{-1}\{f[\sigma_X^2 - (q-2)^2\sigma_X^4/(q|X|^2) + \sigma_Y^2] + q^{-1/2}\hat{J}_B^{-1}(\alpha)\}]_+^{1/2} \tag{3.26}$$

and $\hat{J}_B^{-1}$ is the quantile function of $\hat{J}_B$. Because of (3.24) and (3.25), Theorem 3.2 also holds for D_{SBB}. The improvement in coverage probability achieved by D_{SBB} is describable as follows.

THEOREM 3.3. *For every finite positive c,*

$$\sup_{|\xi|^2 \le qc} |P_\xi(Y \in D_{SBB}) - \alpha| = O(q^{-1}) \tag{3.27}$$

while the corresponding coverage probability errors for D_{SA} and D_{SB} are $O(q^{-1/2})$.

In the special case of confidence sets for ξ, which correspond to $\sigma_Y^2 = 0$, a numerical experiment indicates that D_{SBB} has more accurate coverage probability than either D_{SA} or D_{SB} when $q \ge 5$. See Beran (1993) for further details.

4. Proofs

The following two lemmas are important in our arguments. Let $W_q(X, Y, \xi)$ be a five dimensional random vector whose components are

$$\begin{aligned} W_{q,1} &= q^{-1/2}(|X-\xi|^2 - q\sigma_X^2) \\ W_{q,2} &= q^{-1/2}(|Y-\xi|^2 - q\sigma_Y^2) \\ W_{q,3} &= q^{-1/2}(X-\xi)'(Y-\xi) \\ W_{q,4} &= q^{-1/2}\xi'(X-\xi) \\ W_{q,5} &= q^{-1/2}\xi'(Y-\xi). \end{aligned} \tag{4.1}$$

LEMMA 4.1. *Suppose that* $\{\xi_q \in R^q\}$ *is any sequence such that* $|\xi_q|^2/q \to a\sigma_X^2 < \infty$ *as* $q \to \infty$. *Then, under* P_{ξ_q},

$$W_q(X, Y, \xi) \Rightarrow (2^{1/2}\sigma_X^2 Z_1,\ 2^{1/2}\sigma_Y^2 Z_2,\ \sigma_X\sigma_Y Z_3,\ a^{1/2}\sigma_X^2 Z_4,\ a^{1/2}\sigma_X\sigma_Y Z_5) \tag{4.2}$$

where the $\{Z_i\}$ *are independent standard normal random variables.*

PROOF. By orthogonal equivariance, the distribution of $W_q(X, Y, \xi_q)$ depends on ξ_q only through $|\xi_q|$. Without loss of generality, take each component of ξ_q to be $q^{-1/2}|\xi_q|$. The weak convergence asserted in (4.2) follows from the normal model P_{ξ_q} and the Lindeberg-Feller central limit theorem.

LEMMA 4.2. *Suppose that* $\{\xi_q \in R^q\}$ *is any sequence such that* $|\xi_q|^2/q \to a\sigma_X^2 < \infty$ *as* $q \to \infty$. *Then, under* P_{ξ_q},

$$q^{-1/2}(|\hat{\xi}_{CL}|^2/\sigma_X^2 - |\xi_q|^2/\sigma_X^2) \Rightarrow N(0, 2+4a) \tag{4.3}$$

PROOF. Since

$$|X|^2 = |X-\xi_q|^2 + |\xi_q|^2 + 2\xi_q'(X-\xi_q), \tag{4.4}$$

the desired weak convergence follows from the definition (3.7) of $\hat{\xi}_{CL}$ and from Lemma 4.1.

We will now sketch proofs for the theorems in Section 2 and 3.

PROOF OF THEOREM 2.1. Since X/σ_X has a $N(\xi/\sigma_X, I)$ distribution, the asymptotic minimax bound (2.1) follows from the risk decomposition (1.2) and from Theorem 3.4 in Beran (1992).

Suppose that $\{\xi_q \in R^q\}$ is any sequence such that $|\xi_q|^2/q \to a\sigma_X^2 < \infty$ as q increases. Writing $p = q - 2$, observe that

$$\begin{aligned} |\hat{Y}_S - \xi_q|^2 &= (1 - p\sigma_X^2/|X|^2)^2|X - \xi_q|^2 + p^2\sigma_X^4|\xi_q|^2/|X|^4 \\ &\quad - 2(p\sigma_X^2/|X|^2)(1 - p\sigma_X^2/|X|^2)\xi_q'(X - \xi_q). \end{aligned} \tag{4.5}$$

By the weak law of large numbers or Lemma 4.1, $|X|^2/q \to (1+a)\sigma_X^2$ and therefore $|\hat{Y}_S - \xi_q|^2/q \to [a/(1+a)]\sigma_X^2$, both in P_{ξ_q}-probability. The attainability result (2.3) now follows by a uniform integrability argument and (1.2).

PROOF OF (3.2). Let $\{\xi_q\}$ be as in Lemma 4.1. From (1.6) and the definition (3.1),

$$G_q(D_C, \xi_q) = q^{-1/2}E_{\xi_q}|X - Y| + q^{-1/2}(\sigma_X^2 + \sigma_Y^2)^{1/2}\chi_q^{-1}(\alpha). \tag{4.6}$$

By the central limit theorem and Polya's theorem on uniform convergence of the corresponding c.d.f.'s,

$$\begin{aligned} q^{-1/2}\chi_q^{-1}(\alpha) &= q^{-1/2}[q + (2q)^{1/2}\Phi^{-1}(\alpha) + o(q^{1/2})]^{1/2} \\ &= 1 + O(q^{-1/2}). \end{aligned} \tag{4.7}$$

The first term on the right side of (4.6) converges to $(\sigma_X^2 + \sigma_Y^2)^{1/2}$ because $q^{-1/2}|X - Y| \to (\sigma_X^2 + \sigma_Y^2)^{1/2}$ in P_{ξ_q}-probability and $q^{-1}E_{\xi_q}|X - Y|^2 = \sigma_X^2 + \sigma_Y^2$. The limit (3.2) now follows.

PROOF OF THEOREM 3.1. As above, write $p = q - 2$. Then, from (3.3),

$$\begin{aligned} q^{1/2}T_q(X, Y, \xi_q) &= |\hat{Y}_S - Y|^2 - (q\sigma_X^2 + q\sigma_Y^2 - p^2\sigma_X^4/|X|^2) \\ &= |\hat{Y}_S - \xi_q|^2 - (q\sigma_X^2 - p^2\sigma_X^4/|X|^2) + (|Y - \xi_q|^2 - q\sigma_Y^2) \\ &\quad - 2(1 - p\sigma_X^2/|X|^2)(X - \xi_q)'(Y - \xi_q) \\ &\quad + 2(p\sigma_X^2/|X|^2)\xi_q'(Y - \xi_q). \end{aligned} \tag{4.8}$$

Moreover, from (4.5) and (4.4),

$$\begin{aligned} |\hat{Y}_S - \xi_q|^2 - (q\sigma_X^2 - p^2\sigma_X^4/|X|^2) &= (1 - 2p\sigma_X^2/|X|^2)(|X - \xi_q|^2 - p\sigma_X^2) \\ &\quad - (2p\sigma_X^2/|X|^2)\xi_q'(X - \xi_q) - 2\sigma_X^2. \end{aligned} \tag{4.9}$$

Together, (4.8), (4.9) and Lemma 4.1 imply that

$$\begin{aligned} T_q(X, Y, \xi_q) \Rightarrow &\ [1 - 2/(1+a)]2^{1/2}\sigma_X^2 Z_1 - [2/(1+a)]a^{1/2}\sigma_X^2 Z_4 \\ &+ 2^{1/2}\sigma_Y^2 Z_2 - 2[1 - 1/(1+a)]\sigma_X\sigma_Y Z_3 \\ &+ [2/(1+a)]a^{1/2}\sigma_X\sigma_Y Z_5 \end{aligned} \tag{4.10}$$

under P_{ξ_q}. The theorem assertion follows.

PROOF OF THEOREM 3.2. Suppose that $|\xi_q|^2/q \to a\sigma_X^2 < \infty$. Then, by Lemma 4.2, $|\hat{\xi}_{CL}|^2/q \to a\sigma_X^2$ in P_{ξ_q}-probability. Consequently, $\hat{\tau}_A$ converges to $\tau(a)$ in probability and the distribution of $T_q(X, Y, \xi_q)/\hat{\tau}_A$ converges weakly to the standard normal distribution. This implies (3.19).

From (1.6),

$$G_q(D_{SA}, \xi_q) = q^{-1/2} E_{\xi_q} |\hat{Y}_S - Y| + q^{-1/2} E_{\xi_q} \hat{d}_A. \tag{4.11}$$

By the weak law of large numbers, $q^{-1/2}|\hat{Y}_S - Y| \to r^{1/2}(a)$ and $q^{-1/2}\hat{d}_A \to r^{1/2}(a)$, both in P_{ξ_q}-probability. Moreover, from (1.4), $q^{-1}E_{\xi_q}|\hat{Y}_S - Y|^2 \to r(a)$. A dominated convergence argument completes the proof of (3.20).

The analogous reasoning in the case of D_{SB} relies on (3.12).

PROOF OF THEOREM 3.3. A short heuristic argument for (3.27) and for the corresponding assertions about D_{SA} or D_{SB} parallels the discussion in Section 3 of Beran (1988). A rigorous Edgeworth expansion treatment may be based on the method in Hall (1992).

Acknowledgements

This research was supported in part by NSF grant DMS 9224868. Some of the work was done while I was a guest of the Institut für Angewandte Mathematik at Universität Heidelberg. A conversation with M. Nussbaum threw more light on the local asymptotic minimax bounds in Section 3.1.

REFERENCES

Beran, R. (1988). Prepivoting test statistics: a bootstrap view of asymptotic refinements. *J. Amer. Statist. Assoc.* **83** 687–697.

Beran, R. (1992). Stein estimation in high dimensions and the bootstrap. Preprint.

Beran, R. (1993). Stein confidence sets and the bootstrap. Preprint

Donoho, D. and Johnstone, I. M. (1989). Minimax risk over l_p balls. Technical report.

Hall, P. (1992). *The Bootstrap and Edgeworth Expansion.* Springer-Verlag, New York.

Hasminski, R. Z. and Nussbaum, M. (1984). An asymptotic minimax bound in a regression model with an increasing number of nuisance parameters. *Proc. Third Prague Symp. Asymptotic Statistics* (P. Mandl and M. Hušková, eds.) 275–283. Elsevier, New York.

James, W. and Stein, C. (1961). Estimation with quadratic loss. *Proc. Fourth Berkeley Symp. Math. Statist. Prob.* 1 (J. Neyman, ed.) 361–380. Univ. California.

Olshen, R. A., Biden, E. N., Wyatt, M. P. and Sutherland, D. H. (1989). Gait analysis and the bootstrap. *Ann. Statist.* 71 1419–1440.

Pinsker, M. S. (1980). Optimal filtration of square-integrable signals in Gaussian white noise. *Problems Inform. Transmission* **16** 120–133.

Schou, G. (1978). Estimation of the concentration parameter in von Mises distributions. *Biometrika* **65** 369–377.

Stein, C. (1956). Inadmissibility of the usual estimator for the mean of a multivariate normal distribution. *Proc. Third Berkeley Symp. Math. Statist. Prob.* 1 (J. Neyman, ed.) 197–206. Univ. California.

Stein, C. (1981). Estimation of the mean of a normal distribution. *Ann. Statist.* **9** 1135–1151.

Department of Statistics
University of California
Berkeley, CA 94720
USA

Perpetuities and Random Equations

Paul Embrechts and Charles M. Goldie

ETH, Zürich and

Queen Mary & Westfield College (University of London)

Abstract. In life insurance and finance, stochastically discounted sums of the following type are important:

$$S_n := \sum_{k=1}^{n} Z_1 \cdots Z_k Y_k \qquad (n = 1, \ldots, \infty),$$

where (Z_i) and (Y_i) are independent iid sequences satisfying mild moment conditions. Sums of this type are called perpetuities. The Y_i are interpreted as payments, the Z_i as discount factors. The distributional properties of S_n, for $n < \infty$, can be studied via the following random equation:

$$S_n \stackrel{d}{=} (Y_1 + S_{n-1})Z_1 \qquad (n = 1, 2, \ldots),$$

where $S_0 := 0$. In this paper we discuss some aspects of the behaviour of solutions of this and related random equations, and their applications to perpetuities.

Keywords. Distributional fix-point, finance, insurance, perpetuity, random affine recursion, random equation, stochastic discounting, tail behaviour

AMS 1991 mathematics subject classification: 60H25, 62P05

1. Fix-points of the random affine map.

Any probability law on $\mathbf{R}^2$, say of the pair of real r.v.s (Q, M), defines a random affine map Ψ on $\mathbf{R}$ by

$$\Psi(t) := Q + Mt \qquad (t \in \mathbf{R}). \tag{1.1}$$

In full we would write $\Psi(t, \omega) := Q(\omega) + M(\omega)t$ where the probability space is $(\Omega, \mathcal{A}, P)$ and ω is an element of Ω. Defined thus, $\Psi : \mathbf{R} \times \Omega \to \mathbf{R}$ is automatically product-measurable.

A probability law on the real line, of R say, that satisfies

$$R \stackrel{d}{=} Q + MR, \quad R \text{ independent of } (Q, M), \tag{1.2}$$

is a *fix-point* of Ψ. That is, Ψ gives rise to an action on probability laws on $\mathbf{R}$, that takes the law of R to the law of $\Psi(R) = Q + MR$, where Ψ and R are independent, and our fix-point is a fix-point of that action. We emphasize that the fix-point is a probability law. R is just a convenient, even 'fictional', construct.

Here are criteria, nearly all from (Vervaat 1979), for a fix-point to exist. We put them in terms of $X := \log|M|$. Note that $-\infty \le X < \infty$, and that the event that $X = -\infty$ is the event that $M = 0$.

A visit of Charles Goldie to Zürich was supported by the Mathematisches Forschungsinstitut of the ETH. The hospitality of the ETH is gratefully acknowledged.

Theorem 1.1. *Let* $X := \log |M|$.

(a) *Suppose* $-\infty < EX < 0$. *Then a fix-point of (1.1) exists if and only if* $E\log(1 \vee |Q|) < \infty$.
(b) *Suppose* $P(X = -\infty) > 0$. *Then a fix-point of (1.1) exists.*
(c) *Suppose* $EX = -\infty$ *and* $P(X = -\infty) = 0$. *Then a fix-point of (1.1) exists if* $E\log(1 \vee |Q|) < \infty$.
(d) *Suppose* $0 \leq EX \leq \infty$. *Then a fix-point of (1.1) exists if and only if* $P(Q + cM = c) = 1$ *for some non-random real c.*
(e) *Suppose EX does not exist,* $P(X = -\infty) = 0$ *and*

$$\frac{\int_0^x \Big(P(X > y) - P(X < -y)\Big)\, dy}{xP(X > x)} \not\to \infty \qquad (x \to \infty). \tag{1.3}$$

Then a fix-point of (1.1) exists if and only if $P(Q + cM = c) = 1$ *for some non-random real c.*

Proof: (a) is (Vervaat 1979, Theorem 1.6(b)) and (c) is part of (Vervaat 1979, Theorem 1.6(c)). Part (d) is (Vervaat 1979, Theorem 1.6(a)).

As for (b), set $p := P(M = 0) > 0$, and let N, Q^*, $(\tilde{Q}_1, \tilde{M}_1)$, $(\tilde{Q}_2, \tilde{M}_2)$, ... be independent, where the $(\tilde{Q}_i, \tilde{M}_i)$ each have the conditional distribution of (Q, M) given $M \neq 0$, Q^* has the conditional distribution of Q given $M = 0$, and N has distribution

$$P(N = n) = (1-p)^n p \qquad (n = 0, 1, \ldots).$$

Then it is easy to see that the 'finite perpetuity'

$$Z_f := \sum_{k=1}^{N} \tilde{Q}_k \prod_{j=1}^{k-1} \tilde{M}_j + Q^* \prod_{j=1}^{N} \tilde{M}_j$$

has a distribution that is a fix-point, since (1.1) in this case may be written

$$R \stackrel{d}{=} \begin{cases} Q^* & \text{with probability } p, \\ \tilde{Q} + \tilde{M}R & \text{with probability } 1-p, \end{cases}$$

and N imitates this in that

$$N \stackrel{d}{=} \begin{cases} 0 & \text{with probability } p, \\ 1 + N & \text{with probability } 1-p. \end{cases}$$

In the situation of (e), (Vervaat 1979, Lemma 1.4(d)) says that either

$$\text{Case I: } \textstyle\sum_{k=1}^{n} X_k \stackrel{d}{\to} -\infty,$$

where $X_1, X_2, \ldots$ are iid, each distributed as X, or

$$\text{Case III: } \limsup_{n\to\infty} P\big(\textstyle\sum_{k=1}^{n} X_k > 0\big) > 0.$$

However, Case I is excluded by a recent result (Kesten & Maller 1993, Theorem 2.1), where among other things it is shown that $\sum_{k=1}^{n} X_k \xrightarrow{d} -\infty$ is equivalent to the negation of (1.3). So we are left with Case III, for which in (Vervaat 1979, Theorem 1.5(iii)) it is proved that a fix-point exists only in the degenerate case stated. □

In (Vervaat 1979, Lemma 1.4(a), Theorem 1.5(i)) it is also proved that when a fix-point exists in (a), (b) or (c) it is unique.

Cases (d), (e) are those of a 'tied-down line', where the random line $y = Q + Mx$ always passes through the point (c, c) in the x–y plane. So (1.2) reduces to $R - c \stackrel{d}{=} M(R - c)$ and unless $M = 1$ a.s. the only fix-point is clearly the law degenerate at c. This case is of no further interest.

The following rather arcane cases are those remaining. Vervaat left open the conditions for existence of a fix-point here, and they remain partially open

(f) $E \log |M| = -\infty$, $P(M = 0) = 0$, $E \log(1 \vee |Q|) = \infty$;
(g) $E \log |M|$ does not exist, $P(M = 0) = 0$, and (1.3) does not hold.

With reference to (c) and (f), the condition on Q in (c) is certainly stronger than necessary. In (Kellerer 1992, III, Proposition 8.6) it is proved that for any probability law for $Q \geq 0$ there is a probability law for an independent $0 < M < 1$ such that a fix-point of (1.2) exists. Since we may clearly replace M by any smaller positive r.v. and the fix-point will continue to exist, it follows that we gain a whole class of examples in (f) where the fix-point exists. Examples may also be readily constructed in both (f) and (g) where no fix-point exists (Kellerer 1992, I, Proposition 2.5). No example is known to us in case (g), not of the tied-down line form, where the fix-point can be proved to exist (see Kellerer 1992, III, Theorem 8.5.(a)).

2. Iterations.

We take an iid sequence Ψ_n, defined on iid pairs (Q_n, M_n) by

$$\Psi_n(t) := Q_n + M_n t \qquad (t \in \mathbb{R}), \tag{2.1}$$

and use it to iterate the random affine map (1.1). Note that Q need not be independent of M. We remark that parts of what is known about iid Ψ_n were extended in (Brandt 1986) to ergodic sequences (Ψ_n), and (Bougerol & Picard 1991) later dealt with d–dimensional versions. There had been various partial existence results in d dimensions before their work. We shall, however, scarcely touch further on ergodic non-iid cases or multi-dimensional cases.

There are two ways to iterate:

(a) *Outer iteration.* Start with R_0, independent of (Ψ_n), and form

$$R_{n+1} := \Psi_{n+1}(R_n) = Q_{n+1} + M_{n+1} R_n \qquad (n = 0, 1, 2, \dots), \tag{2.2}$$

that is,

$$R_n = \Psi_n \circ \Psi_{n-1} \circ \cdots \circ \Psi_1(R_0) = \Psi_n(\Psi_{n-1}(\cdots \Psi_1(R_0) \cdots)).$$

(b) *Inner iteration.* Start with $t \in \mathbb{R}$ and form $Z_n : \mathbb{R} \times \Omega \to \mathbb{R}$ by $Z_0(t) := t$ and

$$Z_{n+1}(t) := Z_n \circ \Psi_{n+1}(t) = Z_n(Q_{n+1} + M_{n+1} t) \qquad (n = 0, 1, \dots). \tag{2.3}$$

In full the right-hand side is $Z_n(Q_{n+1}(\omega) + M_{n+1}(\omega)t, \omega)$. Then

$$Z_n(t) := \Psi_1 \circ \Psi_2 \circ \cdots \circ \Psi_n(t) = \Psi_1(\Psi_2(\cdots \Psi_n(t) \cdots)).$$

'Inner' and 'outer' seem to us preferable terms to 'backwards' and 'forwards' iteration, particularly in view of the habit, common in population-modelling settings and for good reasons, of performing iterations backwards *in time*, from the present into the past. Time reversal also has meaning in finance contexts, where discounting is transformed into accumulation and vice versa (Dufresne 1992).

Both iterations are, when $P(M \neq 0) = 1$, random walks on the group of affine maps on $\mathbf{R}$, and in (Vervaat 1979, §2) there are references to the literature of that viewpoint.

Outer iteration gives

$$\begin{aligned} R_n &= Q_n + M_n Q_{n-1} + M_n M_{n-1} Q_{n-2} + \cdots + M_n \cdots M_2 Q_1 + M_n \cdots M_1 R_0 \\ &= R_0 \prod_{j=1}^{n} M_j + \sum_{i=1}^{n} Q_i \prod_{j=i+1}^{n} M_j \end{aligned}$$

(interpret $\prod_{j=n+1}^{n}$ as 1). This iteration gives the simplest way of repeating the Ψ-actions on probability laws: just iterated 'affine convolutions' $\mathcal{L}(R_{n+1}) = \mathcal{L}(\Psi_{n+1}(R_n)) = \mathcal{L}(Q_{n+1} + M_{n+1}R_n)$. Vervaat also proves that in Theorem 1.1 (a)–(c), where there exists a fix-point the probability laws of the iterates R_n converge to it, whatever the law of R_0 (see Lemma 1.4(a), Theorem 1.5(i) and Lemma 1.1 of (Vervaat 1979)).

If there is a fix-point distribution, and R_0 is given that distribution, then the sequence (R_n) is stationary. However under this iteration there is no convergence of the r.v.s R_n in Theorem 1.1 (a)–(c). If M is identically 0, for example, (R_n) reduces to the iid sequence (Q_n).

Inner iteration gives

$$\begin{aligned} Z_n(t) &= Q_1 + M_1 Q_2 + M_1 M_2 Q_3 + \cdots + M_1 \cdots M_{n-1} Q_n + M_1 \cdots M_n t \\ &= \sum_{i=1}^{n} Q_i \prod_{j=1}^{i-1} M_j + t \prod_{1}^{n} M_i. \end{aligned}$$

Under Theorem 1.1 (a)–(c) one has $\prod_1^n M_i \xrightarrow{\text{a.s.}} 0$, and so $Z_n(t)$ converges a.s. to the *perpetuity*

$$Z_\infty := \sum_{i=1}^{\infty} Q_i \prod_{j=1}^{i-1} M_j. \tag{2.4}$$

In (Vervaat 1979, Theorem 1.6) it is proved that then the law of Z_∞ is the fix-point of Ψ.

However inner iteration gives no simple connection between the probability laws of successive iterates, except indirectly by the knowledge that if $R_0 = t$ a.s. they are *marginally* the same laws as for outer iterates and so are connected thereby.

The duality between inner and outer iteration for random recursions is an old idea going back at least to the early days of queueing theory (cf. (Lindley 1952,

p. 279)). G. Letac has built a Principle out of it, for proving existence of fix-points and convergence to them for general random iterations (Letac 1986). When one of the conditions (a)–(c) of Theorem 1.1 is satisfied, ensuring the existence of the unique fix-point law, then as we have noted the probability laws generated by the outer iterations converge to that law, while the random variables generated by the inner iterations converge a.s. to a r.v. having the fix-point law. The idea of Letac's Principle is to prove in more general settings the latter a.s. convergence and deduce from it the former distributional convergence.

With monotonicity or some d-dimensional form of monotonicity the duality between the iterations is a powerful tool, the 'Loynes trick' (Loynes 1962), on which the book (Brandt, Franken & Lisek 1990) is based.

If there is available an iid sequence $((Q_n, M_n))$ with index n running over all integers then the conditions that ensure the existence of a fix-point yield the existence of a process

$$\tilde{R}_n := \sum_{j=-\infty}^{n} Q_j \prod_{i=j+1}^{n} M_i. \tag{2.5}$$

This is a construction of a *stationary* sequence with successive elements linked by outer iteration: $\tilde{R}_{n+1} = \Psi_{n+1}(\tilde{R}_n)$ with Ψ_{n+1} independent of $\tilde{R}_n$. Each $\tilde{R}_n$ is a perpetuity like (2.4) so one might call $(\tilde{R}_n)$ a 'perpetuity process'.

3. Random affine maps in addition-first form.

A random affine map can be formulated as

$$\Phi(t) := M(U + t) \qquad (t \in \mathbf{R}), \tag{3.1}$$

instead of as (1.1). This is less general than (1.1) in that if $P(M = 0) > 0$ then the map (1.1) reduces to a constant-in-t map on that part of the probability space, whereas that cannot occur with (3.1). However some situations to be modelled may present themselves more naturally in the form (3.1). A *fix-point* of (3.1) is any probability law, of S say, that satisfies

$$S \stackrel{\mathrm{d}}{=} M(U + S), \quad S \text{ independent of } (U, M). \tag{3.2}$$

Obviously, (3.1) may be expressed in the form (1.1) by putting $Q := MU$, while if $P(M = 0) = 0$ then (1.1) becomes an instance of (3.1) on setting $U := Q/M$. The extension of Theorem 1.1 to this setting is as follows.

THEOREM 3.1. *Set* $X := \log|M| \in [-\infty, \infty)$.

(a) *Suppose* $-\infty < EX < 0$. *Then a fix-point of (3.1) exists if and only if* $E \log(1 \vee |U|) < \infty$.
(b) *Suppose* $P(X = -\infty) > 0$. *Then a fix-point of (3.1) exists.*
(c) *Suppose* $EX = -\infty$ *and* $P(X = -\infty) = 0$. *Then a fix-point of (3.1) exists if* $E \log(1 \vee |U|) < \infty$.
(d) *Suppose* $0 \leq EX \leq \infty$. *Then a fix-point of (3.1) exists if and only if* $P(M(c + U) = c) = 1$ *for some non-random real* c.
(e) *Suppose* EX *does not exist,* $P(X = -\infty) = 0$, *and (1.3) holds. Then a fix-point of (3.1) exists if and only if* $P(M(c+U) = c) = 1$ *for some non-random real* c.

Proof: $V := \log|U|$ and $X := \log|M|$ take values in $[-\infty, \infty)$ so their sum is always well-defined.

(a) We are assuming $-\infty < EX < 0$, and we apply Theorem 1.1 in the setting that $Q := UM$. So a fix-point exists of (3.1) if and only if $E\log(1 \vee |UM|) < \infty$, that is, $E(V+X)1_{V+X>0} < \infty$. Since $EX1_{V+X>0}$ is finite, from the assumed finiteness of $E|X|$, the criterion reduces to $EV1_{V+X>0} < \infty$.

Now

$$EV1_{V>0} = EV1_{V+X>0} + EV1_{V>0,V+X\leq 0} - EV1_{V\leq 0,V+X>0}, \tag{3.3}$$

and we have

$$0 \leq EV1_{V>0,V+X\leq 0} \leq E(-X)1_{V>0,V+X\leq 0} \leq E|X| < \infty$$

and

$$0 \leq -EV1_{V\leq 0,V+X>0} \leq EX1_{V\leq 0,V+X>0} \leq E|X| < \infty.$$

So the criterion is equivalent to $EV1_{V>0} < \infty$, as claimed.

(b) This follows from Theorem 1.1 (b), with $Q := UM$. In the expansion for the finite perpetuity Q^* is zero identically.

(c) Now we are assuming $EX = -\infty$ and we apply Theorem 1.1's (b) in the setting $Q := UM$. So a fix-point of (3.1) exists if $E(V+X)1_{V+X>0} < \infty$. For the expectation not to be $+\infty$ it suffices that $EV1_{V+X>0} < \infty$ and $EX1_{V+X>0} < \infty$. Since the second of these follows from our assumption $EX = -\infty$, it suffices for a fix-point that $EV1_{V+X>0} < \infty$. But by (3.3), $EV1_{V+X>0} \leq EV1_{V>0}$, so it is enough that $EV1_{V>0} < \infty$, and (b) is thus proved.

(d)–(e) are re-statements of Theorem 1.1's (d)–(e), again for the setting where $Q := MU$. □

In cases (a)–(c) the uniqueness of the fix-point law when it exists, and convergence of the laws of outer iterates $S_{n+1} := \Phi_{n+1}(S_n) = M_{n+1}(U_{n+1} + S_n)$ to it whatever the law of S_0, may be deduced from the results of (Vervaat 1979) similarly to the above proof.

Inner iteration $Z_{n+1}(t) := Z_n(M_{n+1}(U_{n+1} + t))$ now gives

$$Z_n(t) = \sum_{i=1}^{n} U_i \prod_{j=1}^{i} M_j + t \prod_{1}^{n} M_i,$$

which converges a.s., under conditions (a), (b) or (c) of Theorem 3.1, to

$$S_\infty := \sum_{i=1}^{\infty} U_i \prod_{j=1}^{i} M_j. \tag{3.4}$$

This is also known as a perpetuity (Dufresne 1991), and is a construction of a random variable having the fix-point law for (3.1).

4. A duality in the case of independent components.

There are further connections between the fix-points of (1.1) and (3.1) when the pairs (Q, M) and (U, M) each have independent components. Thus let R, Q, M and M' be independent, and $\mathcal{L}(M') := \mathcal{L}(M)$. If R is a fix-point of (1.1) then $S := M'R$ satisfies

$$S = M'R \stackrel{d}{=} M'(Q + MR) \stackrel{d}{=} M(Q + M'R) = M(Q + S),$$

so S is a fix-point of (3.1) with $U := Q$ independent of M.

Conversely, let S, M, U and U' be independent, and $\mathcal{L}(U') := \mathcal{L}(U)$. If S is a fix-point of (3.1) then $R := U' + S$ satisfies

$$R = U' + S \stackrel{d}{=} U' + M(U + S) \stackrel{d}{=} U + M(U' + S) = U + MR,$$

so R is a fix-point of (1.1) with $Q := U$ independent of M.

The above dual relations between fix-points of (1.1) and (3.1) can be expressed via perpetuities. If any of Theorem 1.1's conditions (a)–(c) hold then R_∞, given by (2.4), is a random-variable construction of a fix-point of (1.1). If all Q_i, M_i for $i = 1, 2, \dots$ are mutually independent, and if M_0 is independent of them all and has $\mathcal{L}(M_0) := \mathcal{L}(M)$, then

$$S'_\infty := M_0 R_\infty = \sum_{i=1}^{\infty} Q_i \prod_{j=0}^{i-1} M_i$$

is a random-variable construction of a fix-point of (3.1), with $U_i := Q_i$ for all i. Because of the assumed independence, S'_∞ has the same probability law as $S_\infty := \sum_{i=1}^{\infty} Q_i \prod_{j=1}^{i} M_i$, and we retrieve (3.4), with $U := Q$.

Conversely, if S_∞ given by (3.4) is a fix-point of (3.1) (with $U := Q$ independent of M), and Q_0 is independent of all Q_i, M_i for $i \geq 1$ and has $\mathcal{L}(Q_0) := \mathcal{L}(Q)$, then

$$R'_\infty := Q_0 + S_\infty = \sum_{i=1}^{\infty} Q_{i-1} \prod_{j=1}^{i-1} M_i$$

is a fix-point of (1.1). Again, by the assumed independence of components, R'_∞ has the same probability law as R_∞ given by (2.4).

None of the above content of this section carries over to non-independent components, though it is easy to persuade oneself that it does. We also do not believe there are other simple dualities between fix-points of (1.1) and (3.1), though again it is easy to persuade oneself otherwise by over-confident manipulation of the seductive formulae (1.2) and (3.2). As an instance of how things can go astray, take the remark in (Pakes 1983) where, referring to his formula

$$Y_n = A_n Y_{n-1} + B_n \qquad (n \in \mathbb{N}), \tag{1}$$

the author observes that 'if we define $X_n = Y_n - B_n$, then (1) can be put into the equivalent form

$$X_n = A_n(X_{n-1} + B_{n-1}) \qquad (n \in \mathbb{N}). \qquad (2)\text{'}$$

Now although the formal substitutions work out, (2) is not really equivalent to (1) via this connection at all. For (1) is useful and meaningful only when the 'innovation' (A_n, B_n) is independent of the pre-existing state Y_{n-1}. Otherwise we have some sort of non-causal model, to use more time-series terminology, the dependence structure of which would need further explication. (Of course, $Y_{n-1} = Y_n/A_n - B_n/A_n$ can be run in reverse time, with (A_n, B_n) independent of Y_n; but that would be a completely different process.) In satisfying (2), $X_{n-1} = Y_{n-1} - B_{n-1}$ is not independent of B_{n-1}, so (X_n) satisfies (2) only in a non-useful form.

5. Heavy-tailed perpetuities from light-tailed inputs.

In (Kesten 1973) it was proved that the solution R of (1.2) has under quite general circumstances an upper tail asymptotic to a power. Kesten proved this for a d-dimensional version of (1.2). Extensions of his results to other conditions were later given by (le Page 1983). The specialization to one dimension of Kesten's results is as follows.

THEOREM K. *Let (Q, M) be such that, for some $\kappa > 0$,*

$$E|Q|^\kappa < \infty,$$
$$E|M|^\kappa = 1,$$
$$E|M|^\kappa \log(1 \vee |M|) < \infty,$$

and

$$\log|M| \text{ is not concentrated on the lattice } \{-\infty\} \cup (h\mathbb{Z}), \text{ for any } h > 0.$$

Then there is a unique law for R satisfying (1.2), and

$$P(R > t) \sim C_+ t^{-\kappa} \qquad (t \to \infty), \tag{5.1}$$
$$P(R < -t) \sim C_- t^{-\kappa} \qquad (t \to \infty), \tag{5.2}$$

for some non-negative constants C_+, C_-. Further, $C_+ + C_- > 0$ if and only if

$$\text{for each fixed } c \in \mathbf{R},\ P(Q + Mc = c) < 1.$$

Observe that the last part says that at least one of the tail asymptotics is a genuine tail equivalence, rather than merely the assertion that $P(R > t)$ and $P(R < -t)$ are $o(t^{-\kappa})$, if and only if the 'tied-down line' cases (d), (e) in Theorem 1.1 are outlawed.

In (Goldie 1991) there are formulae for C_+ and C_-, as well as, under extra conditions, rates of approach of $t^\kappa P(R > t)$ and $t^\kappa P(R < -t)$ to their limits. The formulae for C_+ and C_- differ according to whether $M \geq 0$ a.s. or $P(M < 0) > 0$. In the former case they are

$$C_+ = \tfrac{1}{\kappa m} E\big(((Q + MR)^+)^\kappa - ((MR)^+)^\kappa\big),$$
$$C_- = \tfrac{1}{\kappa m} E\big(((Q + MR)^-)^\kappa - ((MR)^-)^\kappa\big),$$

where $m := E|M|^\kappa \log|M|$ (the conditions imply $0 < m < \infty$), and $x^+ := x \vee 0$, $x^- := (-x) \vee 0$. The formulae are inexplicit, involving the unknown R, so in the same paper some bounds are worked out for them and the special cases when the formulae can be made more explicit are detailed.

Here is what Theorem K becomes for the alternative formulation (3.2):

Theorem 5.1. *Let (U, M) be such that, for some $\kappa > 0$, the conditions on M in Theorem K are satisfied, and*

$$E|U|^{\kappa}\mathbf{1}_{M\neq 0} < \infty.$$

Then there is a unique law for S, satisfying (3.2), and

$$P(S > t) \sim C_+ t^{-\kappa} \qquad (t \to \infty), \tag{5.3}$$

$$P(S < -t) \sim C_- t^{-\kappa} \qquad (t \to \infty), \tag{5.4}$$

for some non-negative constants C_+, C_-. Further, $C_+ + C_- > 0$ if and only if

$$\text{for each fixed } c \in \mathbf{R},\ P(M(U + c) = c) < 1.$$

Proof: We shall *not* start from the result of Theorem K, and make deductions for MR, as we saw that for MR to satisfy (3.2) needs independence of U and M, which we do not assume.

Rather, we just use the fact, also previously remarked, that (3.2) is the special case of (1.2) in which $Q := MU$, and the 'solution' R to this special case of (1.2) is our S. Thus all we have to check is that $(Q, M) := (MU, M)$ satisfies the conditions of Theorem K. But that is immediate! For since $E|M|^{\kappa} = 1$, it follows that $E|Q|^{\kappa} = E|U|^{\kappa}\mathbf{1}_{M\neq 0} < \infty$. □

The results in (Goldie 1991), including those on rate of approach, also carry over to the case of (3.2) without trouble. The formulae for C_+ and C_- become, when $M \geq 0$ a.s.,

$$C_+ = \tfrac{1}{\kappa m} E\big(((U + S)^+)^{\kappa} - (S^+)^{\kappa}\big),$$
$$C_- = \tfrac{1}{\kappa m} E\big(((U + S)^-)^{\kappa} - (S^-)^{\kappa}\big),$$

with m as before.

Theorem K has been applied by a number of authors. It was originally needed to apply to random walks and branching processes in random environments, and applications in that direction have continued recently with (Grey & Lu 1992) and (Grey & Lu 1993). It is also natural to use tail estimates as a starting point for an extreme-value analysis, and in (de Haan *et al.* 1989) Theorem K was applied to the one-dimensional case of ARCH processes. ARCH processes are used in econometrics to model exchange-rate yields. The random equation is $R_{n+1} = (\alpha + \lambda R_n)Z_{n+1}^2$ in this case, where the Z_i are independent $N(0,1)$ and $\alpha > 0$, $0 < \lambda < 1$ are constants. Thus it is in fact of the form (3.2). The work of de Haan *et al.* has recently been extended in (Perfekt 1991).

In the extreme-value analyses a dominant role seems to be played by high values of the terms $Q_i \prod_{j=1}^{i-1} M_j$ that make up the perpetuity (2.4). Indeed there is a cluster of such high values, leading to an extremal-index contribution to the tail asymptote. This phenomenon picks up the observation of (Kesten 1973) that the tail of the perpetuity is like that of $\bigvee_{i=1}^{\infty} Q_i \prod_{j=1}^{i} M_j$, but with a multiplicative constant in the exponent. One can also see here instances of the Poisson clumping phenomenon of (Aldous 1989).

We mention at this point that other recent applications in finance have tended to concentrate on continuous-time forms of perpetuity and random recurrence. Thus in (Dufresne 1991) a perpetuity $Z := \int_0^\infty e^{-(\nu t+\sigma W_t)}\,dt$, where ν and σ are positive constants and W is a Wiener process, was treated. (Guegan 1992) and (Yor 1992) are other contributions. See also (De Schepper, De Vylder, Goovaerts & Kaas 1992) and (De Schepper, Goovaerts & Delbaen 1992) for applications in insurance related to the calculation of annuities certain using Feynman integration. More generally, starting from the SDE

$$dr_t = (\delta + \beta r_t)\,dt + \sigma r_t^\gamma\,dW_t,$$

which functions as an analogue of (2.2) or (2.3), in (Geman & Yor 1992, §V) there is derived after some specializations of parameters the 'perpetuity'

$$Z = \lambda \int_0^\infty \exp\left(-\lambda \int_0^u r(\lambda s)\,ds\right)\,du,$$

where r is the square of an Ornstein-Uhlenbeck process.

In a related but differently focussed direction there is a line of work on embedding discrete recurrences of the form (2.2) into continuous-time versions, namely SDEs or stochastic integral equations; cf. (Wolfe 1982), (de Haan & Karandikar 1989).

6. Other tail and distribution results.

Much more common than the phenomenon discussed in §5 is that heavy-tailed output from a stochastic model results simply from heavy-tailed input. This can certainly be the case for the random affine recursion, as was first proved in (Grincevičius 1975). An improved, corrected proof of that result is in (Grey 1993) together with a converse and an application. Multivariate extensions are in (Resnick & Willekens 1991). The conditions of these results make the tail of the perpetuity (2.4) reflect the properties of the distribution of Q in (1.1), in contrast to what occurs in the setting of §5, where the role of M is more prominent.

As one might expect, there are many scattered results where the components of the perpetuity (2.4) and the perpetuity itself are light-tailed. For such cases and in particular those in which Q and/or M has bounded support it is natural to think of the map $\mathcal{L}(R) \mapsto \mathcal{L}(\Psi(R))$, where Ψ and R are independent, as a contraction on some suitably metrized space of probability laws. Under this point of view R can naturally be allowed to live in some general space such as a Banach space. See (Burton & Rösler 1993) or (Rachev & Rüschendorf 1991) for recent instances of this approach.

Rate of convergence in law of the outer iterates is also a natural topic to consider under the latter approach. Some references on rates of convergence are (Grincevičius 1978), (Rachev & Todorovic 1990), (Rachev & Samorodnitsky 1992).

An extremely special case of (1.1) is when M is non-random, when the perpetuity process (2.5) reduces to a weighted moving average of iid random variables. Tail evaluation and extreme-value theory for such processes have been done in great generality in (Rootzén 1986).

Recurrence properties and ergodic behaviour of the outer iterations, for non-negative Q and M, are dealt with in the three-part paper (Kellerer 1992). The restriction to non-negative Q lessens the value of these results for applications and it would be valuable to have it removed. Earlier results on ergodicity and recurrence, from a time-series viewpoint, were in (Feigin & Tweedie 1985).

Some statistical aspects of perpetuities, in particular bootstrap estimation, are considered in (Aebi *et al.* 1992).

Certain other general properties of fix-points are known. These can be considered properties of perpetuities (2.4) or (3.4), as these are constructions for the fix-points of (1.1) and (3.1) respectively. For instance, in (Grincevičius 1974) it is proved that if a fix-point of (1.1) exists and $P(M \neq 0) = 1$ then the fix-point law is either degenerate or continuous (non-atomic). This was extended in (Pakes 1983), where conditions that ensure the law has a density, or a continuous density, are given.

REFERENCES

Aebi, M., Embrechts, P. & Mikosch, T., *Stochastic discounting, aggregate claims and the bootstrap*, Report, 1992, Departement Mathematik, ETH, CH–8092 Zürich, Switzerland; Adv. Appl. Probab. (March 1994) (to appear).

Aldous, D., *Probability Approximations via the Poisson Clumping Heuristic*, Applied Math. Sciences 77, Springer-Verlag, New York, 1989.

Bougerol, Ph. & Picard, N., *Strict stationarity of generalized autoregressive processes*, Ann. Probab. 20 (1992), 1714–1730.

Brandt, A., *The stochastic equation* $Y_{n+1} = A_n Y_n + B_n$ *with stationary coefficients*, Adv. Appl. Probab. 18 (1986), 211–220.

Brandt, A., Franken, P. & Lisek, B., *Stationary Stochastic Models*, Wiley Ser. in Probab. & Math. Stat., Wiley, Chichester, UK, 1990.

Burton, R. M. & Rösler, U., *An* L_2 *convergence theorem for random affine mappings*, Report, 1993, Dept. Math., Oregon State Univ., Corvallis, Oregon 97331, USA.

De Schepper, A., De Vylder, F., Goovaerts, M. & Kaas, R., *Interest randomness in annuities certain*, Insurance: Math. and Econom. 11 (1992), 271-282.

De Schepper, A., Goovaerts, M. & Delbaen, F., *The Laplace transform of annuities certain with exponential time distributions*, Insurance: Math. and Econom. 11 (1992), 291-299.

Dufresne, D., *The distribution of a perpetuity, with applications to risk theory and pension funding*, Scand. Act. J. no. 1–2 (1991), 39–79.

Dufresne, D., *On discounting when rates of return are random*, in Comptes rendus du 24ᵉ Congrés International des Actuaires, Montréal, 1992.

Feigin, P. D. & Tweedie, R. L., *Random coefficient autoregressive processes: a Markov chain analysis of stationarity and finiteness of moments*, J. Time Series Anal. 6 (1985), 1–14.

Geman, H. & Yor, M., *Bessel processes, Asian options and perpetuities*, Report, 1992, Dépt. Finance, ESSEC, 95021 Cergy-Pontoise, France.

Goldie, C. M., *Implicit renewal theory and tails of solutions of random equations*, Ann. Appl. Probab. 1 (1991), 126–166.

Grey, D. R., *Regular variation in the tail behaviour of solutions of random difference equations*, Report, 1993, Dept. Prob. & Stat., Univ. Sheffield, PO Box 597, Sheffield S10 2UN, UK; Ann. Appl. Probab. (to appear).

Grey, D. R. & Lu Zhunwei, *The fractional linear probability generating function in the random environment branching process*, Report, 1992, Dept. Prob. & Stat., Univ. Sheffield, PO Box 597, Sheffield S10 2UN, UK.

Grey, D. R. & Lu Zhunwei, *The asymptotic behaviour of extinction probability in the Smith-Wilkinson branching process*, Adv. Appl. Probab. 25 (1993), 263–289.

Grincevičius, A. K., *On the continuity of the distribution of a sum of dependent variables connected with independent walks on lines*, Theor. Prob. Appl. 19 (1974), 163–168.

Grincevičius, A. K., *One limit distribution for a random walk on the line*, Lithuanian Math. J. 15 (1975), 580–589.

Grincevičius, A. K., *Approximation-in-variation of distributions of products of random linear transformations of a straight line*, Lithuanian Math. J. 18 (1978), 183–190.

Guegan, D., *A continuous time ARCH model*, Report, 1992, Inst. Galilée, Av. Jean Baptiste Clément, 93430 Villetaneuse, France.

Haan, L. de & Karandikar, R. L., *Embedding a stochastic difference equation into a continuous-time process*, Stoch. Proc. Appl. 32 (1989), 225–235.

Haan, L. de, Resnick, S. I., Rootzén, H., Vries, C. G. de, *Extremal behaviour of solutions to a stochastic difference equation, with applications to ARCH processes*, Stoch. Proc. Appl. 32 (1989), 213–224.

Kellerer, H. G., *Ergodic behaviour of affine recursions, I: criteria for recurrence and transience, II: invariant measures and ergodic theorems, III: positive recurrence and null recurrence*, Reports, 1992, Math. Inst. Univ. München, Theresienstrasse 39, D–8000 München, Germany.

Kesten, H., *Random difference equations and renewal theory for products of random matrices*, Acta Math. 131 (1973), 207–248.

Kesten, H. & Maller, R. A., *Infinite limits and infinite limit points of random walks and trimmed sums*, Report, 1993, Dept. Math., Univ. Western Australia, Nedlands, W.A. 6009, Australia.

Letac, G., *A contraction principle for certain Markov chains and its applications*, in Random Matrices and their Applications (Proc. AMS-IMS-SIAM Joint Summer Research Conf. 1984) (J. E. Cohen, H. Kesten, C. E. Newman, eds.), Contemporary Mathematics 50, Amer. Math. Soc., Providence, R.I., 1986, pp. 263–273.

Lindley, D. V., *The theory of queues with a single server*, Proc. Camb. Philos. Soc. 48 (1952), 277–289.

Loynes, R. M., *The stability of a queue with non-independent inter-arrival and service times*, Proc. Camb. Philos. Soc. 58 (1962), 497–520.

le Page, E., *Théorèmes de renouvellement pour les produits de matrices aléatoires; équations aux différences aléatoires*. Séminaire de Probabilités, Rennes 1983, pp. 116. Publ. Sém. Math. I, Univ. Rennes, 1983.

Pakes, A. G., *Some properties of a random linear difference equation*, Austral. J. Statist. 25 (1983), 345–357.

Perfekt, R., *Extremal behaviour of stationary Markov chains with applications*, Report, 1991, Dept. Math. Stat., Univ. Lund, S–22100 Lund, Sweden.

Rachev, S. T. & Rüschendorf, L., *Probability metrics and recursive algorithms*, Report, 1991, Dept. Stat. & Appl. Probab., Univ. California, Santa Barbara, California 93106, USA.

Rachev, S. T. & Samorodnitsky, G., *Limit laws for a stochastic process and random recursion arising in probabilistic modelling*, Report, 1992, School of O.R. & I.E., ETC Bldng., Cornell Univ., Ithaca, NY 14853–3801, USA.

Rachev, S. T. & Todorovic, P., *On the rate of convergence of some functionals of a stochastic process*, J. Appl. Probab. 27 (1990), 805–814.

Resnick, S. I. & Willekens, E., *Moving averages with random coefficients and random coefficient autoregressive models*, Commun. Stat. - Stoch. Models 7 (1991), 511–525.

Rootzén, H., *Extreme value theory for moving average processes*, Ann. Probab. 14 (1986), 612–652.

Vervaat, W., *On a stochastic difference equation and a representation of non-negative infinitely divisible random variables*, Adv. Appl. Probab. 11 (1979), 750–783.

Wolfe, S. J., *On a continuous analogue of the stochastic difference equation* $X_n = \rho X_{n-1} + B_n$, Stoch. Proc. Appl. 12 (1982), 301–312.

Yor, M., *Some aspects of Brownian motion. Part I: Some special functionals*, Lecture Notes in Mathematics, ETH-Zürich, Birkhäuser, Basel, 1992.

Paul Embrechts, Department of Mathematics, ETH-Zentrum, 8092 Zürich, Switzerland

Charles M. Goldie, School of Mathematical Sciences, Queen Mary & Westfield College, Mile End Road, London E1 4NS, UK

ON RECENT DEVELOPMENTS IN THE THEORY OF SET-INDEXED PROCESSES

A unified approach to empirical and partial-sum processes+)

Peter Gaenssler

Math. Institute, University of Munich, D-80333 Munich

Abstract

Two important processes in probability and statistics are the empirical and partial-sum processes. The purpose of this paper is to present a unified approach to both types of processes including their multivariate versions by studying processes $S_n = ((S_n(C))_{C\in\mathscr{C}}$ with $S_n(C)$ being defined by

$$S_n(C) := \sum_{j\le j(n)} 1_C(\eta_{nj})\xi_{nj} \quad , \quad C\in\mathscr{C} ,$$

the η_{nj}'s being random elements (random locations) in an arbitrary sample space X, the ξ'_{nj}s being real-valued random variables (random masses), and where the index set $\mathscr{C}$ is a Vapnik-Chervonenkis class of subsets of X.

Our emphasis is on a asymptotic results (as the sample size n tends to infinity) for the processes S_n such as functional central limit theorems (FCLT) and a uniform law of large numbers (ULLN) (the uniformity being w.r.t. $\mathscr{C}$), containing various results for empirical and partial-sum processes as special cases.

Key words and phrases: Empirical processes, partial-sum processes, set-indexed partial-sum processes with random locations, Vapnik-Chervonenkis classes of sets, metric entropy, maximal inequalities for Rademacher averages, symmetrization inequality, functional central limit theorems, uniform laws of large numbers.
AMS 1980 subject classifications: Primary 60-02, 60F17, 60F25; secondary 60E15, 60G15, 54C70, 62G30.

+) Invited Lecture for the Fifth Prague Symposium on Asymptotic Statistics, 4.-9. Sept., 1993

1. Introduction. In its simplest form, the so-called *multivariate uniform empirical process* $\alpha_n = (\alpha_n(\underline{t}))_{\underline{t}\in I^d}$ (of sample size n) is defined by

$$\alpha_n(\underline{t}) := n^{1/2}(F_n(\underline{t})-F(\underline{t})),\ \underline{t} = (t_1,\dots,t_d)\in I^d = [0,1]^d,\ d\geq 1 .$$

Here F_n denotes the empirical distribution function based on independent random variables (r.v.'s) $\eta_1,\dots,\eta_n$ which are uniformly distributed in the sample space $X = I^d$, i.e. (denoting with 1_A the indicator function of A)

$$F_n(\underline{t}) := n^{-1}\sum_{j\leq n} 1_{[\underline{0},\underline{t}]}(\eta_j),\ \underline{t} = (t_1,\dots,t_d)\in I^d,$$

where $[\underline{0},\underline{t}] = [0,t_1]\times\dots\times[0,t_d]$, and $F(\underline{t}) := t_1\cdot\dots\cdot t_d$.

Concerning partial-sum processes $\zeta_n = (\zeta_n(\underline{t}))_{\underline{t}\in I^d}$ (of sample size n) one defines

$$\zeta_n(\underline{t}) := n^{-d/2} \sum_{\{\underline{j}\in J_n:\,\underline{j}/n\in[\underline{0},\underline{t}]\}} \xi_{\underline{j}},\quad \underline{t}\in I^d,$$

where $J_n = \{1,\dots,n\}^d$, $\underline{j} = (j_1,\dots,j_d)$, and where the r.v.'s $\xi_{\underline{j}}, \underline{j}\in\mathbb{N}^d$, are assumed to be independent and identically distributed (i.i.d.) with $\mathbb{E}(\xi_{\underline{j}}) = 0$ and $\mathbb{E}(\xi_{\underline{j}}^2) = 1$.

Identifying in both cases any $\underline{t}\in I^d$ with the quadrant $C = [\underline{0},\underline{t}]$ one gets the representation

(1.1) $$\alpha_n(C) = n^{1/2}(\nu_n(C) - \nu(C)),\ C\in\mathscr{C},$$

where $\nu_n(C) := n^{-1}\sum_{j\leq n} 1_C(\eta_j)$ is the *empirical measure* based on $\eta_1,\dots\eta_n$,

ν := Lebesgue measure,

(1.2) $$\mathscr{C} := \{[\underline{0},\underline{t}]:\ \underline{t}\in I^d\},$$ and

(1.3) $$\zeta_n(C) = n^{-d/2} \sum_{\underline{j}\in J_n} 1_C(\underline{j}/n)\,\xi_{\underline{j}},\ C\in\mathscr{C},$$

respectively.

Thus, both processes can be considered as *set-indexed processes* (indexed by the class $\mathscr{C}$ of all quadrants as defined by (1.2)).

In addition, both processes are special cases of what we call *partial-sum processes* $S_n = (S_n(C))_{C\in\mathscr{C}}$ (of sample size $j(n)\in\mathbb{N}$) *with random locations*, being normalized and centered at their expectations $\mathbb{E}(S_n(C))$, defined by

(1.4) $$S_n(C) := \sum_{j\leq j(n)} 1_C(\eta_{nj})\xi_{nj},\ C\in\mathscr{C},$$

where $(\eta_{nj})_{1\leq j\leq j(n),n\in\mathbb{N}}$ and $(\xi_{nj})_{1\leq j\leq j(n),n\in\mathbb{N}}$ (with $j(n)\to\infty$ as $n\to\infty$) are given triangular arrays of rowwise independent r.v.'s such that for each $n\in\mathbb{N}$ $(\eta_{nj})_{1\leq j\leq j(n)}$ is independent of $(\xi_{nj})_{1\leq j\leq j(n)}$.

Indeed, taking $\eta_{nj} = \eta_j$ as random locations in $X = I^d$ and $\xi_{nj} \equiv n^{-1}$ as non-random masses, $1\leq j\leq j(n) := n$, one gets

$$n^{1/2}(S_n(C) - \mathbb{E}(S_n(C))) = \alpha_n(C),$$

and, replacing η_{nj} in (1.4) by the non-random locations $\underline{j}/n$ and the random masses ξ_{nj} by $n^{-d}\xi_{\underline{j}}$ where $1\leq j\leq j(n) := n^d$,

$$n^{d/2}(S_n(C) - \mathbb{E}(S_n(C))) = \zeta_n(C).$$

Based on the excellent work of Dudley (1978), Ossiander (1984) and Pyke (1984), we are going to present FCLT's and a ULLN for the processes $S_n = (S_n(C))_{C\in\mathscr{C}}$ as defined by (1.4), where more generally, compared with the both examples mentioned before, the η_{nj}'s are random elements (r.e.'s) in an *arbitrary* sample space X (that is, the η_{nj}'s are assumed to be defined on some basic probability space (p-space) $(\Omega, \mathscr{A}, \mathbb{P})$ with values in X such that for a suitable σ-field $\mathfrak{X}$ of subsets of X each $\eta_{nj}: \Omega \to X$ is $\mathscr{A}$, $\mathfrak{X}$-measurable), and where the index family $\mathscr{C}$ is a so-called *Vapnik-Chervonenkis class* (VCC) of subsets of X (a property which is shared by the class of all quadrants in $X = I^d$).

For this we recall that a class $\mathscr{C}$ of subsets of a set X is said to shatter a finite subset F of X if every $F' \subset F$ is of the form $F \cap C$ for some $C \in \mathscr{C}$. $\mathscr{C}$ is called a VCC if for some $s \in \mathbb{N}$ no s-element subset of X is shattered by $\mathscr{C}$. For example, in $X = \mathbb{R}^d$ the class of all lower left orthants or the class of all rectangles or ellipsoids form a VCC. Any subclass of a VCC is a VCC, and so $\{C \Delta D: C, D \in \mathscr{C}\}$ is a VCC if $\mathscr{C}$ is one. More on VCC's may be found in Vapnik-Chervonenkis (1971), Dudley (1978), Pollard (1984) and in a remarkable paper by Stengle and Yukich (1989).

Given a VCC $\mathscr{C} \subset \mathfrak{X}$ in an arbitrary sample space $X = (X, \mathfrak{X})$ it is known from Dudley ((1978), Lemma 7.13) (see also Assouad (1981)) that there exist universal constants $0 < K < \infty$ and $k \in \mathbb{N}$, depending only on $\mathscr{C}$, such that for any probability measure (p-measure) μ on $\mathfrak{X}$.

(1.5) $\qquad N(\varepsilon, \mathscr{C}, d_\mu) \le K\varepsilon^{-k}$ for all $0 < \varepsilon \le 1/2$,

where $d_\mu(C,D) := \mu(C \Delta D)$, $C, D \in \mathscr{C}$, defines a pseudo-metric on $\mathscr{C}$ and where $N(\varepsilon, \mathscr{C}, d_\mu) := \min\{n \in \mathbb{N}:$ there exist $C_1, \dots, C_n \in \mathscr{C}$ such that for any $C \in \mathscr{C}$ $d_\mu(C, C_i) \le \varepsilon$ for some C_i, $1 \le i \le n\}$.

This, together with a maximal inequality for so-called *Rademacher averages* and a *symmetrization inequality* presented for arbitrary pseudo-metric parameter spaces $T = (T,d)$ (like $\mathscr{C} = (\mathscr{C}, d_\mu)$) in an Appendix below together with simple proofs which may be of independent interest, will be our main tools.

The outline of this paper is as follows: In Section 2 FCLT's are proved for the centered processes $S'_n = (S'_n(C))_{C\in\mathscr{C}}$ with $S'_n(C) := S_n(C) - \mathbb{E}(S_n(C))$ and $S_n(C) := \sum_{j\le j(n)} 1_C(\eta_{nj})\xi_{nj}$, $C \in \mathscr{C}$, yielding various results for empirical and partial-sum processes as special cases. In Section 3 a ULLN in form of a Mean Glivenko-Cantelli result is presented whose proof together with similar consequences for empirical and partial-sum processes will be contained in a separate paper by Gaenssler and Ziegler (1993).

To carry out certain calculations, some further measurability assumptions would be needed: in order to avoid these and also for the sake of simplicity of our presentation we assume that the VCC $\mathscr{C}$ is countable; otherwise we use upper integrals $\mathbb{E}^*$, where for any $f:\Omega \to \mathbb{R}$ $\mathbb{E}^*(f) := \inf\{\mathbb{E}(g): f \le g, g \text{ integrable}\}$.

2. Functional central limit theorems (FCLT) for partial-sum processes with random locations and indexed by a VCC.

As indicated in Section 1 the present context is as follows:

Let $X = (X,\mathfrak{X})$ be an arbitrary measurable space (sample space for locations), $(\eta_{nj})_{1\le j\le j(n),\, n\in\mathbb{N}}$ be a triangular array of r.e.'s in X, $(\xi_{nj})_{1\le j\le j(n),\, n\in\mathbb{N}}$ be a triangular array of r.v.'s, all defined on a common p-space $(\Omega,\mathscr{A},\mathbb{P})$ and $\mathscr{C}\subset\mathfrak{X}$ be a *countable* VCC.

The following basic assumptions on (η_{nj}) and (ξ_{nj}) will be made throughout this section:

Both arrays consists of rowwise independent but not necessarily identically distributed (i.d.) r.e.'s and r.v.'s, respectively, and for each $n\in\mathbb{N}$ $(\eta_{nj})_{1\le j\le j(n)}$ is independent of $(\xi_{nj})_{1\le j\le j(n)}$.

We are going to present a functional central limit theorem (FCLT) for the centered process $S'_n = (S'_n(C))_{C\in\mathscr{C}}$ that is, *weak convergence (convergence in 'law') in the sense of Hoffmann-Jørgensen* (1984) of the processes S'_n (as $n \to \infty$) in the (nonseparable) Banach space

$$l^\infty(\mathscr{C}) := \{x:\mathscr{C} \to \mathbb{R}: \|x\|_{\mathscr{C}} := \sup_{C\in\mathscr{C}} |x(C)| < \infty\}$$

to a mean zero Gaussian process $\mathbb{G} = (G(C))_{C\in\mathscr{C}}$, whose sample paths are contained in

$$U^b(\mathscr{C},d_\nu) := \{x\in l^\infty(\mathscr{C}): x \text{ uniformly } d_\nu\text{-continuous}\}$$

for some p-measure ν on $\mathfrak{X}$.

Note that $U^b(\mathscr{C},d_\nu)$ is a closed subspace of $(l^\infty(\mathscr{C}), \|\cdot\|_{\mathscr{C}})$ being separable iff $(\mathscr{C},d_\nu)$ is totally bounded (cf. Corollary 2 in Gaenssler-Schneemeier (1986)), where the latter follows from (1.5).

In this context, S'_n *converges weakly* to $\mathbb{G}$ ($S'_n \xrightarrow[\text{sep}]{\mathscr{L}} \mathbb{G}$) (where 'sep' indicates the fact that the limiting process $\mathbb{G}$ concentrates on the separable space $U^b(\mathscr{C},d_\nu)$), iff

(2.1) $\lim_{n\to\infty} \mathbb{E}^*(f(S'_n)) = \mathbb{E}(f(\mathbb{G}))$ for all bounded and continuous $f: l^\infty(\mathscr{C}) \to \mathbb{R}$:

(Note that since $\mathbb{G}$ has all its sample paths in the separable subspace $U^b(\mathscr{C},d_\nu)$ of $l^\infty(\mathscr{C})$, it follows that $\mathbb{G}$ (thought of as being also defined on our basic p-space $(\Omega,\mathscr{A},\mathbb{P})$) is $\mathscr{A}$, $\mathscr{B}(l^\infty(\mathscr{C}))$-measurable (where $\mathscr{B}(l^\infty(\mathscr{C}))$ denotes the Borel σ-field in $(l^\infty(\mathscr{C}), \|\cdot\|_{\mathscr{C}})$) and therefore $f(\mathbb{G})$ is $\mathbb{P}$-integrable, whence in

(2.1) $\mathbb{E}(f(\mathbb{G}))$ is well defined. Note also that in this context it is *not* assumed the 'laws' of S'_n to be defined on any particular σ-field in $l^\infty(\mathscr{C})$, i.e. concerning the processes S'_n, $n\in\mathbb{N}$, no measurability has to be assumed; therefore the use of upper integrals (outer expectations) $\mathbb{E}^*$ in (2.1).)

Using the technique in the proof of Theorem 3.1 in Arcones-Gaenssler-Ziegler (1992) one gets, under the basic assumptions of this section as mentioned above, the following result:

2.2 Theorem. Let $S'_n(C) := \sum_{j\le j(n)} (1_C(\eta_{nj})\xi_{nj} - \mathbb{E}(1_C(\eta_{nj})\xi_{nj}))$, $C\in\mathscr{C}$. Assume that there exists a p-measure ν on $\mathfrak{X}$ and finite constants c_i so that with $\nu_{nj} := \mathscr{L}\{\eta_{nj}\}$ (the laws on $\mathfrak{X}$ of the η_{nj}) the following four conditions are fulfilled:

(i) $\sum_{j\le j(n)} \mathbb{E}(|\xi_{nj}|I(|\xi_{nj}| \ge \varepsilon)) \longrightarrow 0$ for all $\varepsilon > 0$

(where here and elsewhere $I(\cdots)$ stands for the indicator function $1_{\{\dots\}}$)

(ii) $\sum_{j\le j(n)} \nu_{nj}(C\cap D)\mathbb{E}(\xi_{nj}^2 I(|\xi_{nj}| \le \varepsilon_1)) \to c_1\nu(C\cap D)$ for all $C,D\in\mathscr{C}$

and some $\varepsilon_1 > 0$

(iii) $\sum_{j\le j(n)} \nu_{nj}(C)\nu_{nj}(D)(\mathbb{E}(\xi_{nj}I(|\xi_{nj}| \le \varepsilon_2)))^2 \to c_2\nu(C)\nu(D)$ for all $C,D\in\mathscr{C}$

and some $\varepsilon_2 > 0$

(iv) $\lim_{\delta\to 0} \limsup_{n\to\infty} \sup_{\{C\in\mathscr{C}:\nu(C)\le\delta\}} \sum_{j\le j(n)} \nu_{nj}(C)\mathbb{E}(\xi_{nj}^2 I(|\xi_{nj}|\le\varepsilon_3)) = 0$

for some $\varepsilon_3 > 0$.

Then $S'_n \xrightarrow[\text{sep}]{\mathscr{L}} \mathbb{G}_\nu$, where $\mathbb{G}_\nu = (G_\nu(C))_{C\in\mathscr{C}}$ is a mean zero Gaussian process with sample paths in $U^b(\mathscr{C},d_\nu)$ and
$\operatorname{cov}(G_\nu(C), G_\nu(D)) = c_1\nu(C\cap D) - c_2\nu(C)\nu(D)$, $C,D\in\mathscr{C}$.

Proof. First we observe that (i) implies that (ii) and (iii) hold for every $\varepsilon>0$, whence in the following we can make use of

(2.3) $$\sum_{j\le j(n)} \nu_{nj}(C\cap D)\mathbb{E}(\xi_{nj}^2 I(|\xi_{nj}| \le \varepsilon)) \to c_1\nu(C\cap D) \text{ for all } C,D\in\mathscr{C} \text{ and all } \varepsilon>0$$

and

(2.4) $$\sum_{j\le j(n)} \nu_{nj}(C)\nu_{nj}(D)(\mathbb{E}(\xi_{nj}I(|\xi_{nj}|\le\varepsilon)))^2 \to c_2\nu(C)\nu(D) \text{ for all } C,D\in\mathscr{C}$$

and all $\varepsilon>0$.

Also, condition (i) implies that there is a sequence $(\tau_n)_{n\in\mathbb{N}}$ of positive real numbers τ_n such that $\tau_n \to 0$ and

(2.5) $$\lim_{n\to\infty} \sum_{j\le j(n)} \mathbb{E}(|\xi_{nj}|I(|\xi_{nj}| \ge \tau_n)) = 0.$$

It follows from (2.5) that

$$\lim_{n\to\infty} \mathbb{P}(\sup_{C\in\mathscr{C}} |S'_n(C) - S''_n(C)| > \rho) = 0 \quad \text{for all } \rho > 0,$$

where $S''_n(C) := \sum_{j\le j(n)} (1_C(\eta_{nj})\xi_{nj}I(|\xi_{nj}| \le \tau_n) - \mathbb{E}(1_C(\eta_{nj})\xi_{nj}I(|\xi_{nj}| \le \tau_n)))$.

Hence, by a Cramér-Slutsky type of argument, it is enough to show that $S''_n \xrightarrow[\text{sep}]{\mathscr{L}} \mathbb{G}_\nu$.

Now, let $\bar{\mathbb{G}}_\nu = (\bar{G}_\nu(C))_{C\in\mathscr{C}}$ be a mean zero Gaussian process having the same covariance structure as $\mathbb{G}_\nu$. Since $(\mathscr{C},d_\nu)$ is totally bounded, in order to prove the assertion that $S''_n \xrightarrow[\text{sep}]{\mathscr{L}} \mathbb{G}_\nu$, it suffices to show (by a result attributed to Hoffmann-Jørgensen (1984), versions of which were proved in other papers, cf. Theorem 3.10 in Gaenssler (1992) and the literature cited there) that

(a) $S''_n \xrightarrow[\text{fidi}]{\mathscr{L}} \bar{\mathbb{G}}_\nu$ (that is, convergence of the finite dimensional marginal distributions *(fidis)* of S''_n to those of $\bar{\mathbb{G}}_\nu$), and

(b) $\lim_{\delta\to 0} \limsup_{n\to\infty} \mathbb{P}(\sup_{\nu(C\Delta D)\le\delta} |S''_n(C) - S''_n(D)| > \varepsilon) = 0$ for alle $\varepsilon > 0$.

As to (a): Let $C_1,\dots,C_N \in \mathscr{C}$ and $a_1,\dots,a_N \in \mathbb{R}$, $N\in\mathbb{N}$, be arbitrary but fixed; then by the Cramér-Wold-Device it suffices to show that

$$\sum_{i=1}^{N} a_i S''_n(C_i) \xrightarrow{\mathscr{L}} \sum_{i=1}^{N} a_i \bar{G}_\nu(C_i) . \tag{2.6}$$

Let $\Phi_{nj} := \sum_{i=1}^{N} a_i 1_{C_i}(\eta_{nj})$ and

$$\zeta_{nj} := \Phi_{nj}\xi_{nj}I(|\xi_{nj}| \le \tau_n) - \mathbb{E}(\Phi_{nj}\xi_{nj}I(|\xi_{nj}| \le \tau_n));$$

then $\sum_{i=1}^{N} a_i S''_n(C_i) = \sum_{j\le j(n)} \zeta_{nj}$.

The Lévy central limit theorem for triangular arrays (cf. Theorem 2.4.7 in Araujo-Giné (1980)) gives us the following sufficient conditions for the convergence in law of $\sum_{j\le j(n)} \zeta_{nj}$ to a centered normally distributed r.v. ζ:

(α) $\sum_{j\le j(n)} \mathbb{P}(|\zeta_{nj}| \ge \varepsilon) \longrightarrow 0$ for all $\varepsilon > 0$

(β) $\sum_{j\le j(n)} \mathbb{E}(\zeta_{nj}^2 I(|\zeta_{nj}| \le \varepsilon))$ converges for some $\varepsilon > 0$

(γ) $\sum_{j\le j(n)} (\mathbb{E}(\zeta_{nj}I(|\zeta_{nj}| \le \varepsilon)))^2$ converges for some $\varepsilon > 0$

(δ) $\sum_{j\le j(n)} \mathbb{E}(\zeta_{nj}I(|\zeta_{nj}| \le \varepsilon)) \longrightarrow 0$ for some $\varepsilon > 0$.

We are going to show (α) - (δ):

As to (α): Since $|\Phi_{nj}| \le \sum_{i=1}^{N} |a_i| := M < \infty$, $\max_{1\le j\le j(n)} |\zeta_{nj}| \le 2M\tau_n$, whence, since $\tau_n \to 0$, there exists for each $\varepsilon > 0$ an $n_o = n_o(\varepsilon) \in \mathbb{N}$ such that for all $n \ge n_o$ $2M\tau_n < \varepsilon$; thus, for all $n \ge n_o$ $\max_{1\le j\le j(n)} \mathbb{P}(|\zeta_{nj}| \ge \varepsilon) = 0$, implying (α).

As to (δ): Since $\max_{1\le j\le j(n)} |\zeta_{nj}| < \varepsilon$ for all $n \ge n_o$, it follows that for all $n \ge n_o$

$\mathbb{E}(\zeta_{nj} I(|\zeta_{nj}| \le \varepsilon)) = \mathbb{E}(\zeta_{nj}) = 0$, $1 \le j \le j(n)$, implying (δ).

As to (γ): As in the proof of (δ) we obtain that

$\sum_{j\le j(n)} (\mathbb{E}(\zeta_{nj} I(|\zeta_{nj}| \le \varepsilon)))^2 \longrightarrow 0$ for some (all) $\varepsilon > 0$.

As to (β): As before, given any $\varepsilon > 0$, we have for all $n \ge n_o = n_o(\varepsilon)$ and $1 \le j \le j(n)$ that

$$\mathbb{E}(\zeta_{nj}^2 I(|\zeta_{nj}| \le \varepsilon)) = \mathbb{E}(\zeta_{nj}^2) =$$

$$\mathrm{Var}(\Phi_{nj}\xi_{nj} I(|\xi_{nj}| \le \tau_n)) = \mathbb{E}(\Phi_{nj}^2 \xi_{nj}^2 I(|\xi_{nj}| \le \tau_n)) - (\mathbb{E}(\Phi_{nj}\xi_{nj} I(|\xi_{nj}| \le \tau_n)))^2 =$$

$$\sum_{i=1}^{N}\sum_{k=1}^{N} a_i a_k \nu_{nj}(C_i \cap C_k)\, \mathbb{E}(\xi_{nj}^2 I(|\xi_{nj}| \le \tau_n)) -$$

$$\sum_{i=1}^{N}\sum_{k=1}^{N} a_i a_k \nu_{nj}(C_i)\, \nu_{nj}(C_k) (\mathbb{E}(\xi_{nj} I(|\xi_{nj}| \le \tau_n)))^2 =:$$

$$\sum_{i=1}^{N}\sum_{k=1}^{N} a_i a_k \nu_{nj}(C_i \cap C_k) A_{nj} - \sum_{i=1}^{N}\sum_{k=1}^{N} a_i a_k \nu_{nj}(C_i)\, \nu_{nj}(C_k)\, B_{nj},$$ say, whence

for all $n \ge n_o$ $\sum_{j\le j(n)} \mathbb{E}(\zeta_{nj}^2 I(|\zeta_{nj}| \le \varepsilon)) =$

$$\sum_{i=1}^{N}\sum_{k=1}^{N} a_i a_k \sum_{j\le j(n)} \nu_{nj}(C_i \cap C_k) A_{nj} - \sum_{i=1}^{N}\sum_{k=1}^{N} a_i a_k \sum_{j\le j(n)} \nu_{nj}(C_i)\, \nu_{nj}(C_k) B_{nj} =:$$

$$\sum_{i=1}^{N}\sum_{k=1}^{N} a_i a_k (D_{nik} - E_{nik}),$$ say.

Now, $\tau_n < \varepsilon$ for all $n \ge n_1(\varepsilon)$, and thus, for all $n \ge n_1(\varepsilon)$ and $1 \le j \le j(n)$,

$$A_{nj} := \mathbb{E}(\xi_{nj}^2 I(|\xi_{nj}| \le \tau_n)) =$$

$$\mathbb{E}(\xi_{nj}^2 I(|\xi_{nj}| \le \varepsilon)) - \mathbb{E}(\xi_{nj}^2 I(\tau_n < |\xi_{nj}| \le \varepsilon)),$$ whence

$$D_{nik} := \sum_{j\le j(n)} \nu_{nj}(C_i \cap C_k) A_{nj} =$$

$$\sum_{j\le j(n)} \nu_{nj}(C_i \cap C_k)\, \mathbb{E}(\xi_{nj}^2 I(|\xi_{nj}| \le \varepsilon)) - \sum_{j\le j(n)} \nu_{nj}(C_i \cap C_k)\, \mathbb{E}(\xi_{nj}^2 I(\tau_n < |\xi_{nj}| \le \varepsilon)),$$ where

$$\sum_{j\le j(n)} \nu_{nj}(C_i \cap C_k)\, \mathbb{E}(\xi_{nj}^2 I(\tau_n < |\xi_{nj}| \le \varepsilon)) \le \varepsilon \sum_{j\le j(n)} \mathbb{E}(|\xi_{nj}| I(|\xi_{nj}| \ge \tau_n)) \to 0$$ by (2.5).

Thus, according to (2.3), $\lim_{n\to\infty} D_{nik} = c_1\nu(C_i \cap C_k)$.

In the same way, for all $n \ge n_1(\varepsilon)$ and $1 \le j \le j(n)$

$$B_{nj} := (\mathbb{E}(\xi_{nj} I(|\xi_{nj}| \le \tau_n)))^2 =$$

$$(\mathbb{E}(\xi_{nj} I(|\xi_{nj}| \le \varepsilon)) - \mathbb{E}(\xi_{nj} I(\tau_n < |\xi_{nj}| \le \varepsilon)))^2 =$$

$$(\mathbb{E}(\xi_{nj} I(|\xi_{nj}| \le \varepsilon)))^2 -$$

$$2\mathbb{E}(\xi_{nj} I(|\xi_{nj}| \le \varepsilon))\,\mathbb{E}(\xi_{nj} I(\tau_n < |\xi_{nj}| \le \varepsilon)) +$$

$$(\mathbb{E}(\xi_{nj} I(\tau_n < |\xi_{nj}| \le \varepsilon)))^2 ,$$ whence

$$E_{nik} := \sum_{j \le j(n)} \nu_{nj}(C_i)\nu_{nj}(C_k) B_{nj} =$$

$$\sum_{j \le j(n)} \nu_{nj}(C_i)\nu_{nj}(C_k)(\mathbb{E}(\xi_{nj} I(|\xi_{nj}| \le \varepsilon)))^2 -$$

$$2\sum_{j \le j(n)} \nu_{nj}(C_i)\nu_{nj}(C_k)\mathbb{E}(\xi_{nj} I(|\xi_{nj}| \le \varepsilon))\mathbb{E}(\xi_{nj} I(\tau_n < |\xi_{nj}| \le \varepsilon)) +$$

$$\sum_{j \le j(n)} \nu_{nj}(C_i)\nu_{nj}(C_k)(\mathbb{E}(\xi_{nj} I(\tau_n < |\xi_{nj}| \le \varepsilon)))^2 ,$$ where

$$\Big|\sum_{j \le j(n)} \nu_{nj}(C_i)\nu_{nj}(C_k)\mathbb{E}(\xi_{nj} I(|\xi_{nj}| \le \varepsilon))\mathbb{E}(\xi_{nj} I(\tau_n < |\xi_{nj}| \le \varepsilon))\Big| \le$$

$$\sum_{j \le j(n)} \mathbb{E}(|\xi_{nj}| I(|\xi_{nj}| \le \varepsilon))\,\mathbb{E}(|\xi_{nj}| I(\tau_n < |\xi_{nj}| \le \varepsilon)) \le$$

$\varepsilon \sum_{j \le j(n)} \mathbb{E}(|\xi_{nj}| I(|\xi_{nj}| \ge \tau_n)) \longrightarrow 0$ by (2.5), and also

$$\sum_{j \le j(n)} \nu_{nj}(C_i)\nu_{nj}(C_k)(\mathbb{E}(\xi_{nj} I(\tau_n < |\xi_{nj}| \le \varepsilon)))^2 \le$$

$\varepsilon \sum_{j \le j(n)} \mathbb{E}(|\xi_{nj}| I(|\xi_{nj}| \ge \tau_n)) \longrightarrow 0$. Thus, according to (2.4),

$\lim_{n\to\infty} E_{nik} = c_2\nu(C_i)\nu(C_k)$. It follows that for each $\varepsilon > 0$

$$\lim_{n\to\infty} \sum_{j \le j(n)} \mathbb{E}(\zeta_{nj}^2 I(|\zeta_{nj}| \le \varepsilon)) =$$

$\sum_{i=1} \sum_{k=1} a_i a_k [c_1\nu(C_i \cap C_k) - c_2\nu(C_i)\nu(C_k)]$. This proves ($\beta$).

Thus, by the cited Lévy CLT, $\sum_{j \le j(n)} \zeta_{nj} \xrightarrow{\mathscr{L}} \zeta$, where

$\mathscr{L}\{\zeta\} = \mathscr{N}(0, \sum_{i=1}^{N} \sum_{k=1}^{N} a_i a_k [c_1\nu(C_i \cap C_k) - c_2\nu(C_i)\nu(C_k)]$. Since

$$\operatorname{Var}\Big(\sum_{i=1}^{N} a_i \bar{G}_\nu(C_i)\Big) = \sum_{i=1}^{N} \sum_{k=1}^{N} a_i a_k [c_1\nu(C_i \cap C_k) - c_2\nu(C_i)\nu(C_k)],$$

this proves (2.6).

As to (b): It is enough to show

(2.7) $$\lim_{\delta\to 0}\ \limsup_{n\to\infty}\ \mathbb{E}\big(\sup_{\nu(C\Delta D)\le\delta} |S''_n(C) - S''_n(D)|\big) = 0.$$

Now, the proof of (2.7) can be carried through in a similar way as in Arcones-Gaenssler-Ziegler (1992) in proving Theorem 3.1 there which proceeds in the present (more general) situation as follows.

According to A.9 (see Appendix)

$$I := \mathbb{E}\big(\sup_{\nu(C\Delta D)\le\delta} |S''_n(C) - S''_n(D)|\big) \le$$

$$2\mathbb{E}\big(\sup_{\nu(C\Delta D)\le\delta} |\sum_{j\le j(n)} \varepsilon_j(1_C(\eta_{nj}) - 1_D(\eta_{nj}))\xi_{nj} I(|\xi_{nj}| \le \tau_n)|\big),$$

where $(\varepsilon_j)_{j\in\mathbb{N}}$ is a Rademacher sequence being independent of both arrays (ξ_{nj}) and (η_{nj}). Let

$$X_n := \sum_{j\le j(n)} \xi_{nj}^2\, I(|\xi_{nj}| \le \tau_n),$$

assuming w.l.o.g. that $X_n > 0$; then

$$I \le 2\mathbb{E}(X_n^{1/2}\, \mathbb{E}_\varepsilon(\sup_{\nu(C\Delta D)\le\delta} |\sum_{j\le j(n)} \varepsilon_j(1_C(\eta_{nj})-1_D(\eta_{nj}))\xi_{nj} I(|\xi_{nj}|\le\tau_n)X_n^{-1/2}|)),$$

where $\mathbb{E}_\varepsilon$ denotes integration w.r.t. (ε_j).

By A.8 (with $p=1$)

$$\mathbb{E}_\varepsilon\big(\sup_{\nu(A\Delta B)\le\delta} |\sum_{j\le j(n)} \varepsilon_j(1_A(\eta_{nj}) - 1_B(\eta_{nj}))\xi_{nj} I(|\xi_{nj}| \le \tau_n)X_n^{-1/2}|\big) \le$$

$$\mathbb{E}_\varepsilon\big(\sup_{d(A,B)\le D} |\sum_{j\le j(n)} \varepsilon_j(1_A(\eta_{nj}) - 1_B(\eta_{nj}))\xi_{nj} I(|\xi_{nj}| \le \tau_n)X_n^{-1/2}|\big) \le$$

$$C_1' \int_0^{D/2} (\log N(\varepsilon,\mathscr{C},d)^{1/2} d\varepsilon,$$

where, for $A,B\in\mathscr{C}$ (assuming w.l.o.g. that $A\Delta B\in\mathscr{C}$ for all $A,B\in\mathscr{C}$)

$$d^2(A,B) := \sum_{j\le j(n)} ((1_A(\eta_{nj}) - 1_B(\eta_{nj}))\xi_{nj} I(|\xi_{nj}| \le \tau_n)X_n^{-1/2})^2 =$$

$$\frac{\sum_{j\le j(n)} 1_{A\Delta B}(\eta_{nj})\xi_{nj}^2\, I(|\xi_{nj}| \le \tau_n)}{\sum_{j\le j(n)} \xi_{nj}^2\, I(|\xi_{nj}| \le \tau_n)}, \quad \text{and}$$

$$D := \sup_{\nu(C)\le\delta} \frac{\big(\sum_{j\le j(n)} 1_C(\eta_{nj})\xi_{nj}^2\, I(|\xi_{nj}| \le \tau_n)\big)^{1/2}}{\big(\sum_{j\le j(n)} \xi_{nj}^2\, I(|\xi_{nj}| \le \tau_n)\big)^{1/2}}$$

Hence, if δ_y denotes the Dirac measure concentrated in y and if Q_n denotes the (random) p-measure on $\mathfrak{X}$ defined by

$$Q_n := \frac{\sum_{j\le j(n)} \delta_{\eta_{nj}} \xi_{nj} I(|\xi_{nj}| \le \tau_n)}{\sum_{j\le j(n)} \xi_{nj}^2 I(|\xi_{nj}| \le \tau_n)},$$

then $d^2(A,B) = d_{Q_n}(A,B)$ $(:= Q_n(A\Delta B))$ and thus by (1.5) it follows that

$$\log N(\varepsilon, \mathscr{C}, d) = \log N(\varepsilon, \mathscr{C}, d_{Q_n}^{1/2}) \le A\varepsilon^{-d}$$

for some universal constants $0 < A < \infty$ and (w.l.o.g.) $0 < d < 2$.

So, denoting by $\mathbb{E}_{\xi,\eta}$ integration w.r.t. $((\xi_{nj}), (\eta_{nj}))$, we obtain that I is bounded by

$$2C_1' \,\mathbb{E}_{\xi,\eta}(X_n^{1/2} \int_0^{D/2} (\log N(\varepsilon, \mathscr{C}, d_{Q_n}^{1/2}))^{1/2} d\varepsilon) \le c\,\mathbb{E}_{\xi,\eta}(X_n^{1/2} D^{1-\frac{d}{2}}) =$$

$$c\,\mathbb{E}_{\xi,\eta}\Big(X_n^{1/2}\Big(\sup_{\nu(C)\le\delta} \frac{(\sum_{j\le j(n)} 1_C(\eta_{nj})\xi_{nj}^2 I(|\xi_{nj}| \le \tau_n))^{1/2}}{\sum_{j\le j(n)} \xi_{nj}^2 I(|\xi_{nj}| \le \tau_n))^{1/2}}\Big)^{1-\frac{d}{2}}\Big) =$$

$$c\,\mathbb{E}_{\xi,\eta}(X_n^{d/4}(\sup_{\nu(C)\le\delta}(\sum_{j\le j(n)} 1_C(\eta_{nj})\xi_{nj}^2 I(|\xi_{nj}| \le \tau_n))^{\frac{1}{2}-\frac{d}{4}}).$$

Thus, by Cauchy and Hölder's inequalities (for $d < 2$) I is bounded by

$$c(\mathbb{E}_{\xi,\eta}(\sum_{j\le j(n)} \xi_{nj}^2 I(|\xi_{nj}|\le\tau_n)))^{d/4} \cdot \mathbb{E}_{\xi,\eta}(\sup_{\nu(C)\le\delta}(\sum_{j\le j(n)} 1_C(\eta_{nj})\xi_{nj}^2 I(|\xi_{nj}|\le\tau_n)))^{\frac{1}{2}-\frac{d}{4}},$$

and, as we have seen so far,

(2.8)
$$\mathbb{E}(\sup_{\nu(A\Delta B)\le\delta} |\sum_{j\le j(n)} \varepsilon_j(1_A(\eta_{nj}) - 1_B(\eta_{nj}))\xi_{nj} I(|\xi_{nj}| \le \tau_n)|) \le$$

$$c/2(\mathbb{E}(\sum_{j\le j(n)} \xi_{nj}^2 I(|\xi_{nj}|\le\tau_n)))^{d/4} \cdot (\mathbb{E}(\sup_{\nu(C)\le\delta}(\sum_{j\le j(n)} 1_C(\eta_{nj})\xi_{nj}^2 I(|\xi_{nj}|\le\tau_n)))^{\frac{1}{2}-\frac{d}{4}}.$$

Now it follows by (ii) (assuming w.l.o.g. that $X \in \mathscr{C}$) that

(2.9)
$$\limsup_{n\to\infty} \mathbb{E}(\sum_{j\le j(n)} \xi_{nj}^2 I(|\xi_{nj}|\le\tau_n)) \le c_1.$$

So it remains to show

(2.10)
$$\lim_{\delta\to 0} \limsup_{n\to\infty} \mathbb{E}(\sup_{\nu(C)\le\delta} \sum_{j\le j(n)} 1_C(\eta_{nj})\xi_{nj}^2 I(|\xi_{nj}| \le \tau_n)) = 0, \text{ where}$$

$$\mathbb{E}(\sup_{\nu(C)\le\delta} \sum_{j\le j(n)} 1_C(\eta_{nj})\xi_{nj}^2 I(|\xi_{nj}| \le \tau_n)) \le$$

$$\mathbb{E}(\sup_{\nu(C)\le\delta} \sum_{j\le j(n)} 1_C(\eta_{nj})\xi_{nj}^2 I(|\xi_{nj}|\le\tau_n) - \mathbb{E}(1_C(\eta_{nj})\xi_{nj}^2 I(|\xi_{nj}| \le \tau_n)))|) +$$

$$\sup_{\nu(C)\le\delta} \mathbb{E}(\sum_{j\le j(n)} 1_C(\eta_{nj})\xi_{nj}^2 I(|\xi_{nj}| \le \tau_n)).$$

Now, by (iv),

$$\lim_{\delta\to 0}\ \limsup_{n\to\infty}\ \sup_{\nu(C)\le\delta} \mathbb{E}\Big(\sum_{j\le j(n)} 1_C(\eta_{nj})\,\xi_{nj}^2 I(|\xi_{nj}|\le\tau_n)\Big) = 0, \text{ and},$$

again by A.9 , the other summand is bounded by

$$2\,\mathbb{E}\Big(\sup_{\nu(C)\le\delta}\Big|\sum_{j\le j(n)} \varepsilon_j 1_C(\eta_{nj})\,\xi_{nj}^2 I(|\xi_{nj}|\le\tau_n)\Big|\Big).$$

By the contraction principle (cf. Lemma 2.12, p. 108 in Araujo and Giné (1980)) this last expression is bounded by

$$4\tau_n\,\mathbb{E}\Big(\sup_{\nu(C)\le\delta}\Big|\sum_{j\le j(n)} \varepsilon_j 1_C(\eta_{nj})\,\xi_{nj} I(|\xi_{nj}|\le\tau_n)\Big|\Big).$$

Finally, by (2.8) and (2.9) (assuming w.l.o.g. that $\emptyset\in\mathscr{C}$),

$$\limsup_{n\to\infty} \mathbb{E}\Big(\sup_{\nu(C)\le\delta}\Big|\sum_{j\le j(n)} \varepsilon_j 1_C(\eta_{nj})\,\xi_{nj} I(|\xi_{nj}|\le\tau_n)\Big|\Big) \le$$

$$\limsup_{n\to\infty} \mathbb{E}\Big(\sup_{\nu(A\Delta B)\le\delta}\Big|\sum_{j\le j(n)} \varepsilon_j (1_A(\eta_{nj})-1_B(\eta_{nj}))\,\xi_{nj} I(|\xi_{nj}|\le\tau_n)\Big|\Big) \le$$

$$c/2\ \limsup_{n\to\infty}\Big(\mathbb{E}\Big(\sum_{j\le j(n)} \xi_{nj}^2 I(|\xi_{nj}|\le\tau_n)\Big)\Big)^{1/2} < \infty ,$$

and thus (2.10) follows since $\tau_n \longrightarrow 0$. □

The proof of Theorem 2.2 shows that also the following result holds true:

2.11 Theorem. Let $S'_n = (S'_n(C))_{C\in\mathscr{C}}$ be as in Theorem 2.2 such that $S'_n \xrightarrow[\text{fidi}]{\mathscr{L}} \bar{\mathbb{G}}$ with a certain process $\bar{\mathbb{G}} = (\bar{G}(C))_{C\in\mathscr{C}}$.

Assume further that the following three conditions are fullfilled:

(i) $\displaystyle\sum_{j\le j(n)} \mathbb{E}(|\xi_{nj}| I(|\xi_{nj}|\ge\varepsilon)) \longrightarrow 0$ for all $\varepsilon>0$

(ii) $\displaystyle\limsup_{n\to\infty}\sum_{j\le j(n)} \mathbb{E}(\xi_{nj}^2 I(|\xi_{nj}|\le\varepsilon_1)) < \infty$ for some $\varepsilon_1>0$

(iii) $\displaystyle\lim_{\delta\to 0}\ \limsup_{n\to\infty}\ \sup_{\nu(C)\le\delta}\sum_{j\le j(n)} \mathbb{E}(1_C(\eta_{nj})\,\xi_{nj}^2 I(|\xi_{nj}|\le\varepsilon_2)) = 0$

for some p-measure ν on $\mathfrak{X}$ and some $\varepsilon_2>0$.

Then $S'_n \xrightarrow[\text{sep}]{\mathscr{L}} \mathbb{G}$, where $\mathbb{G} = (G(C))_{C\in\mathscr{C}}$ is a version of $\bar{\mathbb{G}}$ (that is, $\mathbb{G}$ has the same fidis as $\bar{\mathbb{G}}$) having all its sample paths in $U^b(\mathscr{C}, d_\nu)$.

2.12 Remark. Concerning FCLT's for more general arrays of stochastic processes under random entropy conditions we refer to the forthcoming PhD-thesis by Klaus Ziegler. Independently, and with different proofs, van der Vaart and Wellner obtained similar results which will appear in a forthcoming monograph on 'Weak convergence and empirical processes'.

We are going to show how various results for empirical and partial-sum processes follow from Theorem 2.2 and Theorem 2.11, respectively.

2.13 Corollary (**FCLT for empirical processes indexed by a VCC $\mathscr{C}$ in the non-i.i.d. case**). Let $\xi_{nj} \equiv j(n)^{-1/2}$ for each $1 \le j \le j(n)$ and $n \in \mathbb{N}$ with $j(n) \to \infty$ as $n \to \infty$; then

$$S'_n(C) := \sum_{j \le j(n)} (1_C(\eta_{nj})\xi_{nj} - \mathbb{E}(1_C(\eta_{nj})\xi_{nj})) = j(n)^{1/2}(\nu_n(C) - \bar{\nu}_n(C)),$$

where $\nu_n(C) := j(n)^{-1} \sum_{j \le j(n)} 1_C(\eta_{nj})$ and $\bar{\nu}_n(C) := j(n)^{-1} \sum_{j \le j(n)} \nu_{nj}(C)$, $C \in \mathscr{C}$, with $\nu_{nj} := \mathscr{L}\{\eta_{nj}\}$.

Assume that there exists a p-measure ν on $\mathfrak{X}$ and finite constants c_i so that the following three conditions are fulfilled:

(+) $\qquad \lim\limits_{n\to\infty} j(n)^{-1} \sum\limits_{j \le j(n)} \nu_{nj}(C \cap D) = c_1 \nu(C \cap D)$ for all $C, D \in \mathscr{C}$

(++) $\qquad \lim\limits_{n\to\infty} j(n)^{-1} \sum\limits_{j \le j(n)} \nu_{nj}(C)\nu_{nj}(D) = c_2 \nu(C)\nu(D)$ for all $C, D \in \mathscr{C}$

(+++) $\qquad \lim\limits_{\delta\to 0} \limsup\limits_{n\to\infty} \sup\limits_{\nu(C)\le\delta} j(n)^{-1} \sum\limits_{j \le j(n)} \nu_{nj}(C) = 0.$

Then $(j(n)^{1/2}(\nu_n(C) - \bar{\nu}_n(C)))_{C\in\mathscr{C}} \xrightarrow[\text{sep}]{\mathscr{L}} \mathbb{G}_\nu$, where $\mathbb{G}_\nu = (G_\nu(C))_{C\in\mathscr{C}}$ is a mean zero Gaussian process with sample paths in $U^b(\mathscr{C}, d_\nu)$ and $\operatorname{cov}(G_\nu(C), G_\nu(D)) = c_1\nu(C \cap D) - c_2\nu(C)\nu(D)$, $C, D \in \mathscr{C}$.

Proof. According to Theorem 2.2 we have to verify the conditions (i)-(iv) there.

As to (i): Since $\xi_{nj} \equiv j(n)^{-1/2} \longrightarrow 0$ as $n \to \infty$, there exists for every $\varepsilon > 0$ an $n_o(\varepsilon) \in \mathbb{N}$ such that $j(n)^{-1/2} < \varepsilon$ for all $n \ge n_o(\varepsilon)$, whence $\sum_{j \le j(n)} \mathbb{E}(|\xi_{nj}| I(|\xi_{nj}| \ge \varepsilon)) = 0$ for all $n \ge n_o(\varepsilon)$.

As to (ii): Choosing any $\varepsilon_1 > 0$, $\mathbb{E}(\xi_{nj}^2 I(|\xi_{nj}| \le \varepsilon_1)) = \mathbb{E}(\xi_{nj}^2) = j(n)^{-1}$ for all $n \ge n_o(\varepsilon_1)$, whence for all $n \ge n_o(\varepsilon_1)$ and all $C, D \in \mathscr{C}$

$\sum_{j \le j(n)} \nu_{nj}(C \cap D)\mathbb{E}(\xi_{nj}^2 I(|\xi_{nj}| \le \varepsilon_1)) = j(n)^{-1} \sum_{j \le j(n)} \nu_{nj}(C \cap D) \longrightarrow c_1\nu(C \cap D)$ according to (+).

As to (iii) and (iv): Follows in the same way as (ii) according to (++) and (+++), respectively. □

2.14 Remark (FCLT for empirical processes indexed by a VCC $\mathscr{C}$ in the i.i.d. case). If in Corollary 2.13 the array (η_{nj}) is replaced by a sequence $(\eta_j)_{j\in\mathbb{N}}$ of i.i.d. r.e.'s η_j in $X = (X, \mathfrak{X})$ with law ν on $\mathfrak{X}$, the conditions (+) (with $c_1 = 1$), (++) (with $c_2 = 1$), and (+++) in Corollary 2.13 are trivially fulfilled, whence in this case $(j(n)^{1/2}(\nu_n(C) - \nu(C)))_{C\in\mathscr{C}} \xrightarrow[\text{sep}]{\mathscr{L}} \mathbb{G}_\nu$, where $\nu_n(C) := j(n)^{-1} \sum_{j \le j(n)} 1_C(\eta_j)$ and where $\mathbb{G}_\nu = (G_\nu(C))_{C\in\mathscr{C}}$ is a mean zero Gaussian process with sample paths in $U^b(\mathscr{C}, d_\nu)$ and $\operatorname{cov}(G_\nu(C), G_\nu(D)) = \nu(C \cap D) - \nu(C)\nu(D)$, $C, D \in \mathscr{C}$.

2.15 Corollary (FCLT for partial-sum processes with random locations in the i.i.d. case). Let $\xi_{nj} = j(n)^{-1/2}\xi_j$ for each $1 \le j \le j(n)$ and $n \in \mathbb{N}$ with $j(n) \to \infty$, as $n \to \infty$, the ξ_j's being i.i.d. r.v.'s with $\mathbb{E}(\xi_1) = 0$ and $\mathbb{E}(\xi_1^2) = 1$, and let $(\eta_{nj})_{1 \le j \le j(n),\, n \in \mathbb{N}}$ be a triangular array of rowwise independent (but not necessarily i.d.) r.e.'s in $X = (X, \mathfrak{X})$ which is independent of the sequence $(\xi_j)_{j \in \mathbb{N}}$.

Assume that there exists a p-measure ν on $\mathfrak{X}$ so that the following two conditions are fulfilled:

(*) $$\lim_{n\to\infty} j(n)^{-1} \sum_{j \le j(n)} \nu_{nj}(C \cap D) = \nu(C \cap D) \text{ for all } C, D \in \mathscr{C}$$

(**) $$\lim_{\delta \to 0} \limsup_{n\to\infty} \sup_{\nu(C) \le \delta} j(n)^{-1} \sum_{j \le j(n)} \nu_{nj}(C) = 0.$$

Then $(j(n)^{-1/2} \sum_{j \le j(n)} 1_C(\eta_{nj})\xi_j)_{C \in \mathscr{C}} \xrightarrow[\text{sep}]{\mathscr{L}} \mathbb{G}_\nu$, where $\mathbb{G}_\nu = (G_\nu(C))_{C \in \mathscr{C}}$ is a mean zero Gaussian process with sample paths in $U^b(\mathscr{C}, d_\nu)$ and $\operatorname{cov}(G_\nu(C), G_\nu(D)) = \nu(C \cap D), C, D \in \mathscr{C}$.

Proof. According to Theorem 2.2 we have to verify the conditions (i), (ii) (with $c_1=1$), (iii) (with $c_2=0$), and (iv) there.

As to (i): For each $\varepsilon>0$ we have that $\sum_{j \le j(n)} \mathbb{E}(|\xi_{nj}| I(|\xi_{nj}| \ge \varepsilon)) \le$

$\varepsilon^{-1} \sum_{j \le j(n)} \mathbb{E}(\xi_{nj}^2 I(|\xi_{nj}| \ge \varepsilon)) = \varepsilon^{-1}\mathbb{E}(\xi_1^2 I(|\xi_1| \ge \varepsilon j(n)^{1/2})) \to 0$, since $\mathbb{E}(\xi_1^2) < \infty$.

As to (ii): $\sum_{j \le j(n)} \nu_{nj}(C \cap D)\mathbb{E}(\xi_{nj}^2 I(|\xi_{nj}| \le \varepsilon_1)) =$

$[j(n)^{-1} \sum_{j \le j(n)} \nu_{nj}(C \cap D)]\mathbb{E}(\xi_1^2 I(|\xi_1| \le \varepsilon_1 j(n)^{1/2}))$, where by (*)

$\lim_{n\to\infty} j(n)^{-1} \sum_{j \le j(n)} \nu_{nj}(C \cap D) = \nu(C \cap D)$ and $\lim_{n\to\infty} \mathbb{E}(\xi_1^2 I(|\xi_1| \le \varepsilon_1 j(n)^{1/2})) = 1$ since

$\mathbb{E}(\xi_1^2) = 1$. This proves (ii) with $c_1 = 1$.

As to (iii): Since $\limsup_{n\to\infty} \sum_{j \le j(n)} \nu_{nj}(C)\nu_{nj}(D)(\mathbb{E}(\xi_{nj} I(|\xi_{nj}| \le \varepsilon_2)))^2 \le$

$\limsup_{n\to\infty} \sum_{j \le j(n)} (\mathbb{E}(\xi_{nj} I(|\xi_{nj}| \le \varepsilon_2)))^2 = \lim_{n\to\infty} (\mathbb{E}(\xi_1 I(|\xi_1| \le \varepsilon_2 j(n)^{1/2})))^2 =$

$(\mathbb{E}(\xi_1))^2 = 0$, $\lim_{n\to\infty} \sum_{j \le j(n)} \nu_{nj}(C)\nu_{nj}(D)(\mathbb{E}(\xi_{nj} I(|\xi_{nj}| \le \varepsilon_2)))^2 = 0$ for all $C, D \in \mathscr{C}$,

which shows that (iii) is fulfilled with $c_2 = 0$.

As to (iv): Since $\sum_{j \le j(n)} \nu_{nj}(C)\mathbb{E}(\xi_{nj}^2 I(|\xi_{nj}| \le \varepsilon_3)) =$

$[j(n)^{-1} \sum_{j \le j(n)} \nu_{nj}(C)]\ \mathbb{E}(\xi_1^2 I(|\xi_1| \le \varepsilon_3 j(n)^{1/2}))$, it follows that

$$\sup_{\nu(C)\le\delta} \sum_{j\le j(n)} \nu_{nj}(C)\,\mathbb{E}(\xi_{nj}^2 I(|\xi_{nj}| \le \varepsilon_3)) =$$

$[\sup_{\nu(C)\le\delta} j(n)^{-1} \sum_{j\le j(n)} \nu_{nj}(C)]\ \mathbb{E}(\xi_1^2 I(|\xi_1| \le \varepsilon_3 j(n)^{1/2}))$, whence by (**) and the fact that $\lim_{n\to\infty} \mathbb{E}(\xi_1^2 I(|\xi_1| \le \varepsilon_3 j(n)^{1/2})) = \mathbb{E}(\xi_1^2) = 1$, condition (iv) is also fulfilled. □

2.16 Remark (FCLT for the classical multivariate partial-sum process with non-random locations). Cf. the partial-sum process ζ_n mentioned in the Introduction and the remarks in connection with Example 2.97 in van der Vaart and Wellner (1993).

Replacing the η_{nj}'s in Corollary 2.15 by the non-random locations $\underline{j}/n$, $\underline{j}\in\{1,\dots,n\}^d$, i.e. if ν_{nj}, $1\le j\le j(n) := n^d$, is the collection of Dirac measures located at the knots of the regular grid consisting of the n^d points of $\{\frac{1}{n}, \frac{2}{n},\dots,1\}^d$ in $X := I^d \equiv [0,1]^d$, then the sequence $\bar{\nu}_n := j(n)^{-1} \sum_{j\le j(n)} \nu_{nj}$, $n\in\mathbb{N}$, converges weakly to $\nu = \lambda^d$, where λ^d denotes the d-dimensional Lebesgue measure on $\mathfrak{X} := I^d \cap \mathbb{B}^d$ ($\mathbb{B}^d$ = Borel σ-field in $\mathbb{R}^d$), whence (*) in Corollary 2.15 is fullfilled with $\nu = \lambda^d$ for all $C,D\in\mathscr{C}_o := \{C\in\mathscr{C}: \lambda^d(\partial C) = 0\}$, where ∂C denotes the boundary of C. Restricting the VCC $\mathscr{C}$ further so that $\bar{\nu}_n$ (on $\mathscr{C}_o$) converges uniformly to $\nu = \lambda^d$ (on $\mathscr{C}_o$), then also (**) in Corollary 2.15 is fulfilled, which is the case if for every $\varepsilon>0$ $N_I(\varepsilon,\mathscr{C}_o,\lambda^d) < \infty$, where $N_I(\varepsilon,\mathscr{C}_o,\lambda^d) := \min\{n\in\mathbb{N}$: there exists $C_i^-, C_i^+\in\mathfrak{X}$ with $\lambda^d(\partial C_i^\pm) = 0$, $1\le i\le n$, such that for every $C\in\mathscr{C}_o$ there is some i such that $C_i^- \subset C \subset C_i^+$ and $\lambda^d(C_i^+ \setminus C_i^-) < \varepsilon\}$.

With this in mind, note that both conditions (*) and (**) in Corollary 2.15 are fulfilled for the collection $C = [\underline{0},\underline{t}]$, $\underline{t}\in I^d$; for other classes $\mathscr{C}\subset\mathfrak{X}$ cf. also the result in Alexander and Pyke (1986).

2.17 Remark (FCLT for the sequential uniform empirical process). Cf. Bickel and Wichura (1971). Let $K_n = (K_n(s,t))_{(s,t)\in I^2}$ be the so-called *sequential uniform empirical process* (of sample size n) based on i.i.d. U[0,1] r.v.'s $\eta_1,\dots,\eta_n$, being defined by (cf. Shorack and Wellner (1986), Chapter 3.5)

$$K_n(s,t) := n^{-1/2} \sum_{j=1}^{\langle ns\rangle} (1_{[0,t]}(\eta_j)-t), \quad (s,t)\in I^2 ,$$

where $\langle a\rangle := \sup\{k\in\mathbb{Z}: k\le a\}$ for $a\in\mathbb{R}$ denotes the greatest integer function.

Choosing in this case $X := I^2$, $\mathfrak{X} := X\cap\mathbb{B}^2$, $\mathscr{C} := \{[0,s]\times[0,t]: (s,t\in I^2\}$, $\eta_{nj} := (j/n, \eta_j)$ and $\xi_{nj} := n^{-1/2}$, $1\le j\le j(n) := n$, then K_n can be considered as a set-indexed partial-sum process $S'_n = (S'_n(C))_{C\in\mathscr{C}}$.

Now, $K_n \xrightarrow[\text{fidi}]{\mathcal{L}} \bar{K}$ follows by the multivariate CLT (cf. the proof of (a) within the proof of Theorem 2.2), where $\bar{K}$ is a mean zero Gaussian process (the so-called *Kiefer process*) with $\operatorname{cov}(\bar{K}(s_1,t_1), \bar{K}(s_2,t_2)) = (s_1 \wedge s_2)[(t_1 \wedge t_2) - t_1 t_2]$, $(s_i,t_i) \in I^2$, $i=1,2$. The conditions (i) and (ii) of Theorem 2.11 are obviously fulfilled. As to (iii) there, note that $\nu_{nj} := \mathcal{L}\{\eta_{nj}\} = \delta_{j/n} \times \lambda^1$, whence (cf. Remark 2.16) $\bar{\nu}_n := n^{-1} \sum_{j \le n} \nu_{nj} = (n^{-1} \sum_{j \le n} \delta_{j/n}) \times \lambda^1$ converges weakly to λ^2 as $n \to \infty$. It follows that $\sup_{C \in \mathcal{C}} |\bar{\nu}_n(C) - \lambda^2(C)| \to 0$ as $n \to \infty$ which implies (iii) of Theorem 2.11 with $\nu := \lambda^2$. Thus by Theorem 2.11, $K_n \xrightarrow[\text{sep}]{\mathcal{L}} K$, where $K = (K(s,t))_{(s,t) \in I^2}$ is a version of the Kiefer process with sample paths in $U^b(\mathcal{C}, d_\nu)$ (with $\mathcal{C} = \{[0,s] \times [0,t] : (s,t) \in I^2\}$).

3. A uniform law of large numbers (ULLN) in form of a Mean Glivenko-Cantelli result for partial-sum processes with random locations and indexed by a VCC. The general context is the same as in Section 2 with the only exception that the basic assumptions on the arrays (η_{nj}) and (ξ_{nj}) are weakened to the extent assuming only that for each $n \in \mathbb{N}$ $(\eta_{n1}, \xi_{n1}), \dots, (\eta_{nj(n)}, \xi_{nj(n)})$ is a sequence of independent but not necessarily i.d. pairs of r.e.'s in $X \times \mathbb{R} = (X \times \mathbb{R}, \mathfrak{X} \otimes \mathbb{B})$, where $\mathfrak{X} \otimes \mathbb{B}$ denotes the product σ-field of $\mathfrak{X}$ and the Borel σ-field $\mathbb{B}$ in $\mathbb{R}$; i.e. the components within each pair need not be independent.

The following result improves the Mean Glivenko-Cantelli theorem in Gaenssler and Ziegler (1992) by weakening the rather strong moment conditions on the ξ_{nj}'s. In the *identically distributed (i.d.) case*, that is, when $\xi_{nj} = j(n)^{-1}\xi_j$, $1 \le j \le j(n)$, with $j(n) \to \infty$ as $n \to \infty$, for some i.d. sequence $(\xi_j)_{j \in \mathbb{N}}$, this improvement leads to a Mean Glivenko-Cantelli theorem for partial-sum processes with random locations and indexed by a VCC under the only assumption that $\mathbb{E}(|\xi_1|) < \infty$.

3.1 Theorem. Assume that the following two condition are fulfilled:

(i) $$\lim_{n \to \infty} \sum_{j \le j(n)} \mathbb{E}(|\xi_{nj}| I(|\xi_{nj}| > \delta)) = 0 \quad \text{for all } \delta > 0$$

(ii) $$\sup_{n \in \mathbb{N}} \sum_{j \le j(n)} \mathbb{E}(|\xi_{nj}| I(|\xi_{nj}| \le \delta)) = K' < \infty \quad \text{for some } \delta > 0.$$

Then, for the partial-sum processes $S_n = (S_n(C))_{C \in \mathcal{C}}$ indexed by a countable VCC $\mathcal{C}$ in an arbitrary sample space $X = (X, \mathfrak{X})$, defined by $S_n(C) := \sum_{j \le j(n)} 1_C(\eta_{nj}) \xi_{nj}$, $C \in \mathcal{C}$, one has

$$\lim_{n \to \infty} \mathbb{E}\,(\sup_{C \in \mathcal{C}} |S_n(C) - \mathbb{E}(S_n(C))|) = 0.$$

As mentioned in the Introduction, the proof of Theorem 3.1 will be given in a separate paper by Gaenssler and Ziegler (1993) together with similar consequences for empirical and partial-sum processes as pointed out in Section 2 in connection with FCLT's.

3.2 Remark. Note that the conditions (i) and (ii) in Theorem 3.1 together imply that $\sup_{n\in\mathbb{N}} \sum_{j\le j(n)} \mathbb{E}(|\xi_{nj}|) < \infty$, and thus $\mathbb{E}(S_n(C))$ exists for each $C\in\mathcal{C}$. In the i.d. case mentioned before, we have for each $\delta>0$ that

$$\sum_{j\le j(n)} \mathbb{E}(|\xi_{nj}| I(|\xi_{nj}| > \delta)) = \sum_{j\le j(n)} j(n)^{-1}\mathbb{E}(|\xi_j| I(|\xi_j| > \delta j(n))) =$$

$\mathbb{E}(|\xi_1| I(|\xi_1| > \delta j(n))) \to 0$ as $n \to \infty$ whenever $\mathbb{E}(|\xi_1|) < \infty$, and

$\sup_{n\in\mathbb{N}} \sum_{j\le j(n)} \mathbb{E}(|\xi_{nj}| I(|\xi_{nj}| \le \delta)) \le \sup_{n\in\mathbb{N}} \sum_{j\le j(n)} \mathbb{E}(|\xi_{nj}|) = \mathbb{E}(|\xi_1|)$, whence in the

i.d. case both conditions (i) and (ii) in Theorem 3.1 are fullfilled under the only assumption that $\mathbb{E}(|\xi_1|) < \infty$.

APPENDIX

ON MAXIMAL INEQUALITIES FOR RADEMACHER AVERAGES

Here some basic maximal inequalities used for proving uniform laws of large numbers (ULLN) (Mean Glivenko-Cantelli results) and functional central limit theorems (FCLT), respectively, will be presented. They are based on ideas exposited by Pisier (1983). The present proofs are mainly due to Klaus Ziegler.

The first lemma is a special case of (3.2) in combination with (3.1) in Pollard (1990) proved there for Olicz norms. Since we do not consider Orlicz norms in the present situation, a much simpler proof will be possible.

A.1 Lemma. Given a *Rademacher sequence* $\varepsilon_1,\ldots,\varepsilon_N$, that is a sequence of i.i.d. r.v.'s ε_j taking only the two values ± 1 such that $\mathbb{P}(\varepsilon_j = 1) = \frac{1}{2} = \mathbb{P}(\varepsilon_j = -1)$, and given a finite and non empty subset M of $\mathbb{R}^N$, there exists for each $1\le p<\infty$ a universal constant $0<K_p<\infty$ such that

$$\mathbb{E}^{1/p}(\max_{x\in M} |\sum_{j\le N} \varepsilon_j x_j|^p) \le K_p(1+\log|M|)^{1/2} \max_{x\in M} (\sum_{j\le N} x_j^2)^{1/2} ,$$

where $|M|$ denotes the cardinality of M and $x = (x_1,\ldots,x_N)$.

For the proof the following proposition is used which is but a reformulation of Lemma 1.6 in Pisier (1983) for the present purposes.

A.2 Proposition. Let $\xi_1,\ldots,\xi_n$ be arbitrary nonnegative r.v.'s and Φ be a strictly increasing , nonnegative, convex function on $[0,\infty)$ such that there are $c_i>0$, $1\le i\le n$, and $c>0$ with

$$\mathbb{E}(\Phi(\xi_i/c_i)) \le c \quad \text{for all } i=1,\ldots,n.$$

Then

$$\mathbb{E}(\max_{i\le n} \xi_i) \le \Phi^{-1}(cn) \max_{i\le n} c_i \quad .$$

Proof of A.2. By Jensen's inequality

$$\Phi(\mathbb{E}(\max_{i\le n} \xi_i / \max_{i\le n} c_i)) \le \Phi(\mathbb{E}(\max_{i\le n} (\xi_i/c_i))) \le \mathbb{E}(\Phi(\max_{i\le n} (\xi_i/c_i))) =$$

$$= \mathbb{E}(\max_{i\le n} \Phi(\xi_i/c_i)) \le \mathbb{E}(\sum_{i\le n} \Phi(\xi_i/c_i)) = \sum_{i\le n} \mathbb{E}(\Phi(\xi_i/c_i)) \le cn.$$

Applying Φ^{-1} to both sides yields the assertion.

Proof of A.1. We show at first that the *Rademacher average* on the l.h.s. of the stated inequality in A.1 can be dominated by a so-called *Gaussian average*, to which we then will apply A.2.

For this, let $g_1,\ldots,g_N$ be i.i.d. $\mathcal{N}(0,1)$ r.v.'s being independent of $\varepsilon_1,\ldots,\varepsilon_N$. Note that $\mathcal{L}\{\varepsilon_j|g_j|\} = \mathcal{L}\{g_j\}$, $1\le j\le N$. Let $\mu := \mathbb{E}(|g_1|)$; then, by Jensen's inequality

$$\mathbb{E}(\max_{x\in M} |\sum_{j\le N} \varepsilon_j x_j|^p) = \mu^{-p}\mathbb{E}(\max_{x\in M} |\sum_{j\le N} \varepsilon_j \mathbb{E}(|g_j|)x_j|^p) =$$

$$\mu^{-p}\mathbb{E}(\max_{x\in M} | \sum_{j\le N} \varepsilon_j \mathbb{E}(|g_j| | \varepsilon_1,\ldots,\varepsilon_N) x_j|^p) =$$

$$\mu^{-p}\mathbb{E}(\max_{x\in M} | \mathbb{E} \sum_{j\le N} \varepsilon_j |g_j| x_j | \varepsilon_1,\ldots,\varepsilon_N)|^p) \le$$

$$\mu^{-p}\mathbb{E}(\max_{x\in M} \mathbb{E}(| \sum_{j\le N} \varepsilon_j |g_j| x_j|^p | \varepsilon_1,\ldots,\varepsilon_N)) \le$$

$$\mu^{-p}\mathbb{E}(\mathbb{E}(\max_{x\in M} | \sum_{j\le N} \varepsilon_j |g_j| x_j|^p | \varepsilon_1,\ldots,\varepsilon_N)) \le$$

$$\mu^{-p}\mathbb{E}(\max_{x\in M} |\sum_{j\le N} \varepsilon_j |g_j| x_j|^p) = \mu^{-p}\mathbb{E}(\max_{x\in M} | \sum_{j\le N} g_j x_j|^p) \ .$$

Next, let $\Phi'(u) := \exp(u^{2/p})$, $u\in[0,\infty)$, which ist convex for u larger than $u_p := (p/2)^{p/2}$. Therefore, taking Φ linear on $[0,u_p]$ and equal to Φ' on (u_p,∞), Φ is a strictly increasing, nonnegative, convex function on $[0,\infty)$ with $\Phi \le \Phi'$ and

$\Phi^{-1}(v) \le u_p + (\log v)^{p/2}$ for all $v \in [0,\infty)$. On the other hand, assigning to each $x \in M$ the r.v. $\xi_x := \sum_{j \le N} g_j x_j$, we have for each $x \in M$ that

$$\mathcal{L}\{\xi_x\} = \mathcal{N}(0,c_x) \text{ with } c_x^2 := \sum_{j \le N} x_j^2 \text{ and } \mathbb{E}(\Phi(|\xi_x|^p / 2^p c_x^p)) \le$$

$$\mathbb{E}(\Phi'(|\xi_x|^p / 2^p c_x^p)) = \mathbb{E}(\exp(\xi_x^2 / 4c_x^2)) = \sqrt{2}\,.$$

Hence by A.2 it follows that

$$\mathbb{E}(\max_{x \in M} |\sum_{j \le N} \varepsilon_j x_j|^p) \le \mu^{-p}\mathbb{E}(\max_{x \in M} |\sum_{j \le N} g_j x_j|^p) = \mu^{-p}\mathbb{E}(\max_{x \in M} |\xi_x|^p) \le$$

$$\mu^{-p}\Phi^{-1}(\sqrt{2}\,|M|) \max_{x \in M} (2^p c_x^p) \le \mu^{-p}(u_p + \log(\sqrt{2}\,|M|)^{p/2}) \max_{x \in M} (2^p c_x^p) \le$$

$$K_p^p (1+\log|M|)^{p/2} \max_{x \in M} (\sum_{j \le N} x_j^2)^{p/2}\,. \qquad \square$$

To formulate the other maximal inequalities, let $T = (T,d)$ be a pseudo-metric space. Following Dudley (1984) a set $\{t_1,\ldots,t_n\} \subset T$ is called a *γ-net* (for any given $\gamma > 0$) iff for each $t \in T$ there is some t_i such that $d(t,t_i) \le \gamma$. This gives rise to define so-called *covering numbers of (T,d):*

A.3 Definition. For any $\gamma > 0$ let

$N(\gamma,T,d) := \inf\{n \in \mathbb{N}$: there exists a γ-net $\{t_1,\ldots,t_n\} \subset T\}$, i.e. $N(\gamma,T,d)$ is the minimal number of points in a γ-net.

$H(\cdot,T,d) := \log N(\cdot,T,d)$ is called the *metric entropy* of $T = (T,d)$.[1)]

A closely related concept is concerned with the so-called *packing numbers* of (T,d). For this, given any $\gamma > 0$, let $D(\gamma,T,d)$ denote the largest m such that for some $t_1,\ldots,t_m \in T$ $d(t_i,t_j) > \gamma$ whenever $i \ne j$. The points $t_1,\ldots,t_m$ may be called *γ-distinguishable.*

A.4 Remark. If $\{t_1,\ldots,t_n\}$ is a maximal γ-distinguishable set, then $\{t_1,\ldots,t_n\}$ is a γ-net.

A.5 Proposition. For any $\gamma > 0$

$$D(2\gamma,T,d) \le N(\gamma,T,d) \le D(\gamma,T,d).$$

The following lemma is crucial for proving ULLN; as we shall see, it is an easy consequence of A.1.

A.6 Lemma. Given any $x_j \in \mathbb{R}^T$, $j=1,\ldots,N$, let $d_1(s,t) := \sum_{j \le N} |x_j(s)-x_j(t)|$, $s,t \in T$. Then, for each $1 \le p < \infty$ and for any Rademacher sequence $\varepsilon_1,\ldots,\varepsilon_N$ there exists a universal constant $0 < K_p < \infty$ such that for all $\gamma > 0$

$$\mathbb{E}^{1/p}(\sup_{t \in T} |\sum_{j \in N} \varepsilon_j x_j(t)|^p) \le \gamma + K_p(1+H(\gamma,T,d_1))^{1/2} \sup_{t \in T} (\sum_{j \le N} x_j^2(t))^{1/2}.$$

---

1 If $d = 0$, that is $d(s,t)=0$ for all $s,t \in T$, we put $N(0,T,d) := 1$ (whence in case $d = 0$ $N(\gamma,T,d) = 1$ and therefore $H(\gamma,T,d) = 0$ for all $\gamma \ge 0$).

Proof of A.6. Given any $\gamma>0$, assume w.l.o.g. that $N(\gamma,T,d_1) < \infty$ and choose a finite subset T' of T with $|T'| = N(\gamma,T,d_1)$ such that for any $t\in T$ there is a $u(t)\in T'$ with $d_1(t,u(t)) \le \gamma$. Then, applying Minkowski's inequality and A.1 we get

$$\mathbb{E}^{1/p}(\sup_{t\in T} |\sum_{j\le N} \varepsilon_j x_j(t)|^p) \le$$

$$\mathbb{E}^{1/p}(\sup_{t\in T} |\sum_{j\le N} \varepsilon_j(x_j(t)-x_j(u(t))|^p) + \mathbb{E}^{1/p}(\sup_{t\in T'} |\sum_{j\le N} \varepsilon_j x_j(t)|^p) \le$$

$$\sup_{t\in T} \sum_{j\le N} |x_j(t)-x_j(u(t))| + K_p(1+\log|T'|)^{1/2} \sup_{t\in T} (\sum_{j\le N} x_j^2(t))^{1/2} \le$$

$$\gamma + K_p(1+H(\gamma,T,d_1))^{1/2} \sup_{t\in T} (\sum_{j\le N} x_j^2(t))^{1/2} . \qquad \square$$

Next, on the basis of A.1, we shall apply the so-called *chaining method* (cf. Pollard (1984)) to obtain a *subgaussian inequality* for the special subgaussian process $(\zeta(t))_{t\in T}$ defined by $\zeta(t) := \sum_{j\le N} \varepsilon_j x_j(t)$ (cf. Lemma 2.1 in Arcones-Gaenssler-Ziegler (1992)), being a crucial inequality for proving FCLT.

A.7 Lemma. Let T be an arbitrary index set. Given any $x_j\in\mathbb{R}^T$, $j=1,\dots,N$, and a Rademacher sequence $\varepsilon_1,\dots,\varepsilon_N$, let $t_o\in T$ be arbitrary but fixed. Then, for each $1\le p<\infty$ there exists a universal constant $0<C_p<\infty$ such that

$$\mathbb{E}^{1/p}(\sup_{t\in T} |\sum_{j\le N} \varepsilon_j x_j(t)|^p) \le \sum_{j\le N} |x_j(t_o)| + C_p \int_0^{D/2} (H(\varepsilon,T,d))^{1/2} d\varepsilon, \text{ where}$$

$d^2(s,t) := \sum_{j\le N} (x_j(s)-x_j(t))^2$ and $D := \sup_{s,t\in T} d(s,t)$ denotes the d-diameter of T.

Proof of A.7. At first we will show

$$(+) \quad \mathbb{E}^{1/p}(\sup_{t\in T} |\sum_{j\le N} \varepsilon_j x_j(t)|^p) \le \sum_{j\le N} |x_j(t_o)| + K_p \sum_{k\in\mathbb{Z}_+} D/2^k(1+H(D/2^{k+2},T,d))^{1/2},$$

where K_p is the universal constant occuring in A.1.

Given any $\gamma>0$, assume w.l.o.g. that $N(\gamma,T,d) < \infty$, whence $D<\infty$; for each $k\in\mathbb{Z}_+$, let $\delta_k := D/2^k$, and choose finite subsets T_k of T with $|T_k| \le N(\delta_{k+1},T,d)$ such that for each $t\in T$ there is a $\pi_k(t)\in T_k$ with $d(t,\pi_k(t)) \le \delta_k$. In view of A.4 and A.5 we may also assume w.l.o.g. that $T_k \subset T_{k+1}$ for each $k\in\mathbb{Z}_+$ (choose $T_o=\{t_o\}$ and, for $k\in\mathbb{Z}_+$, T_{k+1} as a maximal δ_{k+1}-distinguishable set of points in T containing T_k).

Put $T_\infty := \bigcup_{k\in\mathbb{N}} T_k$ and $\zeta(t) := \sum_{j\le N} \varepsilon_j x_j(t)$, $t\in T$, noticing that T_∞ is dense in (T,d) by construction of the T_k's and that the process ζ has d-continuous sample paths, whence

$$\mathbb{E}^{1/p}(\sup_{t\in T} |\sum_{j\le N} \varepsilon_j x_j(t)|^p) = \mathbb{E}^{1/p}(\sup_{t\in T} |\zeta(t)|^p) = \mathbb{E}^{1/p}(\sup_{t\in T_\infty} |\zeta(t)|^p) .$$

Now, for each $k\in\mathbb{Z}_+$, $\max_{t\in T_{k+1}} |\zeta(t)| \le \max_{t\in T_{k+1}} |\zeta(\pi_k(t))| + \max_{t\in T_{k+1}} |\zeta(t)-\zeta(\pi_k(t))| \le$

$\max_{t\in T_k} |\zeta(t)| + \max_{t\in T_{k+1}} |\sum_{j\le N} \varepsilon_j (x_j(t)-x_j(\pi_k(t)))|$.

By Minkowski's inequality and A.1 (note that the second maximum on the r.h.s. of the last inequality runs only over at most $N(\delta_{k+2},T,d)$ quantities and that $\sum_{j\le N} (x_j(t)-x_j(\pi_k(t)))^2 = d^2(t,\pi_k(t)) \le \delta_k^2$),

$$\mathbb{E}^{1/p}(\max_{t\in T_{k+1}} |\zeta(t)|^p) \le$$

$$\mathbb{E}^{1/p}(\max_{t\in T_k} |\zeta(t)|^p) + K_p(1+\log|T_{k+1}|)^{1/2} \max_{t\in T_{k+1}} (\sum_{j\le N} (x_j(t)-x_j(\pi_k(t)))^2)^{1/2} \le$$

$$\mathbb{E}^{1/p}(\max_{t\in T_k} |\zeta(t)|^p) + K_p(1+H(\delta_{k+2}, T, d))^{1/2}\delta_k .$$

Since

$$\mathbb{E}^{1/p}(\max_{t\in T_o} |\zeta(t)|^p) = \mathbb{E}^{1/p}(|\sum_{j\le N} \varepsilon_j x_j(t_o)|^p) \le \sum_{j\le N} |x_j(t_o)|,$$

for any m=0,1,...

$$\mathbb{E}^{1/p}(\max_{t\in T_{m+1}} |\zeta(t)|^p) \le \sum_{j\le N} |x_j(t_o)| + K_p \sum_{k=0}^{m} \delta_k(1+H(\delta_{k+2}, T,d))^{1/2}.$$

Since $T_m \uparrow T_\infty$ as $m \to \infty$, by monotone convergence,

$$\mathbb{E}^{1/p}(\sup_{t\in T_\infty} |\zeta(t)|^p) \le \sum_{j\le N} |x_j(t_o)| + K_p \sum_{k\in\mathbb{Z}_+} \delta_k(1+H(\delta_{k+2}, T,d))^{1/2}$$

which proves (+).

From (+) we obtain the assertion of A.7 as follows:
Note first that $N(\gamma,T,d) \ge 2$ for all $0<\gamma<D/2$, whence $H(\delta_{k+2},T,d) \ge \log 2$ for all $k\in\mathbb{Z}_+$. So, by taking $c := 1+(\log 2)^{-1}$, we get the following upper estimate for the series on the r.h.s. of (+):

$$\sum_{k\in\mathbb{Z}_+} \delta_k(1+H(\delta_{k+2},T,d))^{1/2} \le c^{1/2} \sum_{k\in\mathbb{Z}_+} \delta_k(H(\delta_{k+2},T,d))^{1/2} =$$

$$c^{1/2} \sum_{k\in\mathbb{Z}_+} 8(\delta_{k+2}-\delta_{k+3})(H(\delta_{k+2}, T,d))^{1/2} \le 8c^{1/2} \sum_{k\in\mathbb{Z}_+} \int_{\delta_{k+3}}^{\delta_{k+2}} (H(\varepsilon,T,d))^{1/2}d\varepsilon =$$

$8c^{1/2} \int_0^{D/4} (H(\varepsilon,T,d))^{1/2}d\varepsilon$ which proves the assertion of A.7 by taking

$C_p := 8(1+(\log 2)^{-1})^{1/2} K_p$. □

A.8 Corollary. In the situation of A.7, for each $1 \le p < \infty$ there exists another universal constant $0 < C'_p < \infty$ such that for each $\delta > 0$

$$\mathbb{E}^{1/p}\Big(\sup_{d(s,t)\le\delta} \Big| \sum_{j\le N} \varepsilon_j (x_j(s)-x_j(t)) \Big|^p \Big) \le C'_p \int_0^{\delta/2} (H(\varepsilon,T,d))^{1/2} d\varepsilon .$$

Proof of A.8. Apply A.7 to T_δ: $\{(s,t)\in T^2 : d(s,t)\le\delta\}$, $x_j(s,t) := x_j(s)-x_j(t)$ for $(s,t)\in T_\delta$ and with the pseudo-metric e defined on T_δ by $e^2((s,t),(u,v)) := \sum_{j\le N} (x_j(s,t)-x_j(u,v))^2$, noticing that

$$\sup_{(s,t),(u,v)\in T_\delta} e((s,t),(u,v)) \le \sup_{(s,t)\in T_\delta} d(s,t) + \sup_{(u,v)\in T_\delta} d(u,v) \le 2\delta \quad \text{and that}$$

$N(\gamma,T_\delta,e) \le N^2(\gamma/2, T,d)$ for all $\gamma>0$. Therefore,

$$\int_0^{e\text{-diam}(T_\delta)/2} (H(\varepsilon,T_\delta,e))^{1/2} d\varepsilon \le \int_0^{\delta} (H(\varepsilon,T_\delta,e))^{1/2} d\varepsilon \le \int_0^{\delta} 2^{1/2}(H(\varepsilon/2,T,d))^{1/2} d\varepsilon =$$

$= 2^{3/2} \int_0^{\delta/2} (H(\varepsilon,T,d))^{1/2} d\varepsilon$. Since $x_j(w,w)=0$ and $(w,w)\in T_\delta$ for any $w\in T$ the first term on the r.h.s. of the inequalitiy in A.7 vanishes. This proves A.8 with $C'_p := 2^{3/2} C_p$. □

Finally, for proving ULLN and FCLT the following *symmetrization inequality* is used. In its present form, assuming T to be an arbitrary countable index set, it follows from Lemma 2 in Markus (1981).

A.9 Lemma. Let $\xi_1,\dots,\xi_N$ be independent stochastic processes with common parameter space T, viewed as r.e.'s in $(\mathbb{R}^T, \mathbb{B}^T)$, such that for each $1\le j\le N$ $\mathbb{E}(|\xi_j(t)|) < \infty$ for all $t\in T$ and such that $(\xi_1,\dots,\xi_N)$ is independent of a given Rademacher sequence $\varepsilon_1,\dots,\varepsilon_N$. Then

$$\mathbb{E}\Big(\sup_{t\in T} \Big| \sum_{j\le N} (\xi_j(t) - \mathbb{E}(\xi_j(t))) \Big| \Big) \le 2\mathbb{E}\Big(\sup_{t\in T} \Big| \sum_{j\le N} \varepsilon_j \xi_j(t) \Big| \Big) .$$

Note added in proof. Theorem 2.11 is still valid under the weaker basic assumptions of Section 3.

REFERENCES

Alexander, K.S. and Pyke, R. (1986). A uniform central limit theorem for set-indexed partial-sum processes with finite variance. Ann. Probab. 14, 582-597.

Araujo, A. and Giné, E. (1980). The central limit theorem for real and Banach space valued random variables. Wiley, New York.

Arcones, M.A., Gaenssler, P. and Ziegler, K. (1992). Partial-sum processes with random locations and indexed by Vapnik-Červonenkis classes of sets in arbitrary sample spaces. In: Probability in Banach spaces 8; R.M. Dudley, M.G. Hahn and J. Kuelbs (Eds.). Progress in Probability, Vol. 30, Birkhäuser.

Assouad, P. (1981). Sur les classes de Vapnik-Červonenkis. C.R. Acad. Sci. Paris, 292 Sér. I, 921-924.

Bickel, P.J. and Wichura, M.J. (1971). Convergence for multiparameter stochastic processes and some applications. Ann. Math. Statist. 42, 1656-1670.

Dudley, R.M. (1978). Central limit theorems for empirical measures. Ann. Probab. 6, 899-929. (Correction (1979) ibid. 7, 909-911).

Dudley, R.M. (1984). A course on empirical processes, Lecture Notes in Math. 1097, pp. 1-142, Springer, New York.

Gaenssler, P. and Schneemeier, W. (1986). On functional limit theorems for a class of stochastic processes indexed by pseudo-metric parameter spaces (with applications to empirical processes). In: Probability in Banach spaces 6; Proc. of the Sixth Int. Conf., Denmark, 1986. Progress in Probability, Vol. 20, 1990, Birkhäuser.

Gaenssler, P. (1992). On weak convergence of certain processes indexed by pseudo-metric parameter spaces with applications to empirical processes. Trans. 11th Prague Conf. (August 27-31, 1990) pp. 49-77.

Gaenssler, P. and Ziegler, K. (1992). On a mean Glivenko-Cantelli result for certain set-indexed processes. Preprint No. 51, University of Munich.

Gaenssler, P. and Ziegler, K. (1993). Preprint No. 51: Revised version.

Hoffmann-Jørgensen, J. (1984). Stochastic Processes on Polish spaces. Unpublished manuscript. Published in 1991 as Vol. 39 of the Various Publication Series, Matematisk Institut, Aarhus Universitet.

Markus, M.B. (1981). Weak convergence of the empirical characteristic function. Ann. Probab. 9, 194-201.

Ossiander, M. (1984). Weak convergence and a law of the iterated logarithm for processes indexed by points in a metric space. Ph. D. thesis. Univ. of Washington.

Pisier, G. (1983). Some applications of the metric entropy condition to harmonic analysis. Banach Spaces, Harmonic Analysis, and Probability Theory. Lecture Notes in Math. 995, pp. 123-154, Springer, New York.

Pollard, D. (1984). Convergence of Stochastic Processes. Springer, New York.

Pollard, D. (1990). Empirical Processes: Theory and Applications. NSF-CBMS Regional Conference Series in Probability and Statistics, Vol. 2. Institute of Mathematical Statistics, Hayward, California.

Pyke, R. (1984). Asymptotic results for empirical and partial-sum processes: A review. Canadian J. of Statist., Vol. 12, No. 4, pp. 241-264.

Shorack, G.R. and Wellner, J.A. (1986). Empirical Processes with Applications to Statistics. Wiley, New York.

Stengle, G. and Yukich, J.E. (1989). Some new Vapnik-Chervonenkis classes. Ann. Statist. 17, 1441-1446.

van der Vaart, A.W. and Wellner, J.A. (1993). Weak convergence and empirical processes. Forthcoming monograph.

Vapnik, V.N. and Červonenkis, A. Ja. (1971). On the uniform convergence of relative frequencies of events to their probabilities. Theory Prob. Appl. 16, 264-280.

REGRESSION RANK SCORES SCALE STATISTICS AND STUDENTIZATION IN LINEAR MODELS

Jana Jurečková
Department of Probability & Statistics
Charles University
Praha, Czech Republic

and

Pranab K. Sen
Departments of Biostatistics & Statistics
University of North Carolina
Chapel Hill, NC, USA

Abstract. In linear models, regression-invariant and scale-equivariant regression rank scores scale statistics are proposed. Among their potential applications in statistical inference in linear models, studentization and testing for homoscedasticity are considered.

Key words. GM-estimators; homoscedasticity; M-estimators; regression-invariance; regression-equivariance; regression quantiles; scale-equivariance.

AMS Subject Classifications: 62G05, 62J05.

1. Introduction

Scale statistics not only provide estimates for various measures of scale, dispersion or spread of a distribution, but also play an important role in robust and nonparametric inference in linear models. Generally, such estimators of location and regression parameters are nonlinear, and hence, they are not ususally scale-equivariant; a typical example is the classical M-estimators of location or regression parameters [viz., Huber (1981)]. On the otherhand, when studentized by an appropriate scale statistic, such M-estimators become scale-equivariant. Secondly, such scale statistics often provide a criterion which is to be minimized in deriving a location or regression estimator [viz., the usual R-estimators and S-estimators involving the Jaeckel (1972) measure of dispersion]. Thirdly, such scale statistics are also essential in the construction of suitable test statistics for testing plausible null hypotheses in a robust or nonparametric setup. Some other important applications of scale statistics for the location-scale model are discussed in Rousseeuw and Croux (1993).

In the present study, we develop a class of scale statistics in general linear models based on the recently developed concept of regression rank scores. Such regression rank scores, by virtue of their regression-invariance, provide a natural and yet a bit unexplored appeal for the development of scale statistics as will be followed through in this study. Various properties of such regression rank scores scale statistics are studied and their effective role in linear models is critically examined.

Consider the usual linear model:

$$\mathbf{Y} = (Y_1,\ldots,Y_n)' = \mathbf{X}\beta + \mathbf{e};\ \mathbf{e} = (e_1,\ldots,e_n)', \tag{1.1}$$

where $\mathbf{X}$ is an nxp (design) matrix of known regression constants and we denote the ith row of $\mathbf{X}$ by $\mathbf{x}'_i$, i=1,...n, β is a p-vector of unknown regression parameters, and the e_i are independent and identically distributed random variables (i.i.d.r.v.) with a distribution function (d.f.) F, defined on $\mathbb{R}$. The form of F is usually unknown and we impose suitable regularity conditions. Also, without any loss of generality, we may set $x_{i1} = 1$ for every i and $\Sigma_{i=1}^{n} x_{ij} = 0$, for j = 2,...,p.

Let $T_n = T_n(Y)$ be a suitable estimator of β. We say that T_n is regression-equivariant if for every $\mathbf{b} \in R^p$, $\mathbf{Y} \in R^n$,

$$T_n(\mathbf{Y} + \mathbf{X}\mathbf{b}) = T_n(\mathbf{Y}) + \mathbf{b}, \tag{1.2}$$

and T_n is said to be scale-equivariant if for every $c \in \mathbb{R}^+$, $\mathbf{Y} \in \mathbb{R}^n$,

$$T_n(c\mathbf{Y}) = c.T_n(\mathbf{Y}). \tag{1.3}$$

Note that (1.2) and (1.3) are to hold for every $\mathbf{Y}$ and $\mathbf{X}$. Regression and scale equivariance properties are very desirable for estimators of regression (location) parameters. Many estimators including the classical least squares (LSE), trimmed LSE and other L-estimators based on regression quantiles and R-estimators are regression and scale equivariant. However, the usual M-estimators are not generally scale-equivariant. Often this problem is dealt with by a simultaneous estimation of β and the scale parameter or incorporating a studentized score function. The classical M-estimator of $\mathbf{M}_n$ is defined either as

$$\mathbf{M}_n = \text{Arg. min}\ \{\sum_{i=1}^{n} \rho(Y_i - \mathbf{x}'_i\mathbf{b}) : \mathbf{b} \in \mathbb{R}^p\}, \tag{1.4}$$

where $\rho : \mathbb{R} \to \mathbb{R}^+$, or for $\psi = \rho'$ continuous, as a solution of

$$\sum_{i=1}^{n} \mathbf{x}_i\ \psi(Y_i - \mathbf{x}'_i\mathbf{M}_n) = 0. \tag{1.5}$$

Such $\mathbf{M}_n$ satisfes (1.2) but not generally (1.3). The studentized M-estimator is defined as

$$\mathbf{M}_n = \text{Arg. min}\ \{\sum_{i=1}^{n} \rho((Y_i - \mathbf{x}'_i\mathbf{b})/S_n) : \mathbf{b} \in R^p\} \tag{1.6}$$

where $S_n = S_n(Y)$ is an appropriate scale statistics. If S_n is scale-equivariant, i.e., for every $c \in R^+$ and $Y \in R^n$,

$$S_n(cY) = cS_n(Y), \tag{1.7}$$

then M_n is itself scale-equivariant. However, we may remark that M_n defined by (1.6) may be deprived of its regression-equivariance property (1.2) unless S_n is regression-invariant in the sense that

$$S_n(Y + Xb) = S_n(Y), \text{ for every } Y \in R^n,\ b \in R^p. \tag{1.8}$$

Another proposal due to Huber is to prescribe the M-estimator of β as a solution to

$$\sum_{i=1}^{n} x_i \ \psi(Y_i - x_i'M_n)/S_n) = 0. \tag{1.9}$$

where S_n is an estimator of scale which is computed simultaneously with M_n by means of the equation

$$\sum_{i=1}^{n} \chi((Y_i - x_i'M_n)/S_n) = 0, \text{ with } \chi : R \to R. \tag{1.10}$$

Then, M_n satisfies (1.2) and (1.3) only if S_n is scale-equivariant and regression-invariant in the sense of (1.7) and (1.8). Alternatively, one may use adaptive M-estimator of the Huber-type, proposed by Jurečková and Sen (1984), and this satisfies (1.2) and (1.3).

If we intend to have the solution M_n in (1.6) to possess both the regression and scale equivariance properties, we need to solve for (1.6) subject to the additional restraints that

$$S_n\ (Y) \geq 0, \text{ with probability one,} \tag{1.11}$$

and

$$S_n(cY + Xb) = cS_n(Y),\quad c \in R^p \text{ and } Y \in R^n. \tag{1.12}$$

On the other hand, the the solution of (1.9) and (1.10) satisfies (1.2) and (1.3) for M_n and also (1.12) for S_n. The situation is similar for GM-estimator which is prescribed as a solution to

$$\sum_{i=1}^{n} \psi(x_i,\ (Y_i - x_i'M_n)/S_n)x_i = 0, \tag{1.13}$$

where $\psi : R^{p+1} \to R$ and S_n is a scale estimator computed simulatneously with M_n by (1.10).

The population counterpart of S_n is usually a (scale) functional $S(F)$ of the d.f. F, and usually S_n is a consistent estimator of $S(F)$.

However, in the current context, S_n is mainly used to obtain a regression and scale-equivariant M-estimator of β, by means of studentization.

Surprisingly, there are not too many scale-statistics satisfying (1.7) and (1.8), and these have not received much attention either. For the location model, the robust scale statistics proposed in Huber (1981), Hampel et al. (1986), Croux and Rousseeuw (1993) and Rousseeuw and Croux (1992), among others, are translation but not regression invariant. Among classical statistics, the root of the *residual sum of squares*,

$$S_n = \|Y - \hat{Y}\| = \{(Y-\hat{Y})'(Y-\hat{Y})\}^{1/2}, \qquad (1.14)$$

where $\hat{Y} = X\hat{\beta}$ with $\hat{\beta}$ being the LSE, satisfies (1.7) and (1.8); in fact, $Y-\hat{Y}$ is the maximal invariant with respect to the group of transformations $Y* = Y+Xb$, $b \in R^p$. However, such a S_n is closely connected with the normal distribution and is very non-robust.

A broad class of statistics satisfying (1.7) and (1.8) are *L-statistics* based on regression quantiles. For example, the euclidean distance of two regression quantiles,

$$S_n = \|\hat{\beta}_n(\alpha_2) - \hat{\beta}_n(\alpha_1)\|, \; 0<\alpha_1<\alpha_2<1 \qquad (1.15)$$

or the difference of the first components

$$S_n = \hat{\beta}_{n1}(\alpha_2) - \hat{\beta}_{n1}(\alpha_1), \; 0<\alpha_1<\alpha_2<1 \qquad (1.16)$$

of regression quantiles, where $\hat{\beta}_n(\alpha) = (\beta_{n1}(\alpha),\ldots,\hat{\beta}_{np}(\alpha))'$ is the α-regression quantile in the model (1.1), $0<\alpha<1$. The asymptotic properties of these statistics may be derived from theorems in Koenker and Bassett (1978), Ruppert and Carroll (1980) and Gutenbrunner and Jurečková (1992). We shall not consider such scale statistics here.

In the present paper, we shall propose another class of scale statistics, based on *regression rank scores* introduced in Gutenbrunner and Jurečková (1992). Jurečková (1992a) considered such statistics as criteria for estimators based on regression rank scores. These statistics represent an extension of Jaeckel's (1972) rank dispersion of the residuals. However, unlike Jaeckel's dispersion which is only translation-equivariant, our proposed statistics satisfy both (1.11) and (1.12). Hence, the proposed statistics are applicable in studentization in linear models and they are useful in various other contexts. For instance, we shall consider a test of homogeneity of scale parameters based on S_n. The authors believe that such statistics will be also widely applicable in rank tests and R-estimation in linear regression model.

Remark. Let us touch briefly the question of finite-sample breakdown points of scale statistics satisfying (1.11) and (1.12). Note that the regression-invariance in (1.12) may not ensure that these statistics do not react to outliers in the matrix $\mathbf{X}$. For example, if the given $\mathbf{X}$ is equal to $\mathbf{X}^o + \mathbf{E}$ where $\mathbf{X}^o$ is the true matrix and $\mathbf{E}$ consists of outliers and it may not belong to the space spanned by $\mathbf{X}^o$ (denoted by $\mathbf{M}(\mathbf{X}^o)$), then $S_n(c\mathbf{Y} + \mathbf{X}\mathbf{b}) = S_n(c\mathbf{Y} + \mathbf{X}^o\mathbf{b} + \mathbf{E}\mathbf{b})$ may not be equal to $S_n(c\mathbf{Y} + \mathbf{X}^o\mathbf{b})$ unless $\mathbf{E}\mathbf{b} \subset \mathbf{M}(\mathbf{X}^o\mathbf{b})$. Thus, the scope for robustness to the leverage points for such scale statistics is somewhat restrained to the set of $\mathbf{b},\mathbf{E}$, such that $\mathbf{E}\mathbf{b} \subset \mathbf{M}(\mathbf{X}^o\mathbf{b})$. Nevertheless, if $\mathbf{X}$ is nonstochastic (as assumed here), this feature should not be of any serious concern.

2. Scale statistics based on regression rank scores.

Consider the linear regression model (1.1) with the matrix $\mathbf{X} = \mathbf{X}_n$ of order $n \times p$ with the rows $\mathbf{x}_i'$ and satisfying $x_{i1}=1$, $i=1,\ldots,n$. Let $\hat{\mathbf{a}}_n(\alpha) = (\hat{a}_{n1}(\alpha),\ldots,\hat{a}_{nn}(\alpha))'$, $0 \le \alpha \le 1$, be the vector of *regression rank scores,* defined as a solution of the parametric linear programming problem

$$\mathbf{Y}'\hat{\mathbf{a}}(\alpha) = \max, \quad \mathbf{X}'\{\hat{\mathbf{a}}(\alpha) - (1-\alpha)\mathbf{1}_n\} = 0, \qquad (2.1)$$
$$\hat{\mathbf{a}}(\alpha) \in [0,1]^n, \ 0 \le \alpha \le 1.$$

Finite-sample as well as asymptotic properties of regression rank scores (RR) are studied in Gutenbrunner (1986) Gutenbrunner and Jurečková (1992), Jurečková (1992b), Gutenbrunner et al. (1993) and Jurečková (1993). Dodge and Jurečková (1993) proposed an estimator of $1/f(F^{-1}(u))$, $0 < u < 1$, of the Kernel type, based on regression quantiles. These papers also illustrate applications of RR, mainly in estimation and testing in linear regression model. In the location submodel corresponding to $\mathbf{X}=\mathbf{1}_n$ the regression rank scores reduce to

$$a_n^*(R_i,\alpha) = \begin{cases} 1 & \text{if } 0 \le \alpha < (R_i-1)/n, \\ R_i - n\alpha & \text{if } (R_i-1)/n \le \alpha \le R_i/n, \\ 0 & \text{if } R_i/n < \alpha \le 1, \end{cases} \qquad (2.2)$$

where R_i is the rank if Y_i among $Y_1, \ldots, Y_n$, $i=1,\ldots,n$. Such scores where considered for the first time by Hájek and Sidák (1967), Chapter V, as a basis of various rank tests; hence we shall occasionally refer to them as the Hájek-Sidák scores. The function $\hat{a}_{ni}(\alpha)$, $0 \leq \alpha \leq 1$, is a continuous function of bounded variation and is piecewise linear on [0,1], i=1,...,n. Moreover, $\hat{a}(\alpha)$ is regression-invariant in the sense that

$$\hat{a}_n(\alpha, Y+Xb) = \hat{a}_n(\alpha, Y), \quad b \in \mathbb{R}^p. \tag{2.3}$$

This property, which follows directly from (2.1), is in correspondence with the translation- invariance of ranks and hence that of the Hájek-Sidák scores. The most important consequence from our point of view is that every procedure based on RR is also regression-invariant.

We shall define the scale statistic in the following way: Select a score function $\phi:(0,1)\rightarrow\mathbb{R}$, nondecreasing, nonconstant, square-integrable on (0,1) and such that $\phi(\alpha)=-\phi(1-\alpha)$, $0<\alpha<1$. Fix a number α_0, $0<\alpha_0<1/2$ and assume, without loss of generality, that ϕ is standardized so that

$$\int_{\alpha_0}^{1-\alpha_0} \phi^2(\alpha)d\alpha = 1. \tag{2.4}$$

Calculate the regression scores $\hat{b} = (\hat{b}_{n1}, \ldots, \hat{b}_{nn})'$ generated by ϕ in the following way:

$$\hat{b}_{ni} = -\int_{\alpha_0}^{1-\alpha_0} \phi(\alpha)d\hat{a}_{ni}(\alpha), \quad i = 1,\ldots,n. \tag{2.5}$$

We propose the scale statistic S_n of the form

$$S_n = S_n(Y) = n^{-1}\sum_{i=1}^{n} Y_i\hat{b}_{ni} = n^{-1}Y'\hat{b}_n. \tag{2.6}$$

Then S_n is regression-invariant, because

$$S_n(Y+Xd)=n^{-1}(Y+Xd)'\hat{b}_n(Y+Xd)=n^{-1}Y'\hat{b}_n = S_n(Y) \tag{2.7}$$

by (2.3)-(2.5). Moreover, because $\hat{a}_n(\alpha)$ and hence $\hat{b}_n$ are scale-invariant,(2.6) implies that S_n is scale-equivariant, i.e.,

$$S_n(cY) = c\, S_n(Y), \quad c>0. \tag{2.8}$$

Furthermore, because $n^{-1}\Sigma_{i=1}^{n}\hat{b}_{ni} = 0$ and S_n is regression- and hence also translation-invariant , (2.6) also implies that $S_n \geq 0$, and as the e_i are i·d·r·v's with a continuous d·f·F, $S_n>0$ with probability 1.

Typical choices of ϕ are (i) *Wilcoxon scores* with $\phi(\alpha)=\alpha-1/2$, $0<\alpha<1$; (ii) *normal (van der Waerden)* scores with $\phi(\alpha) = \Phi^{-1}(\alpha)$, $0<\alpha<1$, Φ being the standard normal d.f. and (iii) median (sign) scores with $\phi(\alpha) = 1/2\ \mathrm{Sign}\ (\alpha-1/2)$, $0 < \alpha < 1$. Remark: The statistic (2.6) is the first component of the L-statistic of type 2, considered by

Gutenbrunner (1986) and Gutenbrunner and Jurečková (1992); the choice of a nondecreasing score function leads to a measure of scale. If we replace the $\hat{a}_{ni}(\alpha)$ by a_n^* (R_i, α), we arrive at the Jackel (1972) measure of dispersion of the residuals. Computational algorithms for regression quantiles and regression rank scores were elaborated by Koenker and d'Orey (1987, 1990), and they are surprisingly fast. Specially, when using the sign-score function, we have to compute the regression rank scores only for the three values of α (i.e., α_1, 1/2 and α_2) to establish the value of S_n.

3. Asymptotic Theory.

We shall show that, under some regularity conditions on F, ψ and **X**, the statistic S_n converges in probability to the functional $S(F) = \int_{\alpha_0}^{1-\alpha_0} /\Phi(\alpha)F^{-1}(\alpha)d\alpha$ and that $n^{1/2}(S_n-S(F))$ is asymptotically normal distributed as $n \to \infty$.

The asymptotic properties will be proved under the following regularity conditions.

(F.1) We assume that F is absolutely continuous with symmetric, absolutely continuous, positive and bounded density f(x), on $\{n:0 < F(n) < 1\}$.

(X.1) $x_{i1} = 1$, $i=1,\dots,n$.

(X.2) $\lim_{n\to\infty} \mathbf{D}_n = \mathbf{D} =$ where $\mathbf{D}_n = n^{-1}\mathbf{X}'_n\mathbf{X}_n$ and $\mathbf{D}$ is a positive definite (p.d.) pxp matrix

(X.3) $n^{-1}\sum_{i=1}^{n} \|\mathbf{x}_i\|^4 = O(1)$ as $n \to \infty$ $(\Rightarrow \max_{1\le i\le n} \|\mathbf{x}_i\| = o(n^{1/4})$ as $n\to\infty)$.

(C.1) Let ϕ: $(0,1) \to \mathbb{R}$ be a nondecreasing, nonconstant and square-integrable function such that $\phi(\alpha) = -\phi(1-\alpha)$, $0<\alpha<1$ and such that

$$\int_{\alpha_0}^{1-\alpha_0} \phi^2(\alpha)d\alpha = 1, \ 0<\alpha<1/2 \text{ fixed.}$$

The main result of the present section is the following:

THEOREM 3.1. Let S_n be the scale statistic defined in (2.5) and (2.6).

(i) Under the above regularity conditions,

$$S_n \xrightarrow{p} S(F) \text{ as } n \to \infty \tag{3.1}$$

where

$$S(F) = \int_{\alpha_0}^{1-\alpha_0} \phi(\alpha)F^{-1}(\alpha)d\alpha. \tag{3.2}$$

(ii) $n^{1/2}(S_n-S(F))$ is asymptotically $\mathcal{N}(0,\sigma^2)$ where $\sigma^2 = \sigma^2(\phi,F,\alpha_o)$ is given by

$$\int_{F^{-1}(\alpha_o)}^{F^{-1}(1-\alpha_o)} \int_{F^{-1}(\alpha_o)}^{F^{-1}(1-\alpha_o)} \{F(x\wedge y)-F(x)F(y)\}\phi(F(x)\phi(F(y))dxdy. \quad (3.3)$$

(iii) S_n admits the following asymptotic representation

$$S_n-S(F) = n^{-1}\Sigma_{i=1}^{n} \psi(e_i) + o_p(n^{-1/2})$$

$$= \int_{F^{-1}(\alpha_o)}^{F^{-1}(1-\alpha_o)} \phi(F(z))\ \{F(z)-F_n(z)\}dz + o_p(n^{-1/2}) \quad (3.4)$$

where

$$\psi(z) = \int_{\alpha_o}^{1-\alpha_o} \{(\alpha-I[F(z) \le \alpha])/f(F^{-1}(\alpha)\}\phi(\alpha)d\alpha,\ z \in \mathbf{R} \quad (3.5)$$

and $F_n(z)$ is the empirical distribution function of $e_1,\ldots,e_n$.

Proof. The propositions follow from Section 6 in Gutenbrunner and Jurečková (1992), hence we drop the details.

Remark. The scale statistic S_n is robust in the sense that its influence function and hence its global sensitivity are bounded, as one can deduce from (3.4). It is possible to take α_o=0 in the definition of S_n; the asymptotic properties would be retained but S_n would not be robust.

Moreover, we may apply the modified definition of Croux and Rousseeuw (1992) of the finite-sample breakdown point for the scale statistics in the location model:

$$\varepsilon_n^* (S_n,Y) = \min\{\tfrac{m}{n}: \sup_{Y^*} S_n(Y^*) = \infty \text{ or } \inf_{Y^*} S_n(Y^*) = 0\}$$

where Y^* is obtained from the original sample Y by replacing any m observations by arbitrary values. Note that by (1.8), we have $S_n(Y) = S_n(e)$ where e is the vector of errors in the model (1.1). Therefore, we can deduce from the definition of S_n that its breakdown point is at least α_o. However, this breakdown point may not stand valid when there is a contamination in the matrix $\mathbf{X}$ [see the remark at the end of Section 1].

4. Tests for homogeneity of scale parameters.

As a natural extension of the model (1.1), we consider the following:

$$Y_r = X_r\beta_r + e_r \quad r=1,\dots,c\ (\geq 2) \qquad (4.1)$$

where Y_r is an n_r-vector of observations, X_r is a given $n_r \times p_r$ matrix, β_r is an unknown p_rx1 parameter and $e_r = (e_{r1},\dots,e_{rn_r})'$ is a vector of i.i.d errors with the distribution function

$$F_r(e) = F_0(e/\lambda_r),\ e\epsilon\mathbf{R},\ \lambda_r \epsilon \mathbf{R}_1^+,\ 1 \leq r \leq c; \qquad (4.2)$$

$\beta_1,\dots,\beta_c$ are the respective regression parameter vectors (which we shall treat as nuisance parameters), and $\lambda_1,\dots,\lambda_c$ are the scale parameters; the functional form of F_0 is supposed to be known. We are interested in testing the following null hypothesis of homogeneity of scale parameteres:

$$H_0\colon \lambda_1 = \dots = \lambda_c = \lambda \text{ (unknown)} \qquad (4.3)$$

against the alternative that the λ_r are not all equal.

A nonparametric version of the model (4.2) was considered by Koenker and Bassett (1982) and Gutenbrunner (1986), who proposed a test of homoscedasticity against a special heterscedasticity alternative. Here we consider a test of homogeneity of the scale functionals

$$S_r = S(F_r) = \int_{\alpha_0}^{1-\alpha_0} \phi(\alpha)F_r^{-1}(\alpha)d\alpha,\ r=1,\dots c \qquad (4.4)$$

treating the β_r as nuisance parameters. In fact, H_0 is true if and only if $S_1= \dots=S_c$; on the other hand, testing the homogeneity of variances (based on the assumption of normality) may not be appropriate for all distributions while our tests are applicable even for the Cauchy distribution.

Let $S_n^{(r)}$ denote the scale statistic defined in (2.6), based on Y_r, r=1,...,c. Then, by an appeal of Theorem 3.1,

$$n_r^{1/2}\,(S_{n_r}^{(r)} - \lambda_r S(F_0)) \sim \mathcal{N}(0,\ \lambda_r^2\,\sigma_0^2), \qquad (4.5)$$

where $\sigma_0^2 = \sigma^2(\phi,F_0,\alpha_0)$ is given in (3.3). Notice that both F_0 and $S(F_0)$ are known; we shall consider the following (Bartlett-type) test of homogeneity of the λ_r. Assume that all $n_1,\dots,n_c$are sufficiently large and

$$n_r/\Sigma_{i=1}^c n_i \to \rho_r \text{ as } n\to\infty,\ 0<\rho_r<1,\ r=1,\dots,c. \qquad (4.6)$$

We denote by

$$S_n^{o(r)} = S_n^{(r)}/S(F_0),\ 1 \leq r \leq c, \qquad (4.7)$$

and

$$\Sigma_{r=1}^{c} (n_r/n)\log S_{n_r}^{o(r)} = \log \bar{S}_n^o, \tag{4.8}$$

so that

$$\log \bar{S}_n^o = \Sigma_{r=1}^{c} (n_r/n) \log S_{n_r}^{(r)} - \log S(F_o) \tag{4.9}$$
$$= \log \bar{S}_n - \log S(F_o).$$

Finally, let

$$\mathcal{L}_n = \{\Sigma_{r=1}^{c} n_r(\log S_n^{(r)} - \log \bar{S}_n)^2\}\sigma_o^{-2}. \tag{4.10}$$

It follows from (4.5) and (4.10) that under H_0,

$$\mathcal{L}_n \overset{\mathcal{D}}{\to} \chi^2_{c-1}, \tag{4.11}$$

hence the test rejects the hypothesis provided $\mathcal{L}_n$ exceeds the critical value of the χ^2_{c-1} distribution. We could then apply either the Bartlett-type or the bootstrap correction to improve the approximation in (4.11).

Acknowledgment. The authors are grateful to Dr. Cornelius Gutenbrunner for his valuable comments which led to an improvement of the text. The first author's research was supported by the Charles University Grant No. 365.

REFERENCES

CROUX, C. and ROUSSEEUW, P.J. (1993). A class of high-breakdown scale estimators based on subranges. *J. Amer. Statist. Assoc.* 88 (in press).

DODGE, Y. AND JUREČKOVÁ, J. (1993). Estimation of quantile density function based on regression quantiles. Preprint.

GUTENBRUNNER, C. (1986). *zur Asymptotic von Regressions-quantile prozessen und daraus abgeleiten statistiken. Ph.D. Dissertation, Univ. Freiburg, Germany.*

GUTENBRUNNER, C. and JUREČKOVÁ, J. (1992). Regression rank scores and regression quantiles. *Ann. Statist.* 20, 305-330.

GUTENBRUNNER, C. and JUREČKOVÁ, J., KOENKER, R. and PORTNOY, S. (1993). Tests of Linear Hypotheses Based on Regression Rank Scores. *J. Nonparamet. Statist.* 2, 307-331.

HÁJEK, J. and ŠIDÁK, Z. *Theory of Rank Tests.* Academia, Prague.

HAMPEL, F.R., RONCHETTI, E.M., ROUSSEEUW, P.J. and STAHEL, W.A. (1986). *Robust Statistics: the Approach Based on Influence Functions.* J. Wiley, New York.

HUBER, P.J. (1981). *Robust Statistics. J. Wiley, New York.*

JAECKEL, L.A., (1972). Estimating the regression coefficients by minimizing the dispersion of residuals. *Ann. Math. Statist.* **43**, 1449-1458.

JUREČKOVÁ, J. (1992a). Estimation in linear model based on regression rank scores. *J. Nonparamet. Statist.* 1, 197-203.

JUREČKOVÁ, J. (1992b). Uniform asymptotic linearity of regression rank scores process. *Nonparametric Statistics and Related Topics (A.K. Md. E. Saleh, ed.),* pp. 217-228. *Elsevier Science Publishers.*

JUREČKOVÁ, J. (1993). Tests of linear hypotehsis based on regression rank scores. Preprint.

JUREČKOVÁ, J. and SEN, P.K. (1984). On adaptive scale-equivariant M-estimators in linear models. *Statistics & Decisions Supplement Issue* 1, 31-41.

KOENKER, R. and BASSETT, G. (1978). Regression quantiles. *Econometrica* **46**, 33-50.

KOENKER, R. and d'OREY, V. (1987). Computing regression quantiles. *Applied Statist.* **36**, 383-393.

KOENKER, R. and d'OREY, V. (1990). Remark on algorithm 229. Submitted.

MARONNA, R. and YOHAI, V.J. (1991). The breakdown point of simultaneous general M-estimates of regression and scale. *J. Amer. Statist. Assoc.* **86**, 699-716.

ROUSSEEUW, P.J. and CROUX, C. (1992). Explicit scale estimators with high breakdown point. L_1-Statistical Analysis and Related Methods (Y.Dodge, ed.), pp. 77-90. *Elsevier Science Publishers.*

RUPPERT, D. and CARROLL, R. J. (1980). Trimmed least squares estimation in the linear model. *J. Amer. Statist. Assoc.* **75**, 828-838.

ON AN ARGMAX-DISTRIBUTION CONNECTED TO THE POISSON PROCESS

Georg Ch. Pflug
Department of Statistics and OR
University of Vienna

Abstract

Let $\Pi_\gamma(t)$ be a Poisson process with parameter $\gamma < 1$. We determine the distribution of $\text{argmax}_{t\geq 0}\Pi_\gamma(t) - t$. This distribution appears as asymptotic distribution of maximum-likelihood estimates and and other stochastic optimizers. We also consider the distribution of $\text{argmax}_{t\geq 0} t - \Pi_\gamma(t)$ for $\gamma > 1$.
Keywords: Poisson process, argmax distribution, asymptotic distribution.
AMS 1991 Subject Classification: 62E15.

1 Introduction and main result

In this paper, we determine the distribution of the location of the maximum (the argmax) of $\Pi_\gamma(t) - t;\ \ t \geq 0$, where $\Pi_\gamma(t)$ is a Poisson process with Parameter $\gamma < 1$, i.e. $\mathbb{E}(\Pi_\gamma(t)) = \gamma t$.

The motivation for considering this problem stems from the asymptotic theory of statistical experiments and stochastic optimization programs. We give two illustrating examples here.

Example 1. Consider the location parameter family on $\mathbb{R}^1$ with densities $f(x-\theta)$ and suppose that f has a jump at zero:

$$\lim_{x\uparrow 0} f(x) = a > b = \lim_{x\downarrow 0} f(x).$$

It was shown in [2] (cf. also [3]), that if f is smooth enough for $x \neq 0$, then the log-likelihood process

$$\sum_{i=1}^{n} \log f(X_i + t/n) - \log f(X_i),$$

where (X_i) are i.i.d. according to f, converges weakly to the following process:

$$\begin{cases} (\log a - \log b)\Pi_b(t) - t(a-b) & \text{if } t \geq 0 \\ (\log b - \log a)\Pi_a(|t|) - t(a-b) & \text{if } t < 0 \end{cases} \tag{1}$$

where the two appearing Poisson processes are independent. The asymptotic distribution of the maximum likelihood estimate is the argmax distribution of the process (1). In this paper, we consider mainly the argmax-distribution of the positive side. The argmax for $t \le 0$ leads to the calculation of the distribution of $\text{argmax}_{t\ge 0} t - \Pi_\gamma(t)$ for $\gamma > 1$, which is more difficult. Some remarks concerning the latter case are contained in section 3.

Example 2. Consider the stochastic optimization program

$$(P) \quad \left\| \begin{array}{l} c_1 x + c_2 \mathbb{E}(\mathbf{1}_{\{\xi > x\}}) \to \min \\ x \ge a \end{array} \right. \tag{2}$$

An interpretation of (2) is for instance the following: An electricity company has to decide about the capacity $x \ge a$ of a transmission line. The construction costs are proportional to the capacity. If the random demand ξ exceeds the capacity, a breakdown occurs, which causes costs of c_2. Suppose, for simplicity, that $c_1 = c_2 = 1$. If $(\xi_i)_{i=1,\ldots,n}$ is a sequence of i.i.d. data, then the following approximation to (P) is called the empirical program:

$$(P_n) \quad \left\| \begin{array}{l} x + \frac{1}{n}\sum_{i=1}^n \mathbf{1}_{\{\xi_i > x\}} \to \min \\ x \ge a \end{array} \right. \tag{3}$$

Suppose that the density $g(u)$ of ξ is less than 1 for $u \ge a$. In this case, the optimal solution is $x = a$. It was shown in [4], that the limiting program of the sequence (P_n), expressed in local coordinates $t = x \cdot n$ is

$$(P_\infty) \quad \left\| \begin{array}{l} t - \Pi_{g(a)}(t) \to \min \\ x \ge 0 \end{array} \right. \tag{4}$$

which is exactly our problem. In particular, denoting by $\hat{X}_n$ the solution of (P_n), it follows that $n(\hat{X}_n - a)$ converges in distribution to $\text{argmin}_{t\ge 0} t - \Pi_{g(a)}(t)$.

The distribution of $\max_{t\ge 0}(\Pi_\gamma(t) - t)$ is known:

$$\mathbb{P}(\max_t(\Pi_\gamma(t) - t) \ge x) = \sum_{n > x} \frac{(x-n)^n}{n!}(\gamma e^{-\gamma})^n e^{\gamma x}(1-\gamma);$$

see Shorack-Wellner [5], Proposition 1 on page 392, where also some other related formulas may be found.

We start our investigation with the remark that the argmax must be a jump point of the Poisson process. Let (τ_i) be a sequence of i.i.d. Exponential(γ) variables and $S_i := \sum_{j=1}^i (1 - \tau_j)$. Notice that $\text{argmax}_{t\ge 0} \Pi_\gamma(t) - t = I^* - S_{I^*}$, where I^* is such that $S_{I^*} = \max(0, S_1, S_2, \ldots)$.

The distribution of $I^* - S_{I^*}$ can be found using random walk methods: Recall that the first index i for which $S_i \ge 0$ is called ladder index I and the pertaining S_i is called ladder height H (Feller [1]). These random variables are defective, since it may happen that $S_i < 0$ for all $i \ge 1$. In that case, we set $I = \infty$. It is known (see also (6) below) that $\mathbb{P}(I = \infty) = 1 - \gamma$. Since the Poisson process has independent increments,

$$I^* - S_{I^*} = \sum_{i=1}^{\nu} Z_i,$$

where (Z_i) is an i.i.d. sequence of non-defective copies of $I - H$ and ν is a random index counting how many times the defective variables $I - H$ were proper:

$$\mathbb{P}(\nu = i) = (1-\gamma)\gamma^i.$$

Introduce

$$R_i(u) = \mathbb{P}(S_1 < 0, \ldots, S_{i-1} < 0, S_i \leq u).$$

Notice that the jump height of the random walk (S_i) cannot exceed 1 and therefore $R_i(u) = R_i(1)$ for $u \geq 1$.

We will determine $R_i(u)$ for $0 \leq u \leq 1$. Let $T_i := \sum_{j=1}^{i} \tau_j = i - S_i$. The joint density of $(T_1, \ldots, T_n)$ is $\gamma^n e^{-\gamma t_n} \mathbf{1}_{\{0 \leq t_1 \leq \ldots \leq t_n\}}$. Therefore, for $0 \leq u \leq 1$,

$$\begin{aligned}
R_n(u) &= \mathbb{P}(T_1 > 1, T_2 > 2, \ldots, T_n \geq n-u) \\
&= \gamma^n \int_{n-u}^{\infty} \int_2^{\infty} \cdots \int_1^{\infty} \mathbf{1}_{\{t_1 \leq \ldots \leq t_n\}} e^{-\gamma t_n}\, dt_1\, dt_2 \ldots dt_n \\
&= \gamma^n \int_{n-u}^{\infty} \cdots \int_2^{\infty} [t_2 - 1] \mathbf{1}_{\{t_2 \leq \ldots \leq t_i\}} e^{-\gamma t_n}\, dt_2 \ldots dt_n = \\
&= \gamma^n \int_{n-u}^{\infty} \int_3^{\infty} \left[\frac{t_3^2}{2} - t_3\right] \mathbf{1}_{\{t_3 \leq \ldots \leq t_n\}} e^{-\gamma t_n}\, dt_3 \ldots dt_n = \ldots \\
&= \gamma^n \int_{n-u}^{\infty} \left[\frac{t_n^{n-1}}{(n-1)!} - \frac{t_n^{n-2}}{(n-2)!}\right] e^{-\gamma t_n}\, dt_n \qquad (5) \\
&= \gamma^n \left[\gamma^{-n} \sum_{k=0}^{n-1} \frac{(\gamma(n-u))^k}{k!} e^{-\gamma(n-u)} - \gamma^{1-n} \sum_{k=0}^{n-2} \frac{(\gamma(n-u)^k)}{k!} e^{-\gamma(n-u)}\right] \\
&= (1-\gamma) e^{-\gamma(n-u)} \sum_{k=0}^{n-2} \frac{(\gamma(n-u))^k}{k!} + e^{-\gamma(n-u)} \frac{(\gamma(n-u))^{n-1}}{(n-1)!}.
\end{aligned}$$

For $n = 1$ the formula reads

$$R_1(u) = e^{-\gamma(1-u)}.$$

Notice that

$$R_n(0) = R_{n+1}(1) = (1-\gamma) e^{-\gamma n} \sum_{k=0}^{n-2} \frac{(\gamma n)^k}{k!} + e^{-\gamma n} \frac{(\gamma n)^{n-1}}{(n-1)!}.$$

It is easily veryfied that $e^{-\gamma n} \sum_{k=0}^{n-2} \frac{(\gamma n)^k}{k!} \to 1$ and $e^{-\gamma n} \frac{(\gamma n)^{n-1}}{(n-1)!} \to 0$ as $n \to \infty$ and therefore

$$\begin{aligned}
\mathbb{P}(\operatorname{argmax}_{t \geq 0} \Pi_\gamma(t) - t = 0) &= \mathbb{P}(S_i < 0 \text{ for all i}) \\
&= \mathbb{P}(\Pi_\gamma(n) \leq n \quad \text{for all } n) \\
&= \lim_{n \to \infty} R_n(0) = 1 - \gamma \qquad (6)
\end{aligned}$$

which is in accordance with Shorack-Wellner [5], formula (4) on page 389.

The joint distribution of the ladder index I and the ladder height H is

$$\mathbb{P}(I = i, H \leq u) = R_i(u) - R_i(0), \quad 0 \leq u \leq i.$$

These random variables are defective, the defect is given by $1 - \gamma$.

It is remarkable that the marginal distribution of H, conditioned on $\{I < \infty\}$ is uniform on [0,1], which may be proved as follows: By differentiation we get from (5) that the marginal density of H is

$$\sum_{n=1}^{\infty} \frac{d}{du} R_n(u) = \sum_{n=1}^{\infty} \gamma^n e^{-\gamma(n-u)} \frac{(n-u)^{n-1}}{(n-1)!} - \sum_{n=2}^{\infty} \gamma^n e^{-\gamma(n-u)} \frac{(n-u)^{n-2}}{(n-2!} = \gamma. \quad (7)$$

Here we have used the identity

$$\sum_{n=k}^{\infty} \frac{(n-u)^{n-k}}{(n-k)!} \gamma^n e^{-\gamma n} = \gamma^k e^{-\gamma u}/(1-\gamma),$$

which is proved in Shorack-Wellner [5], Lemma 2 on page 397. To get the density conditioned on $\{I < \infty\}$ we have to divide (7) by γ, yielding the density 1 on [0,1].

The random variable $I - H$, which we need, has distribution

$$\begin{aligned} \mathbb{P}(I - H \leq u) &= \sum_{k=1}^{\lfloor u \rfloor} \mathbb{P}(I = k) + \mathbb{P}(I = \lceil u \rceil, H \geq \lceil u \rceil - u) \\ &= 1 - R_{\lfloor u \rfloor}(0) + R_{\lceil u \rceil}(1) - R_{\lceil u \rceil}(\lceil u \rceil - u) = 1 - R_{\lceil u \rceil}(\lceil u \rceil - u). \end{aligned}$$

Here we have used the symbols $\lfloor u \rfloor$ resp. $\lceil u \rceil$ for the floor (entire part) resp. the ceiling of the real number u, if u is integer we set $\lfloor u \rfloor = u$ and $\lceil u \rceil = u + 1$. $I - H$ is clearly also defective: $\mathbb{P}(I - H < \infty) = \gamma$. Let the pertaining not-defective distribution be

$$F(u) = \frac{1}{\gamma} \mathbb{P}(I - H \leq u) = \frac{1}{\gamma}[1 - R_{\lceil u \rceil}(\lceil u \rceil - u)].$$

Again by differentiation, using (5), we get its density

$$f(u) = e^{-\gamma u} \mathbf{1}_{\{0 \leq u < 1\}} + e^{-\gamma u} \sum_{n=2}^{\infty} \gamma^{n-1} \left[\frac{u^{n-1}}{(n-1)!} - \frac{u^{n-2}}{(n-2)!} \right] \mathbf{1}_{\{n-1 \leq u < n\}}.$$

The graph of this density (for $\gamma = 0.6$) is shown below:

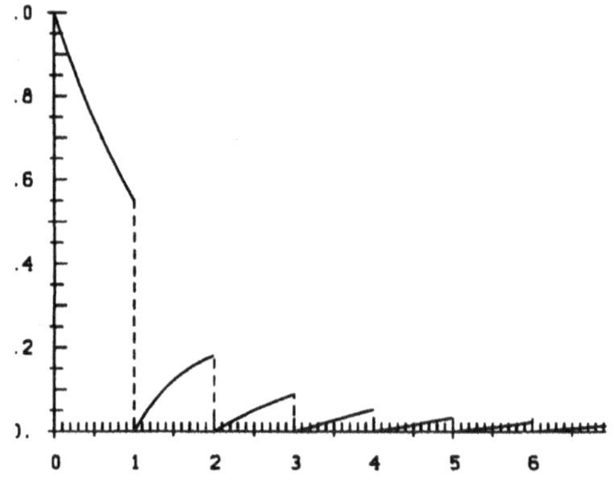

Finally, as already remarked, the distribution of $\operatorname{argmax}_{t\geq 0} \Pi(t) - t$ coincides with the distribution of the random sum

$$\sum_{k=1}^{\nu} Z_i,$$

where (Z_i) is an i.i.d. sequence with common distribution F and ν is a geometric random variable, independent of the $Z_i's$:

$$\mathbb{P}(\nu = i) = (1-\gamma)\gamma^i \qquad i = 0, 1, \ldots$$

($\sum_{i=1}^{0} Z_i$ is set to zero).

2 An upper bound

In this section, we show that the tails of F decrease exponentially fast. This implies that the argmin distribution *possesses moments of all orders*. Moreover, the bound may be used to construct confidence regions.

We start from the remark, that the distribution of $I - H$ is stochastically smaller than that of I alone. Notice that

$$\begin{aligned}
&\mathbb{P}(I > n | I < \infty) = 1 - \frac{1 - R_n(0)}{\gamma} \\
= \;& 1 - \frac{1}{\gamma}\left[e^{-\gamma n}\sum_{k=0}^{\infty}\frac{(\gamma n)^k}{k!} - e^{-\gamma n}\sum_{k=0}^{n-2}\frac{(\gamma n)^k}{k!} + \gamma e^{-\gamma n}\sum_{k=0}^{n-2}\frac{(\gamma n)^k}{k!} - e^{-\gamma n}\frac{(\gamma n)^{n-1}}{(n-1)!}\right] \\
= \;& e^{-\gamma n}\sum_{k=0}^{\infty}\frac{(\gamma n)^k}{k!} - \frac{1}{\gamma}\left[e^{-\gamma n}\sum_{k=n-1}^{\infty}\frac{(\gamma n)^k}{k!} + \gamma e^{-\gamma n}\sum_{k=0}^{n-2}\frac{(\gamma n)^k}{k!} - e^{-\gamma n}\frac{(\gamma n)^{n-1}}{(n-1)!}\right] \\
= \;& (1 - \frac{1}{\gamma})e^{-\gamma n}\sum_{k=n-1}^{\infty}\frac{(\gamma n)^k}{k!} + \frac{1}{\gamma}e^{-\gamma n}\frac{(\gamma n)^{n-1}}{(n-1)!} \\
= \;& e^{-\gamma n}\sum_{k=n-1}^{\infty}\frac{(\gamma n)^k}{k!} - \frac{1}{\gamma}e^{-\gamma n}\sum_{k=n}^{\infty}\frac{(\gamma n)^k}{k!} \\
= \;& e^{-\gamma n}\sum_{k=n}^{\infty}\gamma^{k-1}n^{k-1}\frac{k-n}{k!} = e^{-\gamma n}\frac{\gamma^{n-1}n^{n-1}}{n!}\sum_{l=0}^{\infty} l\gamma^l \frac{n^l}{(n+1)\cdots(n+l)} \\
\leq \;& e^{-\gamma n}\frac{\gamma^{n-1}n^{n-1}}{n!}\sum_{l=0}^{\infty} l\gamma^l = e^{-\gamma n}\frac{\gamma^{n-1}n^{n-1}}{n!}\frac{\gamma}{(1-\gamma)^2} \\
\leq \;& \frac{e^{-\gamma n}\gamma^n e^n}{\sqrt{2\pi}n^{3/2}(1-\gamma)^2} \leq Ce^{-\delta n}
\end{aligned}$$

where $\delta = \gamma - \log\gamma - 1 > 0$ and $C = (1-\gamma)^{-2}(2\pi)^{-1/2}$.

Therefore I and hence $I - H$ is stochastically smaller than Y, where $\mathbb{P}(Y > n) = Ce^{-\delta n}$, i.e.

$$\mathbb{P}(Y = n) = \begin{cases} 0 & n < n_0 \\ 1 - Ce^{-\delta n} & n = n_0 \\ Ce^{-\delta n}(e^{\delta} - 1) & n > n_0 \end{cases}$$

with $n_0 = \lceil (\log C)/\delta \rceil$. Consequently, the argmax distribution is stochastically smaller than

$$\sum_{i=1}^{\nu} Y_i,$$

where (Y_i) are i.i.d. copies of Y.

3 Final remarks

Finally, we consider the problem of determining $\operatorname{argmax}_{t\geq 0} t - \Pi_\gamma(t)$, where $\gamma > 1$. Let $(\tau_i)_{i\geq 0}$ be a sequence of i.i.d. Exponential(γ) variables and $-S_i = \sum_{j=1}^{i}(\tau_i - 1) = T_i - i$. Let further

$$\bar{R}_n(u) = \mathbb{P}(S_1 > 0, S_2 > 0, \ldots, S_n \geq -u) = \mathbb{P}(T_1 < 1, T_2 < 2, \ldots, T_n < n+u). \quad (8)$$

Introduce the following constants

$$\begin{aligned} d_n &= \int_0^1 \int_{t_1}^2 \cdots \int_{t_{n-1}}^n dt_n \ldots dt_1 \\ &= \int_0^n \int_0^{\min(t_n, n-1)} \cdots \int_0^{\min(t_1,1)} dt_1 \ldots dt_n. \end{aligned}$$

Unfortunately, the author was not able to find a nice expression for d_n. Calculations show that $d_1 = 1, d_2 = 3/2, d_3 = 8/3, d_4 = 125/24, \ldots$ The recursive structure is evident: $d_n = \int_0^n h_n \, du$, where $h_{n+1}(u) = \int_0^{\min(u,n)} h_n(v)\, dv$ and $h_1(u) = 1$. Now,

$$\begin{aligned} \bar{R}_n(u) &= \gamma^n \int_0^1 \int_{t_1}^2 \cdots \int_{t_{n-1}}^{n+u} e^{-\gamma t_n} \, dt_n \ldots dt_1 \\ &= \gamma^{n-1} \int_0^1 \int_{t_1}^2 \cdots \int_{t_{n-2}}^{n-1} \left(e^{-\gamma t_{n-1}} - e^{-\gamma(n+u)} \right) dt_{n-1} \ldots dt_1 \\ &= \bar{R}_{n-1}(0) - \gamma^{n-1} e^{-\gamma(n+u)} d_{n-1}. \end{aligned}$$

With

$$\bar{R}_1(u) = 1 - \exp^{-\gamma(1+u)}$$

we get

$$\bar{R}_n(u) = 1 - \sum_{i=1}^{n-1} \gamma^{i-1} e^{-\gamma i} d_{i-1} - \gamma^{n-1} e^{-\gamma u} d_{n-1}.$$

Let

$$\eta(\gamma) = \mathbb{P}(S_i > 0 \quad \text{for all } i) = \lim_{n\to\infty} \bar{R}_n(0) = 1 - \sum_{i=1}^{\infty} \gamma^{i-1} e^{-\gamma i} d_{i-1}.$$

(The limit exists, since $0 \leq \bar{R}_{n+1}(0) \leq \bar{R}_n(0) \leq 1$). Let I resp. H be the ladder index resp. the ladder height of the random walk $(-S_i)$. The random variable is defective, the defect is given by $\eta(\gamma)$. The distribution of $I + H$, which we need, is

$$\bar{F}(u) = \mathbb{P}(I + H \leq u) = \sum_{i=1}^{\lfloor u \rfloor} \mathbb{P}(0 \leq H \leq u - i, I = i) = \sum_{i=1}^{\lfloor u \rfloor} [\bar{R}_i(u-i) - \bar{R}_i(0)].$$

The smallest local maximizer of $t - \Pi_\gamma(t)$ is the first jump point of the Poisson process τ_0, and with probability $\eta(\gamma)$, this is the global maximizer. Therefore the desired argmax distribution is the distribution of

$$\tau_0 + \sum_{i=1}^{\nu} \bar{Z}_i,$$

where τ_0 is Exponential(γ), $\bar{Z}_i$ are i.i.d. with distribution $\bar{F}$ and ν is geometric with parameter $(1 - \eta(\gamma))$.

Acknowledgement: The author is indebted to A. Futschik for his help in carrying out some integrals appearing in this paper.

References

[1] Feller W. An Introduction to Probability Theory and Its Applications. Vol. I, J. Wiley & Sons, 1986 (Third ed.)

[2] Ibragimov, I.A., Hasminski R.Z. Statistical estimation: Asymptotic Theory. Appl. of Math. No 16, Springer Verlag, 1981

[3] Pflug G. Ch. The Limiting Log-Lokelihood Process for Discontinuous Density Families. Z. Wahrscheinlichkeitstheorie verw. Geb. 64, 15-35 (1983)

[4] Pflug G. Ch. Asymptotic stochastic programs. Preprint, Univ. of Vienna (submitted)

[5] Shorack G.R., Wellner J.A. Empirical Processes with Applications to Statistics. J. Wiley & Sons, 1986

ASYMPTOTIC NORMALITY OF REGRESSION M-ESTIMATORS: HADAMARD DIFFERENTIABILITY APPROACHES

BY JIAN-JIAN REN and PRANAB KUMAR SEN
University of Nebraska-Lincoln and University of North Carolina at Chapel Hill

ABSTRACT. Incorporating an alternative derivation of the Jurečková-uniform asymptotic linearity of M-statistics (in regression parameters) based on Hadamard differentiability approaches, asymptotic normality and some other related properties of a general class of M-estimations of regression parameters are studied. A comparative study of the basic regularity conditions pertaining to alternative methods is also made along with.

KEY WORDS: asymptotic normality, Hadamard differentiability, linear models, M-estimation, uniform asymptotic linearity.
AMS 1980 SUBJECT CLASSIFICATION: 62G05, 62G99, 62J05.

1. INTRODUCTION

The linear regression model is one of the most widely used tools in statistical analysis. Due to lack of robustness of the classical least square estimators, other 'robust' estimators have been proposed. These include the Huber (1973) M-estimators of regression. Consider the simple linear regression model:

$$X_i = \underset{\sim}{c}_i^T \underset{\sim}{\beta} + e_i, \qquad i \geq 1 \tag{1.1}$$

where the $\underset{\sim}{c}_i$ are known p-vectors of regression constants, $\underset{\sim}{\beta}=(\beta_1, \cdots, \beta_p)^T$ is the vector of unknown (regression) parameters, $p \geq 1$, and is to be estimated from n observations $X_1, \cdots, X_n$, and e_i are independent and identically distributed random variables (i.i.d.r.v.) with distribution function (d.f.) F. Then the M-estimators of $\underset{\sim}{\beta}$ are defined through minimizing a sum of a less rapidly increasing function ρ of the residuals, i.e.,

$$\sum_{i=1}^n \rho(X_i - \underset{\sim}{c}_i^T \underset{\sim}{\theta}) = \min! \tag{1.2}$$

or, if ρ is convex with derivative ψ, the *M-estimators of Regression* $\hat{\underset{\sim}{\beta}}_n$ of $\underset{\sim}{\beta}$ is equivalently defined by the resulting estimators of the equations

(1.3) $$\sum_{i=1}^{n} \underset{\sim}{c}_i\, \psi(X_i - \underset{\sim}{c}_i^T \underset{\sim}{\theta}) = \underset{\sim}{0},$$

and ψ is called the *score function* of the M-estimators.

To study the asymptotic properties of the M-estimators of regression, various procedures have been considered by various people. Huber (1973) proved the asymptotic normality of the M-estimators of regression, assuming that the score function ψ is continuous, nondecreasing and bounded with a bounded second derivative. Koul (1977) and Jurečková (1977) also proved the asymptotic normality under weaker conditions on ψ, but they both required that the underlying d.f. F has a finite Fisher's information, i.e., $0 < I(f) = \int (f'/f)^2 dF < \infty$, where $F' = f$ is the density function of F. As a corollary of Theorem 4.1 of Bickel (1975), Yohai and Maronna (1979) obtained that asymptotic normality under certain assumptions on ψ and F, one of which is given as below:

(1.4) $$\frac{\psi(u+z)-\psi(u)}{z} \geq d, \qquad \text{if } |u| \leq c,\ |z| \leq b$$

where b, c and d are positive real numbers satisfying $q = F(c) - F(-c) > 0$. For non-convex ρ, Jurečková (1989) derived the asymptotic normality and the weak consistency of the M-estimators in linear models, assuming the existence of the second derivative of F.

However, there exists a class of score functions ψ and error distributions F which do not satisfy the conditions imposed by above authors for the asymptotic normality of regression M-estimators. For instance, if we choose a smooth ψ which is truncated outside of a finite interval, say,

$$\psi_d(x) = \begin{cases} \operatorname{sgn}\{x\}\, x^2, & \text{if } |x| < d \\ \operatorname{sgn}\{x\}\, d^2, & \text{if } |x| \geq d \end{cases}$$

where $d > 0$, and if the error distribution is the mixture of double gamma distributions, say, $F(x) = \lambda F_1(x) + (1-\lambda)F_2(x)$, where $0 < \lambda < 1$ and F_i is the double gamma d.f. with p.d.f.: $f_i(x) = (\beta_i^{\alpha_i}/2\Gamma(\alpha_i))|x|^{\alpha_i - 1} e^{-\beta_i |x|}$, $\alpha_i > 0$, $\beta_i > 0$, i=1, 2, then ψ_d does not have first derivative at $x = \pm d$ and the condition (1.4) does not hold for ψ_d, and moreover f does not have derivative at $x = 0$ and $I(f) = \infty$ for certain α_i and β_i (for example: when $\beta_1 = \beta_2 = 1$, $\alpha_2 = 1$, $1 < \alpha_1 \leq 3/2$). The interest of our current research is to study the asymptotic normality of the M-estimators of regression under weak conditions on the score function ψ and the underlying d.f. F. Our conditions will include the class of score functions ψ and error distributions F mentioned above as a special case.

Since M-estimators are implicitly defined by the estimating equations, it is difficult to study their statistical properties directly. The principle difficulty is the non-linearity of the estimator. The approach presented here is using Hadamard differentiability through the Jurečková- (1971, 1977) uniform asymptotic linearity to study the asymptotic properties of M-estimators of regression. More specifically, setting $Y_i=X_i-\beta^T c_i$ (i.i.d.r.v.'s with d.f. F), and setting the normalized version of the estimating equations in (1.3) as below:

(1.5) $$M_n(u) = \Sigma_{i=1}^n c_{ni}\, \psi(Y_i - c_{ni}^T u)$$

where $C_n=\Sigma_{i=1}^n c_i c_i^T=(r_{nij})_{1\le i,j\le p}$, $C_n^o=\mathrm{Diag}(\sqrt{r_{n11}},\cdots,\sqrt{r_{npp}})$, $u=C_n^o(\theta-\beta)$ and $c_{ni}=(C_n^o)^{-1}c_i=(c_{ni1},\cdots,c_{nip})^T$, $1\le i\le n$, then the following type of (Jurečková-) uniform asymptotic linearity for M-processes:

(1.6) $$\sup_{|u|\le K}\left|M_n(u) - M_n(0) + \gamma_o Q_n u\right| \xrightarrow{P} 0, \qquad \text{as } n\to\infty$$

where K is any finite positive real number, $|\cdot|$ stands for the uniform norm of space R^p, $\gamma_o=\int F'(x)d\psi(x) > 0$ and $Q_n=\Sigma_{i=1}^n c_{ni}c_{ni}^T=(C_n^o)^{-1}C_n(C_n^o)^{-1}$, provides an easy access to the study of the asymptotic properties of M-estimators. Since $M_n(u)$ is a linear functional of the empirical function

(1.7) $$S_n^*(t,u)=\Sigma_{i=1}^n c_{ni} I(Y_i\le F^{-1}(t)+c_{ni}^T u), \qquad t\in[0,1],\ u\in R^p$$

viz.,

$$M_n(u) = \int \psi(F^{-1}(t))\, dS_n^*(t,u),$$

and since the Hadamard derivative is also a linear functional, $M_n(u)$ could be the Hadamard derivative of some appropriate functional τ. The choice of τ may not be unique, and general motivations for this are given in Ren and Sen (1991). We may refer to (3.18) where we shall use a specific τ to facilitate our manipulations. Ren and Sen (1991) showed that if a functional τ is Hadamard differentiable with derivative $M_n(u)$, then

(1.8) $$\sup_{|u|\le K}\left|D_p\{\tau(S_n^*(\cdot,u))-\tau(S_n^*(\cdot,0))-[M_n(u) - M_n(0)]\}\right| \xrightarrow{P} 0,$$

where $D_p = \mathrm{Diag}(\Sigma_{i=1}^n c_{ni1},\cdots,\Sigma_{i=1}^n c_{nip})$ and

$$\tau(S_n^*(\cdot,u)) = \left(\tau\left(\frac{S_{n1}^*(\cdot,u)}{\Sigma_{i=1}^n c_{ni1}}\right),\cdots,\tau\left(\frac{S_{np}^*(\cdot,u)}{\Sigma_{i=1}^n c_{nip}}\right)\right)^T.$$

Therefore (1.6) follows from showing

$$\sup_{|u|\le K}\left|D_p\{\tau(S_n^*(\cdot,u))-\tau(S_n^*(\cdot,0))\} + \gamma_o Q_n u\right| \xrightarrow{P} 0.$$

Our results show that Hadamard differentiability approach provides a good method for the study of the asymptotic properties of M-estimators of regression through the linear approximation of the estimators. Compared with the method of using the convexity property of the criterion function (viz., Pollard, 1991), our method does not depend explicitly on the convexity property. In our approach, Hadamard differentiability (or compact differentiability) properties allows us to derive the asymptotic approximation of the estimating equations $\underset{\sim}{M}_n(\underset{\sim}{u})$ uniformly over a compact set [see (1.8)]. This allows us to bypass the Convexity Lemma (Pollard, 1991) and the concordance-disconcordance condition, which wes used by Jurečková (1977) to establish (1.6).

Some notations along with basic assumptions are presented in Section 2. The asymptotic normality of the M-estimators of regression is considered in Section 3, where none of the assumptions, such as bounded second derivative on ψ, or finite Fisher's information on F, or (1.4), or any existence of the second derivative of F, is required. Some lemmas and theorems needed for the proof of our main results in Section 3 are given in Section 4.

2. PRELIMINARIES

First, we notice that (1.3) is equivalent to

(2.1) $$\underset{\sim}{M}_n(\underset{\sim}{u}) = \underset{\sim}{0} \qquad \text{(with respect to } \underset{\sim}{u}),$$

i.e., $\hat{\underset{\sim}{u}}_n=\underset{\sim}{C}_n^o(\hat{\underset{\sim}{\beta}}_n-\underset{\sim}{\beta})$ is a solution of (2.1). We denote the k-th component of $\underset{\sim}{M}_n(\underset{\sim}{u})$ as

(2.2) $$M_{nk}(t,\underset{\sim}{u})=\Sigma_{i=1}^n c_{nik}\psi(Y_i-\underset{\sim}{c}_{ni}^T\underset{\sim}{u}),$$

the k-th component of $\underset{\sim}{S}_n^*(t,\underset{\sim}{u}))$ as

(2.3) $$S_{nk}^*(t,\underset{\sim}{u})=\Sigma_{i=1}^n c_{nik}I(Y_i\leq F^{-1}(t)+\underset{\sim}{c}_{ni}^T\underset{\sim}{u}),$$

and the expected value of $S_{nk}^*(t,\underset{\sim}{u})$ as

(2.4) $$S_{nk}(t,\underset{\sim}{u})=\Sigma_{i=1}^n c_{nik}F(F^{-1}(t)+\underset{\sim}{c}_{ni}^T\underset{\sim}{u}).$$

We also write for every $1\leq i\leq n$, $1\leq k\leq p$, $t\in[0,1]$ and $\underset{\sim}{u}\in R^p$,

(2.5) $$c_{nik}=c_{nik}^+-c_{nik}^-;\quad c_{nik}^+=\max\{0,\ c_{nik}\},\ c_{nik}^-=-\min\{0,\ c_{nik}\};$$

(2.6) $$\underset{\sim}{c}_{ni}=\underset{\sim}{c}_{ni}^+-\underset{\sim}{c}_{ni}^-;\quad \underset{\sim}{c}_{ni}^+=(c_{ni1}^+,\cdots,c_{nip}^+)^T,\ \underset{\sim}{c}_{ni}^-=(c_{ni1}^-,\cdots,c_{nip}^-)^T;$$

(2.7) $$(d_{nk}^+)^2=\Sigma_{i=1}^n(c_{nik}^+)^2,\qquad (d_{nk}^-)^2=\Sigma_{i=1}^n(c_{nik}^-)^2;$$

(2.8) $$\bar{c}_{nik}^+=c_{nik}^+/d_{nk}^+,\ \text{if } d_{nk}^+>0;\ 0,\ \text{if } d_{nk}^+=0;$$

(2.9) $$\bar{c}_{nik}^-=c_{nik}^-/d_{nk}^-,\ \text{if } d_{nk}^->0;\ 0,\ \text{if } d_{nk}^-=0;$$

(2.10) $$\bar{\underset{\sim}{c}}_{ni}^+=(\bar{c}_{ni1}^+,\cdots,\bar{c}_{nip}^+)^T,\qquad \bar{\underset{\sim}{c}}_{ni}^-=(\bar{c}_{ni1}^-,\cdots,\bar{c}_{nip}^-)^T;$$

(2.11) $S_{nk}^{*+}(t,\underset{\sim}{u})=\sum_{i=1}^{n} c_{nik}^{+} I(Y_i \le F^{-1}(t)+\underset{\sim}{c}_{ni}^{T}\underset{\sim}{u}), \quad S_{nk}^{+}(t,\underset{\sim}{u})=E\{S_{nk}^{*+}(t,\underset{\sim}{u})\};$

(2.12) $S_{nk}^{*-}(t,\underset{\sim}{u})=\sum_{i=1}^{n} c_{nik}^{-} I(Y_i \le F^{-1}(t)+\underset{\sim}{c}_{ni}^{T}\underset{\sim}{u}), \quad S_{nk}^{-}(t,\underset{\sim}{u})=E\{S_{nk}^{*-}(t,\underset{\sim}{u})\};$

so that

(2.13) $S_{nk}^{*}(t,\underset{\sim}{u}) = S_{nk}^{*+}(t,\underset{\sim}{u}) - S_{nk}^{*-}(t,\underset{\sim}{u}), \quad S_{nk}(t,\underset{\sim}{u}) = S_{nk}^{+}(t,\underset{\sim}{u}) - S_{nk}^{-}(t,\underset{\sim}{u}).$

In this paper, we will always consider the D[0,1] space (of right continuous real valued functions with left hand limits) is endowed with the uniform topology. Then, the space C[0,1] of real valued continuous functions, endowed with the uniform topology, is a subspace of D[0,1]. It is easy to see that, for every $\underset{\sim}{u} \in R^p$ and $1 \le k \le p$, $S_{nk}^{*}(\cdot, \underset{\sim}{u})$ is an element of D[0,1]. A convention which we will follow through out the whole paper is: a function f: $R \to R$ is *nondecreasing* if $f(x) \le f(y)$ for $x \le y$, and is *increasing* if $f(x) < f(y)$ for $x < y$. Analogously, for *nonincreasing* and *decreasing*.

Let $\|\cdot\|$ stands for the Euclidean norm. As the following, we present a set of interacting conditions, subsets of which will be incorporated in the main results.

(A1) $c_{nik} \ge 0, \qquad i=1, 2, \dots, n$

(A2) $\overline{\lim}_{n\to\infty} \max_{1 \le i \le n} \|\underset{\sim}{c}_{ni}\|^2 = 0.$

(A3) $\overline{\lim}_{n\to\infty} \max_{1 \le i \le n} \|\underset{\sim}{c}_{ni}^{+}\|^2 = 0, \quad \overline{\lim}_{n\to\infty} \max_{1 \le i \le n} \|\underset{\sim}{c}_{ni}^{-}\|^2 = 0.$

(A4) There exists a positive definite p×p matrix $\underset{\sim}{Q}$ such that $\lim_{n\to\infty} \underset{\sim}{Q}_n=\underset{\sim}{Q}$.

(B) F is absolutely continuous with a derivative F′ which is positive and continuous with limits at $\pm\infty$.

(C1) ψ is a nondecreasing function with a range including positive and negative real numbers.

(C2) $\int \psi \, dF = 0.$

(C3) $0 < \gamma_0 = \int F' \, d\psi < \infty.$

(C4) $0 < \sigma^2 = \int \psi^2 \, dF < \infty.$

(C5) ψ is right continuous and bounded.

REMARK 1. It can be easily checked that for the class of score functions ψ and error distributions F mentioned in Section 1, our conditions (B) and (C) hold.

REMARK 2. It is easy to see that our conditions (A) on $\underset{\sim}{c}_i$ are weaker than those required by Jurečková (1977). The concordance conditions on the $\underset{\sim}{c}_i$ are avoided in our Hadamard differentiability approach.

We should notice that (B) implies that F′ is bounded and uniformly continuous, and that, for any fixed k=1, ⋯ , p, (A1) and (A2) imply that

(2.14) $$\sum_{i=1}^{n} c_{nik} \to \infty, \qquad \text{as } n\to\infty,$$

and (A2) implies

(2.15) $$a_n = \max_{1\le j\le p}\ \max_{1\le i\le n} |c_{nij}| \to 0, \qquad \text{as } n\to\infty.$$

We also should notice that (A3) implies (A2). They are listed separately because we establish (1.6) for the simpler case of (A1) and (A2) first, then generalize it to the case of (A3).

REMARK 3. Note that (C1) does not imply that ρ in (1.2) is convex. We will assume that $\rho' = \psi$ for some function ρ in (1.2) throughout this paper.

REMARK 4. If (C2) and (C3) are replaced by the condition: $\psi = \psi_1 - \psi_2$, where ψ_1 and ψ_2 are two continuous monotone functions, we still have (1.6) in our Theorem 3.3. From Theorem 6.4.2 of Ortega and Rheinboldt (1970) and from (1.6), we can easily show that $\hat{\underset{\sim}{u}}_n = \underset{\sim}{C}_n^o(\hat{\underset{\sim}{\beta}}_n - \underset{\sim}{\beta})$ is bounded in probability using a proof similar to the one of Jurečková (1989). Therefore, the asymptotic normality of $\hat{\underset{\sim}{\beta}}_n$ in our Theorem 3.3 still holds. Such a case of ψ arises with redescending score functions considered in detail by Hampel, Ronchetti, Rousseeuw and Stahel (1986).

3. ASYMPTOTIC NORMALITY

LEMMA 3.1. Assume (A2), (A4), (C2) and (C4). Then,

(3.1) $$\underset{\sim}{M}_n(\underset{\sim}{0}) \xrightarrow{\mathfrak{D}} N_p(\underset{\sim}{0}, \sigma^2 \underset{\sim}{Q}), \qquad \text{as } n \to \infty.$$

If in addition assume (C1) and (C3), and assume, for any K>0, we have

(3.2) $$\sup_{|\underset{\sim}{u}|\le K} \left|\underset{\sim}{M}_n(\underset{\sim}{u}) - \underset{\sim}{M}_n(\underset{\sim}{0}) + \gamma_o \underset{\sim}{Q}_n \underset{\sim}{u}\right| \xrightarrow{P} 0, \qquad \text{as } n\to\infty,$$

then, for large n, $\hat{\underset{\sim}{u}}_n$ exists as a solution of the following equations:

(3.3) $$\underset{\sim}{M}_n(\underset{\sim}{u}) = \underset{\sim}{0} \qquad \text{(with respect to } \underset{\sim}{u}),$$

and

(3.4) $$|\ \underset{\sim}{Q}_n\, \hat{\underset{\sim}{u}}_n - \gamma_o^{-1} \underset{\sim}{M}_n(\underset{\sim}{0})\ | \xrightarrow{P} 0, \qquad \text{as } n\to\infty.$$

Therefore, as $n\to\infty$

(3.5) $$\hat{\underset{\sim}{u}}_n = \underset{\sim}{C}_n^o\ (\ \hat{\underset{\sim}{\beta}}_n - \underset{\sim}{\beta}\ ) \xrightarrow{\mathfrak{D}} N_p(\underset{\sim}{0},\ \sigma^2\gamma_o^{-2}\underset{\sim}{Q}^{-1}).$$

Proof. Recall (1.5), we know

$$\underset{\sim}{M}_n(\underset{\sim}{0}) = \Sigma_{i=1}^n\ \underset{\sim}{c}_{ni}\ \psi(Y_i),$$

where Y_i are i.i.d.r.v's with d.f. F. To prove (3.1), it suffices to show that, for an arbitrary non-zero vector $\underset{\sim}{a}=(a_1, \cdots, a_p)^\top \in R^p$,

(3.6) $$\underset{\sim}{a}^\top \underset{\sim}{M}_n(\underset{\sim}{0}) \xrightarrow{\mathfrak{D}} N(0, \sigma^2 \underset{\sim}{a}^\top \underset{\sim}{Q} \underset{\sim}{a}), \qquad \text{as } n\to\infty.$$

Since

$$\underset{\sim}{a}^\top \underset{\sim}{M}_n(\underset{\sim}{0}) = \Sigma_{i=1}^n\ \underset{\sim}{a}^\top\ \underset{\sim}{c}_{ni}\psi(Y_i),$$

then (3.6) follows from (2.15), (A4) and the theorem of Hájek and Šidák (1967, page 153).

To see the existence of $\hat{u}_n$, we first notice that (C1) implies that there exist α and β such that

$$\psi(\alpha)<0, \quad \text{and} \quad \psi(\beta)>0; \quad \text{for } \alpha<\beta. \tag{3.7}$$

So, for $x<\alpha$ and $y>\beta$, we have

$$\frac{\rho(x)-\rho(\alpha)}{x-\alpha} \le \psi(\alpha) < 0 \quad \text{and} \quad \frac{\rho(y)-\rho(\beta)}{y-\beta} \ge \psi(\beta) > 0.$$

Therefore,

$$\lim_{x\to\pm\infty} \rho(x) = +\infty, \tag{3.8}$$

and ρ has a finite lower bound. Let A_n be a pxp non-singular matrix such that $A_n^T Q_n A_n$ is a diagonal matrix which, by (A4), only has positive eigenvalues for large enough n. Consider a function Φ_n defined as below:

$$\Phi_n(v) = \sum_{i=1}^n \rho(Y_i - c_{ni}^T A_n v), \qquad n\ge 1,\ v\in R^p. \tag{3.9}$$

Then, for any $v\in R^p$, we have, as $\|v\| \to \infty$

$$\sum_{i=1}^n (c_{ni}^T A_n v)^2 = \sum_{i=1}^n v^T A_n^T c_{ni}\, c_{ni}^T A_n v = v^T A_n^T\Big(\sum_{i=1}^n c_{ni}\, c_{ni}^T\Big) A_n v$$
$$= v^T A_n^T Q_n A_n v \ge \lambda_1(Q_n)\|v\|^2 \to \infty,$$

where $\lambda_1(A)$ denotes the smallest eigenvalue of the matrix A. Hence, by (3.8), for any large enough n,

$$\Phi_n(v) \to \infty, \qquad \text{as} \quad \|v\| \to \infty. \tag{3.10}$$

Since ρ is continuous, Φ_n is also a continuous function for $n\ge 1$. Therefore, Φ_n has at least one global minimum, i.e., by the monotone property of ψ, (1.2), equivalent to (1.3), has a finite solution. Therefore, from the equivalent relation between (1.3) and (2.1), (3.3) has a solution for large enough n.

From (3.1), for any $\epsilon>0$, there exist $K_o>0$ and N such that

$$P\{|M_n(0)|>K_o\} < \epsilon, \qquad \text{for } n\ge N. \tag{3.11}$$

Since (3.3) has a solution $\hat{u}_n$ for large n, we have, for arbitrary $\rho>0$ and $K>0$,

$$P\{|\gamma_o Q_n \hat{u}_n - M_n(0)|>\rho\}$$
$$=P\{|\gamma_o Q_n \hat{u}_n - M_n(0)|>\rho,\ \|\hat{u}_n\|\le K\} + P\{|\gamma_o Q_n \hat{u}_n - M_n(0)|>\rho,\ \|\hat{u}_n\|>K\}$$
$$\le P\{\sup_{|u|\le K} \big|M_n(u)-M_n(0)+\gamma_o Q_n u\big|\ge\rho\} + P\{\|\hat{u}_n\|>K\}. \tag{3.12}$$

Consider a function given by

$$h(u, B) = u^T M_n(uB) = \sum_{i=1}^n u^T c_{ni} \psi(Y_i - c_{ni}^T uB), \qquad u\in R^p,\ B\in(0, +\infty). \tag{3.13}$$

Since ψ is nondecreasing, $h(u,B)$ is nonincreasing in B for any $u\in R^p$. Hence, we claim that $\|\hat{u}_n\|>K$ implies that

(3.14) $$h(\underset{\sim}{e}_n,\, K)\geq 0,$$

where $\underset{\sim}{e}_n=\|\hat{\underset{\sim}{u}}_n\|^{-1}\hat{\underset{\sim}{u}}_n\in R^p$. Otherwise, $h(\underset{\sim}{e}_n,\, K)<0$ implies

(3.15) $$h(\underset{\sim}{e}_n,\, \|\hat{\underset{\sim}{u}}_n\|) \leq h(\underset{\sim}{e}_n,\, K) < 0,$$

which, by the definition of $\hat{\underset{\sim}{u}}_n$ and (3.13), contradicts

$$h(\underset{\sim}{e}_n,\, \|\hat{\underset{\sim}{u}}_n\|) = \underset{\sim}{e}_n^T\underset{\sim}{M}_n(\hat{\underset{\sim}{u}}_n) = 0.$$

Therefore, by (3.11), for $n\geq N$,

$$\begin{aligned} P\{\|\hat{\underset{\sim}{u}}_n\|>K\} &\leq P\{\underset{\sim}{e}_n^T\underset{\sim}{M}_n(\underset{\sim}{e}_nK)\geq 0\} \\ &= P\{\underset{\sim}{e}_n^T[\underset{\sim}{M}_n(\underset{\sim}{e}_nK)-\underset{\sim}{M}_n(\underset{\sim}{0})+\gamma_o K\underset{\sim}{Q}_n\underset{\sim}{e}_n]\geq -\underset{\sim}{e}_n^T\underset{\sim}{M}_n(\underset{\sim}{0})+\gamma_o K\underset{\sim}{e}_n^T\underset{\sim}{Q}_n\underset{\sim}{e}_n\} \\ &\leq P\{\underset{\sim}{e}_n^T[\underset{\sim}{M}_n(\underset{\sim}{e}_nK)-\underset{\sim}{M}_n(\underset{\sim}{0})+\gamma_o K\underset{\sim}{Q}_n\underset{\sim}{e}_n]\geq -pK_o+\lambda_1(\underset{\sim}{Q}_n)\gamma_o K\} + \epsilon. \end{aligned}$$

Therefore, by (A4), when n large enough, there exists a positive number λ such that $0<\lambda<\lambda_1(\underset{\sim}{Q})$ and

$$P\{\|\hat{\underset{\sim}{u}}_n\|>K\} \leq P\{\underset{\sim}{e}_n^T[\underset{\sim}{M}_n(\underset{\sim}{e}_nK)-\underset{\sim}{M}_n(\underset{\sim}{0})+\gamma_o K\underset{\sim}{Q}_n\underset{\sim}{e}_n]\geq -pK_o+\lambda\gamma_o K\} + \epsilon.$$

Choose K such that $K>\frac{pK_o}{\gamma_o\lambda}$, then $\beta_1=-pK_o+\lambda\gamma_o K>0$, and by (3.2), we have that, as $n\to\infty$

(3.16) $$P\{\|\hat{\underset{\sim}{u}}_n\|>K\} \leq P\{\sup_{|\underset{\sim}{u}|\leq K} |\underset{\sim}{M}_n(\underset{\sim}{u})-\underset{\sim}{M}_n(\underset{\sim}{0})+\gamma_o\underset{\sim}{Q}_n\underset{\sim}{u}|\geq\beta_1\} + \epsilon \to \epsilon.$$

Therefore, (3.4) follows from (3.12), (3.2) and (3.16). □

THEOREM 3.2. For a fixed $k=1, \cdots, p$, assume (A1), (A2), (B), (C1) through (C3) and (C5). Then, for any $K>0$, as $n\to\infty$

(3.17) $$\sup_{|\underset{\sim}{u}|\leq K} |M_{nk}(\underset{\sim}{u})-M_{nk}(\underset{\sim}{0})+ \gamma_o\Sigma_{i=1}^n c_{nik}\underset{\sim}{c}_{ni}^T\underset{\sim}{u}| \xrightarrow{P} 0,$$

where $\gamma_o = \int F'\, d\psi > 0$.

Proof. Let $M: [0,1]\to R$ be a function defined by $M(t)=\psi(F^{-1}(t))$, $0\leq t\leq 1$. From the assumptions on ψ and F, we have that M is nondecreasing, right continuous and bounded. Consider a functional $\tau: D[0,1]\to R$ defined by

(3.18) $$\tau(G) = \int_0^1 h(G(t))\, e^t dM(t), \qquad G\in D[0,1]$$

where h is defined by

(3.19) $$h(x) = \begin{cases} -e^{-x} & \text{if } x \geq 0; \\ (x-1) & \text{if } x < 0. \end{cases}$$

Then, τ can be expressed as a composition of the following Hadamard differentiable transformations:

$$\gamma_1\colon D[0,1]\to L^1[0,1], \qquad \gamma_1(S)=h\circ S;$$

$$\gamma_2\colon L^1[0,1]\cap D[0,1]\to R, \qquad \gamma_2(S)=\int_0^1 S(t)\, e^t dM(t);$$

and

$$\tau(G) = \gamma_2(\gamma_1(G)).$$

Note that γ_1 is Hadamard differentiable at U by Proposition 6.1.2 of Fernholz (1983) (which is also true for γ: $D[0,1]\to L^q[0,1]$), because h is differentiable everywhere with a bounded and continuous derivative, and that γ_2 is linear and continuous, thus Fréchet differentiable. Since the range of γ_1 is contained by D[0,1], from Proposition 3.1.2 of Fernholz (1983), τ is Hadamard differentiable at U with Hadamard derivative

$$\tau'_U(G) = \int_0^1 h'(t)\, G(t)\, e^t dM(t), \qquad G\in D[0,1]. \tag{3.20}$$

By Theorem 4.1 in Section 4, we have, for any K>0, as $n\to\infty$

$$\sup_{|\underline{u}|\le K}\Big|\sum_{i=1}^n c_{nik}\Big\{\tau\Big(\frac{S^*_{nk}(\cdot,\underline{u})}{\sum_{i=1}^n c_{nik}}\Big)-\tau\big(U(\cdot)\big)\Big\}-\tau'_U\Big(S^*_{nk}(\cdot,\underline{u})-U(\cdot)\sum_{i=1}^n c_{nik}\Big)\Big| \xrightarrow{P} 0.$$

Therefore, as $n\to\infty$

$$\sup_{|\underline{u}|\le K}\Big|\sum_{i=1}^n c_{nik}\Big\{\tau\Big(\frac{S^*_{nk}(\cdot,\underline{u})}{\sum_{i=1}^n c_{nik}}\Big)-\tau\Big(\frac{S^*_{nk}(\cdot,\underline{0})}{\sum_{i=1}^n c_{nik}}\Big)\Big\}-\tau'_U\Big(S^*_{nk}(\cdot,\underline{u})-S^*_{nk}(\cdot,\underline{0})\Big)\Big| \xrightarrow{P} 0. \tag{3.21}$$

Since, $h'(t)=e^{-t}$, for $0\le t\le 1$, for each $\underline{u}$, under (C5),

$$\tau'_U(S^*_{nk}(\cdot,\underline{u})) = \int_0^1 h'(t)S^*_{nk}(t,\underline{u})\, e^t dM(t) = \sum_{i=1}^n c_{nik}\int_{F(Y_i-\underline{c}_{ni}^T\underline{u})}^{1} dM(t)$$

$$= \sum_{i=1}^n c_{nik}\int_{(Y_i-\underline{c}_{ni}^T\underline{u})}^{\infty} d\psi(x) = \psi(\infty)\sum_{i=1}^n c_{nik} - M_{nk}(\underline{u}). \tag{3.22}$$

Therefore (3.17) follows from (3.21), (3.22) and Proposition 4.9. □

In the general case of $\{c_{nik}\}$, $1\le i\le n$, $1\le k\le p$, we have the following theorem.

THEOREM 3.3. Assume (A3), (A4), (B) and (C1) through (C5). Then, we have that, for any K>0 and any $1\le k\le p$,

$$\sup_{|\underline{u}|\le K}\Big|M_{nk}(\underline{u})-M_{nk}(\underline{0})+\gamma_0\textstyle\sum_{i=1}^n c_{nik}\underline{c}_{ni}^T\underline{u}\Big| \xrightarrow{P} 0. \qquad \text{as } n\to\infty. \tag{3.23}$$

Therefore, as $n\to\infty$

$$\sup_{|\underline{u}|\le K}\Big|\underline{M}_n(\underline{u}) - \underline{M}_n(\underline{0}) + \gamma_0\underline{Q}_n\underline{u}\Big| \xrightarrow{P} 0, \tag{3.24}$$

and furthermore,

$$(3.25)\qquad \underset{\sim}{C}_n^o\,(\,\hat{\underset{\sim}{\beta}}_n - \underset{\sim}{\beta}\,) \xrightarrow{\mathfrak{D}} N_p(\underset{\sim}{0}, \sigma^2\gamma_0^{-2}\underset{\sim}{Q}^{-1}).$$

Proof. Consider any $1 \le k \le p$. If $d_{nk}^+=0$ or $d_{nk}^-=0$, (3.23) is the case of Theorem 3.2. If $d_{nk}^+>0$ and $d_{nk}^->0$, using the functional τ given by (3.18) and Theorem 4.2, we have, for any $K>0$ and $1 \le k \le p$, as $n \to \infty$

$$\sup_{|\underset{\sim}{u}|\le K} \Big| \sum_{i=1}^n c_{nik}^+ \tau\Big(\frac{S_{nk}^{*+}(\cdot,\underset{\sim}{u})}{\sum_{i=1}^n c_{nik}^+}\Big) - \sum_{i=1}^n c_{nik}^- \tau\Big(\frac{S_{nk}^{*-}(\cdot,\underset{\sim}{u})}{\sum_{i=1}^n c_{nik}^-}\Big) -$$

$$(3.26)\qquad - \tau(U)\sum_{i=1}^n c_{nik} - \tau_U'\big(S_{nk}^*(\cdot,\underset{\sim}{u}) - U(\cdot)\sum_{i=1}^n c_{nik}\big)\Big| \xrightarrow{P} 0.$$

So, by (3.22), we have, as $n \to \infty$

$$\sup_{|\underset{\sim}{u}|\le K} \Big| \sum_{i=1}^n c_{nik}^+ \Big\{\tau\Big(\frac{S_{nk}^{*+}(\cdot,\underset{\sim}{u})}{\sum_{i=1}^n c_{nik}^+}\Big) - \tau\Big(\frac{S_{nk}^{*+}(\cdot,\underset{\sim}{0})}{\sum_{i=1}^n c_{nik}^+}\Big)\Big\} -$$

$$(3.27)\qquad - \sum_{i=1}^n c_{nik}^- \Big\{\tau\Big(\frac{S_{nk}^{*-}(\cdot,\underset{\sim}{u})}{\sum_{i=1}^n c_{nik}^-}\Big) - \tau\Big(\frac{S_{nk}^{*-}(\cdot,\underset{\sim}{0})}{\sum_{i=1}^n c_{nik}^-}\Big)\Big\} + M_{nk}(\underset{\sim}{u}) - M_{nk}(\underset{\sim}{0})\Big| \xrightarrow{P} 0.$$

Therefore (3.23) follows from (3.26), (3.27) and Lemma 4.10. □

REMARK. Comparing our Theorem 3.2 and Theorem 3.3 with the results available in the literature on the asymptotic normality of M-estimators of regression, we can see that we require weaker conditions on ψ, F and the design matrix by the Hadamard differentiability approach.

4. SOME ASYMPTOTIC RESULTS ON $\underset{\sim}{S}_n^*$

Using a form of the Taylor expansion involving Hadamard derivative of a functional τ defined on D[0,1], we may express $\tau(G)$ for any G in a neighborhood of U, the uniform d.f. on [0,1], as

$$(4.1)\qquad \tau(G) = \tau(U) + \tau_U'(G-U) + \mathrm{Rem}(G-U;\tau)$$

where τ_U' is the Hadamard derivative of τ at U (see Fernholz, 1983, on the definition of the Hadamard derivative) and $\mathrm{Rem}(G-U;\tau)$ is the remainder term in this first order expansion. For $G=S_{nk}^*(\cdot,\underset{\sim}{u})/\sum_{i=1}^n c_{nik}$, we have the following theorems on the remainder term of the expansion (4.1).

THEOREM 4.1. Suppose τ: $D[0,1]\to R$ is a functional and is Hadamard differentiable at U. For any fixed $k=1, 2, \cdots, p$, assume (A1), (A2) and (B). Then, for any $K>0$,

$$(4.2)\qquad \sup_{|\underset{\sim}{u}|\le K}\Big|\sum_{i=1}^{n} c_{nik}\,\mathrm{Rem}\Big(\frac{S^*_{nk}(\cdot,\underset{\sim}{u})}{\sum_{i=1}^{n}c_{nik}} - U(\cdot);\,\tau\Big)\Big| \xrightarrow{P} 0, \qquad \text{as } n\to\infty.$$

Proof. The proof follows along the lines of Theorem 3.1 of Ren and Sen (1991) and hence is omitted. □

THEOREM 4.2. Suppose τ: D[0,1]→R is a functional and is Hadamard differentiable at U. Assume (A3) and (B). Then, for any $1\le k\le p$ and K>0, as $n\to\infty$

$$\sup_{|\underset{\sim}{u}|\le K}\Big|\sum_{i=1}^{n} c^+_{nik}\,\tau\Big(\frac{S^{*+}_{nk}(\cdot,\underset{\sim}{u})}{\sum_{i=1}^{n}c^+_{nik}}\Big) - \sum_{i=1}^{n} c^-_{nik}\,\tau\Big(\frac{S^{*-}_{nk}(\cdot,\underset{\sim}{u})}{\sum_{i=1}^{n}c^-_{nik}}\Big) -$$

$$(4.3)\qquad - \tau(U)\textstyle\sum_{i=1}^{n}c_{nik} - \tau'_U\Big(S^*_{nk}(\cdot,\underset{\sim}{u}) - U(\cdot)\sum_{i=1}^{n}c_{nik}\Big)\Big| \xrightarrow{P} 0.$$

Proof. The proof is similar to the one of Theorem 3.2 of Ren and Sen (1991).□

Let

$$(4.4)\qquad \underset{\sim}{I} = (1, \cdots, 1)^\tau,$$

and

$$(4.5)\qquad E_q = [0,1]^q$$

where q is an arbitrary positive integer, then, for any $0<K<\infty$, $(t,\underset{\sim}{u})\in E_{p+1}$, we write

$$(4.6)\qquad W_{nk}(t,\underset{\sim}{u}) = S^*_{nk}(t, (2\underset{\sim}{u}-\underset{\sim}{I})K) - S_{nk}(t, (2\underset{\sim}{u}-\underset{\sim}{I})K).$$

Also, let f be a function defined on E_q, we denote

$$(4.7)\qquad \omega_f(\delta) = \sup_{|\underset{\sim}{x}-\underset{\sim}{y}|\le\delta} \big|f(\underset{\sim}{x}) - f(\underset{\sim}{y})\big|$$

where $|\underset{\sim}{x}| = |(x_1, \dots, x_q)| = \max_{1\le i\le q} |x_i|$.

LEMMA 4.3. For a fixed k=1, ⋯ , p, assume (A1), (A2) and (B). Then, for any K>0, as $n\to\infty$

$$(4.8)\qquad \sup_{(t,\underset{\sim}{u})\in E_{p+1}} |W_{nk}(t,\underset{\sim}{u}) - W_{nk}(t,\tfrac{1}{2}\underset{\sim}{I})| \xrightarrow{P} 0.$$

Proof. Consider a weighted empirical process $Z_n(t)=\sum_{i=1}^{n}c_{nik}[I(F(Y_i)\le t)-t]$, for $0\le t\le 1$. From the Corollary 2 of Shorack and Wellner (1986, page 109), we have for $\epsilon>0$,

(4.9) $$\lim_{\delta\to 0}\ \overline{\lim_{n\to\infty}}\ P(\omega_{Z_n}(\delta)\geq\epsilon) = 0.$$

This implies that for $\rho>0$, there exist $\delta_o>0$ and N_o such that

(4.10) $$P(\omega_{Z_n}(\delta_o)\geq\epsilon) < \rho, \text{ for } n\geq N_o.$$

Since by (A2) and (B), there exist a positive constant M and a positive integer $N\geq N_o$ such that

$$|\ F(F^{-1}(t)+\underset{\sim}{c}_{ni}^T(2\underset{\sim}{u}-\underset{\sim}{I})K) - F(F^{-1}(t))\ | \leq MpKa_n \leq \delta_o, \qquad \text{for } n\geq N,$$

then for $n\geq N$,

(4.11) $$\sup_{(t,\underset{\sim}{u})\in E_{p+1}} |W_{nk}(t,\underset{\sim}{u}) - W_{nk}(t,\tfrac{1}{2}\underset{\sim}{I})| \leq \omega_{Z_n}(\delta_o).$$

Therefore,

$$P\Big\{\sup_{(t,\underset{\sim}{u})\in E_{p+1}} |W_{nk}(t,\underset{\sim}{u}) - W_{nk}(t,\tfrac{1}{2}\underset{\sim}{I})| \geq \epsilon\Big\} < \rho, \qquad \text{for } n\geq N,$$

follows from (4.10) and (4.11). □

COROLLARY 4.4. For a fixed $k=1, \cdots, p$, assume (A1), (A2) and (B). Then, for any $K>0$, as $n\to\infty$

(4.12) $$\sup\Big\{|[S^*_{nk}(t,\underset{\sim}{u})-S^*_{nk}(t,\underset{\sim}{0})]- \textstyle\sum_{i=1}^n c_{nik}\ \underset{\sim}{c}_{ni}^T\underset{\sim}{u}F'(F^{-1}(t))|;\ t\in[0,1],\ |\underset{\sim}{u}|\leq K\Big\} \xrightarrow{P} 0.$$

Proof. Lemma 4.3 implies that, for any $K>0$, as $n\to\infty$

(4.13) $$\sup\Big\{|[S^*_{nk}(t,\underset{\sim}{u})-S^*_{nk}(t,\underset{\sim}{0})]-[S_{nk}(t,\underset{\sim}{u})-S_{nk}(t,\underset{\sim}{0})]|;\ t\in[0,1],\ |\underset{\sim}{u}|\leq K\Big\} \xrightarrow{P} 0.$$

Hence, it suffices to show that

(4.14) $$\sup\Big\{|[S_{nk}(t,\underset{\sim}{u})-S_{nk}(t,\underset{\sim}{0})]- \textstyle\sum_{i=1}^n c_{nik}\ \underset{\sim}{c}_{ni}^T\underset{\sim}{u}F'(F^{-1}(t))|;\ t\in[0,1],\ |\underset{\sim}{u}|\leq K\Big\} \xrightarrow{P} 0.$$

Since for any $t\in[0,1]$ and $|\underset{\sim}{u}|\leq K$,

$$\begin{aligned}
&|\ [S_{nk}(t,\underset{\sim}{u})-S_{nk}(t,\underset{\sim}{0})]- \textstyle\sum_{i=1}^n c_{nik}\ \underset{\sim}{c}_{ni}^T\underset{\sim}{u}F'(F^{-1}(t))\ |\\
&\quad = |\ \textstyle\sum_{i=1}^n c_{nik}[F(F^{-1}(t)+\underset{\sim}{c}_{ni}^T\underset{\sim}{u})-F(F^{-1}(t))] - \sum_{i=1}^n c_{nik}\underset{\sim}{c}_{ni}^T\underset{\sim}{u}F'(F^{-1}(t))\ |\\
&\quad = |\ \textstyle\sum_{i=1}^n c_{nik}\underset{\sim}{c}_{ni}^T\underset{\sim}{u}[F'(\xi)-F'(F^{-1}(t))]\ |\\
&\quad \leq pK \sup|\ F'(\xi)-F'(F^{-1}(t))\ |,
\end{aligned}$$

where ξ is between $F^{-1}(t)+\underset{\sim}{c}_{ni}^T\underset{\sim}{u}$ and $F^{-1}(t)$, therefore (4.14) follows from (A2) and the uniform continuity of F'. □

COROLLARY 4.5. Assume (A3) and (B). Then, for any $K>0$ and $1\leq k\leq p$, as $n\to\infty$

(4.15) $\sup\left\{|[S^{*+}_{nk}(t,\underset{\sim}{u})-S^{*+}_{nk}(t,\underset{\sim}{0})]-[S^{+}_{nk}(t,\underset{\sim}{u})-S^{+}_{nk}(t,\underset{\sim}{0})]|;\ t\in[0,1],\ |\underset{\sim}{u}|\leq K\right\} \xrightarrow{P} 0,$

and

(4.16) $\sup\left\{|[S^{*-}_{nk}(t,\underset{\sim}{u})-S^{*-}_{nk}(t,\underset{\sim}{0})]-[S^{-}_{nk}(t,\underset{\sim}{u})-S^{-}_{nk}(t,\underset{\sim}{0})]|;\ t\in[0,1],\ |\underset{\sim}{u}|\leq K\right\} \xrightarrow{P} 0.$

Therefore,

(4.17) $\sup\left\{|[S^{*+}_{nk}(t,\underset{\sim}{u})-S^{*+}_{nk}(t,\underset{\sim}{0})]-\Sigma_{i=1}^{n}c^{+}_{nik}\underset{\sim}{c}^{T}_{ni}\underset{\sim}{u}F'(F^{-1}(t))|;\ t\in[0,1],\ |\underset{\sim}{u}|\leq K\right\} \xrightarrow{P} 0,$

and

(4.18) $\sup\left\{|[S^{*-}_{nk}(t,\underset{\sim}{u})-S^{*-}_{nk}(t,\underset{\sim}{0})]-\Sigma_{i=1}^{n}c^{-}_{nik}\underset{\sim}{c}^{T}_{ni}\underset{\sim}{u}F'(F^{-1}(t))|;\ t\in[0,1],\ |\underset{\sim}{u}|\leq K\right\} \xrightarrow{P} 0.$

Proof. Consider any $1\leq k\leq p$. If $d^{+}_{nk}=0$ or $d^{-}_{nk}=0$, it is the case of Corollary 4.4. If $d^{+}_{nk}>0$ and $d^{-}_{nk}>0$, by Lemma 4.3, we have, as $n\to\infty$

(4.19) $\frac{1}{d^{+}_{nk}}\sup\left\{|[S^{*+}_{nk}(t,\underset{\sim}{u})-S^{*+}_{nk}(t,\underset{\sim}{0})]-[S^{+}_{nk}(t,\underset{\sim}{u})-S^{+}_{nk}(t,\underset{\sim}{0})]|;\ t\in[0,1],\ |\underset{\sim}{u}|\leq K\right\} \xrightarrow{P} 0,$

and

(4.20) $\frac{1}{d^{-}_{nk}}\sup\left\{|[S^{*-}_{nk}(t,\underset{\sim}{u})-S^{*-}_{nk}(t,\underset{\sim}{0})]-[S^{-}_{nk}(t,\underset{\sim}{u})-S^{-}_{nk}(t,\underset{\sim}{0})]|;\ t\in[0,1],\ |\underset{\sim}{u}|\leq K\right\} \xrightarrow{P} 0.$

Hence, (4.15) and (4.16) follow from the fact: $0<d^{+}_{nk},\ d^{-}_{nk}<1$. Further, by the similar proof of Corollary 4.4, (4.17) and (4.18) follow from

(4.21) $\sup\left\{|[S^{+}_{nk}(t,\underset{\sim}{u})-S^{+}_{nk}(t,\underset{\sim}{0})]-\Sigma_{i=1}^{n}c^{+}_{nik}\underset{\sim}{c}^{T}_{ni}\underset{\sim}{u}F'(F^{-1}(t))|;\ t\in[0,1],\ |\underset{\sim}{u}|\leq K\right\} \xrightarrow{P} 0,$

and

(4.22) $\sup\left\{|[S^{-}_{nk}(t,\underset{\sim}{u})-S^{-}_{nk}(t,\underset{\sim}{0})]-\Sigma_{i=1}^{n}c^{-}_{nik}\underset{\sim}{c}^{T}_{ni}\underset{\sim}{u}F'(F^{-1}(t))|;\ t\in[0,1],\ |\underset{\sim}{u}|\leq K\right\} \xrightarrow{P} 0,$

respectively. □

LEMMA 4.6. Assume (A2). Then, for any $1\leq k\leq p$,

(4.23) $$\sup_{t\in[0,1]}|S^{*}_{nk}(t,\underset{\sim}{0})-S_{nk}(t,\underset{\sim}{0})| = O_P(1), \qquad \text{as } n\to\infty.$$

Proof. Let a weighted empirical process given by

$$Z_{nk}(t) = S^{*}_{nk}(t,\underset{\sim}{0})-S_{nk}(t,\underset{\sim}{0}) = \Sigma_{i=1}^{n}c_{nik}[I\{F(Y_i)\leq t\}-t],$$

for $0\leq t\leq 1$, then for any $0\leq s,\ t\leq 1$, we have

$$\mathrm{Cov}\{Z_{nk}(s),\ Z_{nk}(t)\} = \Sigma_{i=1}^{n}c^{2}_{nik}\{s\wedge t - st\}=\{s\wedge t - st\}.$$

Hence, by Theorem 1 of Shorack and Wellner (1986, page 109), we know that Z_{nk} weakly converges to some Z on $(D[0,1],\ \mathfrak{D},\ \|\cdot\|)$, where $\|\cdot\|$ is the uniform metric on

D[0,1] and $\mathfrak{D}$ is the σ-field of subsets of D[0,1] generated by the open balls. Therefore, (4.23) follows. □

COROLLARY 4.7. For any fixed $k=1, \cdots, p$, assume (A1), (A2) and (B). Then, for any $K>0$, as $n\to\infty$

$$\sup\{|S^*_{nk}(t, \underset{\sim}{u})-S_{nk}(t,\underset{\sim}{0})|;\ t\in[0,1],\ |\underset{\sim}{u}|\le K\} = O_P(1). \tag{4.24}$$

Therefore, as $n\to\infty$

$$\sup\{|\frac{S^*_{nk}(t, \underset{\sim}{u})}{\sum_{i=1}^n c_{nik}} - t|;\ t\in[0,1],\ |\underset{\sim}{u}|\le K\} \xrightarrow{P} 0. \tag{4.25}$$

Proof. Since for each $t\in[0,1]$, $|\underset{\sim}{u}|\le K$,

$$\begin{aligned} S^*_{nk}(t,\underset{\sim}{u})-S_{nk}(t,\underset{\sim}{0}) &= [S^*_{nk}(t,\underset{\sim}{u})-S_{nk}(t,\underset{\sim}{u})] + [S_{nk}(t,\underset{\sim}{u})-S_{nk}(t,\underset{\sim}{0})] \\ &= \{[S^*_{nk}(t,\underset{\sim}{u})-S_{nk}(t,\underset{\sim}{u})]-[S^*_{nk}(t,\underset{\sim}{0})-S_{nk}(t,\underset{\sim}{0})]\} + \\ &\quad + [S^*_{nk}(t,\underset{\sim}{0})-S_{nk}(t,\underset{\sim}{0})] + [S_{nk}(t,\underset{\sim}{u})-S_{nk}(t,\underset{\sim}{0})], \end{aligned}$$

then, (4.24) follows from (4.13), Lemma 4.6 and (4.14). Furthermore, (4.25) follows from (2.14). □

COROLLARY 4.8. Assume (A3) and (B). Then, for any $K>0$ and $1\le k\le p$, as $n\to\infty$

$$\sup\{|S^{*+}_{nk}(t, \underset{\sim}{u})-S^+_{nk}(t,\underset{\sim}{0})|;\ t\in[0,1],\ |\underset{\sim}{u}|\le K\} = O_P(1), \tag{4.26}$$

$$\sup\{|S^{*-}_{nk}(t, \underset{\sim}{u})-S^-_{nk}(t,\underset{\sim}{0})|;\ t\in[0,1],\ |\underset{\sim}{u}|\le K\} = O_P(1), \tag{4.27}$$

and furthermore, as $n\to\infty$

$$\sup\{|\frac{S^{*+}_{nk}(t, \underset{\sim}{u})}{\sum_{i=1}^n c^+_{nik}} - t|;\ t\in[0,1],\ |\underset{\sim}{u}|\le K\} \xrightarrow{P} 0, \tag{4.28}$$

$$\sup\{|\frac{S^{*-}_{nk}(t, \underset{\sim}{u})}{\sum_{i=1}^n c^-_{nik}} - t|;\ t\in[0,1],\ |\underset{\sim}{u}|\le K\} \xrightarrow{P} 0. \tag{4.29}$$

Proof. Consider any $1\le k\le p$. If $d^+_{nk}=0$ or $d^-_{nk}=0$, it is the case of Corollary 4.7. If $d^+_{nk}>0$ and $d^-_{nk}>0$, from Corollary 4.7, we have, as $n\to\infty$

$$\frac{1}{d^+_{nk}}\sup\{|S^{*+}_{nk}(t, \underset{\sim}{u})-S^+_{nk}(t,\underset{\sim}{0})|;\ t\in[0,1],\ |\underset{\sim}{u}|\le K\} = O_P(1).$$

Since $0<d^+_{nk}<1$ and

$$\frac{S^{*+}_{nk}(t, \underset{\sim}{u})}{\sum_{i=1}^n c^+_{nik}} - t \doteq \frac{[S^{*+}_{nk}(t, \underset{\sim}{u})-S^+_{nk}(t,\underset{\sim}{0})]/d^+_{nk}}{\sum_{i=1}^n c^+_{nik}/d^+_{nk}},$$

and since (A3) implies that

$$\Sigma_{i=1}^{n} c_{nik}^{+}/d_{nk}^{+} \to \infty, \qquad \text{as } n\to\infty$$

(4.26) and (4.28) follow. The proof of (4.27) and (4.29) is similar. □

PROPOSITION 4.9. For a fixed k=1, ⋯ , p, assume (A1), (A2), (B), (C1) and (C5). Let M be a function given by $M(t)=\psi(F^{-1}(t))$, $0\le t\le 1$ and h a function given by $h(x)=-e^{-x}$, if $x\ge 0$; $(x-1)$, if $x<0$. Then, for any $K>0$, as $n\to\infty$

$$\text{(4.30)} \quad \sup_{|\underset{\sim}{u}|\le K} \Big| \sum_{i=1}^{n} c_{nik} \int_0^1 \Big\{ h\Big(\frac{S_{nk}^{*}(t,\underset{\sim}{u})}{\Sigma_{i=1}^{n} c_{nik}}\Big) - h\Big(\frac{S_{nk}^{*}(t,\underset{\sim}{0})}{\Sigma_{i=1}^{n} c_{nik}}\Big)\Big\} e^{t} dM(t) - \gamma_0 \sum_{i=1}^{n} c_{nik} \underset{\sim}{c}_{ni}^{T} \underset{\sim}{u} \Big| \xrightarrow{P} 0.$$

Proof. First, we notice that the function h is continuous and differentiable with a bounded and continuous h′, and that

$$\sum_{i=1}^{n} c_{nik} \int_0^1 \Big\{ h\Big(\frac{S_{nk}^{*}(t,\underset{\sim}{u})}{\Sigma_{i=1}^{n} c_{nik}}\Big) - h\Big(\frac{S_{nk}^{*}(t,\underset{\sim}{0})}{\Sigma_{i=1}^{n} c_{nik}}\Big)\Big\} e^{t} dM(t)$$

$$= \int_0^1 h'(\xi_t)[S_{nk}^{*}(t,\underset{\sim}{u}) - S_{nk}^{*}(t,\underset{\sim}{0})] e^{t} dM(t)$$

$$= \int_0^1 [h'(\xi_t) - h'(t)][S_{nk}^{*}(t,\underset{\sim}{u}) - S_{nk}^{*}(t,\underset{\sim}{0})] e^{t} dM(t) +$$

$$+ \int_0^1 h'(t)\Big\{[S_{nk}^{*}(t,\underset{\sim}{u}) - S_{nk}^{*}(t,\underset{\sim}{0})] - \sum_{i=1}^{n} c_{nik} \underset{\sim}{c}_{ni}^{T} \underset{\sim}{u} F'(F^{-1}(t))\Big\} e^{t} dM(t) +$$

$$\text{(4.31)} \qquad + \int_0^1 h'(t) \sum_{i=1}^{n} c_{nik} \underset{\sim}{c}_{ni}^{T} \underset{\sim}{u} F'(F^{-1}(t)) e^{t} dM(t),$$

where ξ_t is between $S_{nk}^{*}(t,\underset{\sim}{u})/\Sigma_{i=1}^{n} c_{nik}$ and $S_{nk}^{*}(t,\underset{\sim}{0})/\Sigma_{i=1}^{n} c_{nik}$. By Corollary 4.4, Corollary 4.7 and the uniform continuity of $h'(t)$, we have, as $n\to\infty$

$$\text{(4.32)} \quad \sup_{|\underset{\sim}{u}|\le K} \Big| \int_0^1 [h'(\xi_t) - h'(t)][S_{nk}^{*}(t,\underset{\sim}{u}) - S_{nk}^{*}(t,\underset{\sim}{0})] e^{t} dM(t) \Big| \xrightarrow{P} 0.$$

By Corollary 4.4 and the boundedness of h′ and M, we have, as $n\to\infty$

$$\text{(4.33)} \quad \sup_{|\underset{\sim}{u}|\le K} \Big| \int_0^1 h'(t)\Big\{[S_{nk}^{*}(t,\underset{\sim}{u}) - S_{nk}^{*}(t,\underset{\sim}{0})] - \sum_{i=1}^{n} c_{nik} \underset{\sim}{c}_{ni}^{T} \underset{\sim}{u} F'(F^{-1}(t))\Big\} e^{t} dM(t) \Big| \xrightarrow{P} 0.$$

Therefore, (4.30) follows from (4.31), (4.32) and

$$\int_0^1 h'(t) \sum_{i=1}^{n} c_{nik} \underset{\sim}{c}_{ni}^{T} \underset{\sim}{u} F'(F^{-1}(t)) e^{t} \, dM(t) =$$

$$= \sum_{i=1}^{n} c_{nik} \underset{\sim}{c}_{ni}^{T} \underset{\sim}{u} \int_0^1 F'(F^{-1}(t)) \, dM(t) = \gamma_0 \sum_{i=1}^{n} c_{nik} \underset{\sim}{c}_{ni}^{T} \underset{\sim}{u}. \qquad □$$

In the general case of $\{c_{nik}\}$, we have the following lemma.

LEMMA 4.10. Assume (A3), (B), (C1) and (C5). Then, we have that, for any $K > 0$ and $1 \le k \le p$, as $n \to \infty$

$$(4.34) \quad \sup_{|\underline{u}|\le K} \Big| \sum_{i=1}^{n} c_{nik}^{+} \int_0^1 \Big\{ h\Big(\frac{S_{nk}^{*+}(t,\underline{u})}{\sum_{i=1}^{n} c_{nik}^{+}}\Big) - h\Big(\frac{S_{nk}^{*+}(t,\underline{0})}{\sum_{i=1}^{n} c_{nik}^{+}}\Big)\Big\} e^t dM(t) - \gamma_0 \sum_{i=1}^{n} c_{nik}^{+} \underline{c}_{ni}^{T} \underline{u} \Big| \xrightarrow{P} 0,$$

and

$$(4.35) \quad \sup_{|\underline{u}|\le K} \Big| \sum_{i=1}^{n} c_{nik}^{-} \int_0^1 \Big\{ h\Big(\frac{S_{nk}^{*-}(t,\underline{u})}{\sum_{i=1}^{n} c_{nik}^{-}}\Big) - h\Big(\frac{S_{nk}^{*-}(t,\underline{0})}{\sum_{i=1}^{n} c_{nik}^{-}}\Big)\Big\} e^t dM(t) - \gamma_0 \sum_{i=1}^{n} c_{nik}^{-} \underline{c}_{ni}^{T} \underline{u} \Big| \xrightarrow{P} 0.$$

Therefore, as $n \to \infty$

$$(4.36) \quad \sup_{|\underline{u}|\le K} \Big| \sum_{i=1}^{n} c_{nik}^{+} \int_0^1 \Big\{ h\Big(\frac{S_{nk}^{*+}(t,\underline{u})}{\sum_{i=1}^{n} c_{nik}^{+}}\Big) - h\Big(\frac{S_{nk}^{*+}(t,\underline{0})}{\sum_{i=1}^{n} c_{nik}^{+}}\Big)\Big\} e^t dM(t) -$$
$$- \sum_{i=1}^{n} c_{nik}^{-} \int_0^1 \Big\{ h\Big(\frac{S_{nk}^{*-}(t,\underline{u})}{\sum_{i=1}^{n} c_{nik}^{-}}\Big) - h\Big(\frac{S_{nk}^{*-}(t,\underline{0})}{\sum_{i=1}^{n} c_{nik}^{-}}\Big)\Big\} e^t dM(t) - \gamma_0 \sum_{i=1}^{n} c_{nik}\, \underline{c}_{ni}^{T} \underline{u} \Big| \xrightarrow{P} 0.$$

Proof. Consider any $1 \le k \le p$. If $d_{nk}^{+}=0$ or $d_{nk}^{-}=0$, this is just the case of Proposition 4.9, and (4.34) or (4.35) is equal to 0. If $d_{nk}^{+}>0$ and $d_{nk}^{-}>0$, the proof is the same as Proposition 4.9's by using Corollary 4.5 and Corollary 4.8. □

ACKNOWLEDGEMENT

We thank the referee for a very careful reading of the manuscript.

REFERENCES

Bickel, P.J. (1975). One-step Huber estimates in the linear model. *JASA*. 70, 428-433.

Chung, K.L. (1974). *A course in Probability Theory.* Academic Press, INC.

Fernholz, L.T. (1983). *Von Mises Calculus for Statistical Functionals.* Lecture notes in statistics 19. Springer, New York. [1.3f, 2.1b, 2.3a].

Hájek, J. and Šidák, Z. (1967). *Theory of Rank Tests.* Academia, Prague.

Hampel, F.R., Ronchetti, E.M., Rousseeuw, P.J. and Stahel, W.A. (1986). *Robust Statistics.* John Wiley & Sons, New York.

Huber, P.J. (1973). Robust regression: asymptotics, conjectures and Monte Carlo. *Ann. Stat.* 1, 799-821.

Jurečková, J. (1971). Nonparametric estimate of regression coefficients. *AMS.* 42,

1328-1338.

Jurečková, J. (1977). Asymptotic relations of M-estimates and R-estimates in linear regression model. *Ann. Stat.* 5, 464-472.

Jurečková, J. (1989). Consistency of M-estimators of vector parameters. *Proceeding of the Fourth Symposium on Asymptotic Statistics.* Edi. by P. Mandl and M. Hušková, Charles Univ., Prague, 305-312.

Koul, H.L. (1977). Behavior of robust estimators in the regression model with dependent errors. *Ann. Stat.* 5, 681-699.

Ortega, J.M. and Rheinboldt, W.C. (1970). *Iterative Solution of Nonlinear Equations in Several Variables.* Academic Press, New York.

Pollard, D. (1991). Asymptotics for least absolute deviation regression estimators. *Econometric Theory* 7, 186-199.

Ren, J. and Sen, P.K. (1991). On Hadamard differentiability of extended statistical functional. *Multi. Anal.* Vol. 39, No. 1, 30-43.

Shorack, G. and Wellner, J.A. (1986). *Empirical Processes with Application to Statistics.* John Wiley & Sons.

Yohai, V.J. and Maronna, R.A. (1979). Asymptotic behavior of M-estimators for the linear model. *Ann. Stat.* 7, 258-268.

Department of Mathematics & Statistics
University of Nebraska-Lincoln
Lincoln, NE 68588
U.S.A.

Department of Statistics
University of North Carolina
Chapel Hill, NC 27599
U.S.A.

Contributed Papers

ASYMPTOTIC THEORY FOR REGRESSION QUANTILE ESTIMATORS IN THE HETEROSCEDASTIC REGRESSION MODEL

Marc AERTS, Paul JANSSEN and Noël VERAVERBEKE
Limburgs Universitair Centrum, Belgium

ABSTRACT. In this note we consider the asymptotic theory for regression quantile estimators in the heteroscedastic regression model $Y = \mu(x) + \sigma(x)\eta$ where μ and σ are unknown smooth functions and η is a random variable with mean zero and unit variance. We derive asymptotic results for a kernel type estimator for the distribution function of Y and the corresponding estimator for its quantiles.

Key words and Phrases : asymptotic normality; asymptotic representations; heteroscedasticity; kernel estimator; nonparametric regression; regression quantiles.
AMS 1991 subject classification : Primary 62G07, Secondary 62G20.

1 INTRODUCTION

Suppose that $Y_1, \ldots, Y_n$ are observations at fixed design points $0 \le x_1 \le \ldots \le x_n \le 1$, according to the general heteroscedastic regression model

$$Y_i = \mu(x_i) + \sigma(x_i)\eta_i \qquad i = 1, \ldots, n,$$

where μ and σ are some unknown smooth functions defined on $[0,1]$ and $\eta_1, \ldots, \eta_n$ are i.i.d. random variables with $E(\eta_1) = 0$ and $Var(\eta_1) = 1$. If F denotes the unknown distribution function of η_1, then the distribution function of $Y_i (i = 1, \ldots, n)$ is given by

$$F_{x_i}(y) = P(Y_i \le y) = F\left(\frac{y - \mu(x_i)}{\sigma(x_i)}\right), \qquad y \in I\!R.$$

Regression quantiles are quantiles of the unknown distribution function

$$F_x(y) = F\left(\frac{y - \mu(x)}{\sigma(x)}\right)$$

where $0 \le x \le 1$ is fixed. They will be estimated by the corresponding sample regression quantiles, obtained from a kernel type estimator $\widetilde{F}_x$ for F_x, defined by

$$\widetilde{F}_x(y) = \sum_{i=1}^{n} w_{ni}(x; h_n) I(Y_i \le y), \qquad y \in I\!R, \tag{1.1}$$

$$w_{ni}(x;h_n) = \frac{1}{c_n(x;h_n)h_n}\int_{x_{i-1}}^{x_i} K\left(\frac{x-z}{h_n}\right)dz, \qquad i=1,\ldots,n,$$

$$c_n(x;h_n) = \frac{1}{h_n}\int_0^{x_n} K\left(\frac{x-z}{h_n}\right)dz.$$

Here $x_0 = 0$, K is a known probability density function (kernel) and $\{h_n\}$ is a sequence of positive constants (bandwidths) tending to 0 as $n \to \infty$. Note that the weights $w_{ni}(x;h_n)$ sum up to 1, so that $\tilde{F}_x$ is a proper distribution function. The estimator $\tilde{F}_x$ is essentially the same as the one proposed by Stone (1977) in the homoscedastic model ($\sigma(x) = \sigma$ for all $x \in [0,1]$).

For $0 < p < 1$, the p-th regression quantile and p-th sample regression quantile, will be denoted by

$$\xi_p(x) = \inf\{y \in \mathbb{R} : F_x(y) \geq p\}$$
$$\tilde{\xi}_p(x) = \inf\{y \in \mathbb{R} : \tilde{F}_x(y) \geq p\}.$$

Note that $\xi_p(x) = \mu(x) + \sigma(x)\xi_p$, where $\xi_p = \inf\{y \in \mathbb{R} : F(y) \geq p\}$.

For the homoscedastic model, Cheng (1984), Janssen and Veraverbeke (1987), Aerts and Geertsema (1991) used (linear combinations of) regression quantiles for estimation of the regression function $\mu(x)$. Some of their asymptotic results are now reconsidered for the general heteroscedastic model. This is interesting in itself, but more important however, is that these results will turn out to be key tools in showing the validity of a new resampling scheme for heteroscedastic nonparametric regression that is introduced and studied in Aerts, Janssen and Veraverbeke (1992).

Let us introduce some notation. For the design points $0 = x_0 \leq x_1 \leq \ldots \leq x_n \leq 1$, we denote $\delta_n = \min_{1\leq i\leq n}(x_i - x_{i-1})$ and $\Delta_n = \max_{1\leq i\leq n}(x_i - x_{i-1})$. For the kernel K we will use $\|K\|_\infty = \sup_{u\in\mathbb{R}} K(u)$, $\|K\|_2^2 = \int_{-\infty}^{\infty} K^2(u)du$, $\mu_\ell^K = \int_{-\infty}^{\infty} u^\ell K(u)du$ ($\ell = 1,2$).

We will frequently use (some of) the following conditions :

(C1) $x_n \to 1$, $\Delta_n = O(n^{-1})$, $\Delta_n - \delta_n = o(n^{-1})$

(C2) K is a probability density function with finite support $[-L, L]$ for some $L > 0$, $\mu_1^K = 0$, and K is Lipschitz of order 1.

(C3) F, μ and σ are Lipschitz of order 1.

(C4) $\inf_{0\leq x\leq 1} \sigma(x) = \sigma_0 > 0$

(C5) For a fixed $y \in \mathbb{R}$, $\ddot{F}_x(y) = \dfrac{\partial^2 F_x(y)}{\partial x^2}$ exists at the point x of interest.

Note that (C3) and (C4) imply that F_x, as a function of x, is Lipschitz of order 1, i.e. for $0 \leq x, x' \leq 1$:

$$|F_x(y) - F_{x'}(y)| \leq C(y)|x - x'| \tag{1.2}$$

where $C(y)$ does not depend on x, x', but depends on y in the following way : $C(y) = c_1|y| + c_2$ with c_1 and c_2 absolute constants. It also follows that

$$|F_x(y)(1 - F_x(y)) - F_{x'}(y)(1 - F_{x'}(y))| \leq 3\ C(y)|x - x'| \tag{1.3}$$

Also note that for n sufficiently large (depending on x) we can take $c_n(x; h_n) = 1$ since $x_n \to 1$ and K has finite support. Because all results in this paper are asymptotic, we will systematically take $c_n(x; h_n) = 1$ in the proofs.

2 RESULTS FOR THE KERNEL ESTIMATOR

We will give asymptotic expressions for the bias and variance of the kernel estimator $\tilde{F}_x$ in (1.1). The derivation is classical.

LEMMA 1.

Assume (C1) - (C4), $h_n \to 0$, $nh_n^2 \to \infty$. Then, as $n \to \infty$,

$$Var(\tilde{F}_x(y)) = \frac{1}{nh_n} F_x(y)(1 - F_x(y))\ \|K\|_2^2 + o\left(\frac{1}{nh_n}\right) \tag{2.1}$$

If moreover (C5) holds, then

$$E(\tilde{F}_x(y)) - F_x(y) = \frac{1}{2}\ddot{F}_x(y)\mu_2^K h_n^2 + o(h_n^2) \tag{2.2}$$

Remark. The previous lemma allows us to obtain the optimal order of magnitude of the smoothing parameter h_n as a negative power of n. This is done in the usual way, by minimizing the approximate mean squared error. This approximation is provided by Lemma 1 :

$$\begin{aligned} MSE(h_n) &= E(\tilde{F}_x(y) - F_x(y))^2 = Var(\tilde{F}_x(y)) + [E(\tilde{F}_x(y)) - F_x(y)]^2 \\ &\approx \frac{1}{nh_n} F_x(y)(1 - F_x(y))\ \|K\|_2^2 + (\frac{1}{2}\ddot{F}_x(y)\mu_2^K h_n^2)^2. \end{aligned}$$

The optimal choice is then $h_n^{opt} = \left(\frac{F_x(y)(1 - F_x(y))\ \|K\|_2^2}{(\ddot{F}_x(y)\mu_2^K)^2}\right)^{1/2} n^{-1/5}$.

In the next lemma we give the asymptotic normality result for $\tilde{F}_x(y)$. In the (a) part, the differentiability condition (C5) is not needed, but the conditions do not allow for the optimal bandwidth choice. For the optimal choice $h_n = Cn^{-1/5}$, asymptotic normality is given in part (b).

LEMMA 2.

(a) Assume (C1) - (C4), $h_n \to 0$, $nh_n^2 \to \infty$, $nh_n^3 \to 0$. Then, as $n \to \infty$,

$$(nh_n)^{1/2}(\tilde{F}_x(y) - F_x(y)) \xrightarrow{d} N(0; F_x(y)(1 - F_x(y))\|K\|_2^2)$$

(b) Assume (C1) - (C5), $h_n = Cn^{-1/5}$ for some $C > 0$. Then, as $n \to \infty$,

$$(nh_n)^{1/2}(\tilde{F}_x(y) - F_x(y)) \xrightarrow{d} N(\frac{1}{2}\ddot{F}_x(y)\mu_2^K C^{5/2}; F_x(y)(1 - F_x(y))\|K\|_2^2)$$

Proof. A sufficient condition for the asymptotic normality of

$$\frac{\tilde{F}_x(y) - E(\tilde{F}_x(y))}{(Var(\tilde{F}_x(y))^{1/2}} \tag{2.3}$$

is Liapunov's condition

$$\frac{\sum_{i=1}^{n} E|w_{ni}(x; h_n)(I(Y_i \leq y) - F_{x_i}(y))|^3}{(Var(\tilde{F}_x(y)))^{3/2}} \to 0, \qquad \text{as } n \to \infty.$$

Now, the numerator of this expression is bounded by $\sum_{i=1}^{n} w_{ni}^3(x; h_n) \leq \frac{\|K\|_\infty^2 \Delta_n^2}{h_n^2}$ $= O\left(\frac{1}{n^2 h_n^2}\right)$. This, together with (2.1) in Lemma 1, shows that the Liapunov ratio is $O((nh_n)^{-1/2}) = o(1)$.

If we want to replace $E(\tilde{F}_x(y))$ by $F_x(y)$ in (2.3), we need to know the limit of the expression

$$(nh_n)^{1/2}[E(\tilde{F}_x(y)) - F_x(y)]. \tag{2.4}$$

For this we use the decomposition

$$\begin{aligned} E(\tilde{F}_x(y)) - F_x(y) &= \sum_{i=1}^{n} \int_{x_{i-1}}^{x_i} \frac{1}{h_n} K\left(\frac{x-z}{h_n}\right) [F_{x_i}(y) - F_z(y)] dz \\ &+ \int_0^{x_n} \frac{1}{h_n} K\left(\frac{x-z}{h_n}\right) [F_z(y) - F_x(y)] dz. \end{aligned}$$

Without (C5) we readily have that $E(\tilde{F}_x(y)) - F_x(y) = O(n^{-1}) + O(h_n)$, so that the expression in (2.4) tends to zero if $nh_n^3 \to 0$.

With condition (C5), we can use Lemma 1 and find that, for the choice $h_n = Cn^{-1/5}$, the expression in (2.4) tends to $\frac{1}{2}\ddot{F}_x(y)\mu_2^K C^{5/2}$. □

3 RESULTS FOR REGRESSION QUANTILE ESTIMATORS

In this section we establish asymptotic representations for the quantile estimator $\tilde{\xi}_p(x)$. They are of the form (with $f = F'$) :

$$\tilde{\xi}_p(x) = \xi_p(x) + \frac{p - \tilde{F}_x(\xi_p(x))}{f(\xi_p)/\sigma(x)} + \tilde{R}_p(x).$$

For the remainder term $\tilde{R}_p(x)$, we study both the order in probability (Theorem 1) and the almost sure order (Theorem 2). Moreover, in view of further applications, we replace p by p_n, where $\{p_n\}$ is a sequence of numbers $(0 < p_n < 1)$, tending to some fixed p, at a certain rate.

We begin with a representation in probability, as Ghosh (1971) did for the usual quantiles and as Janssen and Veraverbeke (1987) did for regression quantiles in the homoscedastic model. Note that only the (b)-part of Theorem 1 allows for the optimal bandwidth choice $h_n = Cn^{-1/5}$, $C > 0$. That the latter is also the optimal bandwidth for the quantile estimator follows from the mean squared error calculations in Jones and Hall (1990).

THEOREM 1.

Let $0 < p < 1$ and $p_n = p + O((nh_n)^{-1/2})$.

(a) Assume (C1) - (C4), $h_n \to 0$, $nh_n^2 \to \infty$, $nh_n^3 \to 0$. If F is differentiable at ξ_p with $f(\xi_p) = F'(\xi_p) > 0$, then as $n \to \infty$,

$$\tilde{\xi}_{p_n}(x) = \xi_p(x) + \frac{p_n - \tilde{F}_x(\xi_p(x))}{f(\xi_p)/\sigma(x)} + o_P((nh_n)^{-1/2}).$$

(b) Assume (C1) - (C4), $h_n = Cn^{-1/5}$ for some $C > 0$. If F has a bounded second derivative in a neighborhood of ξ_p with $f(\xi_p) > 0$ and if μ and σ are differentiable at x, then as $n \to \infty$, the same conclusion as in (a) holds.

Proof. Put $V_n = (nh_n)^{1/2}(\tilde{\xi}_{p_n}(x) - \xi_p(x) + \frac{p - p_n}{f(\xi_p)/\sigma(x)})$, $W_n = (nh_n)^{1/2}\frac{p - \tilde{F}_x(\xi_p(x))}{f(\xi_p)/\sigma(x)}$.
By verifying the two conditions of Lemma 1 in Ghosh (1971), we will show that $V_n - W_n = o_P(1)$.

From the asymptotic normality results in Lemma 2, it immediately follows that $W_n \doteq O_P(1)$, so that Ghosh's first condition is satisfied. As to his second condition, note that for all t, $\{V_n \le t\} = \{Z_{tn} \le t_n\}$, where

$$\begin{aligned}
Z_{tn} &= (nh_n)^{1/2}\frac{F_x(t_n^*) - \tilde{F}_x(t_n^*)}{f(\xi_p)/\sigma(x)} \\
t_n^* &= \xi_p(x) + \frac{p_n - p}{f(\xi_p)/\sigma(x)} + (nh_n)^{-1/2}t \\
t_n &= (nh_n)^{1/2}\frac{F_x(t_n^*) - p_n}{f(\xi_p)/\sigma(x)}.
\end{aligned}$$

Since $F'_x(\xi_p(x)) = f(\xi_p)/\sigma(x)$, it follows that $t_n \to t$ as $n \to \infty$.
Ghosh's second condition will be satisfied if we show that $E(W_n - Z_{tn})^2 \to 0$ as $n \to \infty$. Now,

$$\begin{aligned} E(W_n - Z_{tn})^2 &= \frac{nh_n}{f^2(\xi_p)/\sigma^2(x)} Var(\widetilde{F}_x(t_n^*) - \widetilde{F}_x(\xi_p(x)) \\ &+ \frac{nh_n}{f^2(\xi_p)/\sigma^2(x)} \{E[\widetilde{F}_x(t_n^*) - \widetilde{F}_x(\xi_p(x))] - [F_x(t_n^*) - F_x(\xi_p(x))]\}^2 \\ &= I_n + II_n. \end{aligned} \tag{3.1}$$

The first term in (3.1) can be bounded by

$$I_n \leq \frac{nh_n}{f^2(\xi_p)/\sigma^2(x)} \frac{\Delta_n \, \|K\|_\infty}{h_n} \sum_{i=1}^n w_{ni}(x;h_n)|F_{x_i}(t_n^*) - F_{x_i}(\xi_p(x))|$$

which is $O((nh_n)^{-1/2})$, using the Lipschitz condition on F and the fact that $t_n^* - \xi_p(x) = O((nh_n)^{-1/2})$.
For the second term in (3.1), we write for any real y,

$$E[\widetilde{F}_x(\xi_p(x) + y)] - F_x(\xi_p(x) + y) = B_1(y) + B_2(y)$$

where

$$B_1(y) = \sum_{i=1}^n \frac{1}{h_n} \int_{x_{i-1}}^{x_i} K\left(\frac{x-z}{h_n}\right) [F_{x_i}(\xi_p(x) + y) - F_z(\xi_p(x) + y)]dz \tag{3.2}$$

$$B_2(y) = \sum_{i=1}^n \frac{1}{h_n} \int_{x_{i-1}}^{x_i} K\left(\frac{x-z}{h_n}\right) [F_z(\xi_p(x) + y) - F_x(\xi_p(x) + y)]dz. \tag{3.3}$$

Then,

$$II_n \leq \frac{nh_n}{f^2(\xi_p)/\sigma^2(x)} \{|B_1(t_n^* - \xi_p(x))| + |B_1(0)| + |B_2(t_n^* - \xi_p(x)) - B_2(0)|\}^2.$$

Using (1.2), it easily follows that for n large enough, $B_1(y) \leq C(\xi_p(x) + y)\Delta_n$. For $B_2(y)$ we obtain different order bounds according to different sets of conditions in (a) or (b). Under (C1) - (C4), we can use (1.2), and this gives, for n large enough,

$$B_2(y) \leq C(\xi_p(x) + y)(\int |u| K(u)du)h_n.$$

With this, $II_n = O(nh_n(n^{-1} + h_n)^2)$. Hence, in the situation of the (a)-part of the theorem we have that $E(W_n - Z_{tn})^2 \to 0$ if $nh_n \to \infty$ and $nh_n^3 \to 0$.
In order to finish the proof under the conditions of the (b)-part of the theorem, note that

$$\begin{aligned} B_2(y) &= \int_{\frac{x-x_n}{h_n}}^{\frac{x}{h_n}} K(z) D_{n,z}(y) dz \\ D_{n,z}(y) &= F_{x-h_n z}(\xi_p(x) + y) - F_x(\xi_p(x) + y), \end{aligned}$$

so that

$$|B_2(y) - B_2(0)| \leq \int_{\frac{x-x_n}{h_n}}^{\frac{x}{h_n}} K(z)|D_{n,z}(y) - D_{n,z}(0)|dz.$$

We have $D'_{n,z}(y) = g_{h_n z}(y) - g_0(y)$, $D''_{n,z}(y) = g'_{h_n z}(y) - g'_0(y)$, where

$$\begin{aligned} g_u(y) &= f\left(\frac{\mu(x) - \mu(x-u) + \sigma(x)\xi_p + y}{\sigma(x-u)}\right)\frac{1}{\sigma(x-u)} \\ g'_u(y) &= f'\left(\frac{\mu(x) - \mu(x-u) + \sigma(x)\xi_p + y}{\sigma(x-u)}\right)\frac{1}{\sigma^2(x-u)}. \end{aligned}$$

By Taylor expansion

$$D_{n,z}(y) - D_{n,z}(0) = y[g_{h_n z}(0) - g_0(0)] + \frac{1}{2}y^2 D''_{n,z}(\theta(y))$$

where $\theta(y)$ is between 0 and y.

Now, $|D''_{n,z}(\theta(y))| \leq K$, some constant, for all y small enough and all $|z| \leq L$, by taking n large and using the conditions that f' is bounded in a neighbourhood of ξ_p and that μ and σ are Lipshitz continuous. Further, since μ' and σ' exist at x, $g_{h_n z}(0) - g_0(0) = \dot{g}_0(0)h_n z + o(h_n z)$, where $\dot{g}_u(0) = \frac{\partial}{\partial u} g_u(0)$.

It follows that, $|B_2(y) - B_2(0)| = O(h_n|y| + y^2)$, which entails that

$$|B_2(t_n^* - \xi_p(x)) - B_2(0)| = O(h_n(nh_n)^{-1/2} + (nh_n)^{-1})$$

and $II_n = O(nh_n(n^{-1} + h_n(nh_n)^{-1/2} + (nh_n)^{-1})^2)$, which tends to zero under the conditions of the (b)-part of the theorem. □

As a corollary of Theorem 1, we obtain the asymptotic normality result for regression quantiles. Here we state the result for fixed p and for the optimal bandwidth choice.

COROLLARY 1.

Let $0 < p < 1$. Assume (C1) - (C4), $h_n = Cn^{-1/5}$ for some $C > 0$. If F has a bounded second derivative in a neighborhood of ξ_p with $f(\xi_p) > 0$ and if μ and σ are twice differentiable at x, then as $n \to \infty$,

$$(nh_n)^{1/2}(\tilde{\xi}_p(x) - \xi_p(x)) \xrightarrow{d} N\left(\frac{-\ddot{F}_x(\xi_p(x))\mu_2^K C^{5/2}}{2f(\xi_p)/\sigma(x)} \, ; \, \frac{p(1-p)\,\|K\|_2^2}{f^2(\xi_p)/\sigma^2(x)}\right). \qquad \square$$

Our next theorem (Theorem 2) will be the asymptotic representation for $\tilde{\xi}_p(x)$, but with an almost sure order bound for the remainder term. The result corresponds to Bahadur's (1966) theorem in the case of classical quantiles. For regression quantiles in the homoscedastic model, see also Cheng (1984), Janssen and Veraverbeke (1987),

Aerts and Geertsema (1990).

THEOREM 2.

Let $0 < p < 1$ and $p_n = p + o((nh_n)^{-1/2}(\log n)^{1/2})$.

Assume (C1) - (C4) and assume that F has a bounded second derivative in a neighborhood of ξ_p with $f(\xi_p) > 0$ and that μ and σ are differentiable at x.

(a) If $h_n \to 0$, $\dfrac{\log n}{nh_n} \to 0$, then as $n \to \infty$,

$$\tilde{\xi}_{p_n}(x) = \xi_p(x) + \frac{p_n - \tilde{F}_x(\xi_p(x))}{f(\xi_p)/\sigma(x)} + o((nh_n)^{-1/2}(\log n)^{1/2}) \text{ a.s.}$$

(b) If $h_n \to 0$, $\dfrac{nh_n^5}{\log n} = O(1)$, then as $n \to \infty$,

$$\tilde{\xi}_{p_n}(x) = \xi_p(x) + \frac{p_n - \tilde{F}_x(\xi_p(x))}{f(\xi_p)/\sigma(x)} + O((nh_n)^{-3/4}(\log n)^{3/4}) \text{ a.s.}$$

The proof of this theorem will be given after a series of three lemmas. These will be the important steps in the proof, but the results are also of independent interest.

LEMMA 3.

Let $0 < p < 1$ and $p_n = p + o((nh)^{-1/2}(\log n)^{1/2})$.

Assume (C1) -(C4), $h_n \to 0$, $\dfrac{\log n}{nh_n} \to 0$. If F is differentiable at ξ_p with $f(\xi_p) > 0$, then as $n \to \infty$,

$$\tilde{\xi}_{p_n}(x) - \xi_p(x) = O((nh_n)^{-1/2}(\log n)^{1/2}) \text{ a.s.}$$

Proof. For any $\epsilon > 0$ (possibly depending on n),

$$P(|\tilde{\xi}_{p_n}(x) - \xi_p(x)| \geq \epsilon) = P(\tilde{\xi}_{p_n}(x) \geq \xi_p(x) + \epsilon) + P(\tilde{\xi}_{p_n}(x) \leq \xi_p(x) - \epsilon).$$

We only deal with the first term since the second one is analogous. As in the proof of Theorem 2.1 in Cheng (1984), we obtain for n sufficiently large

$$P(\tilde{\xi}_{p_n}(x) \geq \xi_p(x) + \epsilon) \leq P(\sum_{i=1}^{n} S_{in} \geq \frac{1}{2}\beta_n)$$

where $S_{in} = w_{ni}(x; h_n)[F_{x_i}(\xi_p(x)+\epsilon) - I(Y_i \leq \xi_p(x)+\epsilon)]$ and $\beta_n = F(\xi_p + \frac{\epsilon}{\sigma(x)}) - p_n$. The condition $f(\xi_p) > 0$ ensures that $\beta_n > 0$ for n sufficiently large. Now since

$$\sum_{i=1}^{n} Var(S_{in}) = \sum_{i=1}^{n} E(S_{in}^2) \leq \frac{\|K\|_\infty \, \Delta_n}{h_n} \sum_{i=1}^{n} E|S_{in}| \leq \frac{\|K\|_\infty \, \Delta_n}{h_n},$$

we have by Bernstein's inequality (see e.g. Serfling (1980), p.95),

$$P(\sum_{i=1}^{n} S_{in} \geq \frac{1}{2}\beta_n) \leq \exp\left\{-\beta_n^2/(8\sum_{i=1}^{n} Var(S_{in}) + \frac{4}{3}\beta_n \frac{\|K\|_\infty \, \Delta_n}{h_n})\right\}.$$

Now let ϵ depend on n, with $\epsilon_n = B(nh_n)^{-1/2}(log\ n)^{1/2}$ for some $B > 0$. Then, since $\beta_n = \epsilon_n f(\xi_p)/\sigma(x) + o(\epsilon_n)$, we obtain that $P(\tilde{\xi}_p(x) > \xi_p(x) + \epsilon_n) = O(n^{-2})$, for B large enough. The result now follows by the Borel-Cantelli theorem. □

LEMMA 4.

Assume (C1) - (C4), $h_n \to 0$, $\dfrac{log\ n}{nh_n} \to 0$. Let $\{d_n\}$ be a sequence of positive constants such that $d_n \sim k_0(nh_n)^{-1/2}(log\ n)^{1/2}$ as $n \to \infty$, for some constant k_0. Then, as $n \to \infty$,

$$\begin{aligned} H_{p_n}(x) &= \sup_{|y|\le d_n} |\tilde{F}_x(\xi_p(x)+y) - \tilde{F}_x(\xi_p(x)) - E[\tilde{F}_x(\xi_p(x)+y) - \tilde{F}_x(\xi_p(x))]| \\ &= O((nh_n)^{-3/4}(log\ n)^{3/4}) \text{ a.s.} \end{aligned}$$

Proof. As in Cheng (1984), let $\{e_n\}$ be a sequence of positive integers such that $e_n \sim k_0\ n^{1/4}(log\ n)^{1/2}$, $n \to \infty$.

For integers $r = -e_n, \ldots, e_n - 1$, put $\eta_{r,n}(x) = \xi_p(x) + d_n e_n^{-1} r$ and

$$\begin{aligned} \alpha_{rn}(x) &= E[\tilde{F}_x(\eta_{r+1,n}(x)) - \tilde{F}_x(\eta_{r,n}(x))] \\ G_{rn}(x) &= |[\tilde{F}_x(\eta_{r,n}(x)) - \tilde{F}_x(\xi_p(x))] - E[\tilde{F}_x(\eta_{r,n}(x)) - \tilde{F}_x(\xi_p(x))]|. \end{aligned}$$

Then,

$$H_{pn}(x) \le L_n(x) + J_n(x)$$

where

$$\begin{aligned} L_n(x) &= \max\{G_{rn}(x) : -e_n \le r \le e_n\} \\ J_n(x) &= \max\{\alpha_{rn}(x) : -e_n \le r \le e_n - 1\}. \end{aligned}$$

Define $\gamma_n = k_1(nh_n)^{-3/4}(log\ n)^{3/4}$, with k_1 to be specified, and

$$Z_{ni} = w_{ni}(x; h_n)[I(Y_i \le \eta_{r,n}(x)) - I(Y_i \le \xi_p(x))].$$

Then,

$$|Z_{ni}| \le \frac{\|K\|_\infty\ \Delta_n}{h_n} = m_n$$
$$\sum_{i=1}^n Var(Z_{ni}) \le m_n \sum_{i=1}^n E|Z_{ni}| \le c\frac{d_n}{nh_n},$$

for some constant $c > 0$ and for n large.

By Bernstein's inequality,

$$P(G_{rn}(x) \ge \gamma_n) = P(|\sum_{i=1}^n (Z_{ni} - E(Z_{ni}))| \ge \gamma_n)$$

$$
\begin{aligned}
&\leq 2\exp\left\{-\frac{\gamma_n^2}{2\sum_{i=1}^{n} Var(Z_{in}) + \frac{2}{3}m_n\gamma_n}\right\} \\
&\leq 2\exp\left\{-\frac{\gamma_n^2}{4c(nh_n)^{-1}d_n}\right\}, \text{ since } \frac{\gamma_n}{d_n} = O((\frac{log\ n}{nh_n})^{1/4}) \to 0, \\
&= O(n^{-2}), \text{ by choice of } k_1.
\end{aligned}
$$

Therefore, $\sum_{n=1}^{\infty} P(L_n(x) \geq \gamma_n) \leq \sum_{n=1}^{\infty} \sum_{r=-e_n}^{e_n} P(G_{rn}(x) \geq \gamma_n) < \infty$ and hence, $L_n(x) = O((nh_n)^{-3/4}(log\ n)^{3/4})$ a.s.

Also,

$$
\begin{aligned}
|\alpha_{rn}(x)| &\leq \sum_{i=1}^{n} |F_{x_i}(\eta_{r+1,n}(x)) - F_{x_i}(\eta_{r,n}(x))| w_{ni}(x) \\
&= O(\frac{d_n}{e_n}) = O(n^{-3/4}h_n^{-1/2}) = o((nh_n)^{-3/4}(log\ n)^{3/4}).
\end{aligned}
$$

□

LEMMA 5.

Assume (C1) - (C4). Let $\{d_n\}$ be any sequence of positive constants such that $d_n \to 0$. If F has a bounded second derivative in a neighborhood of ξ_p and if μ and σ are twice differentiable at x, then as $n \to \infty$,

$$
\begin{aligned}
&\sup_{|y|\leq d_n} |E[\tilde{F}_x(\xi_p(x)+y) - \tilde{F}_x(\xi_p(x))] - [F_x(\xi_p(x)+y) - F_x(\xi_p(x))]| \\
&= O(n^{-1} + d_n h_n + d_n^2). \qquad (3.4)
\end{aligned}
$$

Proof. With B_1 and B_2 as in (3.2) and (3.3), we have as in the proof of Theorem 1 that the left hand side in (3.4) is bounded by

$$
|B_1(0)| + \sup_{|y|\leq d_n} |B_1(y)| + \sup_{|y|\leq d_n} |B_2(y) - B_2(0)||
$$

and this is $O(n^{-1} + d_n h_n + d_n^2)$. □

Proof of Theorem 2.

$$
\begin{aligned}
& F_x(\tilde{\xi}_{p_n}(x)) - F_x(\xi_p(x)) \\
= \ & [\tilde{F}_x(\tilde{\xi}_{p_n}(x)) - p_n] + [p_n - \tilde{F}_x(\xi_p(x))] \\
& -\{[\tilde{F}_x(\tilde{\xi}_{p_n}(x)) - \tilde{F}_x(\xi_p(x))] - E[\tilde{F}_x(\tilde{\xi}_{p_n}(x)) - \tilde{F}_x(\xi_p(x))]\} \\
& -\{E[\tilde{F}_x(\tilde{\xi}_{p_n}(x)) - \tilde{F}_x(\xi_p(x))] - [F_x(\tilde{\xi}_{p_n}(x)) - F_x(\xi_p(x))]\} \\
= \ & A_n + [p_n - \tilde{F}_x(\xi_p(x))] + B_n + C_n.
\end{aligned}
$$

It immediately follows that $|A_n| \leq \max_{1\leq i\leq n} w_{ni}(x; h_n) = O(\frac{1}{nh_n})$ a.s.. Lemma 3, 4 and 5 imply that $B_n = O((nh_n)^{-3/4}(log\ n)^{3/4})$ a.s. and $C_n = O(n^{-1}+h_n(nh_n)^{-1/2}(log\ n)^{1/2}+$

$(nh_n)^{-1}log\ n)$ a.s.

Using a Taylor series expansion and Lemma 3 :

$$F_x(\tilde{\xi}_{p_n}(x)) - F_x(\xi_p(x)) = \frac{f(\xi_p)}{\sigma(x)}(\tilde{\xi}_{p_n}(x) - \xi_p(x)) + O((nh_n)^{-1}log\ n).$$

It follows that $\tilde{\xi}_{p_n}(x) - \xi_p(x) = \dfrac{p_n - \tilde{F}_x(\xi_p(x))}{f(\xi_p)/\sigma(x)} + \tilde{R}_{p_n}(x)$ where a.s.

$$\begin{aligned}\tilde{R}_{p_n}(x) &= O((nh_n)^{-1}) + O((nh_n)^{-3/4}(log\ n)^{3/4}) + O(n^{-1}) \\ &\quad + O(h_n(nh_n)^{-1/2}(log\ n)^{1/2}) + O((nh_n)^{-1}log\ n)\end{aligned}$$

and this is $o((nh)^{-1/2}(log\ n)^{1/2})$ if $\dfrac{log\ n}{nh_n} \to 0$ or $O((nh_n)^{-3/4}(log\ n)^{3/4})$ if $\dfrac{nh_n^5}{log\ n} =$ $O(1)$. □

REFERENCES

Aerts, M. and Geertsema, J. (1990). Bounded length confidence intervals in nonparametric regression. Sequential Analysis 9, 171-192.

Aerts, M., Janssen, P. and Veraverbeke, N. (1992). Bootstrapping regression quantiles. (submitted)

Bahadur, R.R. (1966). A note on quantiles in large samples. Ann. Math. Statist. **37**, 577-580.

Cheng, K.F. (1984). Nonparametric estimation of regression function using linear combinations of sample quantile regression functions. Sankhyā, A, **46**, 287-302.

Ghosh, J.K. (1971). A new proof of the Bahadur representation of quantiles and an application. Ann. Math. Statist. **42**, 1957-1961.

Janssen, P. and Veraverbeke, N. (1987). On nonparametric regression estimators based on regression quantiles. Commun. Statist. -Theory Meth. **16**, 383-396.

Jones, M.C. and Hall, P. (1990). Mean squared error properties of kernel estimates of regression quantiles. Statistics and Probability Letters **10**, 283-289.

Serfling, R.J. (1980). Approximation Theorems of Mathematical Statistics. Wiley, New York.

Stone, C.J. (1977). Consistent nonparametric regression. Ann. Statist. **5**, 595-645.

NONNEGATIVE MOVING-AVERAGE MODELS

JIŘÍ ANDĚL

Charles University, Prague

ABSTRACT. Nonnegative MA(1) and MA(2) models are investigated in the paper. A new method for estimating their parameters is proposed. It is proved that the estimators are strongly consistent. Small-sample properties are demonstrated on a simulation study. Real data on biological oxygen demand are analyzed by the new method.

1990 MATHEMATICS SUBJECT CLASSIFICATION: 62M10

KEY WORDS AND PHRASES: Moving-average models, nonnegative processes, simulations.

1. INTRODUCTION

Let $\{e_t\}$ be a nonnegative strict white noise, i.e. a sequence of nonnegative i.i.d. random variables with $Ee_t^2 < \infty$. Define $F(x) = P(e_t < x)$. Assume that $P(e_t = 0) < 1$. Bell and Smith (1986) investigated a nonnegative AR(1) process $\{X_t\}$ defined by

$$X_t = bX_{t-1} + e_t, \qquad b \in [0,1).$$

They proved that

$$b^* = \min\left\{\frac{X_2}{X_1}, \frac{X_3}{X_2}, \dots, \frac{X_n}{X_{n-1}}\right\} \tag{1.1}$$

is a strongly consistent estimator for b if and only if

$$F(d) - F(c) < 1 \qquad \text{for all} \quad 0 < c < d < \infty. \tag{1.2}$$

Some explicit results were derived under the assumption that e_t have an exponential distribution $Ex(\lambda)$ with the density

$$f(y) = \lambda^{-1}e^{-y/\lambda} \qquad \text{for } y > 0.$$

Anděl (1988) found the distribution of b^* and Turkman (1990) developed Bayesian analysis of the model. The estimation procedure was generalized to AR models of higher order (Anděl 1989). In the present paper we modify the estimator of the type (1.1) to the MA processes.

2. Model MA(1)

Define $X_t = e_t + be_{t-1}$ for all integers t. Assume that the condition (1.2) holds. Since $\{e_t\}$ is a nonnegative strict white noise, the coefficient b must be nonnegative in order that all X_t are also nonnegative. For details see Anděl (1991). In our paper we assume that $0 \le b \le 1$. Let a realization $X_1, \dots, X_n$ of our model be given.

Theorem 2.1. *Define*

$$b^* = \min_{2 \le t \le n-1} \frac{X_{t-1} + X_{t+1}}{X_t}. \tag{2.1}$$

If the condition (1.2) is fulfilled then $b^* \to b$ a.s. as $n \to \infty$.

Proof. We have

$$\frac{X_{t-1} + X_{t+1}}{X_t} = b + Y_t \tag{2.2}$$

where

$$Y_t = \frac{be_{t-2} + (1 - b^2)e_{t-1} + e_{t+1}}{e_t + be_{t-1}}. \tag{2.3}$$

Define $M_n = \min(Y_2, \dots, Y_n)$. If $\lceil a \rceil$ denotes the integer part of a then for every $\varepsilon > 0$ we get

$$\begin{aligned} P(M_n > \varepsilon) &= P(Y_t > \varepsilon \text{ for } t = 2, \dots, n) \le P\left(Y_{2+4k} > \varepsilon \text{ for } k = 0, \dots, \left\lceil \frac{n-2}{4} \right\rceil\right) \\ &\le P\left\{be_{4k} + (1 - b^2)e_{4k+1} + e_{4k+3} > \varepsilon e_{4k+2} \text{ for } k = 0, \dots, \left\lceil \frac{n-2}{4} \right\rceil\right\} \\ &= \left[P\{be_0 + (1 - b^2)e_1 + e_3 > \varepsilon e_2\}\right]^{\lceil \frac{n+2}{4} \rceil}. \end{aligned}$$

Since $0 \le b \le 1$ we have

$$P\{be_0 + (1 - b^2)e_1 + e_3 > \varepsilon e_2\} \le P\{e_0 + e_1 + e_3 > \varepsilon e_2\}.$$

Lemma 7.1 gives that under (1.2) $P(e_0 + e_1 + e_3 > \varepsilon e_2) < 1$ holds for every $\varepsilon > 0$. This implies that

$$\sum_{n=2}^{\infty} P(M_n > \varepsilon) < \infty$$

and according to the Borel–Cantelli lemma with probability one only a finite number of the events $\{M_n > \varepsilon\}$ occur. Thus $M_n \to 0$ a.s. □

Remark 2.2. If the condition (1.2) is not fulfilled then b^* is not a consistent estimator for b. Really, assume that $P(c \le e_t < d) = 1$ for some $0 < c < d < \infty$. Taking into account (2.3) we get that

$$Y_t \ge \frac{bc + (1 - b^2)c + c}{d + bd} \ge \frac{c}{2d} > 0$$

holds with probability one and thus $M_n \ge \frac{c}{2d}$ for all n.

3. Model MA(2)

In this section we consider an MA(2) process

$$X_t = e_t + b_1 e_{t-1} + b_2 e_{t-2} \tag{3.1}$$

with $b_1 \geq 0$, $b_2 \geq 0$ and with a nonnegative strict white noise $\{e_t\}$. We assume that

$$z^2 + b_1 z + b_2 \neq 0 \quad \text{for } |z| > 1. \tag{3.2}$$

It is well known [cf. Box and Jenkins (1970), §3.3.4] that the condition (3.2) is fulfilled if and only if the vector $(b_1, b_2)'$ belongs to the closed triangle with vertices $(-2, 1)$, $(0, -1)$, $(2, 1)$. Since we assume also $b_1 \geq 0$, $b_2 \geq 0$ we deal only with the vectors $(b_1, b_2)' \in K$ where K is the closed tetragon with the vertices (0,0), (1,0), (2,1), (0,1), i.e.

$$K = \{(b_1, b_2)' : b_1 \geq 0, 0 \leq b_2 \leq 1, b_2 \geq -1 + b_1\}.$$

The interior of K corresponds to the invertible MA(2) models (3.1) with positive coefficients b_1, b_2.

Theorem 3.1. *Let a realization $X_1, \ldots, X_n$ of the model (3.1) be given where $(b_1, b_2)' \in K$. Assume that the condition (1.2) is fulfilled. Define*

$$b_1^* = \min_{2 \leq t \leq n-1} \frac{X_{t+1} + 3X_{t-1}}{X_t}, \qquad b_2^* = \min_{3 \leq t \leq n-2} \frac{X_{t+2} + 2X_{t-1} + X_{t-2}}{X_t}.$$

If $n \to \infty$ then $b_1^* \to b_1$ and $b_2^* \to b_2$ a.s.

Proof. We have

$$\frac{X_{t+1} + 3X_{t-1}}{X_t} = b_1 + Z_{t1},$$

$$\frac{X_{t+2} + 2X_{t-1} + X_{t-2}}{X_t} = b_2 + Z_{t2}$$

where

$$Z_{t1} = \frac{e_{t+1} + (3 - b_1^2 + b_2)e_{t-1} + b_1(3 - b_2)e_{t-2} + 3b_2 e_{t-3}}{e_t + b_1 e_{t-1} + b_2 e_{t-2}},$$

$$Z_{t2} = \frac{e_{t+2} + b_1 e_{t+1} + (2 - b_1 b_2)e_{t-1} + (1 + 2b_1 - b_2^2)e_{t-2} + (2b_2 + b_1)e_{t-3}}{e_t + b_1 e_{t-1} + b_2 e_{t-2}} + \frac{b_2 e_{t-4}}{e_t + b_1 e_{t-1} + b_2 e_{t-2}}.$$

It can be easily checked that

$$3 - b_1^2 + b_2 \geq 0, \qquad 3 - b_2 \geq 0, \qquad 2 - b_1 b_2 \geq 0, \qquad 1 + 2b_1 - b_2^2 \geq 0$$

and thus $Z_{t1} \geq 0$, $Z_{t2} \geq 0$ with probability one. Now the remaining part of the proof is quite analogous to that of Theorem 2.1. □

It is also easy to show that if (1.2) is not fulfilled then b_1^* and b_2^* are not consistent estimators.

4. Numerical results in MA(1) model

Consider again the model $X_t = e_t + be_{t-1}$ where $b \in [0,1]$. In this section we assume that e_t has an exponential distribution $Ex(\lambda)$. Then we say that we deal with an MA(1) model with exponentially distributed white noise. In this case it is clear that the condition (1.2) is fulfilled.

We simulated realizations $X_1, \ldots, X_n$ with $e_t \sim Ex(1)$ for different $b \in [0,1]$ and for different n. From each realization the estimate b^* was calculated. The results are summarized in Tab. 1. For each combination of b and n 1 000 simulations were performed. The average of the corresponding 1 000 estimates b^* is introduced in the column with heading "average" whereas the empirical standard deviation calculated from the same estimates is given in the column "sd".

It is clear that the estimator b^* suffers from a large positive bias. Unfortunately, our attempts to derive the distribution of b^* were not successful and so no explicit formula for Eb^* is known. Thus only some approximations to Eb^* were investigated.

Theorem 4.1. *Let $X_t = e_t + be_{t-1}$ where $b \in [0,1]$ and $e_t \sim Ex(1)$. Define*

$$W_t = \frac{X_{t-1} + X_{t+1}}{X_t}.$$

Let $G(x)$ be the distribution function of the random variable W. Then clearly $G(x) = 0$ for $x \le b$. Further it holds:

(i) *If $b = 0$ then*

$$G(x) = \frac{x^2}{(1+x)^2} \qquad \text{for } x > 0.$$

(ii) *If $b \in (0,1)$ then*

$$G(x) = \begin{cases} \dfrac{(x-b)^3}{(1-b)x(x+1-b)(x+1)} & \text{for } b < x \le \dfrac{1}{b}, \\[2ex] \dfrac{(x-b)^2}{x(x+1-b)} + \dfrac{bx-1}{bx}\,\dfrac{b(1+b)x - b^3 - 1}{(x+1-b)(bx+b-1)} & \text{for } x > \dfrac{1}{b}. \end{cases}$$

(iii) *If $b = 1$ then*

$$G(x) = \frac{(x-1)^2(x+2)}{x^3} \qquad \text{for } x > 1.$$

Proof. The distribution function G is calculated by standard elementary methods but the derivation is long and cumbersome and so it is not introduced here. □

We can see that (i) is a special case of (ii) when $b = 0$ and (iii) is also a special case of (ii) when $b = 1$.

Since e_t are independent, the variables W_t and W_s are independent for $|s-t| \ge 4$. One can try to approximate the distribution function of $b^* = \min\{W_2, \ldots, W_{n-1}\}$ by

$$H(x) = 1 - \{1 - G(x)\}^{n-2}$$

TAB. 1. Results of simulations of MA(1) models with exponentially distributed white noise

b	n	average	sd	mean	approx. mean
0.0	20	0.28	0.18	0.28	0.28
	50	0.15	0.09	0.15	0.17
	100	0.10	0.06	0.10	0.12
0.1	20	0.42	0.18	0.42	0.42
	50	0.29	0.10	0.29	0.29
	100	0.23	0.07	0.23	0.24
0.2	20	0.53	0.18	0.55	0.54
	50	0.41	0.10	0.41	0.41
	100	0.35	0.07	0.35	0.35
0.3	20	0.65	0.17	0.66	0.66
	50	0.52	0.10	0.52	0.52
	100	0.47	0.07	0.46	0.45
0.4	20	0.75	0.17	0.77	0.77
	50	0.63	0.10	0.63	0.62
	100	0.57	0.07	0.57	0.56
0.5	20	0.85	0.15	0.86	0.86
	50	0.73	0.10	0.73	0.72
	100	0.67	0.07	0.67	0.65
0.6	20	0.94	0.15	0.95	0.94
	50	0.82	0.09	0.83	0.81
	100	0.77	0.07	0.77	0.75
0.7	20	1.02	0.14	1.03	1.02
	50	0.91	0.09	0.92	0.89
	100	0.86	0.06	0.86	0.84
0.8	20	1.09	0.12	1.09	1.08
	50	0.99	0.08	1.00	0.97
	100	0.95	0.06	0.95	0.92
0.9	20	1.13	0.10	1.14	1.12
	50	1.05	0.06	1.06	1.04
	100	1.02	0.05	1.02	1.00
1.0	20	1.15	0.09	1.15	1.16
	50	1.09	0.05	1.08	1.10
	100	1.06	0.03	1.06	1.07

which is the distribution function of $\min\{W_2, \ldots, W_{n-1}\}$ calculated under the condition that $W_2, \ldots, W_{n-1}$ are independent.

Since $H(x) = 0$ fo $x \leq 0$, the expectation corresponding to this distribution function is

$$\nu = \int_0^\infty \{1 - H(x)\}\, dx = \int_0^\infty \{1 - G(x)\}^{n-2}\, dx. \tag{4.1}$$

Because the formulas for G are introduced in Theorem 4.1, it is easy to calculate ν by numerical integration. The results are given in Tab. 1 in the column "mean". Comparing ν with the averages of simulations we can see that ν can be considered as a good approximation to the true expectation Eb^*.

Using the results of numerical integration it was found that from the practical point of view ν can be approximated by

$$m = b + \frac{1.173 + 1.917b - 2.4b^2}{\sqrt{n}}. \tag{4.2}$$

The values of m are given in Tab. 1 in the column "approximate mean".

If $n > 8$ then we have from (4.2) and from the condition $b \in [0, 1]$ that

$$b = \frac{\sqrt{n} + 1.917 - \sqrt{(\sqrt{n} + 1.917)^2 - 9.6(m\sqrt{n} - 1.173)}}{4.8}.$$

For a given value of b^* define

$$D = (\sqrt{n} + 1.917)^2 - 9.6(b^*\sqrt{n} - 1.173).$$

To reduce the bias of b^* we can introduce a modified estimator

$$b^+ = \begin{cases} \dfrac{\sqrt{n} + 1.917 - \sqrt{D}}{4.8} & \text{for } D \geq 0, \\ 1 & \text{for } D < 0. \end{cases} \tag{4.3}$$

5. Statistical analysis of empirical data

Bell and Smith (1986) considered data from Willamette River, Orego, on biological oxygen demand. They reported the following results. The least squares estimates of the parameters b ans μ in the AR(1) model

$$X_t - \mu = b(X_{t-1} - \mu) + e_t$$

are $\hat{b} = 0.254$ and $\hat{\mu} = 1.09$, respectively. If we use the AR(1) model $X_t = bX_{t-1} + e_t$ with exponentially distributed white noise $\{e_t\}$ then the estimate (1.1) for b is $b^* = 0.307$.

However, the standard statistical analysis shows that this set of data can be a little better described by an MA(1) model than by an AR(1) model. The classical estimated MA(1) model is then $X_t = 6.576 + e_t + 0.271e_{t-1}$ where $\{e_t\}$ is a white noise with $Ee_t = 0$. Considering the model $X_t = e_t + be_{t-1}$ with a nonnegative white noise we have got the estimate (2.1) $b^* = 0.717$. Assuming that the strict white noise $\{e_t\}$ is exponentially distributed, we used (4.3) for improving b^* to $b^+ = 0.522$.

TAB. 2. Results of simulations of MA(2) models with exponentially distributed white noise

b_1	b_2	n	average of b_1^*	average of b_2^*	sd 1	sd 2	mean 1	mean 2
0	0	20	0.50	0.72	0.32	0.37	0.51	0.77
		50	0.27	0.44	0.17	0.21	0.27	0.45
		100	0.18	0.32	0.11	0.15	0.18	0.32
1	0	20	1.80	1.36	0.31	0.44	1.84	1.41
		50	1.55	1.00	0.23	0.33	1.56	1.00
		100	1.41	0.78	0.17	0.24	1.42	0.80
0	1	20	0.83	1.74	0.43	0.32	0.76	1.75
		50	0.52	1.50	0.25	0.19	0.48	1.49
		100	0.39	1.38	0.17	0.14	0.35	1.38
0.5	0.5	20	1.59	1.79	0.42	0.43	1.30†	1.58†
		50	1.28	1.45	0.27	0.29	1.02†	1.70†
		100	1.11	1.27	0.21	0.23	0.88†	1.41†

6. Numerical results in MA(2) model

Consider the model (3.1) such that $(b_1, b_2)' \in K$ where $\{e_t\}$ are independent random variables with a distribution $Ex(1)$. Results of simulations of this process are in Tab. 2. In the columns "average of b_1" and "average of b_2" are averages calculated from 1 000 simulations, whereas the columns "sd 1" and "sd 2" contain the corresponding empirical standard deviations. Also in this case we tried to get some approximations to Eb_1^* and Eb_2^*.

Theorem 6.1. *Define*

$$V_t = \frac{X_{t+1} + 3X_{t-1}}{X_t}, \qquad W_t = \frac{X_{t+2} + 2X_{t-1} + X_{t-2}}{X_t}.$$

Let G_1 and G_2 be the distribution functions of V_t and W_t, respectively. Then $G_1(x) = 0$ for $x \le b_1$ and $G_2(x) = 0$ for $x \le b_2$.

(i) *If $b_1 = 0$, $b_2 = 0$, then*

$$G_1(x) = \frac{x^2}{(1+x)(3+x)} \qquad \text{for } x > 0,$$

$$G_2(x) = 1 + \frac{4}{1+x} - \frac{8}{2+x} - \frac{1}{(1+x)^2} \qquad \text{for } x > 0.$$

† linear approximation

(ii) *If* $b_1 = 1$, $b_2 = 0$, *then*

$$G_1(x) = \begin{cases} 0 & \text{for } x \le 1, \\ \dfrac{(x-1)^3}{2x(x+2)} & \text{for } 1 < x \le 3, \\ \dfrac{x^3 - 17x + 28}{(x-2)x(x+2)} & \text{for } 3 < x, \end{cases}$$

$$G_2(x) = \begin{cases} \dfrac{x^2}{16}\left\{\dfrac{27}{x+3} - \dfrac{19x^2 + 24x + 9}{(x+1)^3}\right\} & \text{for } 0 < x \le 2, \\ \dfrac{27}{8}\dfrac{x^2 + 4x - 6}{(x+1)(x+3)} - \dfrac{(x-2)(19x^2 - 52x + 37)}{8(x-1)^3} & \\ \quad - \dfrac{x(33x^4 - 36x^3 - 54x^2 + 36x + 37)}{8(x-1)^3(x+1)^3} & \text{for } 2 < x. \end{cases}$$

(iii) *If* $b_1 = 0$, $b_2 = 1$, *then*

$$G_1 = 1 - \frac{1}{6(1+x)^2} + \frac{81}{2(3+x)^2} - \frac{256}{3(4+x)^2} \qquad \text{for } x > 0,$$

$$G_2(x) = \begin{cases} 0 & \text{for } x \le 1, \\ (x-1)\left\{\dfrac{5x^2 + 5x - 2}{x^3} - \dfrac{4(x^2 + 4x + 11)}{(x+1)^3}\right\} & \text{for } x > 1. \end{cases}$$

Proof. The formulas follow from standard calculations. □

One can try to approximate the distribution functions of b_1^* and b_2^* by $H_1(x) = 1 - \{1 - G_1(x)\}^{n-2}$ and $H_2(x) = 1 - \{1 - G_2(x)\}^{n-4}$, respectively. The means

$$\nu_1 = \int_0^\infty \{1 - G_1(x)\}^{n-2}\, dx, \qquad \nu_2 = \int_0^\infty \{1 - G_2(x)\}^{n-4}\, dx$$

were computed by numerical integration and can be found in Tab. 2 in the columns "mean 1" and "mean 2", respectively. They approximate the averages of b_1^* and b_2^* quite well.

Unfortunately, the bias of b_1^* and b_2^* cannot be reasonably approximated by a linear function, which is demonstrated by comparing results for $b_1 = 0.5$, $b_2 = 0.5$ in Tab. 2. The problem how to reduce the bias should be further investigated.

7. Appendix

Lemma 7.1. *Let* $Y_0, \ldots, Y_n$ *be nonnegative i.i.d. random variables such that* $P(Y_0 > 0) > 0$. *Then there exists* $\varepsilon > 0$ *such that*

$$P\left(\sum_{k=1}^{n} Y_k > \varepsilon Y_0\right) = 1 \tag{7.1}$$

if and only if there exist numbers $0 < c < d < \infty$ such that $P(c \le Y_0 < d) = 1$.

Proof. Let (7.1) hold for an $\varepsilon > 0$. For every $\delta > 0$ we have

$$0 = P\left(\sum_{k=1}^{n} Y_k \le \varepsilon Y_0\right) \ge P\left(Y_0 \ge \frac{\delta}{\varepsilon}, Y_k \le \frac{\delta}{n} \text{ for } k = 1, \dots, n\right)$$
$$= P\left(Y_0 \ge \frac{\delta}{\varepsilon}\right)\left\{P\left(Y_0 \le \frac{\delta}{n}\right)\right\}^n.$$

There exists $\delta_0 > 0$ such that $P\left(Y_0 \le \frac{\delta_0}{n}\right) > 0$. Then $P\left(Y_0 \le \frac{\delta_0}{\varepsilon}\right) = 0$. Similarly, there exists $\delta_1 > 0$ such that $P\left(Y_0 \ge \frac{\delta_1}{n}\right) > 0$. Then we get $P\left(Y_0 \le \frac{\delta_1}{\varepsilon}\right) = 0$. Now, we can take $c = \frac{\delta_1}{n}$, $d = \frac{\delta_0}{\varepsilon}$.

Let $P(c \le Y_0 < d) = 1$ for some $0 < c < d < \infty$. Then for every $\varepsilon > 0$

$$P\left(\sum_{k=1}^{n} Y_k > \varepsilon Y_0\right) \ge P\left(Y_k \ge \frac{c}{n} \text{ for } k = 1, \dots, n, Y_0 < \frac{c}{\varepsilon}\right)$$
$$\ge P\left(Y_k \ge c \text{ for } k = 1, \dots, n, Y_0 < \frac{c}{\varepsilon}\right)$$
$$= P\left(Y_0 < \frac{c}{\varepsilon}\right)\{P(Y_0 \ge c)\}^n.$$

If we put $\varepsilon = \frac{c}{d}$ then $P\left(Y_0 < \frac{c}{\varepsilon}\right) = P(Y_0 < d) = 1$. Since $P(Y_0 \ge c) = 1$, we have that

$$P\left(\sum_{k=1}^{n} Y_k > \varepsilon Y_0\right) = 1. \qquad \square$$

Remark 7.2. Bell and Smith (1986) proved Lemma 7.1 for $n = 1$. Our proof for $n \ge 1$ is a generalization of their method.

REFERENCES

[1] Anděl J., *On AR(1) processes with exponential white noise*, Commun. Statist. - Theory Meth. **17** (1988), 1481–1495.

[2] Anděl J., *Non-negative autoregressive processes*, Time Ser. Anal. **10** (1989), 1–11.

[3] Anděl M., *Non-negative linear processes*, Apl. mat. **36** (1991), 277–283.

[4] Bell C. B. and Smith E. P., *Inference for non-negative autoregressive schemes*, Commun. Statist. - Theory Meth. **15** (1986), 2267–2293.

[5] Box G. E. P., Jenkins G. M., *Time Series Analysis. Forecasting and control*, Holden Day, San Francisco, 1970.

[6] Turkman M. A. A., *Bayesian analysis of an autoregressive process with exponential white noise*, Statistics **21** (1990), 601–608.

Charles University, Department of Probability and Mathematical Statistics, Sokolovská 83, CZ-186 00 Praha 8 - Karlín, Czech Republic; tel: (+42 2) 231 76 82; fax: (+42 2) 84 76 88; telex: (+42 2) 121 673 mff

E-mail: andel@cspguk11.bitnet

BIASED SAMPLES FROM A SEARCH AND THE ASYMPTOTIC BEHAVIOUR OF BAYESIAN MODELS

J. A. BATHER
University of Sussex, Falmer, Brighton, BN1 9QH, U.K.

ABSTRACT

Consider a hidden population of N objects and a search in which different objects are discovered independently, with probabilities that depend on their values. This paper investigates the distribution of values in the population of discovered objects. It also studies the behaviour of a Bayesian model with a prior process, for the number N and the set of values in the hidden population, which consists of a mixture of Poisson processes. If N is large, the weights on different components in the mixture change in a coherent way during the search.

Key words: Bayesian model; Poisson processes; search.
AMS classification: 90 B 40

1. Introduction

Consider a search for hidden objects, starting at time zero, in which the discovery times for different items are represented by independent exponential random variables. The objects may have several characteristics but, for simplicity, we shall concentrate on a single variable $v \geq 0$, called the value. For example, we can imagine fishing in a pond, with v representing the size of fish. In general, suppose that the hidden population consists of N objects, $0_1, 0_2, \ldots 0_N$, with corresponding values $v_1, v_2, \ldots v_N$. The search is parametrised by a positive function λ, so that the discovery time for an item of value $v \geq 0$ has probability density $\lambda(v) \exp(-\lambda(v)s)$, $s > 0$. In other words, given a population of size N, with values $v_1, v_2, \ldots v_N$ the corresponding discovery times $s_1, s_2, \ldots s_N$ are independent exponential random variables, where s_j has mean $1/\lambda(v_j)$.

For an object of value v and a search of fixed length t, let

$$p(v,t) = 1 - \exp(-\lambda(v)t), \quad q(v,t) = 1 - p(v,t). \tag{1.1}$$

Thus, p(v,t) is the probability of discovery before time t for a single item. The set of objects discovered by time t is a random subset of the original population and, unless the function λ is constant, it will be a biased sample from the population. It might be imagined that the probabilities of inclusion in the sample for objects 0_i and 0_j are in the ratio $p(v_i,t) : p(v_j,t)$, but this only holds asymptotically as $N \rightarrow \infty$.

The next section is concerned with this kind of bias, which arises from discovery rates that depend on the values of the objects. The following result will be established. Imagine an infinite sequence of objects with values $v_1, v_2, \ldots$ and let $p_i = p(v_i,t)$, $i \geq 1$. For each N, consider the population consisting of the first N objects and the random subset of this population discovered after searching until time t. Let $\pi_i(N) = \pi_i(v_1, v_2, \ldots v_N)$ be the probability that the subset is non-empty and that a single object chosen at random from it is 0_i, $i \leq N$. If the sequence $\{v_r, r \geq 1\}$ has the property that Σp_r diverges, then, for any pair of objects, the ratio $\pi_i(N)/\pi_j(N) \to p_i / p_j$ as $N \to \infty$.

It is realistic to treat the population size N and the values $v_1, v_2, \ldots v_N$ as unknown and the rest of the paper is concerned with a Bayesian representation of these quantities. Prior and posterior information about the residual population of undiscovered objects can be represented by Poisson processes. Mixtures of these processes behave in a natural way during the search and also when new discoveries are made. Search models of this type have been discussed in two recent papers: Bather (1992), (1993). The main ideas from the first will be useful here and a summary is given in Section 3. The second paper is about optimal stopping rules for a search based on mixtures with two components. In the present paper, the aim is to study the random behaviour of the weights on different components of a mixture during a search. It will be shown that, if N is large, the weights $w_j(t)$, $j = 1, 2, \ldots, m$, at time t will tend to favour the component that corresponds to the best a-priori description of the hidden population. A more precise statement of this result and its proof are contained in the final section of the paper.

2. Biased sampling

Throughout this section, we suppose there is a fixed but arbitrary sequence of objects with values $v_1, v_2, \ldots$. For any integer N, consider a hidden population of N objects with the values $v_1, v_2, \ldots v_N$ and the random subset of those objects found by searching for the period [0,t]. Object O_i is found with probability $p_i = p(v_i,t)$; it remains hidden with probability $q_i = 1 - p_i$ and events involving different objects are independent, $i = 1, 2, \ldots, N$. Now imagine selecting one item at random from the subset of discovered objects. We defined $\pi_i(N)$ to be the probability that the subset is non-empty and that the object chosen at random is O_i, for $i = 1, 2, \ldots, N$. In general,

(2.1) $$\sum_{i=1}^{N} \pi_i(N) = 1 - q_1 q_2 \cdots q_N \ .$$

For example, when N = 2 and N = 3, we obtain

$$\pi_1(2) = p_1 q_2 + \frac{1}{2} p_1 p_2, \qquad \pi_2(2) = q_1 p_2 + \frac{1}{2} p_1 p_2,$$

$$\pi_1(3) = p_1 q_2 q_3 + \frac{1}{2}\{p_1\ p_2\ q_3 + p_1\ q_2\ p_3\} + \frac{1}{3} p_1 p_2 p_3.$$

The expressions for $\pi_i(N)$ grow in complexity as N increases.

In order to see the pattern that emerges when N is large, it is helpful to use generating functions. Let

(2.2) $$G(N,z) = (p_1 z + q_1)(p_2 z + q_2) \cdots (p_N z + q_N).$$

and consider

$$G_1(N,z) = \frac{p_1}{(p_1 z + q_1)} G(N,z) = p_1 (p_2 z + q_2) \cdots (p_N z + q_N).$$

For r = 1, 2, ..., N, the coefficient of z^{r-1} in this polynomial is the probability that the discovered subset of the population contains 0_1 and exactly r - 1 of the other objects. Since $\int_0^1 z^{r-1} dz = 1/r$, we can express $\pi_1(N)$ as an integral:

$$\pi_1(N) = \int_0^1 G_1(N,z)\, dz.$$

For object 0_i, the corresponding formulae are

(2.3) $$G_i(N,z) = \frac{p_i}{(p_i z + q_i)} G(N,z),$$

(2.4) $$\pi_i(N) = \int_0^1 G_i(N,z)\, dz.$$

The expressions for individual probabilities are still complicated, but we can now determine their relative magnitudes as $N \to \infty$.

Theorem 1. Let $\{v_r, r \geq 1\}$ be any sequence of values and suppose that Σp_r diverges, where $p_r = p(v_r, t)$ is given by (1.1) and $t > 0$ is fixed. Then, for any pair of objects 0_i, 0_j,

$$\frac{\pi_i(N)}{\pi_j(N)} \to \frac{p_i}{p_j} \quad as \quad N \to \infty \ .$$

Proof. Let us restrict attention to $\pi_1(N)$ and $\pi_2(N)$, in order to simplify the notation. On substituting y = 1 -z in (2.4), we obtain

$$\pi_1(N) = \int_0^1 p_1(1-p_2 y)\cdots(1-p_N y)dy$$

and there is a similar expression for $\pi_2(N)$. Now write $H(N,y) = (1-p_3 y)\ldots(1-p_N y)$ and let

$$J(N) = \int_0^1 H(N,y)\,dy, \qquad K(N) = \int_0^1 y H(N,y)\,dy.$$

Then we have

$$\frac{\pi_1(N)}{\pi_2(N)} = \frac{p_1 J(N) - p_1 p_2 K(N)}{p_2 J(N) - p_1 p_2 K(N)} .$$

This ratio converges to p_1/p_2, as required, provided that $K(N)/J(N) \to 0$ as $N \to \infty$.

It is not difficult to verify that, for $0 \le y \le 1$,

$$1 - \sigma_N y \le H(N,y) \le \exp(-\sigma_N y),$$

where $\sigma_N = p_3 + p_4 + \ldots + p_N$. Since $H(N,y) \ge 0$, it follows that

$$J(N) \ \ge \ \int_0^1 \max(1-\sigma_N y, 0)\,dy.$$

When N is large enough, $\sigma_N \ge 1$ and the integral is easily evaluated as $(2\sigma_N)^{-1}$. On the other hand,

$$K(N) \ \le \ \int_0^1 y\exp(-\sigma_N y)\,dy \le \int_0^\infty y\exp(-\sigma_N y)\,dy = \sigma_N^{-2}.$$

Hence $K(N)/J(N) \le 2/\sigma_N$ and this converges to zero, since $\sigma_N \to \infty$ as $N \to \infty$.

Remark (i). According to (2.1), the assumption that Σp_r diverges is also what is required to ensure that $\Sigma\pi_j(N) \to 1$ as $N \to \infty$.

Remark (ii) Perhaps the simplest model for a hidden population of N objects is to regard $v_1, v_2, \ldots v_N$ as a random sample from some probability density function f. In this case, the probability that a typical object is found by time t is

(2.5) $$\gamma(t) = \int_0^\infty p(v,t) f(v)\,dv,$$

where p(v,t) is defined in (1.1). The number of objects discovered is a binomial random variable with parameters N and $\gamma(t)$; their values constitute a random sample from the probability density given by

(2.6) $$g(v,t) = p(v,t) f(v) / \gamma(t) .$$

3. Poisson search models

Consider a Poisson process with intensity $\alpha(v)$, $v \geq 0$. It represents the number and values of objects in a hidden population in the following way. The number of items with values lying in any interval A is a Poisson random variable with mean $\int_A \alpha(v)\,dv$ and the numbers for disjoint intervals are independent of each other.

The population size N has mean $\mu = \int_0^\infty \alpha(v)\,dv$. The values $v_1, v_2, \ldots v_N$ are conditionally independent, given N, and they are equivalent to a random sample from the probability density $\varphi(v) = \alpha(v)/\mu$.

Now consider the effects of searching for these objects. The model is specified by the function λ, as before. Given the values v_j, the corresponding discovery times s_j are independent exponential random variables, where s_j has mean $1/\lambda(v_j)$, $j = 1, 2, \ldots, N$. In fact we have a two-dimensional Poisson process: the points (v_j, s_j) are a realisation of the process with intensity function $\alpha(v)\lambda(v)\exp(-\lambda(v)s)$, for $v \geq 0$, $s \geq 0$. The independence of points generated by this process in disjoint subsets of the positive quadrant makes it easy to obtain posterior distributions. Imagine a search during $[0,t]$ and consider the prospects for the future at time t. The two sets of random points that correspond to past and future discoveries are independent of each other, so previous discoveries cannot affect future prospects. The posterior distribution of the number and values of undiscovered objects is determined by the intensity function:

$$\alpha(v)\lambda(v)\exp(-\lambda(v)s) = \alpha(v)\exp(-\lambda(v)t)\lambda(v)\exp(-\lambda(v)(s-t))$$

for $v \geq 0$ and $s \geq t$. Thus, the original intensity $\alpha(v)$ is replaced by $\alpha(v,t) = \alpha(v)\exp(-\lambda(v)t)$, at time t and the probability density for discovery times is the same as before, except that we are now concerned with the extra waiting time $s-t$

The search model described so far is limited because it is deterministic. However, mixtures of several Poisson processes with different intensity functions are more adaptive. The general model has components with intensities $\alpha_i(v)$, $i = 1, 2, \ldots, m$, and exponential discovery times defined by the positive function $\lambda(v)$. Each individual component behaves deterministically throughout a search, but the weights $w_i(t)$ associated with them change in a way that reflects the likelihoods $L_i(t)$ of events that occur. A detailed description of this behaviour is given in Bather (1992): what follows is a summary of the main features.

For the components, we need the following functions, defined for $v \geq 0$, $0 \leq s \leq t$, $i = 1, 2, \ldots, m$,

(3.1) $$\alpha_i(v,t) = \alpha_i(v)\exp(-\lambda(v)t),$$

(3.2) $$\mu_i(t) = \int_0^{\infty} \alpha_i(v,t)\,dv,$$

(3.3) $$\Psi_i(s,t) = \exp\{\mu_i(t) - \mu_i(s)\}.$$

It is assumed that the integrals in (3.2) are all finite. Note that $\alpha_i(v,t)$ and $\mu_i(t)$ are decreasing in t and that $\mu_i(0) - \mu_i(t)$ is the expected number of random points that fall in the region: $v \geq 0$, $0 \leq s \leq t$. Hence, the probability that no discovery occurs in the interval $[0,t]$, for the i th component of the model, is $\Psi_i(0,t)$.

It is convenient to refer to weights on components of the mixture, rather than probabilities. Suppose we begin with weights $w_i \geq 0$ on the intensities α_i, $i = 1, 2, \ldots, m$, and let $w = \Sigma w_i > 0$. Then the hidden population is described, initially, by a random process with typical realisations constructed in the following way: first choose a component i with probability w_i/w and find N by observing a Poisson random variable with mean $\mu_i(0)$; then take a random sample $v_1, v_2, \ldots, v_N$ from the probability density $\varphi_i(v) = \alpha_i(v)/\mu_i(0)$. We must now consider how this representation changes with time.

By using the independence property of Poisson processes, it can be shown that at any time $t > 0$ during a search, the set of undiscovered objects is represented by a mixture of Poisson processes with intensity functions $\alpha_i(v,t)$, $v \geq 0$. The weights $w_i(t)$ are proportional to $w_i L_i(t)$ where $L_i(t)$ is the likelihood associated with the i th component. Let the data consist of k discoveries, corresponding to points (v_j,s_j) with $0 \leq s_j \leq t$, $j = 1, 2, \ldots, k$. The likelihood $L_i(t)$ is proportional to

$$\Psi_i(0,t)\prod_{j=1}^{k}\{\alpha_i(v_j)\lambda(v_j)\exp(-\lambda(v_j)s_j)\}\quad.$$

It reduces to $\Psi_i(0,t)$ in the case $k = 0$ and, in general, each discovery contributes a factor based on the intensity at that point. Since we are only interested in the relative magnitudes of the weights $w_i(t)$, $i = 1, 2, \ldots, m$, terms in the likelihood that do not depend on i can be ignored and it is enough to define

(3.4) $$L_i(t) = \Psi_i(0,t)\prod_{j=1}^{k}\alpha_i(v_j)\quad,$$

(3.5) $$w_i(t) = w_i \Psi_i(0,t) \prod_{j=1}^{k} \alpha_i(v_j) \quad .$$

It is helpful to separate the effects of searching without success from the immediate consequences of making a discovery. If there are no discoveries during the interval [0,t] the posterior weight associated with the intensity function $\alpha_i(v,t)$ is

(3.6) $$w_i(t) = w_i \Psi_i(0,t) \, .$$

Then the effect of finding an object of value v* at time t is a transition: $w_i(t) \rightarrow w_i^*(t)$, where

(3.7) $$w_i^*(t) = w_i(t)\alpha_i(v^*) \quad .$$

An important question is how the behaviour of this Bayesian model depends on the underlying population of hidden objects. We can make some progress in this direction, in the case of large populations, by following up Remark (ii) in the previous section.

4. Asymptotic behaviour

Equations (3.3) and (3.5) show that

(4.1) $$\log\{w_i(t) / w_i\} = \mu_i(t) - \mu_i(0) + \sum_{j=1}^{k} \log\{\alpha_i(v_j^*)\} \, ,$$

where v_1^*, v_2^*, ..., v_k^* refer to the discovered items from the original population of N objects. From now on, it will be assumed that the population is a random sample of size N from a probability density f. Then the last term in (4.1) has the same probability distribution as

$$\sum_{r=1}^{N} I_r(v_r) \log\{\alpha_i(v_r)\} \quad ,$$

where v_1, v_2, ..., v_N are independent with the density f and the indicator variables $I_r(v_r)$ are conditionally independent, taking the values 1 and 0 with probabilities $p(v_r,t)$ and $q(v_r,t)$, respectively. By (1.1), (3.1) and (3.2), we also have

$$\mu_i(0) - \mu_i(t) = \int_0^\infty \alpha_i(v) p(v,t) \, dv.$$

Hence (4.1) is equivalent to

(4.2) $$\log(w_i(t) / w_i) = \sum_{r=1}^{N} I_r(v_r) \log\{\alpha_i(v_r)\} - \int_0^\infty \alpha_i(v) p(v,t) dv.$$

The independent components of the sum here all have the same moment generating function

(4.3) $$M_i(u,t) = \int_0^\infty \{p(v,t)\alpha_i(v)^u + q(v,t)\} f(v)\,dv$$

and it is easily verified that

(4.4) $$E[\log\{w_i(t)/w_i\}] = N\int_0^\infty \log\{\alpha_i(v)\}p(v,t)f(v)\,dv - \int_0^\infty \alpha_i(v)p(v,t)\,dv \quad ,$$

$i = 1, 2, \cdots, m$.

The Bayesian model with components based on the intensity functions α_i may or may not provide a good match for the underlying population represented by N and f. Any intensity α determines a probability density φ given by

$$\varphi(v) = \alpha(v)/\mu, \qquad \mu = \int_0^\infty \alpha(v)\,dv.$$

Let us define

(4.5) $$R(N,\mu,\varphi) = N\int_0^\infty \log\{\mu\varphi(v)\}p(v,t)f(v)\,dv - \mu\int_0^\infty p(v,t)\varphi(v)\,dv \ ,$$

so that the expectations in (4.4) become $R(N,\mu_i,\varphi_i)$, where $\alpha_i(v) = \mu_i\varphi_i(v)$, $\mu_i = \mu_i(0)$, $i = 1, 2, \ldots, m$. It will be shown that $R(N,\mu,\varphi)$ achieves its maximum with respect to μ and φ when $\mu = N$ and $\varphi = f$. This will provide some guide to the relative magnitudes of the weights $w_i(t)$ when N is large.

Lemma 2. Let $\varphi(v)$ and $f(v)$ be probability densities for $v \geq 0$ and let $p(v,t)$ be defined by (1.1), where $t \geq 0$ is fixed. Then, for any $N \geq 1$ and $\mu > 0$,

$$R(N,\mu,\varphi) \leq R(N,N,f)$$

and equality is attained only when $\mu = N$ and when φ represents the same probability distribution as f.

Proof. Let us simplify the notation here by writing

$$R(N,\mu,\varphi) = N\int\log(\varphi)pf + N\log\mu\int pf - \mu\int p\varphi$$

$$\frac{\partial R}{\partial\mu} = \frac{N}{\mu}\int pf - \int p\varphi .$$

This shows that the maximum with respect to μ is achieved by setting $\mu = N\int pf / \int p\varphi$. By substituting this value and rearranging the result, we obtain

$$R(N,\mu,\varphi) \leq N\{\log N - 1\}\int pf + N\left[\int\log(\varphi)pf - \int pf\log\{\int p\varphi / \int pf\}\right]$$

Now apply Jensen's inequality to $E[\log\{\varphi(v)/f(v)\}]$, where v has the probability density given by (2.6). For this distribution, $E\{\varphi(v)/f(v)\} = \int p\varphi / \int pf$ and we have

$$\int pf\, E[\log\{\varphi(v)/f(v)\}] = \int \log(\varphi)pf - \int \log(f)pf \leq \int pf \log\{\int p\varphi / \int pf\}$$

When this is combined with the previous inequality, it leads to

$$R(N,\mu,\varphi) \leq N\{\log N - 1\}\int pf + N\int \log(f)pf = R(N,N,f).$$

Finally, it is a straightforward matter to check that the maximum we have obtained is essentially unique.

The lemma shows that, in order to provide an effective description of the underlying population when N is large, the mixture of Poisson processes must include components with prior means $\mu_i = \mu_i(0)$ that are of order N. For simplicity, let $\mu_i = Nc_i$ where $c_i > 0$ is constant and consider the asymptotic behaviour of $w_i(t)/w_j(t)$ as $N \to \infty$, for any pair of components $i, j \leq m$. Let

(4.6) $$\Delta_{ij}(t) = \log\{w_i(t)/w_j(t)\} - \log\{w_i/w_j\} .$$

By using (4.2) with $\alpha_i(v) = Nc_i\varphi_i(v)$, $\alpha_j(v) = Nc_j\varphi_j(v)$, we can express this change over the interval [0,t] as a sum of N independent and identically distributed random variables:

(4.7) $$\Delta_{ij}(t) = \sum_{r=1}^{N} I_r(v_r)\log\{c_i\varphi_i(v_r)/c_j\varphi_j(v_r)\}$$
$$- N\int_0^\infty \{c_i\varphi_i(v) - c_j\varphi_j(v)\}p(v,t)\,dv.$$

Further, equations (4.4) and (4.5) show that the mean of this sum is $R(N,Nc_i,\varphi_i) - R(N,Nc_j,\varphi_j)$ and since

$$R(N, Nc, \varphi) = NR(1,c,\varphi) + N\log N\int_0^\infty p(v,t)f(v)\,dv,$$

we have

(4.8) $$E\{\Delta_{ij}(t)\} = N\{R(1,c_i,\varphi_i) - R(1,c_j,\varphi_j)\}.$$

It is clear from (4.6) ....(4.8) and the strong law of large numbers that $w_i(t)/w_j(t) \to 0$ almost surely as $N \to \infty$, if $R(1,c_i,\varphi_i) < R(1,c_j,\varphi_j)$. Hence, the mixture will be dominated, for large N, by the component for which $R(1,c_j,\varphi_j)$ is largest. We also know from Lemma 2 that the maximum possible value of this is R(1,1,f). The results we have established can be summarised in the following way.

Theorem 3. Consider a Poisson search model with intensity functions α_i, $i = 1, 2, \ldots, m$. Suppose that the underlying population of hidden objects is a random sample of size N from the probability density f and that $\alpha_i(v) = Nc_i\varphi_i(v)$ for $v \geq 0$, where φ_i is a probability density and c_i is a positive constant. Let

$$R(1,c,\varphi) = \int_0^\infty \log\{c\,\varphi(v)\}p(v,t)\,f(v)\,dv - c\int_0^\infty p(v,t)\,\varphi(v)\,dv.$$

Then, for fixed $t > 0$, the ratio of weights $w_i(t)/w_j(t) \to 0$ almost surely as $N \to \infty$ if $R(1,c_i,\varphi_i) < R(1, c_j,\varphi_j)$. In particular, if there is a single component with $c_j = 1$ and $\varphi_j = f$, then $w_i(t)/w_j(t) \to 0$ for each $i \neq j$.

REFERENCES

Bather, J. A. (1992). "Search models", *J. Appl. Prob.* **29**, 605-615.

Bather, J. A. (1993). "Poisson search models: Monotone stopping rules", to appear Festschrift for Peter Whittle, Wiley, Chichester.

A MODIFICATION OF LEAST SQUARES WITH HIGH EFFICIENCY AND HIGH BREAKDOWN POINT IN LINEAR REGRESSION

Konrad Behnen
Institut für Mathematische Stochastik
Universität Hamburg, Germany

Abstract

For the linear regression model we define SM-estimators with the following properties:

With high probability the estimator is identical with the least squares estimator, if the data correspond to the assumption of normality. Especially, the estimation procedure has 95% asymptotic efficiency at the normal model.

The procedure will identify groups of data which do not follow the model, as long as more than 50% of the data are in correspondence with the ideal model. Especially, the SM-estimator has breakdown point 1/2.

We evaluate bounds on the supremum asymptotic bias (SAB) for rather big balls around the ideal model, which will prove uniform consistency, too.

AMS 1980 subject classifications. 62F35, 62J05, 62E20.
Key words. linear regression, robust estimate, uniform consistency, high breakdown point, high efficiency, bias bounds, supremum asymptotic bias, diagnostics

1. Introduction. In a graduate course on robust regression [cf. Behnen (1991)] I combined S-estimation [cf. Rousseeuw & Yohai (1984)] and M-estimation [e.g. Huber (1981)] in order to construct a high efficiency and high breakdown point substitute of the least median of squares (LMS) estimator [cf. Rousseeuw & Leroy (1987)]. The resulting SM-estimator (cf. Section 4) turns out to be a smooth version of a MM-estimator [cf. Yohai (1987)]. The smoothness is important for easy evaluation.

By defining a suitable regression adapted distance (cf. Section 1) we are able to derive bias bounds for rather general balls around the ideal model. These balls are much bigger than the usual ϵ-contamination balls and allow an immediate proof of the uniform consistency of the S-estimators of Section 3 and the SM-estimators of Section 4.

Numerical values of these bounds are given in Table 1. Similar bounds for the smaller ϵ-contamination balls are given in Hennig (1993).

2. The balls around the ideal model. As ideal model we consider the multiple linear regression model with normal errors and, in order to have easy and comparable asymptotics, with normally distributed regressors.

Thus, under the ideal model we have i.i.d. observations $(x_1, y_1), \ldots, (x_n, y_n)$ with values in $\mathbb{R}^r \times \mathbb{R}$ such that for some $(\mu_0, \vartheta_0) \in \mathbb{R} \times \mathbb{R}^r$ and some scale $\sigma_0 > 0$ for given $x_1, \ldots, x_n \in \mathbb{R}^r$ the $y_1, \ldots, y_n$ are (conditionally) independent with (conditional) distribution according to

$$\mathcal{L}_0[y_i \mid x_i] \;=\; \mathcal{N}(\mu_0 + \vartheta_0' x_i, \sigma_0^2). \tag{2.1}$$

Here the shift parameter μ_0 (regression with intersection) is separately included in order to get results for the location problem as a special case. For our explicit ideal model results we assume additionally

$$\mathcal{L}_0[x_i] \;=\; \mathcal{N}_r(\eta_0, \Sigma_0) \tag{2.2}$$

with $\eta_0 \in \mathbb{R}^r$ and nondegenerate $r \times r$ covariance matrix Σ_0.

Throughout the paper the ideal model distribution $\mathcal{L}_0[(x_1, y_1)]$ according to (2.2) und (2.1) will be denoted by P_0.

The regression-adapted distance d_1

In order to get rather general bias bounds for the proposed SM-estimator (cf. Section 4), which immediately will imply (uniform) consistency, we define the following distance d_1 on $\mathcal{M}_1 := \mathcal{M}_1(\mathbb{R}^r \times \mathbb{R}, \mathbb{B}^r \otimes \mathbb{B})$, the set of all probability measures on the Borel sets of $\mathbb{R}^r \times \mathbb{R}$:

$$d_1(P,Q) \;:=\; \sup_{B \in \mathcal{C}_0} \mid P(B) - Q(B) \mid, \tag{2.3}$$

where

$$\mathcal{C}_0 \;:=\; \{B(\mu, \vartheta, \tau) : \mu \in \mathbb{R}, \vartheta \in \mathbb{R}^r, \tau > 0\} \tag{2.4}$$

is the set of all closed half spaces

$$B(\mu, \vartheta, \tau) \;:=\; \{(x,y) \in \mathbb{R}^r \times \mathbb{R} : (y - \mu - \vartheta' x)^2 \;\geq\; \tau\} \tag{2.5}$$

in $\mathbb{R}^r \times \mathbb{R}$.

The set $\mathcal{C}_0$ is a Vapnik-Červonenkis class with VC-index $r+3$ [cf. Dudley (1979)]. Thus, using Theorem 1 (Vapnik-Červonenkis) of Chap.26 of Shorack & Wellner (1986), we get the existence of decreasing sequences $0 < \epsilon_n \to 0$ and $0 < \eta_n \to 0$ such that

$$\sup_{P \in \mathcal{M}_1} P^\infty\{z \in (\mathbb{R}^r \times \mathbb{R})^\infty : \max_{m \geq n} d_1(P_m(z, \cdot), P) \geq \epsilon_n\} \;\leq\; \eta_n\,, \tag{2.6}$$

where

$$P_m(z, B) \;:=\; \frac{1}{m} \sum_{i=1}^{m} 1(z_i \in B), \quad B \in \mathbb{B}^r \otimes \mathbb{B}, \tag{2.7}$$

is the empirical distribution of the first m components of the sequence $z = (z_1, z_2, \ldots)$ $\in (\mathbb{R}^r \times \mathbb{R})^\infty$. Note that (2.6) and (2.7) correspond to a sequence of i.i.d. observations (x_i, y_i), $i \in \mathbb{N}$.

Here and in the sequel $1(\cdots)$ denotes the **indicator function** corresponding to condition $(\cdots)$.

Formula (2.6) means that uniformly in $P \in \mathcal{M}_1$ for P^∞- almost all z the empirical distribution $P_n(z, \cdot)$ is close to P for all sufficiently large sample sizes n, if closeness is measured in d_1-distance. This property will guarantee the (uniform) consistency of the estimators defined in Section 3 and Section 4.

Other immediate consequences of definition (2.3) are the inequalities

$$d_1(\ (1-\epsilon)P_0+\epsilon P,\ P_0\ )\ \le\ \epsilon \quad \forall\, P \in \mathcal{M}_1, \tag{2.8}$$
$$d_1(P,P_0)\ \le \|\ P-P_0\ \| := \sup_B \mid P(B)-P_0(B)\mid . \tag{2.9}$$

Thus, given any radius $0<\epsilon<1$ the d_1-ball $\{P \in \mathcal{M}_1 : d_1(P,P_0) \le \epsilon\}$ with radius ϵ and center P_0 contains the corresponding ϵ-contamination ball and the corresponding total variation ball, too.

The bias bounds of the S-estimator of Section 3 and the SM-estimator of Section 4 will be valid for these general d_1-balls with radius $\epsilon \in [0,1/2)$ and the ideal model P_0 as the center.

Now, let $h:[0,\infty) \to [0,1]$ be any (scaling) function with the properties

$$h(t)\ =\ \int_0^{t_1} h'(y)\,1(y \le t)\,dy \quad \forall\, t \ge 0 \quad \text{and} \quad \int_0^{t_1} h'(y)\,dy\ =\ 1, \tag{2.10}$$

for some constant $0<t_1<\infty$ and some measurable function $h' \ge 0$. Especially, the function h is absolutely continuous and nondecreasing on the compact interval $[0,t_1]$ with $h(0)=0$ and $h(t)=1\ \forall\, t \ge t_1$.
The representation (2.10) and Fubini immediately imply the equality

$$\int h(\frac{(y-\mu-\vartheta' x)^2}{s^2})\,dP(x,y)\ =\ \int_0^{t_1} P(B(\mu,\vartheta,s^2 t))\,h'(t)\,dt. \tag{2.11}$$

Therefore the second part of (2.10) proves the inequality

$$\sup_{\mu,\vartheta,s} \mid \int h(\frac{(y-\mu-\vartheta' x)^2}{s^2})\,(dP(x,y)-dQ(x,y))\mid \le\ d_1(P,Q), \tag{2.12}$$

where the sup is taken for all $(\mu,\vartheta) \in \mathbb{R} \times \mathbb{R}^r$ and all $s>0$.

The inequality (2.12) will be central for the derivation of the bias bounds in Section 3 and Section 4.

The scaling function h_1
Throughout this paper we'll use the special scaling function

$$h_1(t)\ :=\ \begin{cases} t\,, & \text{if} \quad t \le 0.95\,, \\ 0.95+g(10t-9.5)/10\,, & \text{if} \quad 0.95<t<1.05\,, \\ 1\,, & \text{if} \quad t \ge 1.05\,, \end{cases} \tag{2.13}$$

with smoothing function

$$g(t)\ :=\ t-t^3+0.5\,t^4\,, \qquad 0\ \le\ t\ \le\ 1. \tag{2.14}$$

It's easily seen that h_1 is two times continuously differentiable with derivative

$$h_1'(t)\ =\ 1(t \le 0.95)+g'(10t-9.5)\ 1(0.95<t<1.05)\ \in [0,1]\ \forall\, t \ge 0 \tag{2.15}$$

and bounded second derivative $h_1'' \le 0$. Especially, this function h_1 has a representation (2.10) with $t_1 = 1.05$.

The function h_1 has been constructed as a smoothed version of the function

$$h_0(t) \;=\; t\,1(t \le 1) + 1(t > 1) \tag{2.16}$$

in order to use iteration methods for easy evaluation of the proposed SM-estimator.

3. S-estimation with breakdown point 1/2. We use the scaling function h_1 defined in (2.13). This function h_1 has the following monotonicity properties ($\forall t \ge 0$):

$$\forall\, 0 < s_1 \le s_2 : \quad h_1(t/s_1) \;\ge\; h_1(t/s_2), \tag{3.1a}$$

$$\forall\, 0 < s_1 \le s_2 : \quad s_1\, h_1(t/s_1) \;\le\; s_2\, h_1(t/s_2). \tag{3.1b}$$

For any $P \in \mathcal{M}_1$ define $s(P,\mu,\vartheta) \ge 0$ as the maximal solution s of

$$\int h_1\Big(\frac{(y-\mu-\vartheta' x)^2}{s^2}\Big)\, dP(x,y) \;=\; 1/2, \tag{3.2}$$

or, if no solution exists, we define $s(P,\mu,\vartheta) = 0$. Then the corresponding S-functional for estimating the scale σ_0 is

$$s(P) \;:=\; \inf\,\{\, s(P,\mu,\vartheta) : \; (\mu,\vartheta) \in \mathbb{R}\times\mathbb{R}^r \,\}. \tag{3.3}$$

Invariance and equivariance of the S-estimator

Definition (3.3) easily implies the invariance of $s(\cdot)$ under nondegenerate affine linear transformations of the x-component and under shift transformations of the y-component (**regression invariance**), and the equivariance of $s(\cdot)$ under scale transformations of the y- component (**scale equivariance**).

Thus, for reason of comparability to other results, we may assume without loss of generality,

$$\mu_0 = 0\,, \quad \vartheta_0 = 0\,, \quad \sigma_0 = 1\,, \quad \eta_0 = 0\,, \quad \Sigma_0 = I_r\,, \tag{3.4}$$

which means that x_i and y_i are independent under the ideal model P_0 with distributions $\mathcal{N}_r(0, I_r)$ and $\mathcal{N}(0,1)$, respectively,

$$P_0 \;=\; \mathcal{N}_r(0, I_r) \otimes \mathcal{N}(0,1). \tag{3.5}$$

Therefore, under P_0, the distribution of $y - \vartheta' x$ is $\mathcal{N}(0, 1+|\vartheta|^2)$ and the distribution of $(y-\mu-\vartheta' x)^2/(1+|\vartheta|^2)$ is a noncentral $\chi_1^2(\mu^2/(1+|\vartheta|^2))$ distribution. This implies

$$\begin{aligned} &\int h_1(\,(y-\mu-\vartheta' x)^2/s^2\,)\, dP_0(x,y) \\ &= \int h_1(\,(y\sqrt{1+|\vartheta|^2}-\mu)^2/s^2\,)\,\varphi(y)\,dy \;=\; E\; h_1\big(\tfrac{1+|\vartheta|^2}{s^2}\cdot \chi_1^2(\tfrac{\mu^2}{1+|\vartheta|^2})\big) \end{aligned} \tag{3.6}$$

Obviously, given any $s > 0$, this integral has a (unique) minimum at $(\mu_0, \vartheta_0) = (0,0) \subset \mathbb{R}\times\mathbb{R}^r$.

As a consequence, for the (unique) positive solution s_0 of $\int h_1(y^2/s^2)\,\varphi(y)\,dy = 1/2$, we get

$$s(P_0) \;=\; s_0 \;=\; 1.0407. \tag{3.7}$$

Therefore the standardized scale functional

$$s_0(P) \;:=\; s(P)/s_0 \tag{3.8}$$

is Fisher-consistent for the scale at the ideal normal model.

Bias bounds

Now let's evaluate bias bounds for $s_0(P)$ with respect to d_1-balls with radius $0 \le \epsilon < 1/2$ and center P_0:

In the sequel let $r_{\mu,\vartheta} : \mathbb{R}^r \times \mathbb{R} \to \mathbb{R}$ denote the residual function

$$r_{\mu,\vartheta}(x,y) \;=\; y - \mu - \vartheta' x\,. \tag{3.9}$$

For any $(\mu,\vartheta) \in \mathbb{R} \times \mathbb{R}^r$ let $s_{01}(\epsilon,\mu,\vartheta)$ denote the (unique) positive solution s of

$$\int h_1(r^2_{\mu,\vartheta}/s^2)\,dP_0 \;=\; \frac{1}{2} + \epsilon, \tag{3.10}$$

and let $s_{02}(\epsilon,\mu,\vartheta)$ denote the (unique) positive solution s of

$$\int h_1(r^2_{\mu,\vartheta}/s^2)\,dP_0 \;=\; \frac{1}{2} - \epsilon. \tag{3.11}$$

Then, because of the h_1 - and P_0 - properties,

$$0 \;<\; s_{01}(\epsilon,\mu,\vartheta) \;\le\; s_{02}(\epsilon,\mu,\vartheta) \;<\; \infty. \tag{3.12}$$

Now, using inequality (2.12) for the d_1-ball with center P_0 and radius ϵ, on one hand we have $(\forall s > 0)$

$$\int h_1(\frac{r^2_{\mu,\vartheta}}{s^2})\,dP_0 - \epsilon \;\le\; \int h_1(\frac{r^2_{\mu,\vartheta}}{s^2})\,dP \;\le\; \int h_1(\frac{r^2_{\mu,\vartheta}}{s^2})\,dP_0 + \epsilon\,, \tag{3.13}$$

on the other hand $s(P,\mu,\vartheta) \ge 0$ is the maximal solution of (3.2). Therefore the monotonicity of h_1 implies

$$0 \;<\; s_{01}(\epsilon,\mu,\vartheta) \;\le\; s(P,\mu,\vartheta) \;\le\; s_{02}(\epsilon,\mu,\vartheta) \;<\; \infty\,. \tag{3.14}$$

Hence, from definition (3.3), we get the bounds

$$s_{01}(\epsilon) \;\le\; s(P) \;\le\; s_{02}(\epsilon)\,, \tag{3.15}$$

where, because of (3.6),

$$s_{01}(\epsilon) \;:=\; \inf\{s_{01}(\epsilon,\mu,\vartheta) : \; (\mu,\vartheta) \in \mathbb{R} \times \mathbb{R}^r\} \tag{3.16}$$

is the (unique) positive solution of

$$\int h_1(y^2/s^2)\,\varphi(y)\,dy \;=\; 1/2 + \epsilon, \tag{3.17}$$

and

$$s_{02}(\epsilon) \ := \ \inf\{s_{02}(\epsilon,\mu,\vartheta) : \ (\mu,\vartheta) \in \mathbb{R}\times\mathbb{R}^r\} \tag{3.18}$$

is the (unique) positive solution of

$$\int h_1(y^2/s^2)\,\varphi(y)\,dy \ = \ 1/2-\epsilon. \tag{3.19}$$

Obviously, $0 < s_{01}(\epsilon) \le s_{02}(\epsilon) < \infty$. Thus, for the standardized scale functional $s_0(P) = s(P)/s_0$, we finally get

$$0 \ < \ s_{01}(\epsilon)/s_0 \ \le \ s_0(P) \ \le \ s_{02}(\epsilon)/s_0 \ < \ \infty. \tag{3.20}$$

Consistency

Because of $s_{01}(\epsilon) \to s_{01}(0) = s_0$ and $s_{02}(\epsilon) \to s_{02}(0) = s_0$, as $\epsilon \to 0$, and because of [cf. formula (2.6)]

$$d_1(P_n(z,\cdot),P) \ \to \ 0 \ \ P^\infty\text{-almost all } z, \ \ \text{as } n\to\infty, \tag{3.21}$$

even uniformly in $P \in \mathcal{M}_1$, formula (3.20) proves the consistency of the scale S-estimator at the ideal model P_0, too,

$$\hat{s}_n(z) \ := \ s_0(P_n(z,\cdot)) \ \to \ \sigma_0 \ \ P_0^\infty\text{-almost all } z, \ \ \text{as } n\to\infty. \tag{3.22}$$

Additionally, the estimator stays close to σ_0 as long as $P_n(z,\cdot)$ stays close to P_0 in d_1 -distance.

The explicit values of the bounds in (3.20) are given in Table 1.

S-estimation of the regression parameter (μ_0,ϑ_0):

In the functional setting

$$(\mu_S(P),\ \vartheta_S(P)) \ = \ \text{argmin}\ \{s(P,\mu,\vartheta) : \ (\mu,\vartheta)\in\mathbb{R}\times\mathbb{R}^r\} \tag{3.23}$$

is the functional for the corresponding S-estimator of (μ_0,ϑ_0).

Note: The existence of $(\mu_S(P),\vartheta_S(P))$ will follow from the bounds given below by a standard compactness argument. If the solution is not unique, we may interprete $(\mu_S(P),\vartheta_S(P))$ as the set of all global minimum points.

From inequalities (3.14) and (3.20) it's obvious that any $(\mu,\vartheta)\in\mathbb{R}\times\mathbb{R}^r$ with the property

$$s_{01}(\epsilon,\mu,\vartheta) \ > \ s_{02}(\epsilon) \tag{3.24}$$

cannot be a global minimum point of $s(P,\cdot)$. Because of the monotonicity of h_1 inequality (3.24) is equivalent to

$$\int h_1(r^2_{\mu,\vartheta}/s^2_{01}(\epsilon,\mu,\vartheta))\,dP_0 \ < \ \int h_1(r^2_{\mu,\vartheta}/s^2_{02}(\epsilon))\,dP_0\,. \tag{3.25}$$

By definition (3.10) of $s_{01}(\epsilon,\mu,\vartheta)$ the left hand side of formula (3.25) equals $1/2+\epsilon$, whereas the right hand side has been evaluated in formula (3.6). Thus, all global minimum points of $s(P,\cdot)$ have to fulfil the condition

$$E\ h_1\Big(\frac{1+|\vartheta|^2}{s^2_{02}(\epsilon)}\cdot\chi_1^2\Big(\frac{\mu^2}{1+|\vartheta|^2}\Big)\Big) \ \le \ \frac{1}{2}+\epsilon. \tag{3.26}$$

Note that $E\ h_1(\alpha^2 \cdot \chi_1^2(\beta^2))$ is continuous and strictly increasing in α^2, for $\beta^2 \geq 0$ fixed, with 1 as limiting value, and continuous and strictly increasing in β^2, for $\alpha^2 > 0$ fixed, with 1 as limiting value, too. The right hand side of (3.26), however, is less than 1.

In order to obtain an explicit compact subset of $\mathbb{R} \times \mathbb{R}^r$ which contains all global minimum points, let's define μ_ϵ^S as the (unique) positive solution μ of

$$E\ h_1\Big(\frac{1}{s_{02}^2(\epsilon)} \cdot \chi_1^2(\mu^2)\Big) \ = \ \frac{1}{2} + \epsilon, \tag{3.27}$$

and d_ϵ^S as the (unique) positive solution d of

$$E\ h_1\Big(\frac{1+d^2}{s_{02}^2(\epsilon)} \cdot \chi_1^2(0)\Big) \ = \ \frac{1}{2} + \epsilon. \tag{3.28}$$

Because of the definition (3.17) of $s_{01}(\epsilon)$ this means

$$d_\epsilon^S \ = \ \sqrt{s_{02}^2(\epsilon)/s_{01}^2(\epsilon) - 1}\,. \tag{3.29}$$

Obviously, no (μ, ϑ) with $|\ \vartheta\ | > d_\epsilon^S$ can fulfil the inequality (3.26), and no (μ, ϑ) with $\mu^2/(1+|\ \vartheta\ |^2) > (\mu_\epsilon^S)^2$ can fulfil inequality (3.26) either. Combining both parts, we've proved that all global minimum points of $s(P, \cdot)$ must lie in the compact set

$$\{\ (\mu, \vartheta) \in \mathbb{R} \times \mathbb{R}^r : \ |\ \mu\ | \leq \ \mu_\epsilon^S \sqrt{1+|\ \vartheta\ |^2},\ |\ \vartheta\ | \leq \ d_\epsilon^S\ \}, \tag{3.30}$$

which obviously shrinks to the point $(\mu_0, \vartheta_0) \ = \ (0,0) \in \mathbb{R} \times \mathbb{R}^r$ if μ_ϵ^S and d_ϵ^S both shrink to zero.

Consistency of the regression S-estimator

Since $s_{02}^2(\epsilon) \rightarrow s_0^2$ as $\epsilon \rightarrow 0$, the definitions (3.27) for μ_ϵ^S, (3.28) for d_ϵ^S, and $\int h_1(y^2/s_0^2)\,\varphi(y)\,dy = 1/2$ immediately imply

$$\mu_\epsilon^S \ \rightarrow 0 \quad \text{and} \quad d_\epsilon^S \ \rightarrow \ 0, \quad \text{as } \epsilon \rightarrow 0, \tag{3.31}$$

which implies the (uniform) consistency of the S-estimator of (μ_0, ϑ_0) at the ideal model P_0.
The explicit values of the bounds μ_ϵ^S and d_ϵ^S are given in Table 1.

Comparability and equivariance

For reasons of comparison note that μ_ϵ^S is a bound for the supremum asymptotic bias (SAB) for estimation of μ_0 in case of the **location model** $(\vartheta_0 = 0)$, whereas d_ϵ^S is a bound for the SAB for estimation of the ϑ_0-part of (μ_0, ϑ_0). But notice that our d_1-balls are much bigger than the usual ϵ contamination balls.

Since the functional $(\mu_S(P), \vartheta_S(P))$ defined in formula (3.23) is equivariant under nondegenerate affine linear transformations of the x-component and under shift transformations of the y-component (**regression equivariance**) and also under scale transformations of the y-component (**scale equivariance**), the special assumptions (3.4) and (3.5) imply no loss of generality, but only normalization of bounds for comparison.

Breakdown point of the regression S-estimator

Since the bounds $\mu_\epsilon^S < \infty$ and $d_\epsilon^S < \infty$ are valid for the d_1 -ball around P_0 with radius $\epsilon \in [0,1/2)$, and since this ball contains the ϵ -contamination ball $\{(1-\epsilon)P_0 + \epsilon Q : Q \in \mathcal{M}_1\}$ the (asymptotic) breakdown point is $1/2$.

4. SM-estimation with breakdown point 1/2 and 95% efficiency. In a first step we evaluate the S-functional $s(P)$ defined in (3.3) for the scaling function h_1 defined in (2.13). Then, in a second step, we use the **same function** h_1 and the S-scale $s(P)$ for the definition of our SM-estimators:

For any $P \in \mathcal{M}_1$ with $s(P) > 0$ we define

$$I_1(P,\mu,\vartheta) \;:=\; \int h_1(\,r^2_{\mu,\vartheta}/(c_1 s(P))^2\,)\,dP\,, \tag{4.1}$$

where $c_1 > 1$ is a constant in order to achieve 95% efficiency at the ideal model P_0 . This means (see below) that c_1 is the (unique) positive solution c of [cf. formula (4.18)]

$$\Big(\int y^2\, h_1'(\,y^2/(cs_0)^2\,)\,\varphi(y)\,dy\Big)^2 \Big/ \Big(\int y^2\,\{h_1'(\,y^2/(cs_0)^2\,)\}^2\,\varphi(y)\,dy\Big) \;=\; 0.95, \tag{4.2}$$

i.e., we get $c_1 s_0 = 2.80$ and $c_1 = 2.69$.

As the (unstandardized) SM-functional for σ_0 -estimation we use

$$\sigma(P) \;:=\; \inf\{\,\sigma(P,\mu,\vartheta) : \; (\mu,\vartheta) \in \mathbb{R} \times \mathbb{R}^r\}\,, \tag{4.3}$$

where

$$\sigma^2(P,\mu,\vartheta) \;=\; \int c_1^2\, s^2(P)\, h_1((\frac{r_{\mu,\vartheta}}{c_1 s(P)})^2\,)\,dP \;=\; c_1^2\, s^2(P)\, I_1(P,\mu,\vartheta)\,, \tag{4.4}$$

if $s(P) > 0$, and $\sigma^2(P,\mu,\vartheta) = 0$, if $s(P) = 0$.

The **standardized** SM-functional for σ_0 -estimation is defined as

$$\sigma_0(P) \;:=\; \sigma(P)/\tau_0, \tag{4.5}$$

where

$$\tau_0^2 \;=\; \sigma^2(P_0) \;=\; \int (c_1 s_0)^2\, h_1(\,y^2/(c_1 s_0)^2\,)\,\varphi(y)\,dy \;=\; 0.9905. \tag{4.6}$$

Finally, the SM-functional for (μ_0, ϑ_0) -estimation is defined as

$$\begin{aligned}(\mu_{SM}(P),\; \vartheta_{SM}(P)) \;&=\; \text{argmin}\,\{\,\sigma(P,\mu,\vartheta) : \; (\mu,\vartheta) \in \mathbb{R} \times \mathbb{R}^r\} \\ &=\; \text{argmin}\,\{\,I_1(P,\mu,\vartheta) : \; (\mu,\vartheta) \in \mathbb{R} \times \mathbb{R}^r\}\,,\end{aligned} \tag{4.7}$$

if $s(P) > 0$, and $(\mu_{SM}(P), \vartheta_{SM}(P)) = (0,0) \in \mathbb{R} \times \mathbb{R}^r$, if $s(P) = 0$.

Bias bounds

Now let's evaluate bias bounds for these SM-estimators, again for d_1 -balls with radius $0 \le \epsilon < 1/2$ and center at the ideal model P_0 :

Firstly, we use (3.2), (3.3), (3.23), the monotonicity of h_1, and $c_1 > 1$ in order to get

$$\inf_{\mu,\vartheta} I_1(P,\mu,\vartheta) \le I_1(\mu_S(P),\vartheta_S(P)) = \int h_1(r^2_{\mu_S(P),\vartheta_S(P)}/(c_1^2 s^2(P)))\,dP$$
$$< \int h_1(r^2_{\mu_S(P),\vartheta_S(P)}/s^2(P))\,dP = \tfrac{1}{2}.$$

Thus,

$$\inf_{\mu,\vartheta} I_1(P,\mu,\vartheta) < \frac{1}{2}, \tag{4.8a}$$

and therefore from inequality (3.15),

$$\sigma(P) \le s_{02}(\epsilon)\,c_1\,/\,\sqrt{2}. \tag{4.9a}$$

Secondly, we use inequality (2.12) for the d_1-ball with center P_0 and radius ϵ, formula (3.6), and $s^2(P) \ge s^2_{01}(\epsilon)$, in oder to get

$$\inf_{\mu,\vartheta} I_1(P,\mu,\vartheta) \le \int h_1(\frac{y^2}{c_1^2 s^2_{01}(\epsilon)})\,\varphi(y)\,dy + \epsilon, \tag{4.8b}$$

and, additionally, $c_1^2 s^2(P) \le c_1^2 s^2_{02}(\epsilon)$ and the monotonicity property (3.1b) in order to get

$$\sigma(P) \le s_{02}(\epsilon)\,c_1 \sqrt{\int h_1(\frac{y^2}{c_1^2 s^2_{02}(\epsilon)})\,\varphi(y)\,dy + \epsilon}. \tag{4.9b}$$

The minimum of the right hand sides of (4.9a) and (4.9b) divided by τ_0 is our upper (bias) bound $\sigma_{02}(\epsilon)$ for $\sigma_0(P)$ with respect to d_1-balls around P_0 with radius $\epsilon \in [0,1/2)$. The explicit values are given in Table 1.

The inequalities (4.8a) and (4.8b) will be used later for establishing bounds for $(\mu_{SM}(P),\vartheta_{SM}(P))$.

Now let's evaluate lower bounds:

Firstly, from (3.1), $0 < s_{01}(\epsilon) \le s(P) \le s(P) \le s(P,\mu,\vartheta)$, $c_1 \ge 1$, and defining equation (3.2) we get

$$\sigma^2(P,\mu,\vartheta) \ge \int s^2(P)\,h_1(r^2_{\mu,\vartheta}/s^2(P))\,dP$$
$$\ge s^2_{01}(\epsilon) \int h_1(r^2_{\mu,\vartheta}/s^2(P,\mu,\vartheta))\,dP \ge s^2_{01}(\epsilon)\,\tfrac{1}{2},$$

and therefore

$$\sigma(P) \ge s_{01}(\epsilon)/\sqrt{2} > 0. \tag{4.10a}$$

Secondly, completely similar to (4.9b) we get

$$\sigma(P) \ge s_{01}(\epsilon)\,c_1 \sqrt{(\int h_1(\frac{y^2}{c_1^2 s^2_{01}(\epsilon)})\,\varphi(y)\,dy - \epsilon)^+}. \tag{4.10b}$$

The maximum of the right hand sides of (4.10a) and (4.10b) divided by τ_0 is our lower (bias) bound $\sigma_{01}(\epsilon)$ for $\sigma_0(P)$ with respect to d_1-balls around P_0 with radius $\epsilon \in [0,1/2)$. The explicit values are given in Table 1.

Consistency, breakdown point, and equivariance of the SM-scale
Since $\tau_0 = s_0 c_1 (\int h_1(y^2/(c_1 s_0)^2)\varphi(y)\,dy)^{1/2}$, and since $s_{01}(\epsilon) \to s_0$ and $s_{02}(\epsilon) \to s_0$, as $\epsilon \to 0$, we get the same (uniform) consistency of the scale SM-estimator as for the scale S-estimator, cf. Section 3.
Especially, the breakdown point is $1/2$, too.
Also the regression invariance and the scale equivariance of $\sigma_0(P)$ is inherited from the S-scale $s(P)$.

Bounds for the regession SM-estimator
Here we use $0 < s(P) \le s_{02}(\epsilon)$ and (3.6) in order to get

$$I_1(P,\mu,\vartheta) \;\ge\; E\,h_1(\frac{1+|\vartheta|^2}{c_1^2 s_{02}^2(\epsilon)}\cdot\chi_1^2(\frac{\mu^2}{1+|\vartheta|^2})) - \epsilon. \tag{4.11}$$

If we define k_ϵ^{SM} as the minimum of the right hand sides of (4.8a) and (4.8b),

$$k_\epsilon^{SM} \;:=\; \min\{\,\frac{1}{2},\; E\,h_1(\frac{1}{c_1^2 s_{01}^2(\epsilon)}\cdot\chi_1^2(0)) + \epsilon\,\}, \tag{4.12}$$

then inequalities $(4.8a), (4.8b)$, and (4.11) imply that any $(\mu,\vartheta) \in \mathbb{R}\times\mathbb{R}^r$ with the property

$$E\,h_1(\frac{1+|\vartheta|^2}{c_1^2 s_{02}^2(\epsilon)}\cdot\chi_1^2(\frac{\mu^2}{1+|\vartheta|^2})) \;>\; k_\epsilon^{SM} + \epsilon \tag{4.13}$$

cannot be a global minimum point of $I_1(P,\cdot)$ or, equivalently, of $\sigma(P,\cdot)$.

Since the right hand side of (4.13) is less than or equal to $1/2 + \epsilon < 1$, we may argue completely similar to Section 3:
Let μ_ϵ^{SM} be the (unique) positive solution μ of

$$E\,h_1(\frac{1}{c_1^2 s_{02}^2(\epsilon)}\cdot\chi_1^2(\mu^2)) \;=\; k_\epsilon^{SM} + \epsilon, \tag{4.14}$$

and d_ϵ^{SM} the (unique) positive solution d of

$$E\,h_1(\frac{1+d^2}{c_1^2 s_{02}^2(\epsilon)}\cdot\chi_1^2(0)) \;=\; k_\epsilon^{SM} + \epsilon. \tag{4.15}$$

Then all global minimum points of $\sigma(P,\cdot)$ must lie in the compact set

$$\{(\mu,\vartheta)\in\mathbb{R}\times\mathbb{R}^r :\; |\mu| \le \mu_\epsilon^{SM}\sqrt{1+|\vartheta|^2},\; |\vartheta| \le d_\epsilon^{SM}\}. \tag{4.16}$$

The explicit values of the bounds μ_ϵ^{SM} and d_ϵ^{SM} are given in Table 1.

Consistency of the regression SM-estimator
For $\epsilon \to 0$ we have $s_{01}(\epsilon) \to s_0$, $s_{02}(\epsilon) \to s_0$, and

$$k_\epsilon^{SM} + \epsilon \;\to\; E\,h_1(\chi_1^2(0)/(c_1 s_0)^2) \;<\; \int h_1(y^2/s_0^2)\,\varphi(y)\,dy \;=\; \frac{1}{2}.$$

Therefore definition (4.14) of μ_ϵ^{SM} and definition (4.15) of d_ϵ^{SM} imply

$$\mu_\epsilon^{SM} \;\to\; 0 \quad\text{and}\quad d_\epsilon^{SM} \;\to\; 0, \quad\text{as}\quad \epsilon \to 0, \tag{4.17}$$

which implies the (uniform) consistency of the SM-estimator of (μ_0, ϑ_0) at the ideal model P_0.

Breakdown point of the regression SM-estimator

Since the bounds $\mu_\epsilon^{SM} < \infty$ and $d_\epsilon^{SM} < \infty$ are valid for the d_1-ball around P_0 with radius $\epsilon \in [0, 1/2)$, and since this ball contains the ϵ-contamination ball $\{(1-\epsilon)P_0 + \epsilon Q : Q \in \mathcal{M}_1\}$ the (asymptotic) breakdown point is $1/2$.

Equivariance of the regression SM-estimator

Because of the definitions (4.1) and (4.7) the regression equivariance and the scale equivariance of the SM-functional $(\mu_{SM}(P), \vartheta_{SM}(P))$ is inherited from the regression invariance and the scale equivariance of the S-functional $s(P)$ of the first step.

Asymptotic normality and efficiency

The smoothness of h_1 and the consistency of the corresponding S- and SM-estimators allow for a simple proof of the asymptotic normality of the SM-estimator $(\hat{\mu}_n, \hat{\vartheta}_n) = (\mu_{SM}(P_n(z,\cdot)), \vartheta_{SM}(P_n(z,\cdot)))$ under the ideal model P_0, cf. Behnen (1991):

$$\mathcal{L}_0\Big[\sqrt{n}\,\big((\hat{\mu}_n, \hat{\vartheta}_n) - (\mu_0, \vartheta_0)\big)\Big] \;\rightarrow\; \mathcal{N}_{r+1}\Big(0, \frac{\beta_1^2}{\beta_2^2} I_{r+1}\Big), \tag{4.18}$$

where

$$\beta_1^2 := \int y^2 \{h_1'(y^2/(c_1 s_0)^2)\}^2 \varphi(y)\, dy \tag{4.19}$$

and

$$\beta_2 := \int y^2 h_1'(y^2/(c_1 s_0)^2)\, \varphi(y)\, dy\,. \tag{4.20}$$

	Table 1 : Bounds for the supremum asymptotic bias (SAB)							
	S-estimator				**SM-estimator**			
ϵ	$\frac{s_{01}(\epsilon)}{s_0}$	$\frac{s_{02}(\epsilon)}{s_0}$	μ_ϵ^S	d_ϵ^S	$\sigma_{01}(\epsilon)$	$\sigma_{02}(\epsilon)$	μ_ϵ^{SM}	d_ϵ^{SM}
0.005	0.988	1.012	0.223	0.225	0.980	1.021	0.368	0.369
0.010	0.976	1.025	0.316	0.323	0.961	1.042	0.526	0.531
0.050	0.881	1.129	0.724	0.802	0.823	1.229	1.299	1.389
0.100	0.768	1.270	1.063	1.316	0.697	1.512	2.087	2.525
0.200	0.560	1.608	1.669	2.691	0.562	2.260	3.834	7.657
0.300	0.366	2.092	2.401	5.619	0.468	3.376	5.397	15.32
0.400	0.181	3.034	3.685	16.71	0.296	5.492	8.377	45.01

Note: The bounds in Table 1 are evaluated for the special case

$$\mu_0 = 0\,,\quad \vartheta_0 = 0\,,\quad \sigma_0 = 1\,,\quad \eta_0 = 0\,,\quad \Sigma_0 = I_r\,.$$

For $\mu_0 \in \mathbb{R}$, $\vartheta_0 \in \mathbb{R}^r$, $\sigma_0 > 0$, $\eta_0 \in \mathbb{R}^r$, and nondegenerate $r \times r$ covariance matrix Σ_0 the SM-functionals $\sigma_0(P)$, $\mu_{SM}(P)$, and $\vartheta_{SM}(P)$ fulfil the inequalities

$$\sigma_0 \cdot \sigma_{01}(\epsilon) \;\leq\; \sigma_0(P) \;\leq\; \sigma_0 \cdot \sigma_{02}(\epsilon), \tag{4.21a}$$

$$\sqrt{(\vartheta_{SM}(P)-\vartheta_0)'\Sigma_0(\vartheta_{SM}(P)-\vartheta_0)} \;\le\; \sigma_0 \cdot d_\epsilon^{SM}, \tag{4.21b}$$

$$\begin{aligned} &|\,(\mu_{SM}(P)-\mu_0)+(\vartheta_{SM}(P)-\vartheta_0)'\eta_0\,| \\ &\qquad \le\; \mu_\epsilon^{SM}\cdot\sqrt{\sigma_0^2+(\vartheta_{SM}(P)-\vartheta_0)'\Sigma_0(\vartheta_{SM}(P)-\vartheta_0)}, \end{aligned} \tag{4.21c}$$

for any $P \in \mathcal{M}_1$ such that $d_1(P,P_0) \le \epsilon$.
The inequalities for the S-functionals $s(P)/s_0$, $\mu_S(P)$, and $\vartheta_S(P)$ are completely similar.

REFERENCES

Behnen, K. (1991). Robuste Statistik: Multiple Lineare Regression. Lecture Notes: Mathem. Stochastik, Univ. Hamburg 1991

Dudley, R.M. (1979). Balls in $\mathbb{R}^k$ do not cut all subsets of $k+2$ points. Adv. Math. 31 (1979), pp. 306-308.

Hennig, Chr. (1993). Efficient high-breakdown-estimators in robust regression. Preprint: Mathem. Stochastik, Univ. Hamburg 1993

Huber, P.J. (1981). Robust Statistics. Wiley: New York 1981.

Maronna, R.A. & Yohai, V.J. (1981). Asymptotic behavior of general M-estimates for regression and scale with random carriers. Z. Wahrsch. verw. Geb. 58 (1981), pp. 7-20.

Martin, R.D., Yohai, V.J., Zamar, R.H. (1989). Min-Max bias robust regression. Ann. Statist. 17 (1989), pp. 1608-1630.

Rousseeuw, P.J. & Leroy, A.M. (1987). Robust Regression and Outlier Detection. Wiley: New York 1987.

Rousseeuw, P.J. & Yohai, V.J. (1984). Robust Regression by Means of S-estimators. In J.Franke, W.Härdle & D.Martin (Eds.), Robust and NonlinearTime Series Analysis, pp. 256-272, Springer: Berlin 1984.

Shorack, G.R. & Wellner, J.A. (1986). Empirical Processes with Applications to Statistics. Wiley: New York 1986.

Yohai, V.J. (1987). High breakdown-point and high efficiency robust estimates for regression. Ann. Statist. 15 (1987), pp. 642-656.

Yohai, V.J. & Zamar, R. (1988). High breakdown point estimates of regression. J. Amer. Statist. Assoc. 83 (1988), pp. 406-413.

Yohai, V.J., Stahel, W.A., Zamar, R.H. (1991). A procedure for robust estimation and inference in linear regression. In W.Stahel & S.Weisberg (Eds.), Directions in Robust Statistics and Diagnostics, Part 2, pp. 365-374, Springer: New York 1991.

SIGNIFICANCE OF DIFFERENCES OF ESTIMATES

Pavel Boček, Jan Ámos Víšek
Academy of Sciences of the Czech Republic

Abstract

A test for evaluating the significance of difference between the estimates of regression model is proposed. The idea is applicable on any estimator having Bahadur's representation, nevertheless for the simplicity it is explained on a concrete example of one-step M-etimates. Performance of the test is illustrated by a numerical example.

Key words: One-step M-estimates, significance of differences between estimates.
AMS classification: Primary 62H15; secondary 62J05.

1 INTRODUCTION

Due to the fact that the robust statistics has developed in the last twenty or thirty years a lot of procedures for solution of the same problem, e.g. for estimating regression coefficients, we may expect that the results of the different procedures may differ. Sometimes they may differ considerably. The practical experiences have confirmed that (Víšek (1992 b)). To make the problem more transparent let us assume some special type of estimators. Since the illustrating example had been computed for the one-step M-estimators, and because it has brought in addition also some interesting experience, we shall consider in what follows one-step M-estimators in the form given by Jurečková and Portnoy (1988). Because of a (wide) range of the possible initial estimators as well as of the ψ-functions we obtain (at least) a few different, competing estimates of the model. We use as the initial estimator usually the asymptotically consistent ones, which are possibly in some sense asymptotically optimal. Nevertheless the one-step estimates evaluated by means of different ψ-functions for the same initial estimate may differ a lot. It does not imply that the difference is statistically significant. As we shall see below the numerical illustration offers also rather surprising fact that the one-step M-estimates (as defined in Jurečková and Portnoy (1988)) may depend also on the order of the regressors in the design matrix, although, of course, the initial estimates - LS and LMS - had appeared to be the same for different permutations of the columns of the design matrix.

The present paper proposes a simple test for the differences between the estimates of the linear model. Naturally, such a test is only a diagnostic, helping tool for a solution of the problem stated above.

2 NOTATION AND PRELIMINARIES

Let N denote the set of all positive integers, R the real line and R^ℓ the ℓ-dimensional Euclidean space. We shall consider the linear regression model

$$Y^{(n)} = X^{(n)}\beta + \varepsilon^{(n)} \tag{1}$$

where for every $n \in N$ and some (fixed) $p \in N$ we have $Y^{(n)} = (Y_1, Y_2, ..., Y_n)$, $X^{(n)} = (x_{ij})_{i=1,2,...,n}^{j=1,2,...,p}$, $\beta^0 = (\beta_1^0, \beta_2^0, ..., \beta_p^0)^T$ and $\varepsilon^{(n)} = (\varepsilon_1, \varepsilon_2, ..., \varepsilon_n)^T$ where "T" indicates the transposition. We assume that the sequence $\{\varepsilon_i\}_{i=1}^{\infty}$ consists of independent and identically distributed random variables being defined on a probability space $(\Omega, \mathcal{B}, P)$. and denote the distribution function of ε_1 by F. The carriers x_{ij}'s (design matrix) are assumed to be fix (nonrandom) and known while β^0 (a "true" value of the regression coefficients) is unknown but also fix. Of course, we will assume $\beta^0 \neq 0$. Sometimes we will use the notation $X_i = (x_{i1}, x_{i2}, ..., x_{ip})^T$, so that $X_i^T\beta = \sum_{j=1}^{p} x_{ij}\beta_j$ for any $\beta \in R^p$. We shall use the result of Jurečková and Portnoy (1988), namely the asymptotic representation of one-step M-estimator. The representation has been derived under the following conditions (see Jurečková and Portnoy (1988)).

CONDITIONS A.

i) For some $\tau \in \left(\frac{1}{4}, \frac{1}{2}\right]$ there is an n^τ-consistent regression invariant estimate S_n of the scale $S(F)$ of residuals.

ii) $\psi : R \to R$ may be written as $\psi_1 + \psi_2$ where ψ_ℓ $(\ell = 1, 2)$ are monotone, bounded, and for $\max\{|t|, |u|\} < \delta_1$, $\delta_1 > 0$, and some $K_1 > 0$

$$\int_{-\infty}^{\infty} \left[z\psi_\ell'(e^{-u}\,(z - t)/S(F))\right]^2 \, \mathrm{d}F(z) < K_1.$$

iii) ψ_ℓ' is absolutely continuous on intervals $(r_0, r_1), (r_1, r_2), ..., (r_k, r_{k+1})$ with $-\infty = r_0 < r_1 < ... < r_k < r_{k+1} = \infty$; denote $\psi_\ell'' = \frac{\mathrm{d}\psi_\ell'}{\mathrm{d}_z}$, $\ell = 1, 2$. ψ' be bounded on intervals $(r_1 - \delta_2, r_1), (r_1, r_2), ..., (r_{k-1}, r_k), (r_k, r_k + \delta_2)$, $\delta_2 > 0$, $\ell = 1, 2$. The d. f. F is Lipschitz of the first order in neighbourhoods of the points $r_1 \cdot S(F), r_2 \cdot S(F), ..., r_k \cdot S(F)$. Moreover, for $\max\{|t|, |u|\} < \delta_3$, $\delta_3 > 0$ and some $K_2 > 0$

$$\int_{-\infty}^{\infty} |\psi_\ell''(e^{-u}\,(z - t)/S(F))| \, \mathrm{d}F(z) < K_2.$$

iv) Denoting $Q_n = \frac{1}{n}\left[X^{(n)}\right]^T X^{(n)} = \left\{q_{ij}^{(n)}\right\}_{i=1,2,...,p}^{j=1,2,...,p}$, we have $\lim_{n\to\infty} Q_n = Q$ where Q is a positively definite $(p \times p)$-matrix.

v) For $j = 1, 2, ..., p$

$$n^{-1}\sum_{i=1}^{n} x_{ij}^4 = O(1) \quad \text{and} \quad \left\{\max_{1\le k\le n} x_{kj}^2\right\}\left[\sum_{i=1}^{n} x_{ij}^2\right]^{-1} \to 0 \quad \text{as} \quad n \to \infty.$$

vi) $\tilde{\beta}^{(n)}$ is n^τ-consistent initial estimator of β^0.

vii)

$$\gamma = S^{-1}(F)\int_{-\infty}^{\infty} \psi'(z/S(F))\,\mathrm{d}F(z) \neq 0 \qquad (2)$$

and

$$\int_{-\infty}^{\infty} z\psi'(z/S(F))\,\mathrm{d}F(z) = 0.$$

REMARK 1 *Let D^Q be the determinant of the matrix Q and $\alpha = \max_{1\leq i,j\leq p}|q_{i,j}|$. Since we have assumed that Q is regular (and hence positive definite) $D^Q > 0$. Select any $\delta \in (0, \frac{1}{2}D^Q\,[p!]^{-1}\,\alpha^{1-p})$. Under A.iv we may find $n_0 \in N$ such that for any $n > n_0$ $\max_{1\leq i,j\leq p}|q_{i,j} - q_{i,j}^{(n)}| < \delta$. But it implies that also the determinants D^{Q_n} are positive starting with n_0.*

Following Jurečková, Portnoy (1988) let us put

$$\hat{\gamma}^{(n)} = \frac{1}{2}n^{\tau-1}\sum_{i=1}^{n} x_{i1}\left[\psi((Y_i - X_i^T\tilde{\beta}^{(n)} + n^{-\tau}X_i^T q^{(1)})/S_n)\right. \qquad (3)$$

$$\left. -\psi((Y_i - X_i^T\tilde{\beta}^{(n)} - n^{-\tau}X_i^T q^{(1)})/S_n)\right]$$

where $q^{(1)}$ is the first column (or the first row) of the matrix Q_n and the (column) vector X_i denotes i-th row of design matrix. Let, moreover,

$$\mathcal{W}_n = n^{-1}Q_n^{-1}\sum_{i=1}^{n} X_i\psi((Y_i - X_i^T\tilde{\beta}^{(n)})/S_n)$$

and fixing $a > 0$ (arbitrary constant) let us define the one-step estimator

$$\hat{\beta}^{(n)} = \begin{cases} \tilde{\beta}^{(n)} + \left[\hat{\gamma}^{(n)}\right]^{-1}\mathcal{W}_n & \text{if } \left\|[\hat{\gamma}(n)]^{-1}\mathcal{W}_n\right\| \leq a, \\ \tilde{\beta}^{(n)} & \text{otherwise} \end{cases} \qquad (4)$$

LEMMA 1 *(Jurečková, Portnoy (1988)) Let Conditions A hold. Then $\hat{\gamma}^{(n)} - \gamma = O_p(n^{-\tau})$ and*

$$\sqrt{n}(\hat{\beta}^{(n)} - \beta^0) = n^{-\frac{1}{2}}\gamma^{-1}Q_n^{-1}\sum_{i=1}^{n} X_i\psi((Y_i - X_i^T\beta^0)/S(F)) + O_p(n^{\frac{1}{2}-2\tau}).$$

Proof: For the proof of the first assertion of the lemma see Jurečková, Portnoy (1988), the relation (3.4). The second assertion follows immediately from Corollary 2.1 and the last line of the proof of Theorem 3.1 of the same paper. □

3 NUMERICAL ILLUSTRATION

To demonstrate the fact mentioned in the introduction we will use "Engine Knock Data". They may be found in Mason, Gunst and Hess (1989) (and also in Hettmansperger and Sheather (1992) or in Rubio, Víšek (1993, in these proceedings)).

EXHIBIT 1. Least Squares and Least Median of Squares Analysis of Engine Knock Data

method	intercept	Spark	Air	Intake	Exhaust
$\hat{\beta}_{LS}$	12.01258	1.09504	2.18894	0.93487	-0.00190
$\hat{\beta}_{LMS}$	30.04044	0.14413	3.07847	0.46004	-0.00694

$\hat{\beta}_{LMS}$ was evaluated by a method based on simplex technique, see Boček and Lachout (1993).

In the next two exhibits the one-step estimates of the regression model for Engine Knock Data with the initial LS-estimate and LMS-estimate are presented. The subindex of $\hat{\beta}$ (see the first columns of Exhibit 2 and 3) indicates which regressor was in the design matrix as the first column, i.e. for instance in the line denoted as $\hat{\beta}_{intercept}$ of Exhibit 2 the one step-estimate with the initial LS-estimate and with the design matrix containing the intercept as the first regressor is given.

EXHIBIT 2. One-step M-estimates for Engine Knock Data with $\hat{\beta}_{LS}$ as initial estimate

method	intercept	Spark	Air	Intake	Exhaust
$\hat{\beta}_{intercept}$	12.73064	1.18772	2.04005	0.98721	-0.00397
$\hat{\beta}_{Spark}$	12.42510	1.14828	2.10341	0.96494	-0.00308
$\hat{\beta}_{Air}$	11.81530	1.06957	2.22985	0.92049	-0.00133
$\hat{\beta}_{Intake}$	12.99157	1.22140	1.98595	1.00623	-0.00472
$\hat{\beta}_{Exhaust}$	11.73219	1.05885	2.24708	0.91443	-0.00109

EXHIBIT 3. One-step M-estimates for Engine Knock Data with $\hat{\beta}_{LMS}$ as initial estimate

method	intercept	Spark	Air	Intake	Exhaust
$\hat{\beta}_{intercept}$	-50.49829	1.63484	7.09455	1.71014	-0.05865
$\hat{\beta}_{Spark}$	-28.21173	1.22234	5.98323	1.36422	-0.04434
$\hat{\beta}_{Air}$	38.37274	-0.01009	2.66298	0.33071	-0.00159
$\hat{\beta}_{Intake}$	47.52891	-0.17957	2.20641	0.18859	0.00429
$\hat{\beta}_{Exhaust}$	19.10427	0.34655	3.62381	0.62979	-0.01396

Comparing the values of the one-step LMS-estimates (Exhibit 3) one may ask wheather they are signifficantly different or not. In the next section we shall propose a test for answering this question. Numerical results may be then found in SECTION 5.

4 THE SIGNIFICANCE OF THE DIFFERENCE BETWEEN TWO ONE-STEP M-ESTIMATES

Let $\tilde{\beta}^{(n,j)}$ and $\psi_{(j)}$, $j = 1, 2$ be two (not necessarily distinct) initial estimators and functions, respectively, both of them fulfilling Conditions A. Moreover, let us denote by $\hat{\beta}^{(n,j)}$ the corresponding one-step M-estimates of the regression coefficints obtained according to (4), and $\gamma_{(j)}$ and $\hat{\gamma}_{(j)}^{(n)}$ the corresponding constants and their estimates, respectively (see (2) and (3)). To be able to derive the asymptotic distribution of below proposed test we would need to assume only that $\mathsf{E}_F\psi_{(1)}(\varepsilon_1/S(F)) = \mathsf{E}_F\psi_{(2)}(\varepsilon_1/S(F))$. We will assume a little more, namely that $\mathsf{E}_F\psi_{(j)}(\varepsilon_1/S(F)) = 0$, $j = 1, 2$, i.e. we will assume Fisher consistency of $\hat{\beta}^{(n,j)}$, $j = 1, 2$. Let us further denote by $r_i^{(n,j)}$ the i-th residual with respect to the j-th estimate, i.e. $r_i^{(n,j)} = Y_i - X_i^T \hat{\beta}^{(n,j)}$. One easy verifies that the following assertion holds.

ASSERTION 1 *Under Conditions A we have*

$$r_\ell^{(n,1)} - r_\ell^{(n,2)} = n^{-1} X_\ell^T Q_n^{-1} \sum_{i=1}^{n} X_i \left\{ \gamma_{(2)}^{-1} \psi_{(2)}((Y_i - X_i^T \beta^0)/S(F)) \right. \tag{5}$$

$$\left. - \gamma_{(1)}^{-1} \psi_{(1)}((Y_i - X_i^T \beta^0)/S(F)) \right\} + h_\ell.$$

where $h_\ell = X_\ell^T \cdot \xi$ *with* $\|\xi\| = O_p(n^{-2\tau})$.

Proof: We have

$$r_\ell^{(n,j)} - \varepsilon_\ell = Y_\ell - X_\ell^T \hat{\beta}^{(n,j)} - Y_\ell + X_\ell^T \beta^0$$

$$= -X_\ell^T (\hat{\beta}^{(n,j)} - \beta^0) = -n^{-1} \gamma_{(j)}^{-1} X_\ell^T Q_n^{-1} \sum_{i=1}^{n} X_i \psi_{(j)}((Y_i - X_i^T \beta^0)/S(F)) + h_\ell^*$$

where $h_\ell^* = X_\ell^T \cdot \xi^{(j)}$ with $\|\xi^{(j)}\| = O_p(n^{-2\tau})$. It implies immediately the assertion.

□

Let us denote

$$\hat{\sigma}_n^2 = \left[\hat{\gamma}_{(1)}^{(n)}\right]^{-2} \hat{\sigma}_{(n,1,1)}^2 - \left[\hat{\gamma}_{(1)}^{(n)} \hat{\gamma}_{(2)}^{(n)}\right]^{-1} \hat{\sigma}_{(n,1,2)}^2 + \left[\hat{\gamma}_{(2)}^{(n)}\right]^{-2} \hat{\sigma}_{(n,2,2)}^2$$

with

$$\hat{\sigma}_{(n,j,\ell)}^2 = \frac{1}{n} \sum_{i=1}^{n} \left[\psi_{(j)}(r_i^{(n,j)}/S_n) - \bar{\psi}_{(j)}\right] \cdot \left[\psi_{(\ell)}(r_i^{(n,\ell)}/S_n) - \bar{\psi}_{(\ell)}\right] \qquad j, \ell = 1, 2,$$

where

$$\bar{\psi}_{(k)} = \frac{1}{n} \sum_{i=1}^{n} \psi_{(k)}(r_i^{(n,k)}/S_n) \qquad k = 1, 2.$$

Then a straightforward consequence of Assertion 1 is the following theorem:

□

THEOREM 1 *Let Conditions A hold and let moreover* $\mathsf{E}_F \psi_{(j)}(\varepsilon_1/S(F)) = 0,\ j = 1, 2$. *Then*

$$T_n^2 = \hat{\sigma}_n^{-2} \sum_{i=1}^{n} (r_i^{(n,1)} - r_i^{(n,2)})^2$$

has asymptotically χ^2*-distribution with* p *degrees of freedom.*

Proof: Let us denote

$$\Psi_n(\varepsilon) = \left(\frac{\psi_{(2)}(\varepsilon_1/S(F))}{\gamma_{(2)}} - \frac{\psi_{(1)}(\varepsilon_1/S(F))}{\gamma_{(1)}}, \frac{\psi_{(2)}(\varepsilon_2/S(F))}{\gamma_{(2)}} - \frac{\psi_{(1)}(\varepsilon_2/S(F))}{\gamma_{(1)}}, \ldots, \right.$$

$$\left. \frac{\psi_{(2)}(\varepsilon_n/S(F))}{\gamma_{(2)}} - \frac{\psi_{(1)}(\varepsilon_n/S(F))}{\gamma_{(1)}} \right)^T$$

and

$$\kappa^{(in)} = (\kappa_1^{(in)}, \kappa_2^{(in)}, ..., \kappa_n^{(in)})^T$$

with $\kappa_j^{(in)} = n^{-\frac{1}{2}} \sum_{\ell=1}^{p} \sum_{k=1}^{p} x_{jk} \tilde{q}_{k\ell}^{(n)} x_{i\ell}$ where $\tilde{q}_{k\ell}^{(n)} = (Q_n^{-1})_{k\ell}$. Then we may write (5) in the form

$$r_i^{(n,1)} - r_i^{(n,2)} = n^{-\frac{1}{2}} \Psi^T(\varepsilon) \cdot \kappa^{(in)} + h_i$$

(for h_i see Asertion 1). Now we have

$$\sum_{i=1}^{n} h_i \Psi^T(\varepsilon) \kappa^{(in)} = \sum_{k=1}^{p} \sum_{\ell=1}^{p} \sum_{r=1}^{p} [n^{-1} \sum_{i=1}^{n} x_{ir} x_{i\ell} \tilde{q}_{k\ell}^{(n)}] \cdot \left[n^{-\frac{1}{2}} \sum_{j=1}^{n} x_{jk} [\Psi(\varepsilon)]_j \right] \cdot n^{\frac{1}{2}} \xi_r.$$

Since

$$\sum_{\ell=1}^{p} \left[n^{-1} \sum_{i=1}^{n} x_{ir} x_{i\ell} \right] \tilde{q}_{k\ell}^{(n)} = \delta_{rk} O(1),$$

$$n^{-\frac{1}{2}} \sum_{j=1}^{n} x_{jk} [\Psi(\varepsilon)]_j = O_p(1) \quad \text{and} \quad n^{\frac{1}{2}} \xi_r = O_p(n^{\frac{1}{2} - 2\tau})$$

we have

$$n^{-\frac{1}{2}} \sum_{i=1}^{n} h_i \Psi^T(\varepsilon) \kappa^{(in)} = O_p(n^{-2\tau}) \quad \text{and} \quad \sum_{i=1}^{n} h_i^2 = O_p(n^{1-4\tau}).$$

Since for $\tau \in (\frac{1}{4}, \frac{1}{2}]$ $1 - 4\tau > -2\tau$, we obtain

$$\sum_{i=1}^{n} (r_l^{(n,1)} - r_l^{(n,2)})^2 = n^{-1} \Psi^T(\varepsilon) \sum_{i=1}^{n} \left\{ \kappa^{(in)} \left[\kappa^{(in)} \right]^T \right\} \Psi(\varepsilon) + \Delta_n$$

with $\Delta_n = O_p(n^{1-4\tau})$. First of all, let us show that

$$n^{-1} \sigma_n^{-2} \Psi^T(\varepsilon) \sum_{i=1}^{n} \left\{ \kappa^{(in)} \left[\kappa^{(in)} \right]^T \right\} \Psi(\varepsilon)$$

with

$$\sigma_n^2 = \left[\gamma_{(1)} \right]^{-2} \text{var} \psi_{(1)}(\varepsilon_1 / S(F)) - \left[\gamma_{(1)} \gamma_{(2)} \right]^{-1} \text{cov} \left\{ \psi_{(1)}(\varepsilon_1 / S(F)), \psi_{(2)}(\varepsilon_1 / S(F)) \right\}$$
$$+ \left[\gamma_{(2)} \right]^{-2} \text{var} \psi_{(2)}(\varepsilon_1 / S(F))$$

has asymptotically χ^2-distribution with p degrees of freedom. Let us observe that

$$\left[\sum_{i=1}^{n} \left\{ \kappa^{(in)} \left[\kappa^{(in)} \right]^T \right\} \right]_{jk} = n^{-1} \sum_{i=1}^{n} \sum_{\ell_1=1}^{p} \sum_{\ell_2=1}^{p} x_{i\ell_1} \tilde{q}_{\ell_1 \ell_2}^{(n)} x_{j\ell_2} \sum_{\ell_3=1}^{p} \sum_{\ell_4=1}^{p} x_{i\ell_3} \tilde{q}_{\ell_3 \ell_4}^{(n)} x_{k\ell_4}$$

$$= \sum_{\ell_1=1}^{p} \sum_{\ell_2=1}^{p} x_{j\ell_1} \tilde{q}_{\ell_1 \ell_2}^{(n)} x_{k\ell_2}.$$

Now due to the fact that (starting with some n_0) Q_n^{-1} is symmetric and positive definite it may be written as $V_n \cdot V_n^T$ where V_n is a regular $(p \times p)$-matrix, and denoting $U_n = X^{(n)} V_n$ we find that $\sum_{i=1}^{n} \kappa^{(in)} \left[\kappa^{(in)} \right]^T = U_n \cdot U_n^T$. Let us consider

$V_n \left[n\mathcal{I}_p - V_n^T \left[X^{(n)}\right]^T X^{(n)} V_n \right] V_n^T$ where $\mathcal{I}_p$ is the $(p \times p)$ identity matrix. We easy find that this expression is equal to $\mathbf{0}$ and hence $n^{-1}U_n^T \cdot U_n = \mathcal{I}_p$. Finally putting $W_n = 1/(\sqrt{n}\sigma_n)\Psi^T(\varepsilon)U_n$, and using Ljapunov theorem, we find that W_n is asymptotically $N_p(\mathbf{0}, \mathcal{I}_p)$. Now, let Z_n be a random vector distributed according to $N_p(\mathbf{0}, \mathcal{I}_p)$. Then $Z_n^T Z_n$ is distributed as χ^2 with p-degrees of freedom. Applying Slucky theorem we find that $Z_n^T Z_n - W_n^T W_n \to 0$ in probability. Making use of the standard technique of approximation by means of the law of large numbers we can prove that we may substitute σ_n^2 by $\hat{\sigma}_n^2$, which concludes the proof. □

5 CONCLUSION

The result of the paper shows that the test for differences between the estimates of regression models may be based on the sum of squared differences of residuals and that the asymptotic distribution of the test statistic is χ^2 with p degrees of freedom (where p is the dimension of model). It means that it is easy to carry out a test whether two estimates of the regression model differ significantly or not. (Let us recall that $\chi_5(5\%) = 11.07$.)

EXHIBIT 4. Testing one-step M-estimates for Engine Knock Data with $\hat{\beta}_{LMS}$ as initial estimate

estimate tested aganist estimate	$\hat{\beta}_{intercept}$ $\hat{\beta}_{Spark}$	$\hat{\beta}_{intercept}$ $\hat{\beta}_{Air}$	$\hat{\beta}_{intercept}$ $\hat{\beta}_{Intake}$	$\hat{\beta}_{intercept}$ $\hat{\beta}_{Exhaust}$
T_n^2	38.99	619.95	754.29	575.91

estimate tested against estimate	$\hat{\beta}_{Spark}$ $\hat{\beta}_{Air}$	$\hat{\beta}_{Spark}$ $\hat{\beta}_{Intake}$	$\hat{\beta}_{Spark}$ $\hat{\beta}_{Exhaust}$
T_n^2	384.00	450.29	266.14

estimate tested against estimate	$\hat{\beta}_{Air}$ $\hat{\beta}_{Intake}$	$\hat{\beta}_{Air}$ $\hat{\beta}_{Exhaust}$	$\hat{\beta}_{Intake}$ $\hat{\beta}_{Exhaust}$
T_n^2	6.58	44.14	96.05

Keep in mind that both estimators of the model parameters in any tested pair are under the Conditions A asymptotically consistent and hence the estimates should be "near" to the "true" model, and of course "near" each to other. In other words, if the result of test indicates that the difference is not insignifficant, it implies that either some of the assumptions under which the consistency, and possibly some other asymptotic properties had been derived, have been distorted, or the sample size may be insufficient so that the asymptotics need not yet work for given data. The matter is also tightly connected with the problem of discriminability between different models. It is clear that for small and mediate sample sizes we have only a limited chance to recognize collinearity. It may however cause that we may have two models, with considerably different coefficients and rather different laws of random fluctuations and both of them may describe the data very well. Then two different estimating methods may "prefer" for our data different "theoretical" models although both methods may be asymptotically consistent in each of given models. Nevertheles it does not guarantee that it cannot happen that the asymptotic has started to work for the first method in the first model (much) earlier than in the second one. And for the second method the situation may be "symmetric". Maybe that it is in this context useful to recall that the most of the robust methods use, at least implicitly, some robustified likelihood, and it partially compels the method to assume some special type of the law of random fluctuations.

An objection may appear that, it may happen that the two estimates are very close each to other but far from a "reasonable" model. It hints that the diagnostic process has to contain also methods for rejecting the estimates which are not appropriate for given data. The methods may be based e. g. on the idea of estimate subsample stability (Víšek (1992 a)) or on a comparison of residual density estimates for some (complementary) subsamples (Rubio and Víšek (in this volume) or Víšek (1993)). Since such methods usually need intensive computations, it may be preferable to test goodness of fit of an etimate only once. If we find that given estimate is not so bad to be rejected by a "goodness-of-fit" test, we may then compare it with the other estimates.

REFERENCES

Boček, P., Lachout, P. (1993), "Linear programming approach to LMS-estimation", to appear in *Memorial volume of Comput. Statist. & Data Analysis devoted to T. Havránek.*

Hettmansperger, T. P., Sheather, S. J. (1992), "A Cautionary Note on the Method of Least Median Squares", *The American Statistician 46, 79–83.*

Jurečková, J., Portnoy, S. (1988): Asymptotics for one-step M-estimators in regression with application to combining efficiency and high breakdown point. *Comumunications in Statist. A 16, pp. 2187 - 2199.*

Mason, R. L., Gunst, R. F., Hess, J. L. (1989), *Statistical Design and Analysis of Experiments,* New York: John Wiley.

Rousseeuw, P.J. (1984): Least median of square regression. *Journal of Amer. Statist. Association 79, pp. 871-880.*

Rubio, A. M., Víšek, J.Á. (1993): Diagnostics of regression model: Test of goodness-of-fit. In these proceedings.

Víšek, J. Á. (1992 a): Adaptive estimation in linear regression model and test of symmetry of residuals. *Contributions to Statistics: Model-Oriented Data Analysis, Springer Verlag (1992), eds. V. Fedorov, W. G. Müller and I. N. Vuchkov, 133–149.*

Víšek, J. Á. (1992 b): Stability of regression model estimates with respect to subsamples. *Computational Statistics 7, 183–203.*

Víšek, J. Á. (1993): A goodness of fit test of the linear regression models based on the weighted Hellinger distance. Submitted to print.

Authors address:

Department of Stochastic Informatics, Institute of Information Theory and Automation, Academy of Sciences the Czech Republic, Pod vodárenskou věží 4, 182 08 Prague, Czech Republic, Phone: 42-2-6605-2019 (or 2466) Fax: 42-2-0774-452, e-mail: VISEK@CSPGAS11.BITNET

Adaptiveness in Time Series Models

FEIKE C. DROST, CHRIS A.J. KLAASSEN, AND BAS J.M. WERKER*
Tilburg University, University of Amsterdam, and Tilburg University

Abstract

Many time series are modelled by stationary processes via innovations that are independent of the past. Suppose the parameter of interest is unrelated to the distribution of the innovations. We compare the estimation problem of this parameter of interest within a purely parametric framework to estimation within a semiparametric model where the shape of the distribution of the innovations appears as a non-Euclidean nuisance parameter. Typically the asymptotic estimation problem is equally hard in both models, mainly due to the independence of innovations and past. We illustrate this phenomenon in several well-known time series models.

KEYWORDS: time series, adaptiveness, linear regression model, ARMA, GARCH, TAR.
AMS 1991 SUBJECT CLASSIFICATION: 62F12, 62G20, 62M10.

1 Introduction

Let $(\Omega, \mathcal{S})$ be a measurable space and let $\mathcal{P} = \{P_{(\theta,g)} : \theta \in \Theta \subset \mathbf{R}^k,\ g \in G\}$ be a set of probability measures on $(\Omega, \mathcal{S})$. The Euclidean (finite dimensional) parameter $\theta^T = (\nu^T, \eta^T)$ is split into two parts with $\nu \in N \subset \mathbf{R}^m$ the parameter of interest and $\eta \in H \subset \mathbf{R}^{k-m}$ a finite dimensional nuisance parameter. The parameter $g \in G$ serves as infinite dimensional nuisance. To estimate in this semiparametric framework we are given observations from a real valued discrete time process $X = \{X_t : t \in \mathbf{N}\}$ on $(\Omega, \mathcal{S})$. The measure on $(\mathbf{R}^n, \mathcal{B}^n)$ induced by $(X_1, \ldots, X_n)$ under some P on $(\Omega, \mathcal{S})$ will be denoted by $P^{(n)}$.

Fix θ_0 in the interior of Θ, $g_0 \in G$, and write $P_0 = P_{\theta_0, g_0}$. Let E_0 denote the expectation under P_0. Consider a parametric submodel $\mathcal{Q} = \{Q_{(\theta,\zeta)} : \theta \in \Theta,\ \zeta \in Z \subset \mathbf{R}^l\}$ of $\mathcal{P}$ containing P_0. Often $\mathcal{Q}$ will satisfy a Local Asymptotic Normality (LAN) condition at P_0 as follows. For $\theta_0^T = (\nu_0^T, \eta_0^T)$ there exists a stationary martingale

*Dept. of Econometrics, Tilburg University, P.O. Box 90153, 5000 LE Tilburg, The Netherlands and Dept. of Mathematics, University of Amsterdam, Plantage Muidergracht 24, 1018 TV Amsterdam, The Netherlands. The first author is research fellow of the Royal Netherlands Academy of Arts and Sciences (K.N.A.W.).

difference sequence $\{\dot{l}_{(t)} : t \in \mathbb{N}\}$ only depending on P_0 such that, under P_0, for all sequences $\theta_n^T = (\nu_n^T, \eta_n^T)$, $\nu_n = \nu_0 + u/\sqrt{n}$, $\eta_n = \eta_0 + v/\sqrt{n}$, and $\zeta_n = \zeta_0 + w/\sqrt{n}$

$$\log \frac{dQ^{(n)}_{(\theta_n,\zeta_n)}}{dQ^{(n)}_{(\theta_0,\zeta_0)}}(X_1,\ldots,X_n) = (u^T,v^T,w^T)\frac{1}{\sqrt{n}}\sum_{t=1}^{n}\dot{l}_{(t)} - \frac{1}{2}\sigma^2 + o_p(1)$$
$$\to_d N(-\frac{1}{2}\sigma^2,\sigma^2),$$

where $\sigma^2 = E_0\left((u^T,v^T,w^T)\dot{l}_{(1)}\right)^2$. The stationarity condition on $\{\dot{l}_{(t)} : t \in \mathbb{N}\}$ is not necessary to deduce the convolution theorem (to be discussed below), but it allows us to simplify notation a lot by only considering $\dot{l}_{(1)} = \dot{l} = (\dot{l}_1^T, \dot{l}_2^T, \dot{l}_3^T)^T$.

By $\mathcal{P}_{12}$ we denote the special parametric submodel of $\mathcal{P}$ with $g = g_0$ fixed. By the convolution theorem of Hájek and Le Cam a best locally regular (efficient) estimator $\hat{\nu}_n$ of ν at P_0 in the model $\mathcal{Q}$ is asymptotically normal with covariance matrix $\left(E_0 l_1^* l_1^{*T}\right)^{-1}$, where l_1^* is the so-called efficient score function for ν in $\mathcal{Q}$. Hence, under P_0,

$$\sqrt{n}\,(\hat{\nu}_n - \nu_0) \to_d N\left(0, \left(E_0 l_1^* l_1^{*T}\right)^{-1}\right), \quad n \to \infty.$$

In case $\mathcal{Q} = \mathcal{P}_{12}$, the efficient score function equals

$$l_1^* = \dot{l}_1 - I_{12} I_{22}^{-1} \dot{l}_2, \tag{1.1}$$

where $I_{ij} = E_0 \dot{l}_i \dot{l}_j^T$. Typically l_1^* depends on $\mathcal{Q}$ and P_0. However, at some instances the efficient score function does not depend on $\mathcal{Q}$, but only on P_0. In those cases, regular estimation of ν at P_0 in the model $\mathcal{P}$ is as difficult asymptotically as estimation of ν in the parametric model $\mathcal{P}_{12}$, at least in principle. If this is the case and if there exists a locally Gaussian regular sequence of estimators $\{\hat{\nu}_n : n \in \mathbb{N}\}$ of ν at P_0, we say that $\{\hat{\nu}_n\}$ is adaptive at P_0. A detailed treatment of adaptiveness, especially for the i.i.d. case, is given by Bickel et al. (1993), henceforth referred to as BKRW (1993); see Definition 3.4.2, page 75, in particular.

Let $L_2^0(P_0)$ denote the Hilbert-space of all random variables with mean zero and finite variance under P_0 and define $\dot{\mathcal{P}}_i = [\dot{l}_i]$ $(i = 1,2)$ the closed linear span of the components of $\dot{l}_i$ in $L_2^0(P_0)$. $\dot{\mathcal{P}}_3$ is the closed linear span of the components of all possible $\dot{l}_3$. Similarly define $\dot{\mathcal{P}}_{ij} = [\dot{l}_i, \dot{l}_j]$. By $\prod(\cdot|\dot{\mathcal{P}}_i)$ we denote the projection operator in $L_2^0(P_0)$ projecting onto $\dot{\mathcal{P}}_i$. As explained in detail in Section 2.4 of BKRW (1993), the efficient score function for ν in $\mathcal{P}_{12}$ (c.f. (1.1)) may be viewed as the residual of a projection,

$$l_1^* = \dot{l}_1 - \prod\left(\dot{l}_1|\dot{\mathcal{P}}_2\right),$$

and a necessary condition for adaptiveness (cf. Proposition 3.4.3 of BKRW (1993)) is

$$\dot{l}_1 - \prod\left(\dot{l}_1|\dot{\mathcal{P}}_2\right) \perp \dot{\mathcal{P}}_3. \tag{1.2}$$

Whether adaptive estimators actually exist and, if so, how to construct them, is outside the scope of this paper.

Many models are described by a k-dimensional parameter κ that does not admit a partition $\kappa^T = (\nu^T, \eta^T)$ such that (1.2) holds (with θ replaced by κ). In the remainder we will study when a reparametrization $\kappa = \kappa(\theta)$ exists such that $\theta^T = (\nu^T, \eta^T)$ does satisfy (1.2). Such a reparametrization might yield a more natural parametrization of the model. Moreover, it exactly identifies which functions of the Euclidean parameters can be estimated adaptively. A seemingly related problem is discussed in Cox and Reid (1987), but they consider orthogonality between ν and η within the parametric model. In Section 2 we will explore the geometry of the situation. In Section 3 we will introduce a class of models for which the theory is further developed. Finally, in Section 4, we will give several examples, mainly of time series models, fitting our framework.

2 Some Geometry

As indicated in the introduction, score functions and their closed linear spans in $L_2^0(P_0)$ play an important role in semiparametric estimation. In this section, therefore, we will consider an arbitrary Hilbert-space H. Let H_1 and H_2 be (possibly infinite dimensional) closed subspaces of H. With $\prod$ we denote projection operators. They either act on elements of H giving the projection of that element, or they act on closed subspaces H_0 of H giving the linear subset of all projections of elements in H_0. Note that in general this does not yield a closed subspace (see, e.g., Weidmann (1980), Exercise 4.32). The following definition is inspired by (1.2).

Definition 2.1 *Let H_1 and H_2 be closed subspaces of the Hilbert-space H. A closed subspace H_0 of H_1 is called (H_1, H_2)-orthogonalizing if*

$$\forall h \in H_1 : h - \prod(h|H_0) \perp H_2. \tag{2.1}$$

H_0 is called minimal (H_1, H_2)-orthogonalizing if it is (H_1, H_2)-orthogonalizing and it does not contain a proper (H_1, H_2)-orthogonalizing closed subspace.

Observe that H_1 itself is always (H_1, H_2)-orthogonalizing. Moreover, $H_1 \perp H_2$ if and only if $H_0 = \{0\}$ is minimal (H_1, H_2)-orthogonalizing.

With this terminology the adaptiveness condition (1.2) translates into: $\dot{\mathcal{P}}_2$ is $(\dot{\mathcal{P}}_{12}, \dot{\mathcal{P}}_3)$-orthogonalizing. Of course, we would like $\dot{\mathcal{P}}_2 = [\dot{l}_2]$ to be as small as possible. Hence, it is important to know the minimal orthogonalizing closed subspaces. In fact, there is just one such subspace as the following result points out.

Proposition 2.2 *Let H_1 and H_2 be closed subspaces of the Hilbert-space H.*

(i) *The closed subspace H_0 of H_1 is (H_1, H_2)-orthogonalizing if and only if*

$$\overline{\prod(H_2|H_1)} \subset H_0. \tag{2.2}$$

(ii) *$\overline{\prod(H_2|H_1)}$ is the unique minimal (H_1, H_2)-orthogonalizing closed subspace.*

Proof. Let H_0 be any closed subspace of H_1. Since $\{h - \prod(h|H_0) : h \in H_1\} = H_1 \cap H_0^\perp$, condition (2.1) can be written as $H_1 \cap H_0^\perp \subset H_2^\perp$ or, equivalently,

$$H_2 \subset H_1^\perp \oplus H_0. \tag{2.3}$$

Now, (2.3) and $H_0 \subset H_1$ imply (2.2). Since $H_2 \subset \prod(H_2|H_1) \oplus H_1^\perp$, the inclusion (2.2) implies (2.3). Claim (i) is proved which yields (ii) directly. □

For the situation of Section 1 this proposition means that, even after reparametrization, $k - \dim(\Pi(\dot{\mathcal{P}}_3|\dot{\mathcal{P}}_{12}))$ is the maximum number of components of the Euclidean parameter one can estimate adaptively, a priori. Moreover, if

$$\Pi\left(\dot{\mathcal{P}}_3\middle|\dot{\mathcal{P}}_{12}\right) = \dot{\mathcal{P}}_2, \tag{2.4}$$

then the adaptiveness condition (1.2) holds.

3 Reparametrizing the Euclidean Parameter

In this section we study models where $\dot{l}_{12} = (\dot{l}_1^T, \dot{l}_2^T)^T$, introduced in the LAN condition in Section 1, has a specific structure. Often this score function $\dot{l}_{12}$ is not orthogonal to the tangent space $\dot{\mathcal{P}}_3$ induced by g. Therefore, we are interested in the minimal orthogonalizing subspace of $\dot{\mathcal{P}}_{12} = [\dot{l}_{12}]$ with respect to $\dot{\mathcal{P}}_3$. The specific structure of $\dot{l}_{12}$, which is quite common as we shall see in the examples of Section 4, allows determination of the unique minimal $(\dot{\mathcal{P}}_{12}, \dot{\mathcal{P}}_3)$-orthogonalizing subspace via straightforward application of Proposition 2.2.

Theorem 3.1 *Suppose*

$$\dot{l}_{12} = \psi(\varepsilon)W, \tag{3.1}$$

where $\psi : \mathbf{R} \to \mathbf{R}$ is measurable, ε a random variable and W a random k-vector. Assume that ε and W are independent, that EWW^T is non-singular, and that

$$\dot{\mathcal{P}}_3 \subset L_2^0(\varepsilon) = \{\lambda(\varepsilon) : E\lambda^2(\varepsilon) < \infty,\ E\lambda(\varepsilon) = 0\}. \tag{3.2}$$

If $\psi(\varepsilon)$ and $\dot{\mathcal{P}}_3$ are not orthogonal, then

$$\Pi\left(\dot{\mathcal{P}}_3\middle|\dot{\mathcal{P}}_{12}\right) = [a^T\psi(\varepsilon)W], \text{ with } a = \left(EWW^T\right)^{-1}EW. \tag{3.3}$$

In particular, if the covariance matrix of W is singular then

$$\Pi\left(\dot{\mathcal{P}}_3\middle|\dot{\mathcal{P}}_{12}\right) = [\psi(\varepsilon)]. \tag{3.4}$$

If $\psi(\varepsilon) \perp \dot{\mathcal{P}}_3$ or $EW = 0$, then

$$\Pi\left(\dot{\mathcal{P}}_3\middle|\dot{\mathcal{P}}_{12}\right) = \{0\} \text{ and } \dot{\mathcal{P}}_3 \perp \dot{\mathcal{P}}_{12}. \tag{3.5}$$

PROOF. If $E\psi^2(\varepsilon) > 0$, then for each $\lambda(\varepsilon) \in \dot{\mathcal{P}}_3$,

$$\Pi\left(\lambda(\varepsilon)\middle| [\psi(\varepsilon)W]\right) = \frac{E\lambda(\varepsilon)\psi(\varepsilon)}{E\psi^2(\varepsilon)} EW^T \left(EWW^T\right)^{-1} \psi(\varepsilon)W.$$

This immediately yields (3.3) and (3.5).

Furthermore, note that singularity of $E(W - EW)(W - EW)^T$ implies the existence of some non-zero vector $b \in \mathbf{R}^k$ with $b^T(W - EW) = 0$ a.s. Since EWW^T is non-singular, $b^T EW = b^T W$ cannot vanish, and for $a = b/(b^T EW)$, we have $a^T W = 1$ a.s. which implies $EWW^T a = EW$. Hence (3.3) reduces to (3.4). □

In the situation of Section 1 and Theorem 3.1 with (3.5) valid at all P_0, full adaptive estimation is likely to be possible. If (3.3) or (3.4) holds with $\prod(\dot{\mathcal{P}}_3|\dot{\mathcal{P}}_{12}) = \dot{\mathcal{P}}_2$, the adaptation condition (1.2) is satisfied. In case this does not happen at all P_0, it might be possible to reparametrize the Euclidean parameter in such a way that it does, i.e.

$$\prod(\dot{\mathcal{P}}_3|\dot{\mathcal{P}}_{12}) = [EW^T(EWW^T)^{-1}\psi(\varepsilon)W] = \dot{\mathcal{P}}_2. \tag{3.6}$$

In order for such a reparametrization to be feasible it may not depend on the unknown non-Euclidean nuisance parameter g.

To study the issue of reparametrization more carefully, let us consider models as in Theorem 3.1 parametrized by the nuisance parameter $g \in G$ and a Euclidean parameter $\kappa \in K$ in some natural or traditional way with $K \subset \mathbf{R}^k$ open. So $\dot{l}_{12} = \psi(\varepsilon)W$ now denotes the vector of score functions with respect to κ. Generally, this parametrization is not such that the adaptation condition (1.2) is satisfied, i.e. the minimal othogonalizing subspace is not generated by a sub-parameter of κ (c.f. (2.4)). Therefore, let us reparametrize κ by $\theta = (\nu^T, \eta)^T \in \Theta \subset \mathbf{R}^k$ such that $\kappa = \kappa(\theta)$ is differentiable with differentiable inverse ($\kappa(\cdot)$ is a diffeomorphism) and $\partial\kappa(\theta)/\partial\eta$ the vector of derivatives $\partial\kappa_i(\theta)/\partial\eta$, $i = 1, \dots, k$. Here $\kappa(\cdot)$ is understood to be independent of g.

Our reparametrization leads to (3.6) if

$$\left[EW^T(EWW^T)^{-1}\dot{l}_{12}\right] = \left[\left(\frac{\partial\kappa(\theta)}{\partial\eta}\right)^T \dot{l}_{12}\right], \tag{3.7}$$

i.e. if $(EWW^T)^{-1}EW$ and $\partial\kappa(\theta)/\partial\eta$ have the same direction viewed as vectors in $\mathbf{R}^k$. Consequently, the desired reparametrization exists only if, for all $\kappa \in K$

$$d_g(\kappa) = a/|a|, \text{ with } a = \left(EWW^T\right)^{-1} EW, \tag{3.8}$$

does not depend on g. Clearly, this holds if the distribution of W depends only on κ and not on g, or more generally, if the first two moments of W depend only on κ and do not depend on g. If the covariance matrix of W is singular, this occurs typically. Then (3.4) applies, the minimal orthogonalizing subspace is a subset of $L_2^0(\varepsilon)$ and the parameter generating $\psi(\varepsilon)$ as score is a parameter of the distribution of ε. From a practical point of view this is important since characteristics of this distribution are often of less importance than parameters governing the other behavioural characteristics of the model. Consequently, this will often lead to quite natural parametrizations.

Now assume $d_g(\kappa) = d(\kappa)$ is independent of g. Then we might reparametrize κ by choosing $\kappa(\theta)$ such that

$$\frac{\partial\kappa(\theta)}{\partial\eta} = d(\kappa(\theta)). \tag{3.9}$$

If in this case $d(\kappa) = d$ does not depend on κ either, and without loss of generality d_k does not vanish, we may reparametrize e.g. via

$$\kappa_k(\theta) = d_k\eta, \; \kappa_i(\theta) = \theta_i + d_i\eta, \; i = 1, \dots, k-1. \tag{3.10}$$

4 Linear Regression and Time Series

We present here four well-known models fitting the framework of Section 3. As an example with independent observations we consider the linear regression model. However our chief examples are some time series models. We will consider ARMA (Auto-Regressive Moving Average), GARCH (Generalized Auto-Regressive Conditional Heteroskedastic) and TAR (Threshold Auto-Regressive) models.

EXAMPLE 4.1. Consider the linear regression model with observations from i.i.d. copies of (X, Y) with

$$Y = X^T \kappa + \varepsilon. \tag{4.1}$$

Here, X is a random k-vector independent of the error ε. The distribution of X is assumed to be known, but ε has unknown density g with respect to Lebesgue measure on $(\mathbf{R}, \mathcal{B})$. If g is absolutely continuous with derivative g', if the Fisher information for location $I(g) = \int (g'/g)^2 g$ is finite and if EX^TX is finite, it can be shown that the LAN condition of Section 1, (3.1) and (3.2) hold with

$$\psi(\varepsilon) = -g'/g(\varepsilon), \; W = X, \text{ and } \dot{\mathcal{P}}_3 = L_2^0(\varepsilon); \tag{4.2}$$

cf., e.g., Example 4.2.2., pp. 89,95,96 of BKRW (1993).

Let EXX^T be non-singular. If it would be singular, there would be a $b \neq 0$ with $b^TX = 0$ a.s. and $X^T\kappa = X^T(\kappa + \alpha b)$ a.s., $\alpha \in \mathbf{R}$, would show that κ is unidentifiable. If $EX = 0$, (3.5) applies and adaptation is possible. If $EX \neq 0$, the vector $d = a/|a|$ with $a = (EXX^T)^{-1}EX$ characterizes the one-dimensional orthogonalizing subspace. Since d does not depend on any of the parameters in the model a suitable reparametrization is given by the device of (3.10). As a special case consider the model with a constant term, say $X_k = 1$ a.s. Then the covariance matrix of X is singular, (3.4) holds and in fact, $\kappa_k = \eta$ corresponds to a location parameter for ε. Note that one has to restrict G in this case to avoid identifiability problems. Doing this by assuming g to be symmetric about zero, $\dot{\mathcal{P}}_3$ becomes $\{\lambda(\varepsilon) \in L_2^0(\varepsilon) : \lambda \text{ even}\}$, $\psi(\varepsilon) \perp \dot{\mathcal{P}}_3$ and (3.5) applies again; c.f. Example 3.2.2, p. 53 of BKRW (1993). For references about adaptive estimators in the general linear model see Example 4.2.2, pp. 95-96 of BKRW (1993). □

In the time series models we will study, the innovations ε, ε_t, $t \in \mathbf{Z}$, are assumed to be i.i.d from some sufficiently smooth distribution with mean zero and density g. Again the density will act as the infinite dimensional nuisance parameter and hence the tangent space with respect to this parameter is given by (cf. Example 3.2.3, p. 53 and Example 5.3.2, p. 193 of BKRW (1993))

$$\dot{\mathcal{P}}_3 = \{\lambda(\varepsilon) \in L_2^0(\varepsilon) : \; E\varepsilon\lambda(\varepsilon) = 0\}. \tag{4.3}$$

EXAMPLE 4.2. The first time series model we examine is the ARMA(p,q) model. For $\alpha \in \mathbf{R}^p$ and $\beta \in \mathbf{R}^q$ it is described by

$$X_t = \sum_{i=1}^{p} \alpha_i X_{t-i} + \varepsilon_t + \sum_{j=1}^{q} \beta_j \varepsilon_{t-j}, \; t \in \mathbf{Z}. \tag{4.4}$$

Here the stationarity and invertibility conditions are satisfied, i.e. $1 - \alpha_1 z - \ldots - \alpha_p z^p$ and $1 + \beta_1 z + \ldots + \beta_q z^q$ have their roots outside the unit circle in the complex plane.

Furthermore these polynomials do not have common roots. Set $\kappa = (\alpha^T, \beta^T)^T$. The structure (3.1) is valid with $\psi(\varepsilon)$ the score for location $-g'/g(\varepsilon)$ as in (4.2) of the linear regression example, and $W = \chi(X_0, X_{-1}, X_{-2}, \ldots)$ with $\chi : \mathbf{R}^\infty \to \mathbf{R}^k$ such that $EW = 0$. This may be deduced from Theorem 3.1, Theorem 3.5 and Example 3.6 of Kreiss (1987a) under his conditions: g is positive on $\mathbf{R}$, it is absolutely continuous with finite Fisher information for location and it has finite variance, $\int x^2 g(x) dx < \infty$. Since $EW = 0$, the orthogonality (3.5) holds. Adaptive estimators have been constructed by Kreiss (1987a,1987b).

Extending this model with a location parameter by replacing every X_s in (4.4) by $X_s - \mu$ we obtain

$$X_t = \mu + \sum_{i=1}^{p} \alpha_i (X_{t-i} - \mu) + \varepsilon_t + \sum_{j=1}^{q} \beta_j \varepsilon_{t-j}, \; t \in \mathbf{Z}. \tag{4.5}$$

Set $\kappa = (\alpha^T, \beta^T, \mu)^T$. Now (3.1) is still valid with ψ as above, but $W = \chi(X_0, X_{-1}, \ldots)$ is replaced by

$$W = \begin{pmatrix} \chi(X_0 - \mu, X_{-1} - \mu, \ldots) \\ 1 - \sum_{i=1}^{p} \alpha_i \end{pmatrix}. \tag{4.6}$$

Clearly, the covariance matrix of W is singular and (3.4) applies. Note that we have no identifiability problems here since $E\varepsilon g'/g(\varepsilon) = -1$ implies $\psi(\varepsilon) \notin \dot{\mathcal{P}}_3$ by (4.3). Also note that reparametrization is not necessary.

So, for estimation of the parameters α and β the parametric model with $g = g_0$ known and the semiparametric model with g unknown are equally difficult asymptotically. □

EXAMPLE 4.3. GARCH(p, q) models are used a lot in modelling financial data. See, e.g., the survey paper of Bollerslev, Chou and Kroner (1992). These models capture the stylized fact in financial data series that there is persistence in volatility, i.e. big shocks are clustered together. Nelson (1990) proves results on stationarity and ergodicity for GARCH(1,1) models. Usually the GARCH(p,q) model is parametrized as follows

$$\begin{aligned} X_t &= \sqrt{Y_t}\varepsilon_t, \qquad\qquad (4.7) \\ Y_t &= \psi + \alpha_1 X_{t-1}^2 + \ldots + \alpha_q X_{t-q}^2 + \beta_1 Y_{t-1} + \ldots + \beta_p Y_{t-p}. \end{aligned}$$

Here $\alpha_1, \ldots, \alpha_q, \beta_1, \ldots, \beta_p, \psi > 0$, $\alpha_1 + \ldots + \alpha_q + \beta_1 + \ldots + \beta_p < 1$ and it is assumed that $1 - \alpha_1 z - \ldots - \alpha_q z^q$ and $\beta_1 z + \ldots + \beta_p z^p$ have no common roots. To warrant identifiability of the model we have to add the extra condition that the innovations ε_t and ε have unit variance. This restricts the tangent space (4.3) to

$$\dot{\mathcal{P}}_3 = \{\lambda(\varepsilon) \in L_2^0(\varepsilon) : \; E\varepsilon\lambda(\varepsilon) = 0, \; E\varepsilon^2\lambda(\varepsilon) = 0\}. \tag{4.8}$$

Set $\kappa = (\alpha_1, \ldots, \alpha_q, \beta_1, \ldots, \beta_p, \psi)^T$. Again $\dot{l}_{12} = \psi(\varepsilon)W$, but $\psi(\varepsilon) = -\frac{1}{2}(1 + \varepsilon g'/g(\varepsilon))$ is the score for scale now. Furthermore, $W_i = \frac{\partial}{\partial \kappa_i} \log Y_1$ and $EW \neq 0$. This is proved in Drost and Klaassen (1993) under conditions such that $\log g$ is twice differentiable and g has finite Fisher information for scale $E\psi^2(\varepsilon)$. Note that $\psi(\varepsilon)$ is not orthogonal to $\dot{\mathcal{P}}_3$ except for some special g. So, (3.5) does not apply generally. In fact, the covariance matrix of W is singular. In search for a such that $a^TW = 1$ observe that for such an a we have

$$\begin{aligned}
Y_1 &= Y_1W^Ta \\
&= \begin{bmatrix} X_0^2 & \cdots & X_{1-q}^2 & Y_0 & \cdots & Y_{1-p} & 1 \end{bmatrix} a \\
&\quad +\beta_1\left(\frac{\partial}{\partial\kappa}Y_0\right)a + \ldots + \beta_p\left(\frac{\partial}{\partial\kappa}Y_{1-p}\right)a \\
&= \begin{bmatrix} X_0^2 & \cdots & X_{1-q}^2 & Y_0 & \cdots & Y_{1-p} & 1 \end{bmatrix} a \\
&\quad +\beta_1Y_0\left(\frac{\partial}{\partial\kappa}\log Y_0\right)a + \ldots + \beta_pY_{1-p}\left(\frac{\partial}{\partial\kappa}\log Y_{1-p}\right)a \\
&= \begin{bmatrix} X_0^2 & \cdots & X_{1-q}^2 & Y_0 & \cdots & Y_{1-p} & 1 \end{bmatrix} a \\
&\quad +\beta_1Y_0 + \ldots + \beta_pY_{1-p} \\
&= \begin{bmatrix} X_0^2 & \cdots & X_{1-q}^2 & Y_0 & \cdots & Y_{1-p} & 1 \end{bmatrix} a \\
&\quad +Y_1 - \alpha_1X_0^2 - \ldots - \alpha_qX_{1-q}^2 - \psi
\end{aligned}$$

This yields $a^T = \begin{bmatrix} \alpha_1 & \cdots & \alpha_q & 0 & \cdots & 0 & \psi \end{bmatrix}$. It may be verified that $a^TW = 1$ indeed. So (3.4) holds and in GARCH models the scale of the innovations generates the most difficult one-dimensional regular subproblem for all the parameters $\alpha_1, \ldots, \alpha_q, \beta_1, \ldots, \beta_p$. The reparametrization induced by a is already given in Linton (1993) for ARCH models (with $p = 0$) and in Drost and Klaassen (1993) for the present GARCH model. See Linton (1993) and Drost and Klaassen (1993) for the construction of efficient estimators.

To find a suitable reparametrization, define $\gamma \in \mathbf{R}^q$, $\delta \in \mathbf{R}^p$, $\eta \in \mathbf{R}$, $\theta = (\gamma^T, \delta^T, \eta)^T$ and let $\kappa : \mathbf{R}^{q+p+1} \to \mathbf{R}^{q+p+1}$ be a diffeomorphism. By (3.8) and (3.9) κ may be chosen to satisfy $\frac{\partial}{\partial\eta}\kappa(\theta)^T = \begin{bmatrix} \alpha^T & 0^T & \psi \end{bmatrix}$. Solving this differential equation and completing κ to obtain a diffeomorphism yields, e.g.,

$$\alpha = e^\eta\gamma,\ \beta = \delta,\ \psi = e^\eta.$$

□

EXAMPLE 4.4. Finally we consider the so-called threshold auto-regressive models (TAR). See, for example, Tong and Lim (1980) for a discussion of these models. Let $\mathcal{A} = (A_1, \ldots, A_k)$ be a partition of $\mathbf{R}$ and put $\mu = (\mu_1, \ldots, \mu_k)^T$, $\rho = (\rho_1, \ldots, \rho_k)^T$ and $\kappa^T = (\mu^T, \rho^T)$. The TAR model is defined by

$$Y_t = \sum_{i=1}^{k}[\mu_i + \rho_iY_{t-1}]I_{A_i}(Y_{t-1}) + \varepsilon_t.$$

See Chan et al. (1985) for results on stationarity and ergodicity of TAR models. Put $I_{\mathcal{A}} = (I_{A_1}, \ldots, I_{A_k})^T$. The scores for this model are of the form $\psi(\varepsilon_t(\kappa))W_t(\kappa)$ where ψ is the score for location and $W_t(\kappa) = [I_{\mathcal{A}}^T(Y_{t-1}), Y_{t-1}I_{\mathcal{A}}^T(Y_{t-1})]^T$. As before, the tangent space generated by the unknown distribution g of the innovations is given by (4.3).

If $a = [1, \ldots, 1, 0, \ldots, 0]^T$ then $a^TW(\kappa) = 1$. Since a does not depend on κ adaptive estimation is possible if the overall location parameter $a^T\mu = \sum_{i=1}^k \mu_i$ is taken as one-dimensional nuisance parameter.

If it is at the forehand known that $\mu_1 = \mu_2 = \ldots = \mu_k = 0$, Theorem 3.1 shows that $a(\kappa) \in \mathbf{R}^k$ with $a_i = EYI_{A_i}(Y)/EY^2 I_{A_i}(Y)$ generates the $([\dot{l}_\kappa], \dot{\mathcal{P}}_3)$-orthogonalizing subspace. In this case a probably depends on the infinite dimensional nuisance parameter g and hence the suggested reparametrization will also depend on g. Then the reparametrization of this model cannot be handled by our techniques. A special case occurs if the process and each of the cells A_i are symmetric about zero. Then $a(\kappa) = 0$ and full adaptive estimation will be possible. □

References

BICKEL, P.J., KLAASSEN, C.A.J., RITOV, Y., and WELLNER, J.A. (1993), *Efficient and Adaptive Statistical Inference for Semiparametric Models*, John Hopkins University Press, Baltimore.

BOLLERSLEV, T., CHOU, R.Y., and KRONER, K.F. (1992), ARCH Modeling in Finance: a Review of the Theory and Empirical Evidence, *Journal of Econometrics*, **52**, 5–59.

CHAN, K.S., PETRUCCELLI, J.D., TONG, H., and WOOLFROD, S.W. (1985), A Multiple-Threshold AR(1) Model, *Journal of Applied Probability*, **22**, 267–279.

COX, D.R. and REID, N. (1987), Parameter Orthogonality and Approximate Conditional Inference, *Journal of the Royal Statistical Society Series B*, **49**, 1–39.

DROST, F.C. and KLAASSEN, C.A.J. (1993), Adaptivity in Semiparametric GARCH models, *Tilburg University*, preprint.

KREISS, J.-P. (1987a), On Adaptive Estimation in Stationary ARMA Processes, *The Annals of Statistics*, **15**, 112–133.

KREISS, J.-P. (1987b), On Adaptive Estimation in Autoregressive Models when there are Nuisance Functions, *Statistics and Decisions*, **5**, 59–76.

LINTON, O. (1993), Adaptive Estimation in ARCH Models, *Econometric Theory*, forthcoming.

NELSON, D.B. (1990), Stationarity and Persistence in the GARCH(1,1) Model, *Econometric Theory*, **6**, 318–334.

TONG, H. and LIM, K.S. (1980), Threshold Autoregression, Limit Cycles and Cyclical Data, *Journal of the Royal Statistical Society, Series B*, **42**, 245–292.

WEIDMANN, J. (1980), *Linear Operators in Hilbert Spaces*, Springer-Verlag, New York.

KERNEL ESTIMATORS OF INTEGRATED SQUARED DENSITY DERIVATIVES IN NON-SMOOTH CASES

A.J. VAN ES AND A.J. HOOGSTRATE

University of Amsterdam and University of Limburg

ABSTRACT. Kernel type estimators of integrated squared density derivatives are studied from a point of view of robustness against non–smoothness. Their behaviour for samples from densities with discontinuities or with discontinuities in their derivatives is investigated. Expansions of the expectation and asymptotic normality are derived.

1. Introduction and results.

Let $X_1, \ldots, X_n$ denote a sample from a distribution having probability density f. Estimators of integrated squared derivatives of f have recently enjoyed a renewed interest because of their importance in plug-in bandwidth selection procedures for kernel density estimators. The integrated squared density derivatives are introduced by

$$R(f^{(m)}) = \int_{-\infty}^{\infty} \{f^{(m)}(x)\}^2 dx \qquad m = 1, 2, \ldots. \tag{1.1}$$

Several authors have introduced and studied estimators of this functional which are based on the kernel estimation technology. We mention Levit (1978), Hall and Marron (1987a), Bickel and Ritov (1988) and Jones and Sheather (1991). These papers all assume smoothness of the density f up to the order of derivative to be estimated and often still more. We will extend these results to certain classes of non–smooth densities. For reasons of completeness and comparison we will also state the results for the smooth cases as obtained in the earlier mentioned papers.

Our motivation for this type of research originates from a point of view of robustness against non–smoothness. Suppose that one is applying a certain bandwidth selection procedure, or any other procedure where integrated squared density derivatives have to be estimated, and that one is not sure that the required smoothness assumptions are satisfied. In such situations one would like to use a procedure which warns one of the presence of non–smoothness. Studying the behaviour of the integrated squared density derivatives in non–smooth cases serves as a first step in understanding the behaviour of, for instance, bandwidth selection procedures in such cases. For more details we refer to Van Es and Hoogstrate (1993), where several popular plug-in bandwidth selection procedures are investigated.

1991 *Mathematics Subject Classification.* Primary 62G05; secondary 62E20.

Key words and phrases. Density estimation, kernel estimators, integrated squared density derivatives.

1.1. Assumptions and definitions.

Firstly we introduce some classes of densities f of different smoothness. We impose the following conditions.

Condition F. *We say that a probability density function f satisfies Condition F for some integer s if*

(1) *f has bounded support.*
(2) *All derivatives of f up to the s–th order, denoted by $f^{(i)}$, $i = 1, 2, \ldots, s$, exist at every point of the real line except at a finite set of points which we denote by D. In these points we give $f^{(i)}$, $i = 1, 2, \ldots, s$, arbitrary values.*
(3) *The functions $f^{(i)}$, $i = 0, 1, \ldots, s$, have finite left and right limits at the points in D.*
(4) *The left and right derivatives $f^{(i)}$, $i = 1, 2, \ldots, s$, of f are finite at points in D.*
(5) *The function $f^{(s)}$ is continuous on the complement of D.*

Under these condions we allow $f^{(i)}$, $i = 0, 1, \ldots, s$, to have jumps in points d in a finite set D. This leads to several distinct classes of non–smooth densities.

Definition 1.1.1. *Let f be a probability density function satisfying Condition F. For every point d in D we define*

$$\delta^{(i)}(d) := f^{(i)}(d+) - f^{(i)}(d-), \qquad i = 0, 1, \ldots, s.$$

We further define for $D \neq \emptyset$

$$\nu := \min\{i : \delta^{(i)}(d) \neq 0, \text{ for some } d \in D\}.$$

If $D = \emptyset$ we take $\nu = s + 1$.

The above definition of ν yields $s + 2$ distinct classes of non–smooth densities, corresponding to the values of ν, i.e. $\nu = 0, \ldots, s+1$. Whether a density f satisfying Condition F belongs to a certain class is determined by the lowest derivative which has a jump. For instance, $\nu = 0$ corresponds to densities with a finite number of jumps. For results on the performance of kernel estimators of the density f itself, under similar conditions, we refer to Van Eeden (1985), Swanepoel (1987), Liebscher (1990), Cline and Hart (1991) and Van Es (1991).

Next we turn to conditions on the kernels. Our results allow higher order kernels. The order of a kernel is defined as follows.

Definition 1.1.2. *The function G will be said to have order k when*

$$\int_{-\infty}^{\infty} x^j G(x)dx = \begin{cases} C_1, & j = 0, \\ 0, & j = 1, \ldots k-1, \\ C_2, & j = k, \end{cases}$$

with $-\infty < C_1 < \infty$ and $-\infty < C_2 < \infty$, $C_2 \neq 0$. If G is a probability density function we have $C_1 = 1$.

When we use the kernel type estimators to estimate $R(f^{(m)})$ we need stronger conditions on the kernel than for estimation of the density itself. We therefore introduce the following condition.

Condition G. *The function G satisfies Condition G if*

(1) *G is a bounded symmetric measurable function of order k with bounded support $[-1,1]$, which integrates to one.*
(2) *G has $2m$ derivatives $G^{(1)},\dots,G^{(2m)}$ which are equal to zero at ± 1.*
(3) *$G^{(2i)}(0) \neq 0$ for $i = 0,1,\dots,m$.*

We further introduce some functions which play an important role in non-smooth kernel density estimation.

Definition 1.1.3. *For G a bounded symmetric measurable function with bounded support, the functions b^G, b_i^G $i = 0,1,\dots$ and $l = 0,1,\dots$ are defined by*

$$b^G(x,h,l) = \frac{1}{h}\int_{-\infty}^{\infty} G\left(\frac{x-u}{h}\right) f^{(l)}(u)du - f^{(l)}(x)\int_{-\infty}^{\infty} G(u)du,$$

$$b_i^G(t) = \begin{cases} \displaystyle\int_{-\infty}^{t} (t-u)^i G(u)du, & \text{for} \quad t < 0, \\ \displaystyle -\int_{t}^{\infty} (t-u)^i G(u)du, & \text{for} \quad t \geq 0. \end{cases}$$

Notice that, when G is a kernel of a kernel estimator, the function $b^G(x,h,0)$ equals the bias of the estimator, since we have $b^G(x,h,0) = Ef_{nh}(x) - f(x)$. We will often use the notation $b^G(x,h)$ for $b^G(x,h,0)$ and will, whenever convenient, denote f by $f^{(0)}$.

Finally we introduce the class of estimators of $R(f^{(m)})$. For a motivation of these estimators we refer to Hall and Marron (1987).

Definition 1.1.4. *Let G satisfy Condition G. We define the estimator $\theta_m(G,h)$ by*

$$\theta_m(G,h) = (-1)^m n^{-1}(n-1)^{-1}h^{-2m-1}\sum_{i=1}^{n}\sum_{j=1}^{n} G^{(2m)}\left(\frac{X_i - X_j}{h}\right).$$

Considering $\theta_m(G,h)$ more closely we see that it contains a known deterministic bias term which originates from the diagonal terms,

$$\begin{aligned}\theta_m(G,h) =&(-1)^m(n-1)^{-1}h^{-2m-1}G^{(2m)}(0)+ \\ &(-1)^m n^{-1}(n-1)^{-1}h^{-2m-1}\sum_{i\neq j}^{n} G^{(2m)}\left(\frac{X_i - X_j}{h}\right). \end{aligned} \qquad (1.2)$$

Since it is a known term, in principle, it can also be left out. Notice, however, that we have left this deterministic term in, which is in accordance to Jones and Sheather (1991). They show that this bias term can be used to improve the asymptotic results, obtaining a smaller minimum MSE than the estimator without this bias term.

1.2. Results.

Our main results describe the asymptotic behaviour of the estimator $\theta_m(G,h)$. Theorem 1.2.2 concerns the expectation of the estimator and Theorem 1.2.3 establishes asymptotic normality.

Firstly we show that the expectation of $\theta_m(G,h)$ can be written as the sum of the known bias and an integral of the bias function $b^{G^{(2m)}}(x,h)$ introduced in Definition 1.1.3.

Lemma 1.2.1. *Let G satisfy condition G and $b^{G^{(2m)}}$ be as defined in Definition 1.1.3, then, with $J_m(G) := (-1)^m G^{(2m)}(0)$, we have*

$$E\theta_m(G,h) = (n-1)^{-1}h^{-2m-1}J_m(G) + (-1)^m h^{-2m}\int b^{G^{(2m)}}(x,h)f(x)dx.$$

Proof. The result follows immediately from (1.2) and

$$h^{-2m-1}\int\int G^{(2m)}\left(\frac{x-y}{h}\right)f(y)dyf(x)dx = h^{-2m}\int b^{G^{(2m)}}(x,h)f(x)dx,$$

which holds because of $\int G^{(2m)}(u)du = 0$ □

It follows that the behaviour of the bias function $b^{G^{(2m)}}(x,h)$ under the various smoothness conditions determines the behaviour of the expectation of $\theta_m(G,h)$. Keeping this in mind we arrive at the following theorem.

Theorem 1.2.2. *Let G satisfy Condition G and be of order k and let the probability density function f satisfy Condition F, with $s = m + \frac{k}{2}$, $m > 0$. Then, with $J_l(G) = (-1)^l G^{(2l)}(0)$, for $l = 0,1,2,\dots$, we have for $h \to 0$*

$$E\theta_m(G,h) = (n-1)^{-1}h^{-2m-1}J_m(G) +$$

(1) *for* $\nu < m$,

$$J_{m-\nu-1}(G)h^{2(\nu-m)+1}\sum_{d\in D}\delta^{(\nu)}(d)^2 + O(h^{2(\nu-m)+2}),$$

(2) *for* $\nu = m$,

$$R(f^{(m)}) - \sum_{d\in D}\delta^{(m)}(d)f^{(m-1)}(d) + O(h),$$

(3) *for* $m < \nu < m + \frac{k}{2}$,

$$R(f^{(m)}) + 2(-1)^{(\nu-m-1)}\frac{h^{2(\nu-m)}}{(2(\nu-m)-1)!}\int_0^1 b^G_{2(\nu-m)-1}(t)dt$$

$$\cdot\sum_{d\in D}\delta^{(\nu)}(d)f^{(\nu-1)}(d) + O(h^{2(\nu-m)+1}),$$

(4) *for* $\nu = m + \frac{k}{2}$,

$$R(f^{(m)}) + (-1)^{\frac{k}{2}} \frac{h^k}{k!} \int_{-1}^{1} u^k G(u)du \int_{-\infty}^{\infty} \{f^{(m+\frac{k}{2})}(x)\}^2 dx$$
$$+ 2(-1)^{(\frac{k}{2}-1)} \frac{h^k}{(k-1)!} \int_0^1 b_{k-1}^G(t)dt \sum_{d\in D} \delta^{(m+\frac{k}{2})}(d) f^{(m+\frac{k}{2}-1)}(d) + o(h^k),$$

(5) *for* $\nu > m + \frac{k}{2}$,

$$R(f^{(m)}) + (-1)^{\frac{k}{2}} \frac{h^k}{k!} \int_{-1}^{1} u^k G(u)du \int_{-\infty}^{\infty} \{f^{(m+\frac{k}{2})}(x)\}^2 dx + o(h^k).$$

□

Remark. We have taken s equal to $m + \frac{k}{2}$ in all cases, but it will become clear from the proof that we can relax this assumption for the cases (1), (2) and (3) to get the same results.

Theorem 1.2.3. *Assume that G satisfies Condition G and that the probability density function f satisfies Condition F with $s = 2m$. Further let the bandwidths h_n be positive and satisfy $h_n \to 0$. Then for $0 \le \lambda < \infty$*

$$d_n(\theta_m(G,h_n) - E(\theta_m(G,h_n))) \xrightarrow{D}$$

(1)	$N(0, 2\sigma^2 + 4\lambda\sigma_s^2)$	*if* $\nu \ge 2m$, $d_n = nh_n^{\frac{4m+1}{2}}$ and $nh_n^{4m+1} \to \lambda$,
(2)	$N(0, 4\sigma_s^2)$	*if* $\nu \ge 2m$, $d_n = n^{\frac{1}{2}}$ and $nh_n^{4m+1} \to \infty$,
(3)	$N(0, 2\sigma^2 + 4\lambda\sigma_\nu^2)$	*if* $\nu < 2m$, $d_n = nh_n^{\frac{4m+1}{2}}$ and $nh_n^{2\nu+2} \to \lambda$,
(4)	$N(0, 4\sigma_\nu^2)$	*if* $\nu < 2m$, $d_n = n^{\frac{1}{2}} h_n^{\frac{4m-2\nu-1}{2}}$ and $nh_n^{2\nu+2} \to \infty$.

The variances σ^2, σ_ν^2 and σ_s^2 are given by

$$\sigma^2 - \int_{-\infty}^{\infty} \{G^{(2m)}(v)\}^2 dv \int_{-\infty}^{\infty} f^2(x)dx,$$

$$\sigma_s^2 = \left(\int_{-\infty}^{\infty} \{f^{(2m)}(x)\}^2 f(x)dx - \left(\int_{-\infty}^{\infty} f^{(2m)}(x) f(x)dx \right)^2 \right),$$

$$\sigma_\nu^2 = (\nu!)^{-2} \int_0^1 \{b_\nu^{G^{(2m)}}(t)\}^2 dt \sum_{d\in D} \delta^{(\nu)}(d)^2 (f(d+) + f(d-)).$$

□

Remark. Notice that σ_s^2 equals $\text{var}_f(f^{(2m)}(X))$. Levit (1978) and Bickel and Ritov (1988) show that this variance is the efficient variance when the density f is sufficiently smooth.

1.3. Discussion.

In Theorem 1.2.2 the first thing which catches the eye is the fact that in case $\nu \leq m$ the estimator does not, ignoring the bias term, estimate the functional $R(f^{(m)})$ consistently. In fact, if $\nu < m$ and the bandwidth tends to zero, as usual in kernel estimation, the expectation of the estimator becomes infinite! However, this does not mean that bandwidth selectors which make use of this estimator will completely break down. Since the estimated integrated squared second derivative appears in the denominator of the optimal bandwidth for smooth densities, the estimated bandwidths will get smaller for less smooth densities. This means that they adapt to non–smoothness in the right way. The optimal bandwidths for non–smooth densities are indeed smaller than the optimal bandwidths for smooth densities. For details see Van Es and Hoogstrate (1993).

We also note that the condition G3 plays a very important role in Theorem 1.2.2. For instance we could have changed this condition to the other extreme, i.e. $G^{(2i)}(0) = 0$, for $i = 0, 1, ..., m$. Then, for $\nu < m$, the bias, which now tends to infinity, reduces to a constant bias term.

From Theorem 1.2.3 we learn, that in case nh^{4m+1} tends to λ or infinity and $\nu \geq 2m$, we always have a convergence rate of $n^{\frac{1}{2}}$, and the estimator is $\sqrt{n}$–consistent. However if $\nu < 2m$, we obtain, in case $nh^{2\nu+2}$ tends to λ, a rate of $n^{1-\frac{4m+1}{4\nu+4}}$. This means that in case $\nu < m$, apart from the fact that the expectation tends to infinity, the variance also explodes. For $\nu = m$ the variance vanishes, but there is a constant bias. If $m < \nu < 2m$ we get consistency but not $\sqrt{n}$–consistency. If $nh^{2\nu+2} \to \infty$ the question of consistency depends on the rate at which $nh^{2\nu+2}$ tends to infinity. For instance, let us take $h^* = cn^{-\frac{1}{2m+k+1}}$, the optimal bandwidth for estimating $R(f^{(m)})$ in the smooth case, see Jones and Sheather (1991). Then, for $\nu < m + \frac{k-1}{2}$, noting that then $nh^{*2\nu+2} \to \infty$, we have $d_n = n^{\frac{-2m+k+2\nu+2}{2(2m+k+1)}}$. This means that the variance does not vanish asymptotically for $\nu \leq m - \frac{k}{2} - 1$. In case of equality we have a constant asymptotic variance. This happens for instance if we have $k = 2$, $m = 2$ and $\nu = 0$. Note that for these choices of k and m we do have consistency for all other cases of ν.

Comparing our results to those of Hall and Marron (1987) and Jones and Sheather (1991) we have to be aware of the more general notion of smoothness they use. Their smoothness classes are induced by a continuous parameter p. Our smoothness class ν is a subset of their class $p = \nu$, so the results can be compared. Our results agree with their results when $p = \nu$ and $\nu > m$. However, we have also considered the cases $\nu < m$ for the expectation as well as the variance of $\theta_m(G, h)$. We have further derived the asymptotic distribution of the estimator and explicit constants for the expectation for all cases $\nu = 0, 1, ...$ and equally for the variance.

2. Proofs.

2.1. Proof of theorem 1.2.2

Before we give the proof of Theorem 1.2.2 we have to pay some attention to the function $G^{(2m)}$ and the bias functions as defined in Definition 1.1.3. The results obtained for these functions play a crucial role in the proof.

Firstly, the assumptions on the kernels G, imposed by Condition G, lead to the following result.

Lemma 2.1.1. *Assume that G satisfies Condition G then, for* $m = 1, 2, \ldots,$

$$\int_{-1}^{1} u^i G^{(2m)}(u)du = \begin{cases} 0, & \text{for } i = 0, 1, 2, \ldots, 2m-1, \\ (2m)!, & \text{for } i = 2m. \end{cases}$$

Proof. The result follows by repeated partial integration and the fact that the functions $G, G^{(1)}, \ldots, G^{(2m)}$ vanish at ± 1. □

We shall repeatedly split up an area of integration into two parts, one containing the singular points and one containing points at least at a distance h of D.

Definition 2.1.2. *We define the set D_h, as the set of points which have at least distance h to the singular points $d \in D$, i.e.*

$$D_h = (-\infty, \infty) \setminus \cup_{d \in D} [d - h, d + h].$$

The following lemma gives the asymptotic behaviour of the functions introduced in Definition 1.1.3. It is an extention of Theorem 2.3 in Van Es (1991). For a proof, similar to the proof in Van Es (1991), we refer to Van Es and Hoogstrate (1993).

Lemma 2.1.3. *Let G be a bounded measurable symmetric function of order k with bounded support $[-1, 1]$. Assume that f satisfies Condition F with $s = k + l$ for some integer l, $l \geq 0$. Furthermore let (h'_n) be a vanishing sequence of positive real numbers.*

(a) Then

$$b^G(x, h, l) = \frac{h^k}{k!} f^{(k+l)}(x) \int_{-1}^{1} u^k G(u)du + r(x, h) \tag{2.1}$$

where the remainder satisfies

$$\lim_{n \to \infty} \sup_{0 \leq h \leq h'_n} \sup_{x \in D_h \cap [-M, M]} h^{-k} |r(x, h)| = 0, \tag{2.2}$$

for every positive M.

(b) For a fixed point x_0 we have

$$b^G(x_0 + th, h, l) = \sum_{i=0}^{k} \frac{h^i}{i!} b_i^G(t) \delta^{(i+l)}(x_0) \tag{2.3}$$

$$+ \frac{h^k}{k!} \int_{-1}^{1} u^k G(u)du \{ f^{(k+l)}(x_0 -) I_{(-\infty, 0)}(t) + f^{(k+l)}(x_0 +) I_{(0, \infty)}(t) \}$$

$$+ r_k(t, h),$$

where the remainder satisfies

$$\lim_{n \to \infty} \sup_{0 \leq h \leq h'_n} \sup_{-M \leq t \leq M, t \neq 0} h^{-k} |r_k(t, h)| = 0 \tag{2.4}$$

for every positive M. □

Remark. We note that an alternative for (b) is given by the following. If $\delta^{(i+l)}(x_0) = 0$ for $i + l = 0, 1, 2, ..., i_0 + l - 1$ and $\delta^{(i_0+l)} \neq 0$ then $b^G(x + th, h, l) = \frac{h^{i_0}}{i_0!}\delta^{(i_0+l)} + o(h^{i_0})$ when f satisfies the weaker Condition F with $s = i_0 + l$.

As we have already shown in Lemma 1.2.1, the asymptotic behaviour of the expectation of the estimator $\theta_m(G, h)$ is determined by an integral of a bias function. In the following lemma we will derive the asymptotic behaviour of the integral of this bias function over the area D_h. The requirement $s = m + \frac{k}{2}$ is suggested by the same requirement in Theorem 1.2.2.

Lemma 2.1.4. *Let the probability density function f satisfy Condition F with $s = m + \frac{k}{2}$, $m > 0$ and let G satisfy Condition G, then*

$$(-1)^m h^{-2m} \int_{D_h} b^{G^{(2m)}}(x, h) f(x) dx =$$

$$\int_{D_h} \{f^{(m)}(x)\}^2 dx + (-1)^{\frac{k}{2}} \frac{h^k}{k!} \int_{-1}^{1} u^k G(u) du \int_{D_h} \{f^{(m+\frac{k}{2})}(x)\}^2 d(x) + o(h^k),$$

with D_h as in Definition 2.1.2.

Proof. We proceed as in the proof of Lemma 3.1 Hall and Marron (1987). Recall that $b^{G^{(2m)}}(x, h, 0) = b^{G^{(2m)}}(x, h)$ by definition. We get by Definition 1.1.3 and the $m + \frac{k}{2}$ times differentability of f on D_h as well as the conditions on G

$$\begin{aligned}
&(-1)^m h^{-2m} \int_{D_h} b^{G^{(2m)}}(x, h, 0) f(x) dx \\
&= (-1)^m h^{-2m-1} \int_{D_h} \left(\int_{\mathbf{R}} G^{(2m)} \left(\frac{x-y}{h} \right) f(y) dy \right) f(x) dx \\
&= (-1)^m h^{-2m} \int_{D_h} \left(\int_{\mathbf{R}} G^{(2m)}(u) f(x) f(x - hu) du \right) dx \\
&= (-1)^m h^{-m+\frac{k}{2}} \int_{D_h} \left(\int_{\mathbf{R}} G^{(m-\frac{k}{2})}(u) f(x) f^{(m+\frac{k}{2})}(x - hu) du \right) dx \\
&\overset{*}{=} (-1)^{\frac{k}{2}} h^{-m+\frac{k}{2}} \int_{D_h} \left(\int_{\mathbf{R}} G^{(m-\frac{k}{2})}(u) f^{(m+\frac{k}{2})}(x) f(x - hu) du \right) dx \\
&= (-1)^{\frac{k}{2}} \int_{D_h} \left(\int_{\mathbf{R}} G(u) f^{(m+\frac{k}{2})}(x) f^{(m-\frac{k}{2})}(x - hu) du \right) dx \\
&= (-1)^{\frac{k}{2}} \int_{D_h} \left\{ b^G(x, h, m - \frac{k}{2}) + f^{(m-\frac{k}{2})}(x) \int_{\mathbf{R}} G(u) du \right\} f^{(m+\frac{k}{2})}(x) dx \\
&= \int_{D_h} \{f^{(m)}(x)\}^2 dx + (-1)^{\frac{k}{2}} \int_{D_h} b^G(x, h, m - \frac{k}{2}) f^{(m+\frac{k}{2})}(x) dx.
\end{aligned}$$

The equality * holds by the Fubini's Theorem and partial integration with respect to x. Now by Lemma 2.1.3 part (a) we get the stated result. Note that we allow $m - \frac{k}{2}$ to be negative. In these cases $G^{(m-\frac{k}{2})}(x)$ should be interpreted as the $m - \frac{k}{2}$ order primitive of G. The density f can be treated similarly. □

Without proof we mention the following properties of the functions $b_i^{G^{(2m)}}(t)$. For a proof we refer to Van Es and Hoogstrate (1993).

Lemma 2.1.5. *Let G satisfy Condition G and let $b_i^{G^{(2m)}}(t)$ be defined as in Definition 1.1.3. Then we have, for $i, j \in \mathbb{N}$, such that $i + j < 2m - 1$,*

$$(a) \int_{-1}^{1} b_i^{G^{(2m)}}(t)t^j dt = 0,$$

$$(b) \int_{0}^{1} b_i^{G^{(2m)}}(t)t^j dt = (-1)^{j+1} i!\, j!\, G^{(2m-(i+j)-2)}(0), \text{ if } i + j \text{ is even,}$$

$$(c) \int_{0}^{1} b_i^{G^{(2m)}}(t)t^j dt = 0, \text{ if } i + j \text{ is odd,}$$

$$(d) \int_{0}^{1} b_m^{G^{(2m)}}(t)t^{m-1} dt = (-)^{m+1}\frac{1}{2} m!(m-1)!.$$

□

Having obtained the lemmas above we are now settled for the proof of Theorem 1.2.2.

Proof of Theorem 1.2.2. In Lemma 1.2.1 we have already established the separation of the expectation of $\theta_m(G, h)$ into the known bias term and an integral over a bias function,

$$E\theta_m(G, h) = (n-1)^{-1} h^{-2m-1} J_m(G) + (-1)^m h^{-2m} \int_{\mathbb{R}} b^{G^{(2m)}}(x, h) f(x) dx. \quad (2.5)$$

Now we have to deal with the second part of (2.5) which describes the behaviour of the stochastic components of our estimator.

To prove case (5) if $\nu > m + \frac{k}{2}$ we apply Lemma 2.1.4 with $s = m + \frac{k}{2}$, $D = \emptyset$ and hence $D_h = \mathbb{R}$, which leads directly to the stated result.

If $\nu \leq m$ we first divide the integration area into a part where f is smooth and a part where f has the singular points. We have, for h small enough,

$$\begin{aligned}(-1)^m h^{-2m} \int_{\mathbb{R}} b^{G^{(2m)}}(x, h) f(x) dx = \\ (-1)^m h^{-2m} \int_{D_h} b^{G^{(2m)}}(x, h) f(x) dx \\ + (-1)^m h^{-2m} \sum_{d \in D} \int_{d-h}^{d+h} b^{G^{(2m)}}(x, h) f(x) dx.\end{aligned} \quad (2.6)$$

Now we can handle the first term on the right, the integral, by Lemma 2.1.4 whereafter we replace D_h by $\mathbb{R}$ at the cost of an $O(h)$ error term, to obtain

$$(-1)^m h^{-2m} \int_{D_h} b^{G^{(2m)}}(x, h) f(x) dx = \int_{\mathbb{R}} \{f^{(m)}(x)\}^2 dx + O(h). \quad (2.7)$$

Next we treat the sum over the intervals containing the singular points. Using $\delta^{(i)}(d) = 0$ for $i = 0, 1, ..., \nu - 1$, Lemma 2.1.3 part (b) and the remark following

Lemma 2.1.3, necessary because $G^{(2m)}(x)$ is a function of order $2m$, we get,

$$
\begin{aligned}
&(-1)^m h^{-2m} \sum_{d\in D} \int_{d-h}^{d+h} b^{G^{(2m)}}(x,h) f(x)dx \\
&\quad = (-1)^m h^{-2m+1} \sum_{d\in D} \int_{-1}^{1} b^{G^{(2m)}}(d+th,h) f(d+th)dx \\
&\quad = (-1)^m \frac{h^{-2m+1+\nu}}{\nu!} \sum_{d\in D} \delta^{(\nu)}(d) \int_{-1}^{1} \{b_\nu^{G^{(2m)}}(t) + R_{\nu+1}(t,h)\} f(d+th)dt,
\end{aligned}
$$

where $R_{\nu+1}(t,h) = O(h)$, uniformly for t in bounded intervals.

In case $\nu < m$ we construct a $\nu - th$ order Taylor expansion of $f(d+th)$ and note that $b_\nu^{G^{(2m)}}(t)t^\nu$ is an odd function, which gives us for the above

$$
(-1)^m \frac{h^{2(\nu-m)+1}}{(\nu!)^2} \sum_{d\in D} \delta^{(\nu)}(d)^2 \int_0^1 b_\nu^{G^{(2m)}}(t)t^\nu dt + O(h^{2(\nu-m)+2}). \tag{2.8}
$$

By combining (2.7) and (2.8) and substituting it into (2.6) , using Lemma 2.1.5, we get statement (1) of the theorem.

In case $\nu = m$ we have to be more careful because of the fact that according to Lemma 2.1.5 we have $\int_0^1 b_\nu^{G^{(2m)}}(t)t^{\nu-1}dt \neq 0$. In this case we get, noting that $f^{(\nu-1)}$ is continuous in d, and $b_\nu^{G^{(2m)}}(t)t^{\nu-1}$ is even,

$$
2\frac{(-1)^m}{(\nu!)(\nu-1)!} \sum_{d\in D} \delta^{(\nu)}(d) f^{(\nu-1)}(d) \int_0^1 b_\nu^{G^{(2m)}}(t)t^{\nu-1}dt + O(h), \tag{2.9}
$$

which gives us statement (2).

Now we are left with the proof for $m < \nu \leq m + \frac{k}{2}$. Assume $\nu = m + \gamma$ with $0 < \gamma \leq \frac{k}{2}$, $\gamma \in \mathbb{N}$. In the same way as in the proof of Lemma 2.1.4 with $\frac{k}{2}$ replaced by $\gamma - 1$ and $D_h = \mathbb{R}$, we now get

$$
\begin{aligned}
&(-1)^m h^{-2m} \int_{\mathbb{R}} b^{G^{(2m)}}(x,h) f(x)dx = \\
&\int_{-\infty}^{\infty} \{f^{(2m)}(x)\}^2 dx + (-1)^{\gamma-1} \int_{-\infty}^{\infty} b^G(x,h,m-\gamma+1) f^{(m+\gamma-1)}(x)dx. \qquad (2.10)
\end{aligned}
$$

Note that the above is possible since the derivatives of f are continuous up to the $\nu - 1$–th order. We again divide the integration area into an area where f is smooth and an area containing the singular points. For small enough h we have

$$
\begin{aligned}
&(-1)^{\gamma-1} \int_{-\infty}^{\infty} b^G(x,h,m-\gamma+1) f^{(m+\gamma-1)}(x)dx \\
&\quad = (-1)^{\gamma-1} \int_{D_h} b^G(x,h,m-\gamma+1) f^{(m+\gamma-1)}(x)dx \qquad (2.11) \\
&\quad + (-1)^{\gamma-1} \sum_{d\in D} \int_{d-h}^{d+h} b^G(x,h,m-\gamma+1) f^{(m+\gamma-1)}(x)dx.
\end{aligned}
$$

Now we see that if we apply Lemma 2.1.3 part (a) to the first term on the right side of (2.10) we see that it is of order $o(h^k)$. For the second part we get, by Lemma 2.1.3 part (b), noting that $\delta^{(i)}(d) = 0$, for $i = 0, 1, \ldots, \nu - 1$ and $2\gamma - 1 < k$

$$(-1)^{\gamma-1} \sum_{d\in D} \int_{d-h}^{d+h} b^G(x, h, m - \gamma + 1) f^{(m+\gamma-1)}(x) dx =$$

$$(-1)^{\nu-m-1} \sum_{d\in D} \frac{h^{2(\nu-m)}}{(2(\nu - m) - 1)!} \tag{2.12}$$

$$\cdot\, \delta^{(\nu)}(d) \int_{-1}^{1} \{b^G_{2(\nu-m)-1}(t) + R_{2\gamma}(t, h)\} f^{(\nu-1)}(d + th) dt,$$

with $R_{2\gamma}(t, h) = O(h)$, uniformly for t in bounded intervals. Now , noting that $f^{(\nu-1)}$ is continuous as well as the fact that $b^G_{2(\nu-m)-1}(t)$ is an even function, we get

$$2(-1)^{(\nu-m-1)} \frac{h^{2(\nu-m)}}{(2(\nu - m) - 1)!} \int_0^1 b^G_{2(\nu-m)-1}(t) dt \sum_{d\in D} \delta^{(\nu)}(d) f^{(\nu-1)}(d) +$$
$$O(h^{2(\nu-m)+1}). \tag{2.13}$$

To obtain statement (3), the case in which $m < \nu < m + \frac{k}{2}$, we combine (2.13) with (2.11), (2.10) and (2.5).

In case $\nu = m + \frac{k}{2}$ we need to be a little more careful. In this case we see that the last expression (2.13), derived for the second part on the right hand side of (2.11), is of the same order, $O(h^k)$, as the integral over D_h. Applying Lemma 2.1.3 part (a) we get for the integral over D_h

$$(-1)^{\frac{k}{2}} \int_{D_h} b^G(x, h, m - \frac{k}{2}) f^{(m+\frac{k}{2})}(x) dx$$
$$= (-1)^{\frac{k}{2}} \frac{h^k}{k!} \int_{-1}^{1} u^k G(u) du \int_{D_h} \{f^{(m+\frac{k}{2})}(x)\}^2 d(x) + o(h^k). \tag{2.14}$$

Now substituting (2.13), with $\nu = m + \frac{k}{2}$, and (2.14) into (2.11) gives us statement (4). This concludes the proof of Theorem 1.2.2. □

2.2. Some remarks on the proof of Theorem 1.2.3

Returning to Definition 1.1.4. we note that, apart from normalization, the estimator $\theta_m(G, h)$ is in fact a U–statistic of order two, with a kernel depending on h, and thus on the samplesize. Central limit theorems for such statistics can be derived by decomposing them, by a Hoeffding decomposition, in a degenerate order two U–statistic and a linear term. Balancing the variances of the two components then yields the order of the variance of the statistic. The order of the variance of the degenerate U–statistic is usually independent of the smoothness of the density f. However, in our case, the variance of the linear term is proportional to $\mathrm{var}(b^{G^{(2m)}}(X_1, h))$. The dependence of this variance on the smoothness of f is demonstrated by the following lemma, the proof of which follows by the same techniques based on the expansions of Lemma 2.1.3. as previously applied in this paper.

Lemma 2.2.1. *Let G satisfy Condition G and the probability density function f Condition F with $s = 2m$ then*

(1) *if $\nu \geq 2m$*

$$\mathrm{var}(b^{G^{(2m)}}(X_1, h)) = h^{4m}\sigma_s^2 + o(h^{4m});$$

(2) *if $\nu < 2m$*

$$\mathrm{var}(b^{G^{(2m)}}(X_1, h)) = h^{2\nu+1}\sigma_\nu^2 + o(h^{2\nu+1}),$$

where σ_s^2 and σ_ν^2 are defined in Theorem 1.2.3.

Combining these variances with the variance of the degenerate U–statistic proves the theorem. For the details of the proof we refer to Van Es and Hoogstrate (1993).

References

Bickel, P.J. and Ritov, Y., *Estimating integrated squared density derivatives: Sharp best order of convergence estimates*, Sankhyā Ser. **A 50** (1988), 381–393.

Cline, D.B.H. and Hart J.D., *Kernel estimation of densities with discontinuities or discontinuous derivatives*, Statistics **22** (1991), 69–84.

Van Eeden, C., *Mean integrated squared error of kernel estimators when the density and its derivatives are not necessarily continuous*, Ann. Inst. Statist. Math. **37 Part A** (1985), 461–472.

Van Es, A.J., *Aspects of Nonparametric Density Estimation*, CWI tract 77, CWI Amsterdam, The Netherlands, 1991.

Van Es, A.J. and Hoogstrate. A.J., *Properties of plug-in bandwidth selection procedures in non–smooth cases*, Technical Report Department of Mathematics and Computer Science, University of Amsterdam (1993).

Hall, P. and Marron, J.S., *Estimation of integrated squared density derivatives*, Statist. Probab. Lett. **6** (1987), 109–115.

Jones, M.C. and Sheather, S.J., *Using nonstochastic terms to advantage in kernel based estimation of integrated squared density derivatives*, Statist. Probab. Lett. **11** (1991), 511–514.

Levit, B.Ya., *Asymptotically efficient estimation of nonlinear functionals*, Problems Inform. Transmiss. **14** (1978), 65-72.

Liebscher, E., *Kernel estimators for probability densities with discontinuities*, Statistics **21** (1990), 185–196.

Swanepoel, J.H., *Optimal kernels when estimating non–smooth densities*, Commun. Statist. Theory Meth. **16** (1987), 1835–1848.

Department of Mathematics and Computer Science, University of Amsterdam, Plantage Muidergracht 24, 1018TV Amsterdam, The Netherlands

E-mail address: vanes@fwi.uva.nl

Department of Economics, University of Limburg, P.O. Box 616, 6200MD Maastricht, The Netherlands

E-mail address: A.Hoogstrate@KE.RULimburg.NL

CURVE SELECTION: A NONPARAMETRIC APPROACH

A. Futschik
University of Vienna

Abstract

Considerable effort has been devoted to the construction of procedures for subset selection in parametric models. In this paper we will propose a procedure to select a subset from a set of k unknown curves $g_1, \ldots, g_k$ that are only assumed to be smooth. The best curve is defined as the curve that has maximal value at a certain point. A sequence of sets of curves is given on which our procedure includes the best curve asymptotically and uniformly with probability $1-\alpha$. Proposals for the choice of the kernel and the bandwidth are given from a large deviation point of view.

Keywords: Nonparametric regression, subset selection, large deviation.
AMS Classification: 60F10, 62F07, 62G07.

1. Introduction.

Many experiments in different fields yield a set of curves that lie in a nonparametric family. (See e.g. Kneip and Gasser (1992) for examples from biomedicine. Hall and Hart (1990) consider environmental data.) In the analysis of such curves the researcher might be interested in comparing the curves.

Many questions concerning the differences between curves can be answered by doing hypothesis testing. (For examples in nonparametric regression, see e.g. King (1989), Hall and Hart (1990), Härdle and Marron (1990). Mammen (1992) (section 3) considers tests in the density estimation setting.)

But often one not only wants to test hypotheses concerning the curves, but also wants to select functions that are best (worst) in some sense. For many parametric models the practical relevance of this question has been pointed out a long time ago. Goel and Rubin (1977) write for example:

> During the late fourties and early fifties, it was pointed out by several researchers, e.g. Paulson or Bahadur, that testing the homogeneity of population means or variances is not a satisfactory solution to a comparison of the performance of several populations with respect to their performance measured by a parameter value. (One would, generally, want to either rank them according to their performance or select one or more from among them for future use or further evaluation. These problems are known as ranking and selection problems.)

In linear regression selection methods have been applied to choose the population with the highest regression value at point t. (For details and practical applications see

Gibbons, Olkin, and Sobel (1977).) Therefore it seems worthwile to consider selection procedures in nonparametric regression models too. In this paper we will consider k unknown smooth curves, combined in the vector valued function $\mathbf{g} := (g_1, \ldots, g_k)$. We want to select a (small) set $C \subset \{1, \ldots, k\}$ that contains the index of any curve g_l satisfying $g_l(t) = \max_{1 \leq j \leq k} g_j(t)$ with high probability.

We will adapt the classical subset selection procedure of Gupta (1965) to our problem and will give a set of curves on which our procedure has asymptotically probability of correct selection (PCS) $1 - \alpha$. Suggestions concerning the choice of the bandwidth and kernel will be derived from a large deviation approximation to the expected number of nonbest populations that are included in the subset.

2. Classical Subset Selection

The problem of subset selection has first been investigated by Gupta (1965). He proposed the following now classical procedure for selecting a subset containing the population that has the largest mean:

Consider k normal populations π_i, $i = 1, \ldots, k$ and assume that the population π_i has mean μ_i and variance σ^2. Let $\bar{X}_i$, $i = 1, \ldots, k$, denote the sample means (all based on samples of size n) of these populations. Let $\mu_{(1)} < \cdots < \mu_{(k)}$ denote the ordered means. We want to define a set $C \subset \{1, \ldots, k\}$ such that C contains the index l of the population with (the largest) mean $\mu_{(k)}$ with a prescribed probability $1 - \alpha$. The set C is obtained by the following rule (R_μ):

$$\text{Include } j \text{ in } C \Leftrightarrow \bar{X}_j \geq \max_{1 \leq i \leq k} (\bar{X}_i) - d_n \tag{1}$$

where $d_n = q_{1-\alpha} \frac{\sigma}{\sqrt{n}}$. Since we have that, with Φ the standard normal distribution,

$$\begin{aligned}
\mathbf{P}(l \in C | R_\mu) &= \mathbf{P}(\bar{X}_l \geq \max_{1 \leq i \leq k} (\bar{X}_i) - q_{1-\alpha} \frac{\sigma}{\sqrt{n}}) \\
&= \mathbf{P}(\frac{\sqrt{n}}{\sigma} \bar{X}_l \geq \max_{1 \leq i \leq k} \left(\frac{\sqrt{n}}{\sigma} \bar{X}_i \right) - q_{1-\alpha}) \\
&\geq \int_{\mathbb{R}} \Phi^{k-1}(q_{1-\alpha} + x)\, d\Phi(x);
\end{aligned}$$

the value of $q_{1-\alpha}$ is obtained as the solution of

$$\int_{\mathbb{R}} \Phi^{k-1}(q_{1-\alpha} + x)\, d\Phi(x) = 1 - \alpha. \tag{2}$$

For practical implementation we replace σ in (1) by a consistent estimator.

Selection procedures have afterwards been proposed for many other situations. (See e.g. the books by Gibbons, Olkin, and Sobel (1977), Gupta and Panchapakesan (1979), or Rizvi (1985, 1986).)

3. Model and Methods

We consider the following model (fixed design regression model):

$$Y_{ij} = g_i(t_{ij}) + \varepsilon_{ij}, \quad 1 \leq i \leq k,\ 1 \leq j \leq n.$$

$Y_{ij} \in \mathbb{R}$ are observations at fixed (i.e. nonrandom) points $t_{ij} \in [0, 1]$. For all i the t_{ij} are assumed to be ordered (i.e. $t_{i,j-1} \leq t_{ij}$, for $2 \leq j \leq n$) and asymptotically equidistant, by which we mean

$$\max_{i,j} |t_{ij} - t_{i,j-1} - \frac{1}{n}| = O(\frac{1}{n^\delta}), \quad \delta > 1,\ n \to \infty. \tag{3}$$

ε_{ij} $(1 \leq i \leq k,\ 1 \leq j \leq n)$ are i.i.d. error terms satisfying $\mathbf{E}\varepsilon_{ij} = 0$ and

$$\sigma^2 := \mathbf{Var}\varepsilon_{ij} < \infty.$$

Note that in the asymptotic considerations the errors ε_{ij} and the design points form triangular arrays. For notational simplicity an additional index n will be omitted.

$\mathbf{g} := (g_1, \ldots, g_k)$: $[0, 1] \to \mathbb{R}^k$ is assumed to have $r+1$ $(r \geq 1)$ continuous derivatives.

Obviously the component functions g_i can be estimated separately. For this purpose we will use the kernel estimator proposed by Gasser and Müller (1979):

$$\hat{g}_{i,n}(t) := \frac{1}{b_n} \sum_{j=1}^{n} \int_{s_{i,j-1}}^{s_{ij}} K(\frac{t-s}{b_n})\, ds\, Y_{ij} \tag{4}$$

Here

$K(x)$ is a bounded kernel with support $[-1, 1]$, satisfying $\int_{-1}^{1} K(x)\, dx = 1$. K is also assumed to be Lipschitz continuous on its support.

$(s_{ij})_{j=0}^{n}$ are taken for all i and $1 \leq j \leq n-1$ as $s_{ij} := \frac{t_{ij}+t_{i,j+1}}{2}$. Furthermore $s_{i0} := 0$, $s_{in} := 1$.

We assume that $b_n \to 0$ and $nb_n \to \infty$.

For smooth functions g_i the use of higher order kernels improves the asymptotic (MSE) behavior. K is called kernel of order r, if

- $\int_{-1}^{1} K(x)\, dx = 1$,
- $\int_{-1}^{1} K(x)x^j\, dx = 0$, for $1 \leq j \leq r-1$,
- $\int_{-1}^{1} K(x)x^r\, dx \neq 0$.

(For other estimation methods in nonparametric regression see e.g. Härdle (1990).)

We will investigate the following selection rule (R_c)

$$\text{Include } j \text{ in } C \Leftrightarrow \hat{g}_j(t) \geq \max_i \hat{g}_i(t) - d_n \tag{5}$$

where $d_n = q_{1-\alpha}S$, with $q_{1-\alpha}$ being the solution of (2), and S^2 is a variance estimator satisfying (for all i)

$$\sqrt{nb_n}(S - \sqrt{\mathbf{Var}(\hat{g}_{i,n})(t)}) \to 0, \quad \text{in prob.} \tag{6}$$

(Such estimators exist, see e.g. Rice (1984).)

4. Probability of Correct Selection

In this section we will introduce conditions on $\mathbf{g}$ such that the PCS of procedure (5) is asymptotically bounded by $1-\alpha$. Intuitively, the conditions require that the component functions have values that are either not too close at point t, or, if $g_i(t)$ and $g_j(t)$ are close for some i, j, then $g_i(t)$ and $g_j(t)$ are also locally close.

At first we will cite some well known properties of the Gasser–Müller estimator that will be needed. The conditions of the previous section apply.

Bias: (Gasser and Müller (1979).) Let g_i have a finite $(r+1)$th derivative $g^{(r+1)}$ in some open interval around t and K be a kernel of order r. Then for $t \in (0, 1)$

$$\mathbf{E}\hat{g}_{i,n}(t) - g_i(t) = \frac{(-1)^r}{r!} b_n^r \Big[\int_{-1}^{1} x^r K(x)\,dx\Big] g_i^{(r)}(t) + O(\frac{1}{n}) + O(b_n^{r+1}). \tag{7}$$

The remainder terms depend on $\mathbf{g}$ only via the the size of $r+1$-th derivative in an arbitrary open interval containing t.

Variance: (Gasser and Müller (1979).)

$$\mathbf{Var}\hat{g}_{i,n}(t) = \frac{\sigma^2}{nb_n}\int_{-1}^{1} K^2(x)\,dx + O(\frac{1}{(nb_n)^2}) + O(\frac{1}{n^\delta b_n}). \tag{8}$$

Asymptotic normality: (Gasser and Müller (1984).)

$$\frac{\hat{g}_{i,n}(t) - \mathbf{E}\hat{g}_{i,n}(t)}{\sqrt{\mathbf{Var}\hat{g}_{i,n}(t)}} \xrightarrow{\mathcal{D}} N(0,1) \tag{9}$$

THEOREM 1 *For any $r \geq 1$, let*

$$|g_j^{(r)}(t) - g_i^{(r)}(t)| \leq \beta_n |g_j(t) - g_i(t)| + \gamma_n, \quad \text{for } 1 \leq i, j \leq k; \tag{10}$$

$$\text{and} \quad \|\mathbf{g}^{(r+1)}(x)\| \leq c, \quad \text{for } x \in (t-\varepsilon, t+\varepsilon) \tag{11}$$

with

$$\beta_n = b_n^{-r} \frac{r!}{\int_{-1}^{1} x^r K(x)\,dx}, \quad \gamma_n = o(n^{-1/2} b_n^{-(2r+1)/2}).$$

Let furthermore $nb_n^{2r+3} \to 0$ and let l denote the index of any curve with (the highest) value $g_{(k)}(t)$. Then

$$\lim_{n\to\infty} \mathbf{P}(l \in C \mid R_c) = \lim_{n\to\infty} \mathbf{P}(\hat{g}_{l,n} \geq \max_{1\leq j\leq k} \hat{g}_{j,n} - d_n) \geq 1-\alpha.$$

Proof:

Using formula (7) for the asymptotic bias, we obtain for any $\mathbf{g}$ whose component functions satisfy our assumptions

$$\begin{aligned}
&\mathbf{E}\hat{g}_{i,n} - \mathbf{E}\hat{g}_{l,n} = \\
&\quad g_i(t) - g_l(t) + \frac{(-b_n)^r}{r!}\Big[\int_{-1}^{1} x^r K(x)\,dx\Big][g_i^{(r)}(t) - g_l^{(r)}(t)] + O(\frac{1}{n}) + O(b_n^{r+1}) \\
&\quad \leq g_i(t) - g_l(t) + |g_i(t) - g_l(t)| + o((nb_n)^{-1/2}) + O(\frac{1}{n}) + O(b_n^{r+1}) \\
&\quad = o((nb_n)^{-1/2}) + O(b_n^{r+1}) \quad \text{since } g_i(t) - g_l(t) \leq 0.
\end{aligned}$$

Therefore

$$\mathbf{P}(\hat{g}_{(k),n} - \hat{g}_{l,n} \geq d_n) \leq \mathbf{P}(S^{-1}[\max_{1\leq i\leq k}(\hat{g}_{i,n}(t) - \mathbf{E}\hat{g}_{i,n}(t)) - \\ - (\hat{g}_{l,n}(t) - \mathbf{E}\hat{g}_{l,n}(t))] > S^{-1}[d_n - o((nb_n)^{-1/2}) - O(b_n^{r+1})]). \tag{12}$$

It remains to show that the limit of the right hand side probability does not exceed α. But, using (9) and assumption (6) we see that

$$S^{-1}(\max[\hat{g}_{i,n} - \mathbf{E}\hat{g}_{i,n}] - [\hat{g}_{l,n} - \mathbf{E}\hat{g}_{l,n}]) \xrightarrow{\mathcal{D}} \mathcal{L}(\max_{1\leq i\leq k} Z_i - Z_k),$$

where Z_i are $N(0, 1)$ distributed and $\mathcal{L}(\max_{1\leq i\leq k} Z_i - Z_k)$ is the distribution of $\max_{1\leq i\leq k} Z_i - Z_k$. Also $S^{-1}[d_n - o((nb_n)^{-1/2}) - O(b_n^{r+1})]$ converges to d in probability. Therefore, by Slutzky's Lemma, the limit of the right hand side of inequality (12) converges to

$$\mathbf{P}\left(\max_{1\leq i\leq k} Z_i - Z_k \geq d\right) = \alpha.$$

□

5. Large Deviation Analysis

Given now a fixed $\mathbf{g}$ that satisfies (10) and (11) we might ask how the bandwidth and the kernel should be chosen such that the expected number of selected nonbest curves (i.e. curves satisfying $g_i(t) < g_{(k)}(t)$) is small.

The moment generating function of the error terms is denoted by $\mathbf{E}e^{t\epsilon_{ij}} =: M(t)$. We make the following assumptions. (In addition to those of section 3.)

(A1) $\mathbf{g}$ has a unique component function g_l with maximal value at t. (I.e. $g_l(t) > g_j(t)$ for all $j \neq l$.)

(A2) For later specified i and j there is a unique minimum $\tau_{i,j} > 0$ (in x) of

$$2\int_{-1}^{1} \log M(K(u)x)\,du - (g_j(t) - g_i(t))x. \tag{13}$$

(A3) $M(t) < \infty$ on some open interval I that contains

$$[\tau_{i,j}\min_x(K(x)),\ \tau_{i,j}\max_x(K(x))].$$

(A4) $\sigma^2 > 0$.

Let l^* denote the index of any (second best) curve satisfying $g_{l^*}(t) = g_{(k-1)}(t)$.

Then we obtain an exponential approximation for the expected number of selected nonbest curves:

THEOREM 2 *Under the above assumptions (choose $i = l^*$ and $j = l$ in (13))*

$$\sum_{i \neq l} P(\max_{1 \leq j \leq k} \hat{g}_{j,n}(t) - \hat{g}_{i,n}(t) \leq d_n) = \exp\Big(nb_n[2\int_{-1}^{1} \log(M(K(u)\tau_{l^*,l}))\,du$$

$$- (g_{(k)}(t) - g_{(k-1)}(t))\tau_{l^*,l}]\Big)\exp(-o(nb_n)).$$

A proof, together with the needed auxiliary results, will be given at the end of the section.

According to Theorem 2 we note that if our aim is to minimize the expected number of selected nonbest curves and when there is a unique optimal curve, we have to choose b_n as large as possible. But if we want to guarantee the constraint that the PCS is not smaller than $1-\alpha$ on a quite large set of curves (the curves satisfying (10) and (11)) we have to choose $b_n = o(n^{-\frac{1}{2k+3}})$. Compared to the MSE-optimal bandwidth of $cn^{-\frac{1}{2k+1}}$ (see Gasser and Müller (1979)), we thus should undersmooth and multiply the estimated MSE-optimal bandwidth by a factor slightly smaller than $n^{2/[(2k+1)(2k+3)]}$. Simulations will have to be carried out to confirm these theoretical conclusions.

It is possible to find the optimal kernel for specific moment generating functions. E.g. for $\varepsilon_i \sim N(0, \sigma^2)$ we have

$$M(t) = \exp(\frac{t^2\sigma^2}{2})$$

and straightforward calculations show that it is asymptotically optimal (w.r.t. the expected size) to choose

$$K(x) = \frac{1}{2}.$$

This kernel is known as minimum variance kernel of order 2. (See Gasser, Müller, Mammitsch (1985).)

The auxiliary results needed for the proof of the Theorem will be stated now. The proof uses the following (slightly modified) Theorem by Bahadur and Ranga Rao. (See Bahadur (1971).)

LEMMA 1 *Let $(Z_n)_n$ be a sequence of random variables.*

a) *If $\rho_n = \inf_{t \geq 0} M_{Z_n}(t)$, then*

$$P(Z_n \geq 0) \leq \rho_n.$$

b) *Assume that $M_{Z_n}(\tau_n) = \rho_n$ for a unique τ_n. Let $\sigma_n^2 := \int z^2\,dQ_n$ with*

$$Q_n := \frac{e^{\tau_n x}}{\rho_n} P^{Z_n}(x) \qquad \textit{(exponential centering).}$$

For $F_n := Q_n((-\infty, \sigma_n x])$ assume that $F_n \to F$ in distribution for $n \to \infty$, where F is strictly increasing on $[0, c]$ for some $c > 0$ and that $\tau_n\sigma_n \to \infty$. Then as $n \to \infty$

$$\frac{1}{\tau_n\sigma_n}[\log P(Z_n \geq 0) - \log \rho_n] = o(1).$$

The next Lemma gives a large deviation approximation for the difference of two curves.

LEMMA 2 *Apart from the conditions specified at the beginning of the section, let $a_n \to 0$ as $n \to \infty$. Define $\tau_{i,j}$ as in (13). If $g_j(t) - g_i(t) > 0$, then*

$$\frac{1}{nb_n}\left(\log \mathbf{P}(\hat{g}_{i,n}(t) - \hat{g}_{j,n}(t) - a_n \geq 0) - 2\int_{-1}^{1} \log(M(K(u)\tau_{i,j}))\,du - (g_j(t) - g_i(t))\tau_{i,j}\right)$$

is of order $o(1)$.

Proof: The proof makes use of Lemma 1. Note first that

$$\begin{aligned}&\mathbf{P}\left(\hat{g}_{i,n}(t) - \hat{g}_{j,n}(t) - a_n \geq 0\right) \qquad (14)\\ &\quad = \mathbf{P}\left((\hat{g}_{i,n}(t) - \mathbf{E}\hat{g}_{i,n}(t)) - (\hat{g}_{j,n}(t) - \mathbf{E}\hat{g}_{j,n}(t)) - (a_n - \mathbf{E}\hat{g}_{i,n}(t) + \mathbf{E}\hat{g}_{j,n}(t)) \geq 0\right).\end{aligned}$$

Denoting $c_n := a_n - \mathbf{E}\hat{g}_{i,n}(t) + \mathbf{E}\hat{g}_{j,n}(t)$ and

$$Z_n := nb_n[(\hat{g}_{i,n}(t) - \mathbf{E}\hat{g}_{i,n}(t)) - (\hat{g}_{j,n}(t) - \mathbf{E}\hat{g}_{j,n}(t)) - c_n].$$

(14) may be written as $\mathbf{P}(Z_n \geq 0)$. To approximate this probability we need the limit of ρ_n from Lemma 1. By substituting the definition of the kernel estimator (4) we get

$$Z_n = \sum_{k=1}^{n} nb_n w_{ik}\varepsilon_{ik} - \sum_{k=1}^{n} nb_n w_{jk}\varepsilon_{jk} - nb_n c_n$$

with

$$w_{ik} = \frac{1}{b_n}\int_{s_{i,k-1}}^{s_{ik}} K\left(\frac{t-s}{b_n}\right) ds$$

and w_{jk} defined in analagous way.

Let $U(\tau)$ be a closed neighborhood of $\tau := \tau_{i,j}$ such that for all $x \in U(\tau)$

$$[x \min_x K(x),\ x \max_x K(x)] \subset I.$$

We will derive first an approximation to the moment generating function of Z_n, called $M_{Z_n}(x)$, on $U(\tau)$. For this note that

$$M_{Z_n}(x) = e^{-nb_n c_n x} \prod_{k=1}^{n} \left(M(w_{ik}nb_n x)M(w_{jk}nb_n x)\right) \qquad (15)$$

and therefore

$$\frac{1}{nb_n}\log M_{Z_n}(x) = \frac{1}{nb_n}\sum_{k=1}^{n}\log M(w_{ik}nb_n x) + \frac{1}{nb_n}\sum_{k=1}^{n}\log M(w_{jk}nb_n x) - c_n x. \qquad (16)$$

Looking at w_{ik}, we notice

$$\begin{aligned} w_{ik} &= \frac{1}{b_n}\int_{s_{ik-1}}^{s_{ik}} K\left(\frac{t-s}{b_n}\right) ds = \int_{(t-s_{ik})/b_n}^{(t-s_{ik-1})/b_n} K(u)\,du \\ &= K(\xi_{ik})\frac{s_{ik} - s_{ik-1}}{b_n} = \frac{K(\xi_{ik})}{nb_n} + o((nb_n)^{-1}) \end{aligned}$$

for a $\xi_{ik} \in \left(\frac{t-s_{ik-1}}{b_n}, \frac{t-s_{ik}}{b_n}\right)$. Note that $w_{ik} = 0$ if $\xi_{ik} \notin [-1,1]$, and that the above $o()$ term is uniform in i and k. For w_{jk} an analagous transformation can be carried out. Resubstituting in (16) we obtain for $x \in U(\tau)$

$$\frac{1}{nb_n} \sum_{\substack{1\le k\le n \\ K(\xi_{ik})\neq 0}} \log M(K(\xi_{ik})x + o(1)) + \frac{1}{nb_n} \sum_{\substack{1\le k\le n \\ K(\xi_{jk})\neq 0}} \log M(K(\xi_{jk})x + o(1)) - c_n x$$

$$= \frac{1}{nb_n} \sum_{\substack{1\le k\le n \\ K(\xi_{jk})\neq 0}} \log M(K(\xi_{ik})x) + \frac{1}{nb_n} \sum_{\substack{1\le k\le n \\ K(\xi_{jk})\neq 0}} \log M(K(\xi_{jk})x) - cx + o(1)$$

where $c := \lim_{n\to\infty} c_n = g_j(t) - g_i(t)$. (This follows since not more than $O(nb_n)$ terms are not equal to zero and $\log \circ M$ is Lipschitz continuous on $U(\tau)$.) The above quantity may now be written as Rieman-sum

$$\sum_{k=1}^{n} \log M(K(\xi_{ik})x)\frac{s_{ik} - s_{ik-1}}{b_n} + \sum_{k=1}^{n} \log M(K(\xi_{jk})x)\frac{s_{jk} - s_{jk-1}}{b_n} - cx + o(1)$$

that converges to

$$2\int_{-1}^{1} \log(M(K(u)x))\, du - cx =: h(x).$$

In the next step we will show that $\tau_n = \operatorname{argmin}(M_{Z_n}(x))$ converges to τ. According to our assumptions, h has a unique minimum τ. We obtain therefore for any ε such that $[\tau-\varepsilon, \tau+\varepsilon] \subset U(\tau)$

$$h(\tau-\varepsilon) > h(\tau), \quad h(\tau+\varepsilon) > h(\tau).$$

But then also

$$\frac{1}{nb_n}\log M_{Z_n}(\tau-\varepsilon) > \frac{1}{nb_n}\log M_{Z_n}(\tau), \quad \frac{1}{nb_n}\log M_{Z_n}(\tau+\varepsilon) > \frac{1}{nb_n}\log M_{Z_n}(\tau),$$

for n large enough. Since log is strictly increasing and M_{Z_n} is convex, we obtain that $\tau_n \in [\tau-\varepsilon, \tau+\varepsilon]$ for large n. (Note that τ_n is unique, since M_{Z_n} is strictly convex because $P(Z_n = 0) < 1$.) Since ε has been chosen arbitrarily, it follows that $\tau_n \to \tau$.

Furthermore

$$\frac{1}{nb_n}\log M_{Z_n}(\tau_n) - h(\tau) \to 0.$$

This may be seen as follows. Since $\tau_n \to \tau$, the above statement is true, if $\frac{1}{nb_n}\log M_{Z_n}$ are equicontinuous on $U(\tau)$. But this can be verified by recalling the definition of M_{Z_n}.

$$|\frac{1}{nb_n}[\log M_{Z_n}(x)]'| =$$

$$= |\frac{1}{nb_n} \sum_{\substack{1\le k\le n \\ w_{ik}\neq 0}} \frac{M'(w_{ik}nb_nx)}{M(w_{ik}nb_nx)} w_{ik}nb_n + \frac{1}{nb_n} \sum_{\substack{1\le k\le n \\ w_{jk}\neq 0}} \frac{M'(w_{jk}nb_nx)}{M(w_{jk}nb_nx)} w_{jk}nb_n|$$

$$\le c \max_{x\in U(\tau)} |\frac{M'(x)}{M(x)}|.$$

The last inequality follows, since there are only $O(nb_n)$ nonzero terms in the sum, and since for any $\delta > 0$ and large enough n

$$\min_x K(x) - \delta \le w_{ik} n b_n \le \max_x K(x) + \delta.$$

Therefore the rate ρ_n in Lemma 1 is

$$\frac{1}{nb_n} \log \rho_n = 2 \int_{-1}^{1} \log(M(K(u)\tau))\, du - c\tau + o(1).$$

To complete the proof, we still have to check the condition in part b) of Lemma 1. Let $Q_n = \frac{e^{\tau_n u}}{\rho_n} \mathbf{P}^{Z_n}(u)$. Then Q_n has moment generating function

$$\begin{aligned} M_{Q_n}(x) &= \frac{\int_{-\infty}^{\infty} e^{xu} e^{\tau_n u}\, dP^{Z_n}(u)}{\rho_n} \\ &= \frac{M_{Z_n}(x+\tau_n)}{M_{Z_n}(\tau_n)} \\ &= e^{-nb_n c_n x} \prod_{k=1}^{n} \left(\frac{M(w_{ik} n b_n (x+\tau_n))}{M(w_{ik} n b_n \tau_n)} \frac{M(w_{jk} n b_n (x+\tau_n))}{M(w_{jk} n b_n (\tau_n))} \right). \end{aligned}$$

Now this is the moment generating function of

$$\sum_{k=1}^{n} [\varepsilon_{ik}^* + \varepsilon_{jk}^*] - nb_n c_n =: V_n,$$

where ε_{ik}^* are random variables with moment generating functions

$$\frac{M(w_{ik} n b_n (x+\tau_n))}{M(w_{ik} n b_n \tau_n)}.$$

(ε_{jk}^* are defined in analogous way.)

To approximate $\sigma_n^2 = \mathbf{Var}(V_n)$ note that for large n

$$\begin{aligned} \mathbf{Var}\varepsilon_{ik}^* &= \frac{w_{ik}^2 n^2 b_n^2 M''(w_{ik} n b_n \tau_n)}{M(w_{ik} n b_n \tau_n)} - (\mathbf{E}\varepsilon_{ik}^*)^2 \\ &\le \max_x |K(x)|^2 \max_{x \in U(\tau)} \frac{M''(x)}{M(x)} + o(1), \end{aligned}$$

which might be bounded by a constant not depending on i and k. Since $O(nb_n)$ terms have nonzero variance we obtain $\sigma_n^2 = O(nb_n)$.

To see that there is a constant $c_1 > 0$ such that $\sigma_n^2 > c_1 n b_n$, note that there is a $c_2 > 0$, such that $w_{ik} n b_n > \varepsilon$ for a set of more than $c_2 n b_n$ indices k. (If $\varepsilon > 0$ is chosen sufficiently small.) For any k in this set ε_{ik}^* has distribution

$$\frac{e^{\tau_n u}}{M(w_{ik} n b_n \tau_n)} \mathbf{P}_{\varepsilon_{ik}} \left(\frac{u}{w_{ik} n b_n} \right).$$

From this it follows that $\mathbf{Var}(\varepsilon_{ik}^*) > c_3 > 0$. (Otherwise it would follow that ε_{ik} has zero variance which is a contradiction to our assumptions.)

Finally we will show that F_n, the distribution of

$$S_n := \frac{\sum_{k=1}^n \varepsilon_{ik}^* + \sum_{k=1}^n \varepsilon_{jk}^* - nb_nc_n}{\sigma_n},$$

converges to the normal distribution. Since $\mathbf{E}S_n = 0$ we may write S_n also as

$$S_n = \frac{\sum_{k=1}^n(\varepsilon_{ik}^* - \mathbf{E}\varepsilon_{ik}^*) + \sum_{k=1}^n(\varepsilon_{jk}^* - \mathbf{E}\varepsilon_{jk}^*)}{\sigma_n},$$

Since ε_{ik}^*, ε_{jk}^* are independent for all i, j, k a sufficient condition for $S_n \xrightarrow{\mathcal{D}} N(0,1)$ is

$$\sum_{k=1}^n \mathbf{E}(\varepsilon_{ik}^* - \mathbf{E}\varepsilon_{ik}^*)^\nu + \sum_{k=1}^n \mathbf{E}(\varepsilon_{jk}^* - \mathbf{E}\varepsilon_{jk}^*)^\nu = o(\sigma^\nu)$$

for some $\nu > 2$. (See Serfling (1980), p. 32.) We choose $\nu = 4$ and calculate $\mathbf{E}\tilde{\varepsilon}_{ik}^4$, where $\tilde{\varepsilon}_{ik} := \varepsilon_{ik}^* - \mathbf{E}\varepsilon_{ik}^*$. Note that

$$\begin{aligned}
\mathbf{E}\tilde{\varepsilon}_{ik}^r &= M_{\tilde{\varepsilon}_{ik}}^{(r)}(0) \\
&= \left(\frac{M(w_{ik}nb_n(x+\tau_n))}{M(w_{ik}nb_n\tau_n)}\exp(-\mathbf{E}\varepsilon_{ik}^* x)\right)^{(r)}\Bigg|_{x=0} \\
&= \sum_{m=0}^r \binom{r}{m} \frac{(w_{ik}nb_n)^m M^{(m)}(w_{ik}nb_n\tau_n)(-\mathbf{E}\varepsilon_{ik}^*)^{n-r})}{M(w_{ik}nb_n\tau_n)}.
\end{aligned}$$

Now $M_{\tilde{\varepsilon}_{ik}}^{(r)}(0) \le c_r$ for some c_r because

$$\mathbf{E}\varepsilon_{ik}^* = \frac{M'(nb_nw_{ik}\tau_n)nb_nw_{ik}}{M(nb_nw_{ik}\tau_n)}$$

and

$$\frac{(w_{ik}nb_n)^m M^{(m)}(w_{ik}nb_n\tau_n)}{M(w_{ik}nb_n\tau_n)} \le (\max_x |K(x)|)^m \max_{x\in U(\tau)} \frac{M^{(m)}(x)}{M(x)} + o(1).$$

Therefore $\mathbf{E}(\varepsilon_{jk} - \mu_{jk})^4$ is bounded by a constant and

$$\sum_{k=1}^n \mathbf{E}(\varepsilon_{ik}^* - \mathbf{E}\varepsilon_{ik}^*)^4 + \sum_{k=1}^n \mathbf{E}(\varepsilon_{jk}^* - \mathbf{E}\varepsilon_{jk}^*)^4 = O(nb_n) = o(\sigma^4).$$

So all the conditions of Lemma 1 have been verified.

□

LEMMA 3 *Assume that $X_1, \dots, X_k$ are independent. Then for any d*

$$\prod_{i=1}^k \mathbf{P}(X_i - X_j < d) \le \mathbf{P}(\max_i X_i - X_j \le d) \le \mathbf{P}(X_l - X_j \le d).$$

Proof: The right hand side inequality is obvious. The left hand side inequality can be derived by splitting

$$\begin{aligned}\mathbf{P}(\max_i X_i - X_j \le d) &= \\ &= \mathbf{P}(X_l - X_j \le d)\mathbf{P}(X_{l^\bullet} - X_j \le d | X_l - X_j \le d)\cdots \\ &\cdots \mathbf{P}(X_{l_1} - X_j \le d | X_l - X_j \le d, \ldots, X_{l_2} - X_j \le d),\end{aligned}$$

where l_1 and l_2 are the indices corresponding to the smallest and second smallest means respectively.

Now the conditional probabilities are not smaller than the corresponding unconditional ones.

□

Proof: (Of Theorem 2.) Let $I_1 = \{i | g_i(t) = g_{(k-1)}(t)\}$. ($g_{(k-1)}(t)$ is the second largest element of $\mathbf{g}(t)$). Choose the $j_0 \in I_1$ that has largest expectation $\mathbf{E}\hat{g}_{j,n}(t)$. ($j_0$ may depend on n.) Then (by Lemma 3)

$$\mathbf{P}(\max_i \hat{g}_{i,n}(t) - \hat{g}_{j_0,n}(t) \le d_n) \ge \tag{17}$$

$$\begin{aligned}&\ge \prod_{i=1}^{k} \mathbf{P}(\hat{g}_{i,n}(t) - \hat{g}_{j_0,n}(t) \le d_n) = \\ &= \mathbf{P}(\hat{g}_{l,n}(t) - \hat{g}_{j_0,n}(t) \le d_n) \prod_{i \ne l} \mathbf{P}(\hat{g}_{i,n}(t) - \hat{g}_{j_0,n}(t) \le d_n)\end{aligned} \tag{18}$$

Since $\hat{g}_{i,n}(t), \hat{g}_{j_0,n}(t)$ have asymptotic normal distribution, and $\mathbf{E}\left(\hat{g}_{i,n}(t) - \hat{g}_{j_0,n}(t)\right) \le 0$ for $i \ne l$ and large n, and $d_n \ge 0$, we obtain

$$\mathbf{P}(\hat{g}_{i,n}(t) - \hat{g}_{j_0,n}(t) \le d_n) \ge c > 0$$

Therefore the product term in (18) is bounded away from 0 for large n. According to Lemma 2

$$\mathbf{P}(\hat{g}_{l,n}(t) - \hat{g}_{j_0,n}(t) \le d_n) = \exp\Big(2nb_n[\int_{-1}^{1} \log(M(K(u)\tau_{l^\bullet,l}))\, du \tag{19}$$

$$- \left(g_{(k)}(t) - g_{(k-1)}(t)\right)\tau_{l^\bullet,l}]\Big) \exp(-o(nb_n))$$

and together we notice that (17) is not smaller than (19). But since

$$\mathbf{P}(\max_i \hat{g}_{i,n}(t) - \hat{g}_{j_0,n}(t) \le d_n) \le \mathbf{P}(\hat{g}_{l,n}(t) - \hat{g}_{j_0,n}(t) \le d_n)$$

(19) is also an upper bound for (17).

For $j \in \{1, \ldots, k\} \setminus \{l, j_0\}$. we may bound

$$\mathbf{P}(\max_i \hat{g}_{i,n}(t) - \hat{g}_{j,n}(t) \le d_n) \le \mathbf{P}(\hat{g}_{l,n}(t) - \hat{g}_{j,n}(t) \le d_n)$$

which is not larger than $\mathbf{P}(\hat{g}_{l,n}(t) - \hat{g}_{j_0,n}(t) \le d_n)$ for large n. Together the statement follows. □

References

[1] Bahadur, R. R. (1971) Some limit theorems in statistics. Regional conference series in applied mathematics 4. SIAM, Philadelphia.

[2] Gasser Th. and Müller H. G. (1979) Kernel estimation of regression functions. In: Smoothing techniques for curve estimation. Lecture notes in mathematics 757. 23–68. Ed. Th. Gasser and M. Rosenblatt.

[3] Gasser Th. and Müller H. G. (1984) Estimating regression functions and their derivatives by the kernel method. Scand. J. Stat. 11. 171–185.

[4] Gasser Th., Müller H. G. and Mammitzsch V. (1985) Kernels for nonparametric curve estimation. J. R. Statist. Soc. 47. 238–252.

[5] Goel, P. K. and Rubin H. (1977). On selecting a subset containing the best population–A Bayesian approach. Ann. Statist. 5. 969–983.

[6] Gibbons, J. D., Olkin, I. and Sobel, M. (1977) Selecting and Ordering Populations. Wiley, New York.

[7] Gupta, S. S. (1965) On some multiple decision (selection and ranking) rules, Technometrics 7. 225–245.

[8] Gupta, S. S. and Panchapakesan, S. (1979) Multiple Decision Procedures: Theory and methodology of selecting and ranking populations. Wiley, New York.

[9] Hall, P. and Hart, J. D. (1990). Bootstrap test for difference between means in nonparametric regression. JASA 85. 1039–1049.

[10] Härdle, W. (1990) Applied nonparametric regression. Cambridge University Press.

[11] Härdle, W. and Marron, J. S. (1990) Semiparametric comparison of regression curves. Ann. Statist. 18. 63–89.

[12] King, E. C. (1989). A test for the equality of two regression curves based on kernel smoothers. Ph.D. dissertation, Dept. of Statistics, Texas A&M University.

[13] Kneip, A. and Gasser, Th. (1992). Statistical tools to analyze data representing a sample of curves. Ann. Statist. 20. 1266–1305.

[14] Mammen, E. (1992) When does bootstrap work? Lecture notes in statistics 77. Springer, New York.

[15] Rice, J. (1984) Bandwidth choice for nonparametric regression. Ann. Stat. 12. 1215–1230.

[16] Rizvi, M. H. (1985, 1986) Modern Statistical Selection I and II. American Sciences Press, Columbus, Ohio.

[17] Serfling, R. J. Approximation Theorems of Mathematical Statistics. Wiley (1980).

COMPLETE CONVERGENCE

Allan Gut, Department of Mathematics, Uppsala University,
Box 480, S-751 06 Uppsala, Sweden.

The concept complete convergence was introduced in 1947 by Hsu and Robbins, who proved that the sequence of arithmetic means of i.i.d. random variables converges completely to the expected value of the variables provided their variance is finite. In this paper we summarize a number of extensions and generalizations of that result.

AMS 1980 subject classifications

Primary: 60F15, 60G50 Secondary: 60G40

Key words and phrases: Complete convergence, sums of i.i.d. random variables, law of large numbers, multidimensional index set, subsequences, rowwise independent arrays, weighted sums of i.i.d. random variables, Cesàro summation.

1. Introduction

The concept complete convergence was introduced by Hsu and Robbins (1947) as follows.

DEFINITION. A sequence $\{U_n,\, n \geq 1\}$ of random variables *converges completely* to the constant θ if

$$\sum_{n=1}^{\infty} P(|U_n - \theta| > \varepsilon) < \infty \quad \text{for all} \quad \varepsilon > 0. \qquad \square$$

In the same paper the authors prove that the sequence of arithmetic means of i.i.d. random variables converges completely to the expected value of the variables provided their variance is finite. Erdős (1949, 1950) proved the converse. Thus, complete convergence in the classical law of large numbers holds iff the variance is finite. In comparison, the strong law itself holds iff the mean is finite. This implies, for example, that if $\{X_k,\, k \geq 1\}$ is a sequence of i.i.d. random variables with partial sums $\{S_n,\, n \geq 1\}$ and we only suppose that the mean is finite $= 0$, say, then the sequence of arithmetic means converges almost surely but *not* completely to 0. In particular,

$$P(|S_n| > n\varepsilon \text{ i.o.}) = 0 \quad \text{and/but} \quad \sum_{n=1}^{\infty} P(|S_n| > n\varepsilon) = \infty \quad \text{for all} \quad \varepsilon > 0.$$

This thus provides an example where only finitely many events occur a.s. and/but the Borel-Cantelli sum diverges. On the other hand, complete convergence always implies almost sure convergence in view of the (first) Borel-Cantelli lemma. We also note that complete convergence and almost sure convergence are equivalent for sequences of independent random variables.

The purpose of this paper is to give a survey on results on complete convergence related to strong laws. In Section 2 we recall the Hsu-Robbins-Erdős law, sketch the ideas of the original proof and, then, present an alternative proof, which turns out to be very efficient and suitable for generalizations and extensions, some of which are described in the remaining sections. Very little will be proved in detail, since proofs can be found in the sources referred to.

2. The Hsu-Robbins-Erdős law

THEOREM HRE. *Let $\{X_k,\ k \geq 1\}$ be a sequence of i.i.d. random variables with partial sums $\{S_n,\ n \geq 1\}$. Then*

$$\sum_{n=1}^{\infty} P(|S_n| > n\varepsilon) < \infty \quad \text{for all} \quad \varepsilon > 0$$

iff $EX = 0$ and $EX^2 = \sigma^2 < \infty$. □

A first naive attempt to prove the sufficiency would be to use Chebyshev's inequality, which, however, (only) yields $\sum_{n=1}^{\infty} P(|S_n| > n\varepsilon) \leq \varepsilon^{-2}\sigma^2 \sum_{n=1}^{\infty} n^{-1}$.

The classical proof of the sufficiency runs, essentially, as follows. Let $2^i \leq n < 2^{i+1}$ and set

$$A_n^{(1)} = \{|X_k| > 2^{i-2} \text{ for at least one } k \leq n\}$$
$$A_n^{(2)} = \{|X_{k_1}| > n^{\gamma}, |X_{k_2}| > n^{\gamma} \text{ for at least two } k_i \leq n\}$$
$$A_n^{(3)} = \{|\sum{}' X_k| > 2^{i-2}\},$$

where γ is "suitably" chosen and where $\sum'$ denotes that summation is over those indices which are not among the first two sets. After this one observes that

$$\{|S_n| > n\varepsilon\} \subset A_n^{(1)} \cup A_n^{(2)} \cup A_n^{(3)}.$$

A fair amount of computations yield the desired result.

A more efficient way to prove the result is to use (standard) symmetrization/desymmetrization methods (the details of which we omit) and the following inequality due to Kahane and Hoffmann-Jørgensen; see Hoffmann-Jørgensen (1974), page 164.

LEMMA KHJ. *Let $Y_1, Y_2, \ldots$ be i.i.d. symmetric random variables vith partial sums $\{V_n,\ n \geq 1\}$ and let $t > 0$. Then*

$$P(|V_n| > 3t) \leq P(\max_{1\leq k\leq n} |Y_k| > t) + 4\big(P(|V_n| > t)\big)^2.$$
$$\leq nP(|Y_1| > t) + 4\big(P(|V_n| > t)\big)^2.$$ □

With the aid of this lemma and Chebyshev's inequality one obtains, in the symmetric case,

$$\sum_{n=1}^{\infty} P(|S_n| > 3n\varepsilon) \leq$$
$$\leq \sum_{n=1}^{\infty} nP(|X_1| > n\varepsilon) + 4\sum_{n=1}^{\infty} (P(|S_n| > n\varepsilon))^2 \leq$$
$$\leq cEX^2 + 4\sum_{n=1}^{\infty}\Big(\frac{\operatorname{Var} S_n}{n^2\varepsilon^2}\Big)^2 = c\sigma^2 + 4\varepsilon^{-4}\sigma^4\pi^2/6 < \infty.$$

It is the squaring of the original probability that does the job in that $\sum_{n=1}^{\infty} n^{-1}$ (which is divergent) is replaced by $\sum_{n=1}^{\infty} n^{-2}$ (which is convergent).

As for a proof of the converse one first notes that, in the symmetric case,

$$2P(|S_n| > n\varepsilon) \geq P(\max_{1\leq k\leq n} |S_k| > n\varepsilon) \geq P(\max_{1\leq k\leq n} |X_k| > 2n\varepsilon) = (1-(1-P(|X_1| > 2n\varepsilon)))^n,$$

(where the first inequality is a consequence of Lévy's inequality). This, together with the fact that the strong law holds and, hence, the mean exists and equals 0, shows that

$$2P(|S_n| > n\varepsilon) \geq \tfrac{1}{2}nP(|X_1| > 2n\varepsilon) \text{ for } n \text{ large.}$$

Summation establishes the finiteness of the variance and, hence, the theorem (in the symmetric case).

3. Convergence rates in the strong law

The following result is due to Baum and Katz (1965).

THEOREM BK. *Let $\{X_k, k \geq 1\}$ be a sequence of i.i.d. random variables with partial sums $\{S_n, n \geq 1\}$ and let α and r be positive reals, such that $\alpha r \geq 1$ and $\alpha > 1/2$. The following are equivalent:*

a) $E|X_1|^r < \infty$ and, *if* $r \geq 1$, $EX = 0$

b) $\sum_{n=1}^{\infty} n^{\alpha r-2} P(|S_n| > n^\alpha \varepsilon) < \infty$ *for all* $\varepsilon > 0$

c) $\sum_{n=1}^{\infty} n^{\alpha r-2} P(\max_{1\leq k\leq n} |S_k| > n^\alpha \varepsilon) < \infty$ *for all* $\varepsilon > 0$.

If $\alpha r > 1$ *and* $\alpha > 1/2$ *the above are also equivalent to*

d) $\sum_{n=1}^{\infty} n^{\alpha r-2} P(\sup_{k\geq n} |S_k/k^\alpha| > \varepsilon) < \infty$ *for all* $\varepsilon > 0$. □

For $r = 2, \alpha = 1$ the equivalence between a) and b) is, of course, Theorem HRE. For $r = 1, \alpha = 1$ the equivalence is due to Spitzer (1956). For $r \geq 1, \alpha = 1$, see Katz (1963).

The classical proof of a) ⇒ b) is based on the HRE-method. An alternative proof is obtained via the alternative approach described above with the inequalities of Marcinkiewicz and Zygmund (1937, 1938) replacing Chebyshev's inequality. In some cases the inequality KHJ has to be replaced by an iteration due to Jain (1975), see also Gut (1978, 1993). For details, see Gut (1978). The converse implication follows as above. Implication b) ⇒ c) is Lévy's inequality (immediately in the symmetric case and with a little more work in the general case) and the converse is immediate. Implication b) ⇒ d) follows in the symmetric case via dividing the positive integers into blocks bounded by powers of 2 and Lévy's inequality, see Baum and Katz (1965). Again, the converse is trivial.

Note, as claimed in the title of this section, that the various sums can be interpreted as convergence rates in the law of large numbers (Kolmogorov's law for $\alpha = 1$ and the Marcinkiewicz law otherwise).

Finally in this section we mention the existence of the random variables $L(\varepsilon) = \sup\{n : |S_n| > n\varepsilon\}$ and $N(\varepsilon) = \text{Card}\{n : |S_n| > n\varepsilon\}$. The Kolmogorov strong law is equivalent to the statement that $L(\varepsilon) < \infty$ almost surely and the relation $\{\sup_{k\geq n} |S_k/k| > \varepsilon\} = \{L(\varepsilon) \geq n\}$ and Theorem BK can be used to find conditions for the finiteness of moments of $L(\varepsilon)$. For some results in a more general setting, see Gut (1980).

4. Multidimensional index sets

Several classical limit theorems have been extended to the case where the index set is, e.g., the positive integer d-dimensional lattice points, $\mathbf{Z}^d$, for $d \geq 2$, with coordinatewise partial ordering, $\prec$. The points in the lattice are denoted $\mathbf{k}$, $\mathbf{n}$ etc. and $|\mathbf{n}|$ is the product of the coordinates of $\mathbf{n}$, i.e. the number of points "preceding" $\mathbf{n}$. The results above can all be generalized to this case. As a sample we state the following extension of Theorem HRE.

THEOREM 4.1. *Let* $\{X_{\mathbf{k}}, \mathbf{k} \in \mathbf{Z}^d\}$ *be i.i.d. random variables with partial sums* $\{S_{\mathbf{n}}, \mathbf{n} \in \mathbf{Z}^d\}$. *Then*

$$\sum_{\mathbf{n}\in\mathbf{Z}^d} P(|S_{\mathbf{n}}| > |\mathbf{n}|\varepsilon) < \infty \quad \textit{for all} \quad \varepsilon > 0$$

iff $EX = 0$ *and* $EX^2(\log^+|X|)^{d-1} < \infty$. □

Remark. $\log^+ x = \max\{1, \log x\}$. □

The proof of the sufficiency follows by observing that the KHJ inequality, Chebyshev's inequality (as well as the Marcinkiewicz-Zygmund inequalities) do not depend on the index set, just on the number of summands in the sums. For details (and an extension of Theorem BK), see Gut (1978).

As for the necessity we only remark that the proof follows the same pattern as for $d = 1$ (see above), and from the fact that

$$\sum_{\mathbf{n}\in\mathbf{Z}^d} |\mathbf{n}|P(|X_{\mathbf{n}}| > |\mathbf{n}|) < \infty \quad \Longleftrightarrow \quad EX^2(\log^+|X|)^{d-1} < \infty.$$

Generalizations of the counting variables mentioned at the end of the previous section have been investigated in Gut (1980). Note also that the theorem reduces to Theorem HRE for $d = 1$.

5. Randomly indexed sums

Another way to extend limit theorems for random walks is to let the index be random (i.e. to consider stopped random walks). The following is a complete convergence analog to Theorem HRE.

THEOREM 5.1. *Let* $\{X_k, k \geq 1\}$ *be a sequence of i.i.d. random variables with mean 0 and finite variance and let* $\{S_n, n \geq 1\}$ *be their partial sums. Further, suppose that* $\{N_n, n \geq 1\}$ *is a sequence of positive integer valued random variables, such that*

$$\sum_{n=1}^{\infty} P(|N_n - nN| > n\delta) < \infty \quad \textit{for some} \quad \delta > 0,$$

where N is a positive random variable.

a) *If* $P(N \geq A) = 1$ *for some* $A > \delta > 0$, *then*

$$\sum_{n=1}^{\infty} P(|S_{N_n}| > N_n\varepsilon) < \infty \quad \textit{for all} \quad \varepsilon > 0.$$

b) *If* $P(N \le B) = 1$ *for some* $B > 0$, *then*

$$\sum_{n=1}^{\infty} P(|S_{N_n}| > n\varepsilon) < \infty \quad \text{for all} \quad \varepsilon > 0. \qquad \square$$

To prove a) one first observes that the assumptions imply that

$$P(|S_{N_n}| > N_n\varepsilon) \le P(\sup_{k \ge n(A-\delta)} |S_k/k| > \varepsilon) + P(|N_n - nN| > n\delta).$$

This and Theorem BK d) prove that part. For part b) one argues similarly, but exploits Theorem BK c).

A first theorem on complete convergence was given in Szynal (1972). The above is part of a more general result, which also includes convergence rates (also with respect to the Marcinkiewicz law); recall Theorem BK. For details, see Gut (1983). Łagodowski and Rychlik (1985) provide results for randomly indexed partial sums with multidimensional indices.

6. Subsequences

Let, again, $\{X_k,\ k \ge 1\}$ be a sequence of i.i.d. random variables with partial sums $\{S_n,\ n \ge 1\}$. In this section we investigate the usual HRE-sum, *however*, with the summation extending over a subset of the positive integers only.

As a first example, consider $\sum_{k=1}^{\infty} P(|S_{2^k}| > 2^k\varepsilon)$ for an arbitrary $\varepsilon > 0$ under the sole assumption that the mean of the summands exists and equals 0. The strong law then, obviously, yields

$$\frac{S_{2^k}}{2^k} \xrightarrow{a.s.} 0 \quad \text{as} \quad k \to \infty,$$

from which it follows that

$$\left|\frac{S_{2^{k+1}} - S_{2^k}}{2^k}\right| \le 2\left|\frac{S_{2^{k+1}}}{2^{k+1}}\right| + \left|\frac{S_{2^k}}{2^k}\right| \xrightarrow{a.s.} 0 \quad \text{as} \quad k \to \infty.$$

Moreover, since the random variables $\{S_{2^{k+1}} - S_{2^k},\ k \ge 1\}$ are independent it follows from the (second) Borel-Cantelli lemma that

$$\sum_{k=1}^{\infty} P(|S_{2^{k+1}} - S_{2^k}| > 2^k\varepsilon) < \infty \quad \text{for all} \quad \varepsilon > 0.$$

Finally, since $S_{2^{k+1}} - S_{2^k} \overset{d}{=} S_{2^k}$, we conclude that

$$\sum_{k=1}^{\infty} P(|S_{2^k}| > 2^k\varepsilon) < \infty \quad \text{for all} \quad \varepsilon > 0,$$

hence, under the only proviso that the mean exists (and equals 0). The assumption can also be shown to be necessary. This suggests that for sufficiently sparse subsequences one might be able to weaken the assumption of finite variance, which is necessary in the HRE theorem. The following findings taken from Gut (1985) affirm this suggestion.

Let $\{n_k,\, k \geq 1\}$ be a strictly increasing subsequence of the positive integers with inverse ψ defined by

$$\psi(0) = 0 \quad \text{and} \quad \psi(x) = \text{Card}\{k : n_k \leq x\} \text{ for } x > 0$$

and set

$$M(x) = \sum_{k=1}^{[x]} n_k, \quad x > 0 \quad \text{and} \quad \beta_k = \frac{n_{k+1}}{M(k)}, \quad k \geq 1.$$

LEMMA 6.1. *For any random variable* X,

$$\sum_{k=1}^{\infty} n_k P(|X| > n_k) = E\, M(\psi(|X|)),$$

and if

$$0 < \liminf_{k\to\infty} \beta_k \leq \limsup_{k\to\infty} \beta_k < \infty,$$

then

$$E\, M(\psi(|X|)) < \infty \quad \Longleftrightarrow \quad E|X| < \infty. \qquad \square$$

The proof of the first part follows from the fact that $\{|X| \geq n_k\} = \{\psi(|X|) \geq k\}$ and partial summation.

THEOREM 6.1. *Let* $\{X_k,\, k \geq 1\}$ *be a sequence of i.i.d. random variables with partial sums* $\{S_n,\, n \geq 1\}$.

a) *Suppose that*

$$\liminf_{k\to\infty} \beta_k > 0.$$

If $E|X| < \infty$ *and* $E\, X = 0$, *then*

$$\sum_{k=1}^{\infty} P(|S_{n_k}| > n_k \varepsilon) < \infty \quad \textit{for all} \quad \varepsilon > 0.$$

If the sum is finite for all $\varepsilon > 0$, *then* $E\, M(\psi(|X|)) < \infty$.

b) *If, in addition,*

$$\limsup_{k\to\infty} \beta_k < \infty,$$

then the sum is finite for all $\varepsilon > 0$ *iff* $E|X| < \infty$ *and* $E\, X = 0$. $\square$

Typical examples are subsequences which increase at least geometrically.

The following is a companion result for subsequences which increase less rapidly.

THEOREM 6.2. *Let* $\{X_k,\, k \geq 1\}$ *be a sequence of i.i.d. random variables with partial sums* $\{S_n,\, n \geq 1\}$ *and suppose that* $x^{-r} M(\psi(x)) \to \infty$ *as* $x \to \infty$ *for some* $r > 1$. *Then*

$$\sum_{k=1}^{\infty} P(|S_{n_k}| > n_k \varepsilon) < \infty \quad \textit{for all} \quad \varepsilon > 0$$

iff $E\, M(\psi(|X|)) < \infty$ *and* $E\, X = 0$. $\square$

For $n_k = k^2$, for example, the moment assumption is $E|X|^{3/2} < \infty$. For $n_k = k$ one rediscovers Theorem HRE (of course).

The proof of Theorem 6.1 is similar to the derivation of the special case above. The proof of Theorem 6.2 is, again, based on the KHJ and the Marcinkiewicz-Zygmund inequalities.

One earlier and one more recent reference in the area are Asmussen and Kurtz (1980) and Kuczmaszewska and Szynal (1988), respectively.

7. Arrays

Another way to generalize the HRE theorem is to consider arrays of random variables. In the following we consider an array $\{(X_{ni},\ 1 \le i \le n),\ n \ge 1\}$ of rowwise independent random variables and let $S_n = \sum_{i=1}^n X_{ni}$, $n \ge 1$, denote the row sums. Thus, no independence *between* the rows is assumed. We also assume (without loss of generality) that $E\,X_{ni} = 0$ whenever $E|X_{ni}| < \infty$, $1 \le i \le n$, $n \ge 1$. Following is the main result of Hu, Móricz and Taylor (1989) – note that the summands in each row are independent, but not necessarily i.i.d. random variables.

THEOREM HMT. *Let $\{(X_{ni},\ 1 \le i \le n),\ n \ge 1\}$ be an array of rowwise independent random variables, such that $E\,X_{ni} = 0$ for all i and n and set $S_n = \sum_{i=1}^n X_{ni}$, $n \ge 1$. Further, suppose that the random variables are uniformly dominated in the sense that there exists a random variable X, such that*

$$P(|X_{ni}| > x) \le P(|X| > x) \quad \text{for all } x > 0 \text{ and for all } i \text{ and } n.$$

If $E|X|^{2p} < \infty$ for some $p \in [1,2)$, then

$$\sum_{n=1}^{\infty} P(|S_n| > n^{1/p}\varepsilon) < \infty \quad \text{for all} \quad \varepsilon > 0. \qquad \square$$

The authors also mention that the moment assumption is sufficient for more general arrays $\{(X_{ni},\ 1 \le i \le k_n),\ n \ge 1\}$, where $k_n \to \infty$ as $n \to \infty$.

The proof is very long and involved. With the previous sections in mind it is natural to suggest that

(i) it should be possible (worth a try) to give another proof via the alternative method
(ii) if the length of the rows grows (very) rapidly one might, in fact, sharpen the result in that less integrability might be required. That this is indeed the case is shown next.

THEOREM 7.1. *Let $\{(X_{ni},\ 1 \le i \le k_n),\ n \ge 1\}$ be an array, subject to the condition*

$$\liminf_{n\to\infty} \frac{k_{n+1}}{\sum_{k=1}^n k_i} > 0,$$

of rowwise independent random variables with mean 0 whenever the mean exists, finite, and set $S_{k_n} = \sum_{i=1}^{k_n} X_{ni}$, $n \ge 1$. Further, suppose that the random variables are mean dominated in the sense that there exists a random variable X, such that, for some $\gamma > 0$,

$$\frac{1}{k_n}\sum_{i=1}^{k_n} P(|X_{ni}| > x) \le \gamma P(|X| > x) \quad \text{for all} \quad x > 0 \quad \text{and all} \quad n.$$

If $E|X|^p < \infty$ *for some* $p \in (0,2)$, *then*

$$\sum_{n=1}^{\infty} P(|S_{k_n}| > k_n^{1/p}\varepsilon) < \infty \quad \text{for all} \quad \varepsilon > 0. \qquad \square$$

The proof of this result does not follow the pattern of the proof of Theorem 6.1 (where the i.i.d. assumption was crucial), rather the "standard" pattern of proving the classical law of large numbers. Note also that, in addition to a weakening of the moment assumption in Theorem HMT, the domination condition is weaker.

By using the alternative method described in Section 2 the way it can be applied to prove Theorem 6.2 above it is possible to prove an analog of that result for arrays. In particular, an HMT-theorem holds under the (weaker) mean domination assumption and for $p \in (0,2)$. We leave the details, referring the reader to Gut (1992), Section 4.

Results of this kind have been used in branching processes. Two references in this direction are Athreya and Kaplan (1978) and Nerman (1981).

As immediate corollaries we have corresponding results for *sequences* of independent random variables. In particular, generalizations of (the sufficiency parts of) the results in Section 6 can be obtained.

One application of the results in this section is to weighted sums of i.i.d. random variables, where one considers a *sequence* $\{Y_i, i \geq 1\}$ of i.i.d. random variables and an *array* $\{(a_{ni}, 1 \leq i \leq n), n \geq 1\}$ of weights. The relevant array $\{(X_{ni}, 1 \leq i \leq n), n \geq 1\}$ is obtained by identifying $a_{ni}Y_i$ with X_{ni}. For more on this, see Gut (1992), Section 7, where also further references can be found.

We close this section with a remark on the relation between complete convergence and a.s. convergence. As mentioned before, the former always implies the latter.

Suppose that $\{(X_{ni}, 1 \leq i \leq n), n \geq 1\}$ is an i.i.d. array, i.e. that *all* X'es are i.i.d. random variables. Further, suppose that the strong law holds; $\frac{S_n}{n} \xrightarrow{a.s.} 0$, say, as $n \to \infty$. Thus,

$$P(|S_n| > n\varepsilon \text{ i.o.}) = 0 \quad \forall \varepsilon > 0.$$

From this (one can show that) it follows that

$$P(|X_{ni}| > n\varepsilon \text{ i.o.}) = 0 \quad \forall \varepsilon > 0,$$

which, by independence and the Borel-Cantelli lemma, is equivalent to

$$\sum_{n=1}^{\infty} \sum_{i=1}^{n} P(|X_{ni}| > n\varepsilon) < \infty \quad (\forall \varepsilon > 0),$$

which, due to stationarity, is the same as

$$\sum_{n=1}^{\infty} nP(|X_{11}| > n\varepsilon) < \infty \quad (\forall \varepsilon > 0),$$

which, in turn, is equivalent to $E|X_{11}|^2 < \infty$. From the sufficiency part it now follows that complete convergence holds.

For i.i.d. arrays complete convergence and a.s. convergence thus are equivalent. Recall, however, that this was not the case for sequences.

8. Cesàro summation

Summation methods for divergent series of real numbers have also been studied in the context of (i.i.d.) random variables. In this section we present a result on complete convergence related to Cesàro summation.

Let $\{x_n,\, n \geq 0\}$ be a sequence of real numbers and set, for $\alpha > -1$,

$$A_n^\alpha = \frac{(\alpha+1)(\alpha+2)\cdots(\alpha+n)}{n!}, \quad n = 1,2,\ldots \quad \text{and} \quad A_0^\alpha = 1$$

(and $A_n^{-1} = 0$ for $n \geq 1$ and $A_0^{-1} = 1$). The sequence $\{x_n,\, n \geq 0\}$ is (C,α)-*summable* iff $\frac{1}{A_n^\alpha}\sum_{k=0}^{n} A_{n-k}^{\alpha-1} x_k$ converges as $n \to \infty$; see e.g. Hardy (1949) and Zygmund (1968).

We now replace the sequence of reals by a sequence $\{X_k,\, k \geq 0\}$ of i.i.d. random variables with the aim of studying convergence properties of

$$\frac{1}{A_n^\alpha}\sum_{k=0}^{n} A_{n-k}^{\alpha-1} X_k \quad \text{as} \quad n \to \infty$$

when $0 < \alpha \leq 1$. A first natural result is a strong law, which states that the sequence $\{X_k,\, k \geq 0\}$ is a.s. (C,α)-summable iff $E|X|^{1/\alpha} < \infty$ (for $\alpha = 1$ this reduces to the ordinary strong law of large numbers). The proof of this result was completed only in Déniel and Derrienic (1988), where it was shown how to handle the remaining case, $\alpha = 1/2$. In their paper they also provide proofs for $0 < \alpha < \frac{1}{2}$ and $\frac{1}{2} < \alpha < 1$, respectively, and note that in the former case they actually prove complete convergence. The following result provides necessary and sufficient conditions for all $\alpha \in (0,1]$, (where, for $\alpha = 1$, the theorem reduces to the HRE theorem).

THEOREM 8.1. *For the sequence* $\{(A_n^\alpha)^{-1}\sum_{k=0}^{n} A_{n-k}^{\alpha-1} X_k,\, n \geq 0\}$ *to converge completely to* μ *under the above setup, it is necessary and sufficient that* $E|X_1|^{1/\alpha} < \infty$ *for* $0 < \alpha < \frac{1}{2}$, $E\,X_1^2 \log^+ |X_1| < \infty$ *for* $\alpha = \frac{1}{2}$ *and* $E\,X_1^2 < \infty$ *for* $\frac{1}{2} < \alpha \leq 1$ *and that* $EX_1 = \mu$. □

The proof follows the usual pattern. For details, see Gut (1993), where also a convergence rate result can be found.

Let us also mention that there are connections with results for weighted sums of i.i.d. random variables (and, thus, also with arrays). One point, however, is that here we have rather precise information about the asymptotics of the weights, namely, $A_n^\alpha \sim \frac{n^\alpha}{\Gamma(\alpha+1)}$ as $n \to \infty$, where $\sim$ denotes that the limit as $n \to \infty$ of the ratio between the members on either side equals 1.

9. Further extensions

Several further extensions are of course possible. One is to consider vector valued random variables. Fazekas (1992) investigates the problems of Section 7 in the case of Banach space valued random variables.

Another generalization is to drop the i.i.d. assumption. One reference in this direction is Plachky (1988), where exchangeable random variables are considered. Yu (1990) considers weighted sums of martingale differences.

A further variant on the theme is Spătaru (1990).

One can also study the same kind of problems with respect to the law of the iterated logarithm. Here, however, it turns out that the corresponding "HRE-sum" diverges for all ε, see Slivka (1969). A positive result can only be obtained for $\sum_{n=1}^{\infty} n^{-1} P(|S_n| > \varepsilon\sqrt{n \log\log n})$. We refer to Davis (1968), to Gut (1980), where also the multidimensional index case and analogs of the counting variables introduced at the end of Section 3 are studied and where further references can be found, to Gut (1983), where the random index case is investigated and to Gut (1986), where subsequences are considered.

References

[1] ASMUSSEN, S. AND KURTZ, T.G. (1980). Necessary and sufficient conditions for complete convergence in the law of large numbers. *Ann. Probab.* **8** 176–182.

[2] ATHREYA, K.B. AND KAPLAN, N. (1978). Additive property and its applications in branching processes. In: *Advances in Probability and Related Topics* (A. Joffe and P. Ney eds.) Marcel Dekker, New York

[3] BAUM, L.E. AND KATZ, M. (1965). Convergence rates in the law of large numbers. *Trans. Amer. Math. Soc.* **120** 108–123.

[4] DAVIS, J.A. (1968). Convergence rates for the law of the iterated logaritm. *Ann. Math. Statist.* **39** 1479–1485.

[5] DÉNIEL, Y. AND DERRIENNIC, Y. (1988). Sur la convergence presque sure, au sens de Cesàro d'ordre α, $0 < \alpha < 1$, de variables aléatoires indépendantes et identiquement distribuées. *Probab. Th. Rel. Fields* **79** 629–636.

[6] ERDŐS, P. (1949). On a theorem of Hsu and Robbins. *Ann. Math. Statist.* **20** 286–291.

[7] ERDŐS, P. (1950). Remark on my paper "On a theorem of Hsu and Robbins". *Ann. Math. Statist.* **21** 138.

[8] FAZEKAS, I. (1992). Convergence rates in the law of large numbers for arrays. *Publ. Math. Debrecen* **41** 53–73.

[9] GUT, A. (1978). Marcinkiewicz laws and convergence rates in the law of large numbers for random variables with multidimensional indices. *Ann. Probab.* **6** 469–482.

[10] GUT, A. (1980). Convergence rates for probabilities of moderate deviations for sums of random variables with multidimensional indices. *Ann. Probab.* **8** 298–313.

[11] GUT, A. (1983). Complete convergence and convergence rates for randomly indexed partial sums with an application to some first passage times. *Acta Math. Acad. Sci. Hungar.* **42** 225–232; Correction, *ibid.* **45** (1985) 235–236.

[12] GUT, A. (1985). On complete convergence in the law of large numbers for subsequences. *Ann. Probab.* **13** 1286–1291.

[13] GUT, A. (1986). The law of the iterated logarithm for subsequences. *Probab. Math. Statist.* **VII** 27–58.

[14] GUT, A. (1992). Complete convergence for arrays. *Period. Math. Hungar.* **25** 51–75.

[15] GUT, A. (1993). Complete convergence and Cesàro summation for i.i.d. random variables. *Probab. Th. Rel. Fields* **97** 169–178.

[16] HARDY, G.H. (1949). *Divergent series.* Oxford University Press.

[17] HOFFMANN-JØRGENSEN, J. (1974). Sums of independent Banach space valued random variables. *Studia Math.* **LII** 159–186.

[18] HSU, P.L. AND ROBBINS, H. (1974). Complete convergence and the law of large numbers. *Proc. Nat. Acad. Sci. USA* **33** 25–31.

[19] HU, T.-C., MÓRICZ, F. AND TAYLOR, R.L. (1989). Strong laws of large numbers for arrays of rowwise independent random variables. *Acta Math. Acad. Sci. Hungar.* **54** 153–162.

[20] JAIN, N.C. (1975). Tail probabilities for sums of independent Banach space valued random variables. *Z. Wahrscheinlichkeitstheorie verw. Gebiete* **33** 155–166.

[21] KATZ, M. (1963). The probability in the tail of a distribution. *Ann. Math. Statist.* **34** 312–318.

[22] KUCZMASZEWSKA, A. AND SZYNAL, D. (1988). On the Hsu-Robbins law of large numbers for subsequences. *Bull. Acad. Sci. Polon. Ser. Math.* **36** 69–79.

[23] ŁAGODOWSKI, Z.A. AND RYCHLIK, Z. (1985). Complete convergence and convergence rates for randomly indexed sums of random variables with multidimensional indices. *Bull. Acad. Sci. Polon. Ser. Math.* **32** 219–223.

[24] MARCINKIEWICZ, J. AND ZYGMUND, A. (1937). Sur les fonctions indépendantes. *Fund. Math.* **29** 60–90.

[25] MARCINKIEWICZ, J. AND ZYGMUND, A. (1938). Quelques théorèmes sur les fonctions indépendantes. *Studia. Math.* **7** 104–120.

[26] NERMAN, O. (1981). On the convergence of supercritical general (C-M-J) branching processes. *Z. Wahrsch. verw. Gebiete* **57** 365–395.

[27] PLACHKY, D. (1988). Complete convergence of the sample mean for exchangeable random variables. *Proc. 5th Pannonian Symp. on Math. Stat.*, (eds. W. Grossmann, J. Mogyoródi, I. Vincze, W. Wertz) 191–197. Reidel.

[28] SLIVKA, J. (1969). On the law of the iterated logarithm. *Proc. Nat. Acad. Sci. USA* **63** 289–291.

[29] SPĂTARU (1990). Strenghtening the Hsu-Robbins-Erdős theorem. *Rev. Roumaine Math. Pures et Appl.* **35** 463–465.

[30] SPITZER, F.L. (1956). A combinatorial lemma and its application. *Trans. Amer. Math. Soc.* **82** 323–339.

[31] SZYNAL, D. (1972). On almost complete convergence for the sum of a random number of independent random variables. *Bull. Acad. Sci. Polon. Ser. Math. Astronom. Phys.* **20** 571–574.

[32] YU, K.F. (1990). Complete convergence of weighted sums of martingale differences. *J. Theoret. Probab.* **3** 339–347.

[33] ZYGMUND, A. (1968). *Trigonometric series.* Cambridge University Press.

TESTS FOR HETEROSCEDASTICITY BASED ON REGRESSION QUANTILES AND REGRESSION RANK SCORES

C. Gutenbrunner
University of Marburg, FRG

We propose a new class of tests for heteroscedasticity in semiparametric linear models with independent errors. The test statistics are a sum of two asymptotically independent parts, one based on regression quantiles, the other on regression ranks. Those components of heteroscedasticity that are correlated to covariates in the model are detected by the regression quantile part of the statistics, whereas components orthogonal to the model are detected by its regression rank part. The proposed tests are a natural application of the concept of regression ranks and regression quantiles. They may be considered as canonical generalizations of rank tests for scale and scale tests based on L-statistics from the two- or p- sample case to a general semiparametric linear model setting. Moreover, the test statistics have desirable invariance properties with respect to the nuisance parameters of the model and do not require symmetry of the error distribution. Their relationship to the likelihood ratio statistic for the corresponding parametric model (with known form of the error distribution) is analogous to that of corresponding rank- and L-statistics to likelihood ratio statistics in the two sample scale testing problem. We also suggest, how generalizations to hypotheses on other departures from i.i.d. errors than classical heteroscedasticity could be obtained.

1. Introduction

Consider the linear model

$$\mathbf{Y}_n = \mathbf{X}_n \beta_n + \mathbf{V}_n \tag{1.1}$$

with independent errors $V_{n1}, \ldots, V_{nn}$. In this paper, the regression parameter β_n is considered as a nuisance parameter, whereas the focus is on the null hypothesis that the errors are i.i.d. against some alternative hypothesis preserving independence, but permitting non-identical distributions of the errors. As we shall see, in the semiparametric context, i.e. when the error density f is not assumed to belong to a parametric class of distributions, regression quantiles and regression ranks provide a natural tool to treat such testing problems. Assuming a semiparametric family of densities $f(\cdot, \tau)$ with $f(\cdot, 0) = f$, for example $f(x, \tau) = \exp(-\tau) f(x \exp(-\tau))$, and $V_{ni} \sim f(\cdot, \tau_{ni})$, we might test

$$H_0 : \tau_n = 0 \tag{1.2}$$

against

$$H_1 : \quad \tau_n' \mathbf{X}_{n2}' \mathbf{X}_{n2} \tau_n + \tau_n' \mathbf{X}_{n3}' \mathbf{X}_{n3} \tau_n > 0, \tag{1.3}$$

where $\mathbf{X}_n = (\mathbf{1}_n | \mathbf{X}_{n1} | \mathbf{X}_{n2})$ and $\mathbf{X}_{n3}$ represents another set of covariates, not contained in $\mathbf{X}_n$.

In the sequel we shall treat only the case $f(x, \tau) = \exp(-\tau) f(x \exp(-\tau))$ in detail. The extension to other models is outlined in section 5.

Example: As an introductory example we shall consider a two sample problem, in which regression quantiles and ranks are simple functions of ordinary quantiles and ranks computed in both groups (samples) separately. We might test then (1.2) against $H_1 : (\bar{\tau}_{n1}-\bar{\tau}_{n2})^2+(\sum_i x_{n3i}\tau_{ni})^2+(\sum_j x_{n3j}\tau_{nj})^2 > 0$, where summation over i and j is summation over the first and second groups respectively and x_{n3i}, x_{n3j} are the values of a covariate which is suspected to be associated with heteroscedasticity within groups, normalized such that $\sum_i x_{n3i} = \sum_j x_{n3j} = 0, \sum_i x_{n3i}^2 = \sum_j x_{n3j}^2 = 1$. A robust test statistic may be constructed using order statistics $Y_{1n:i}, Y_{2n:j-n_1}$ and ranks R_{1ni}, R_{2nj} computed separately in both groups, as

$$(1.4) \qquad T_n = \hat{\sigma}_n(\nu)^{-2}\frac{n_1 n_2}{n_1+n_2}\Big(\sum_i \nu_{ni}Y_{1n:i} - \sum_j \nu_{nj}Y_{2n:j-n_1}\Big)^2 + \Big(\sum_i x_{n3i}b(n_1, R_{1ni})\Big)^2 + \Big(\sum_j x_{n3j}b(n_2, R_{2nj})\Big)^2.$$

Here, ν_{ni}, ν_{nj} are appropriate weights which for example could be generated by a signed measure satisfying (W.1) in section 3, via $\nu_{ni} = \nu[(i-1)/n_1, i/n_1)$, ν_{nj} analogously, so that $\sum_i \nu_{ni} = \sum_j \nu_{nj} = 0$, $\sum_{i=1}^k \nu_{ni} \le 0$ for $k \le n_1$, $\sum_{j=n_1+1}^k \nu_{nj} \le 0$ for $k \le n_2$. The rank scores $b(n,r)$ should be generated by a score function that is u-shaped w.r.t. $\frac{1}{2}$, and satisfy $\sum_i b(n_1,i) = \sum_j b(n_2,j) = 0$, $\sum_i b(n_1,i)^2 = \sum_j b(n_2,j)^2 = 1$. $\hat{\sigma}_n^2(\nu)$ estimates the ν-dependent component of the variance of the L-statistic (linear combination of order statistics) and may be defined as in (3.3). Under the conditions of the theorem in section 3, T_n is asymptotically χ^2 with 3 degrees of freedom. The crucial fact we want to demonstrate with this example is that T_n avoids pre-alignment of observations by capturing heteroscedasticity between groups via order statistics and heteroscedasticity within groups via rank scores. The same principle will be used in section 3 treating the general case. □

Turning back to heteroscedasticity in general linear models, we shall for convenience assume orthogonality between the different parts of the design matrix,

$$(1.5) \qquad \mathbf{1}_n \perp \mathbf{X}_{n1} \perp \mathbf{X}_{n2} \perp \mathbf{X}_{n3},$$

denoting the column dimension of $\mathbf{X}_{n1}, \mathbf{X}_{n2}, \mathbf{X}_{n3}$ by p,q and r respectively (p, q or r may be zero i.e. not all 3 components must really be present). By definition $\mathbf{X}_{n1}$ contains the covariates influencing only location, $\mathbf{X}_{n2}$ those influencing location and scale and $\mathbf{X}_{n3}$ those influencing only scale. The corresponding partition of the $1+p+q$-dimensional regression parameter is

$$(1.6) \qquad \beta_n = (\beta_{n0} /\!/ \gamma_{n1} /\!/ \gamma_{n2}),$$

where $/\!/$ denotes vertical concatenation of vectors or matrices.
Introducing the scale factors $\sigma_{ni} = \exp(\tau_{ni})$, $\mathbf{\Sigma}_n = \mathrm{diag}(\sigma_{ni})$, $U_{ni} = \sigma_{ni}^{-1}V_{ni}$ (U_{ni} i.i.d. $\sim f$ also under the alternatives), we arrive at the explicit model equation

$$(1.7) \qquad \mathbf{Y}_n = \mathbf{X}_n\beta_n + \mathbf{\Sigma}_n\mathbf{U}_n$$

$$(1.8) \qquad = \beta_{n0}\mathbf{1}_n + \mathbf{X}_{n1}\gamma_{n1} + \mathbf{X}_{n2}\gamma_{n2} + \mathbf{\Sigma}_n\mathbf{U}_n.$$

2. Short review of regression quantiles and regression rank scores

According to Koenker and Bassett (1978) the (unweighted) u-th *regression quantile* (RQ) $\widehat{\beta}_n(u)$ is defined as the solution of the linear program

$$
(2.1) \qquad \begin{aligned} & u\mathbf{1}_n^{'}\mathbf{r}^+ + (1-u)\mathbf{1}_n^{'}\mathbf{r}^- = \min \\ & \mathbf{X}_n\widehat{\beta} + \mathbf{r}^+ - \mathbf{r}^- = \mathbf{Y}_n \\ & (\widehat{\beta}, \mathbf{r}^+, \mathbf{r}^-) \in \mathbb{R}^p \times \mathbb{R}_+^n \times \mathbb{R}_+^n \end{aligned}
$$

whereas the centered *regression rank scores* (RR)$\widehat{\mathbf{\Delta}}_n(u) \in \mathbb{R}^n$ introduced in Gutenbrunner and Jurečková (1992) are the solutions of the corresponding dual program

$$
(2.2) \qquad \begin{aligned} & \mathbf{Y}_n^{'}\widehat{\mathbf{\Delta}} = \max \\ & \mathbf{X}_n^{'}\widehat{\mathbf{\Delta}} = \mathbf{0} \\ & \widehat{\mathbf{\Delta}} \in [u-1, u]^n. \end{aligned}
$$

The functions $u \mapsto \widehat{\beta}_n(u)$ resp. $u \mapsto \widehat{\mathbf{\Delta}}_n(u)$ are piecewise constant resp. piecewise linear, with only a finite number of jumps resp. corners. A fast algorithm to compute both functions at the same time is given in Koenker and d'Orey (1987,1990). The properly standardized functions, called *regression quantile process* and *regression rank score process*, have similar properties as the ordinary quantile process and Hájek's rank score process respectively (see e.g. Gutenbrunner and Jurečková (1992)). In accordance with the notation $\beta = (\beta_0 /\!\!/ \gamma_1 /\!\!/ \gamma_2)$, we shall decompose the regression quantile vector as $\widehat{\beta}_n = (\widehat{\beta}_{n0} /\!\!/ \widehat{\gamma}_{n1} /\!\!/ \widehat{\gamma}_{n2})$.

The test statistics proposed in section 3 are made up from weighted averages of the regression quantile and regression rank score processes, like

$$
(2.3) \qquad \mathbf{L}_n = \int \widehat{\gamma}_{n2} d\nu \quad \text{(L-statistic)}
$$

and

$$
(2.4) \qquad \mathbf{R}_n = \mathbf{X}_{n3}^{'} \int \widehat{\mathbf{\Delta}}_n db \quad \text{(regression rank statistic).}
$$

Note, that the integrals in (2.3) and (2.4) actually are finite sums and may be computed very fast, because of the jump character of the regression quantile and rank processes (see section 4). As in Gutenbrunner and Jurečková (1992), we assume here that the finite signed measure ν and the totally bounded measure generating function b satisfy

(W) ν and the signed measure genenerated by b have compact support in $(0,1)$, $\int b(u)du = 0$, $\int b(u)^2 du = 1$.

In Gutenbrunner et al. (1993), regression rank statistics with unbounded b were investigated, but under more restrictive assumptions on f than are assumed in the present paper.

In order to prove asymptotic results we shall make the further assumptions

(F.1) $I_1(f) = \int (t[f'(t)/f(t)] + 1)^2 \, dF(t) < \infty$ (finite Fisher information for scale)

(F.2) $f > 0$ on $\{0 < F < 1\}$

(F.3) $F(0) = \frac{1}{2}$ (w.l.o.g.)

(X.1) $\mathbf{X}'_{nl}\mathbf{1}_n = 0, \mathbf{X}'_{nl}\mathbf{X}_{nm} = 0, \quad l, m = 1, 2, 3, \quad l \neq m$ (w.l.o.g.)

(X.2) $\|\mathbf{X}_{nl}\|_\infty \to 0$ as $n \to \infty$

(X.3) $\mathbf{X}'_{nl}\mathbf{X}_{nl} \to \mathbf{Q}_l$ positive definite, $l = 1, 2, 3$

Either

(T.1) $\mathbf{X}'_{nl}\tau_n \to \eta_l$, $l = 1, 2, 3$ and

(T.2) $\|\tau_n\|_\infty = o(1), \quad \|\tau_n\|_2 = O(1)$ as $n \to \infty$

or

($\tilde{\text{T}}$.1) $\tau_n = \mathbf{X}_{n2}\delta_2 + \mathbf{X}_{n3}\delta_3$ for some $\delta_2 \in I\!R^q$ and $\delta_3 \in I\!R^r$.

It should be stressed, that we do not require symmetry of f. Procedures based on pre-alignment of observations need symmetry of f at least if they can detect the within-model part (correlated to $\mathbf{X}_{n2}$) of heteroscedasticity.

The standardisation (X.3) of the design matrices differs from the one used in other papers on regression quantiles, like e.g. Gutenbrunner and Jurečková (1992) and Koenker and Bassett (1978) by a factor $n^{-1/2}$, implying that in the present paper the slope components of regression parameters and hence also regression quantiles differ by the inverse factor $n^{1/2}$ from corresponding quantities in the cited papers. Our present standardisation is useful when dealing with asymptotic tests and local (contiguous) alternatives, but would be confusing when dealing mainly with estimation problems.

From Gutenbrunner and Jurečková (1992) under $H_0 : \tau_n = 0$ we have representations

(2.5) $$\mathbf{L}_n = (\mathbf{X}'_{n2}\mathbf{X}_{n2})^{-1}\mathbf{X}'_{n2}\mathbf{\Psi}_{\nu,F}(\mathbf{U}_n) + o_p(1) \underset{\mathcal{D}}{\to} N(0, \sigma^2(\nu, F)\mathbf{Q}_2^{-1}),$$

(2.6) $$\mathbf{R}_n = \mathbf{X}'_{n3}\mathbf{b}\circ\mathbf{F}(\mathbf{U}_n) + o_p(1) \underset{\mathcal{D}}{\to} N(0, \mathbf{Q}_3),$$

where $\mathbf{\Psi}_{\nu,F}(\mathbf{U}_n) = (\psi_{\nu,F}(U_{n1}), \ldots, \psi_{\nu,F}(U_{nn}))'$, $\psi_{\nu,F}$ is the well known influence function of the functional

(2.7) $$F \mapsto \int F^{-1} d\nu = \mu(\nu, F),$$

i.e.

(2.8) $$\psi_{\nu,F}(t) = \int \frac{u - I[F(t) \leq u]}{f\circ F^{-1}(u)} d\nu(u),$$

(2.9) $$\sigma^2(\nu, F) = \iint \frac{u \wedge v - uv}{f\circ F^{-1}(u)\, f\circ F^{-1}(v)} d\nu(u) d\nu(v) = \mathrm{Var}[\psi_{\nu,F}(U_{n1})]$$

and where $\mathbf{b}\circ\mathbf{F}(\mathbf{t}) = (b\circ F(t_1), \ldots, b\circ F(t_n))'$.

3. Tests for Heteroscedasticity

For testing (1.2) against (1.3), we propose test statistics

(3.1)
$$T_n = \widehat{\sigma}_n(\nu)^{-2}\mathbf{L}_n'\mathbf{X}_{n2}'\mathbf{X}_{n2}\mathbf{L}_n + \mathbf{R}_n'(\mathbf{X}_{n3}'\mathbf{X}_{n3})^{-1}\mathbf{R}_n$$

and

(3.2)
$$S_n = \breve{\sigma}_n(\nu)^{-2}(\int \breve{\gamma}_{n2}d\nu)'\mathbf{X}_{n2}'\mathbf{X}_{n2}(\int \breve{\gamma}_{n2}d\nu) + \breve{\sigma}_n(\nu_2)^{-2}(\int \breve{\gamma}_{n3}d\nu_2)'\mathbf{X}_{n3}'\mathbf{X}_{n3}(\int \breve{\gamma}_{n3}d\nu_2).$$

T_n contains $\mathbf{L}_n$ and $\mathbf{R}_n$ from (2.3) and (2.4) (computed w.r.t. model (1.8)) and an estimate $\widehat{\sigma}_n^2(\nu)$ of $\sigma^2(\nu, F)$, which is defined as

(3.3)
$$\widehat{\sigma}_n^2(\nu) = -\int_0^1\int_0^v (\widehat{\beta}_{n0}(u) - \widehat{\beta}_{n0}(v))^2 \, d\tilde{J}(u)dJ(v) - (\int \widehat{\beta}_{n0}(u)\, d\tilde{J}(u))^2,$$

if ν has a "signed density" J of bounded variation w.r.t. Lebesgue measure and where we set $\tilde{J}(u) = uJ(u)$. If ν has no such density, we define

(3.4)
$$\widehat{\sigma}_n^2(\nu) = -\int\int(u \wedge v - uv)\hat{q}_n(u)\hat{q}_n(v)\, d\nu(u)d\nu(v),$$

where $\hat{q}_n(u) = (2s_n)^{-1}(\widehat{\beta}_{n0}(u + s_n) - \widehat{\beta}_{n0}(u - s_n)) \underset{p}{\rightarrow} (f \circ F^{-1})^{-1}$ uniformly on compact sets if $s_n^{-1} = o(n^{1/2})$. The reason for this distinction of cases is that it may be shown that (3.3) converges to $\sigma^2(\nu, F)$ with rate $O_p(n^{-1/2})$, while for (3.4) the rate is lower, and, still more important, the bandwidth s_n introduces some arbitrariness to the test statistic. Thus, we recommend use of ν with signed density J as mentioned above, and $\widehat{\sigma}_n^2(\nu)$ according to (3.3). The estimator given in (3.3) is closely related to the first part of an estimator given in Welsh (1987) (v_n on page 631 bottom). Indeed, if in that formula G_n would be replaced by the Bassett-Koenker regression quantile based empirical cdf $\hat{F}_n$ ((3.3) in Bassett and Koenker (1982)), both estimators would coincide. Our representation (3.3) avoids the inversion of $\widehat{\beta}_{n0}(\cdot)$ and is particularily useful for J of bounded variation.

In S_n, as is indicated by the cups $\breve{}$ instead of hats $\widehat{}$, we compute regression quantiles w.r.t. the larger model obtained from (1.8) by adding a term $\mathbf{X}_{n3}\gamma_{n3}$.

To make (3.1) and (3.2) useful, the weighting signed measures ν, ν_2 and the function b can not be arbitrary of course. ν should have property

(W.1)
$$\nu(0,1) = 0, \quad \nu \neq 0, \quad \nu(0,u] \leq 0 \ \forall u \in (0,1).$$

i.e. ν has total charge zero and has a negative part ν^- that is stochastically smaller than its positive part ν^+. (W.1) yields location (regression) invariance of T_n and S_n and guarantees $\mu(\nu, F) > 0$ for all f satisfying (F.2).
ν_2 must not have total charge zero, but if it has not, it should satisfy

(W.2)
$$\nu^-(1/2, 1) = \nu^+(0, 1/2) = 0,$$

i.e. its measure generating function should be u-shaped w.r.t. 1/2. (W.2) guarantees $\mu(\nu, F) > 0$ for all f satisfying (F.2+3). Note, that under the side condition $\nu(0,1) = 0$, (W.2) is more restrictive than (W.1). In correspondence to (W.2), b should satisfy

(W.3) b u-shaped w.r.t. 1/2,

implying for all f with (F.2+3)

(3.5)
$$\mu_R(b, F) = \int F^{-1} f o F^{-1} db > 0.$$

Examples for useful ν (given by their density J) and b are

(1) $$J(u) = \Phi^{-1}\left(\frac{u-\epsilon}{1-2\epsilon}\right) I\left[\Phi^{-1}\left(\frac{u-\epsilon}{1-2\epsilon}\right)^2 \leq k^2\right],$$

(2) $$b(u) = \Phi^{-1}\left(\frac{u-\epsilon}{1-2\epsilon}\right)^2 \wedge k^2 \quad \text{(up to standardisation)}$$

and

(3) $$J(u) = (u-0.5)I[\epsilon \leq u < 1-\epsilon],$$

(4) $$b(u) = (u-0.5)^2 \wedge (\epsilon - 0.5)^2 \quad \text{(up to standardisation)},$$

where Φ denotes the normal cdf and $k > 0$, $0 < \epsilon < 0.5$ are constants. (1) and (2) correspond to the ψ_f^* (3.7) generated by the distribution minimizing Fisher information for scale in the contamination neighbourhood of the normal distribution, cf. Huber (1981) pp. 118-122. (3) and (4) were choosen because of their simplicity.

Usually nonparametric statistics are asymptotically equivalent to $2\log L_n^*$, where L_n^* is the likelihood ratio statistic for a corresponding parametric problem with known f and nuisance parameters. This is true also for T_n and S_n, as we shall see from their asymptotic representation. In fact, the relation is exactly as in the case of two-sample scale tests based on ordinary L-statistics and rank statistics. In our case the likelihood ratio statistic is given by

(3.6) $$\begin{aligned} 2logL_n^* = I_1(f)^{-1}\big[&\mathbf{\Psi}_f^*(\mathbf{Y}_n - \mathbf{X}_n\beta_n)'\mathbf{X}_{n2}(\mathbf{X}_{n2}'\mathbf{X}_{n2})^{-1}\mathbf{X}_{n2}'\mathbf{\Psi}_f^*(\mathbf{Y}_n - \mathbf{X}_n\beta_n) \\ &+ \mathbf{\Psi}_f^*(\mathbf{Y}_n - \mathbf{X}_n\beta_n)'\mathbf{X}_{n3}(\mathbf{X}_{n3}'\mathbf{X}_{n3})^{-1}\mathbf{X}_{n3}'\mathbf{\Psi}_f^*(\mathbf{Y}_n - \mathbf{X}_n\beta_n)\big] + o_p(1) \end{aligned}$$

where

(3.7) $$\psi_f^*(t) = -t[f'(t)/f(t)] - 1, \quad \mathbf{\Psi}(\mathbf{t}) = (\psi(t_1), \ldots, \psi(t_n))'.$$

A different type of tests for heteroscedasticity also based on regression quantiles was proposed by Koenker and Bassett (1982). These tests are designed for another type of heteroscedasticity: For a fixed set of numbers $0 < u_1 < \ldots < u_m < 1$, parallelism of the corresponding regression quantiles is tested against all possible deviations from it, leading to statistics asymptotically χ^2-distributed with a number of degrees of freedom depending on the number m of considered regression quantiles. This different approach was generalized to infinitely many regression quantiles by Portnoy (1992).

Theorem: Assume that (F.1-2), (X.1-3), (W) and alternatively (T.1-2) or ($\tilde{\text{T}}$.1) are true and that (W.1) holds for ν, (W.3) for b and either (W.1) or (W.2) holds for ν_2. Then we have

(a) $T_n \underset{\mathcal{D}}{\to} \chi^2_{q+r}\left(\frac{\mu^2(\nu,F)}{\sigma^2(\nu,F)}\eta_2'\mathbf{Q}_2^{-1}\eta_2 + \mu_R^2(b,F)\eta_3'\mathbf{Q}_3^{-1}\eta_3\right)$

(b) $S_n \underset{\mathcal{D}}{\to} \chi^2_{q+r}\left(\frac{\mu^2(\nu,F)}{\sigma^2(\nu,F)}\eta_2'\mathbf{Q}_2^{-1}\eta_2 + \frac{\mu^2(\nu_2,F)}{\sigma^2(\nu_2,F)}\eta_3'\mathbf{Q}_3^{-1}\eta_3\right)$,

$\mu(\nu,F)$,$\sigma^2(\nu,F)$ and $\mu_R(b,F)$ according to (2.7),(2.8) and (3.5) are positive, and in case of ($\tilde{\text{T}}$.1), we have $\eta_l = \mathbf{Q}_l\delta_l, l = 2,3$.

(c) $S_n - T_n = o_p(1)$ iff $d\nu_2/db = f\circ F^{-1}$.

(d) $S_n = T_n + o_p(1) = 2logL_n^* + o_p(1)$ if $d\nu/d\lambda = \psi_f^{*'}oF^{-1}$, $b = \psi_f^* oF^{-1}$ and $\nu_2 = \nu$ (L_n^* acc. to (3.6), λ Lebesque measure).

(e) T_n and S_n are invariant, $T_n(\sigma\mathbf{Y}_n + \mathbf{X}_n\beta) = T_n(\mathbf{Y}_n)$, S_n analogously, for all $\sigma > 0$, $\beta \in \mathbb{R}^{1+p+q}$.

Remark: If $\psi(t) = b\circ F(t)$, then $\mu_R^2(b,F) = E[U\psi'(U)]/\text{Var}[\psi(U)]$, if $\psi(t) = \psi_{\nu,F}(t)$, then $\mu^2(\nu,F)/\sigma^2(\nu,F) = E[U\psi'(U)]/\text{Var}[\psi(U)]$ and if $d\nu_2/db = f\circ F^{-1}$, then $\mu_R^2(b,F) = \mu^2(\nu,F)/\sigma^2(\nu,F)$. Hence the asymptotic relative efficiency relations between T_n, S_n and M-statistic based procedures are analogous to those between one dimensional scale tests based on rank statistics, L-statistics and M-statistics.

Proof of the theorem: We shall prove first

$$\widehat{\sigma}_n^2(\nu) = \sigma^2(\nu,F) + O_p(n^{-1/2}) \tag{3.8}$$

in the case where $\widehat{\sigma}_n^2(\nu)$ is defined via (3.3). Note, that

$$\begin{aligned}\sigma^2(\nu,F) &= \iint \frac{u\wedge v - uv}{foF^{-1}(u)\, foF^{-1}(v)}d\nu(u)d\nu(v)\\ &= \iint(u\wedge v - uv)J(u)J(v)dF^{-1}(u)dF^{-1}(v)\\ &= 2\iint_{u\leq v} \tilde{J}(u)J(v)dF^{-1}(u)dF^{-1}(v) - (\textstyle\int \tilde{J}dF^{-1})^2\\ &= -\iint_{s\leq t}(F^{-1}(s) - F^{-1}(t))^2 d\tilde{J}(s)dJ(t) - (\textstyle\int F^{-1}d\tilde{J})^2\end{aligned} \tag{3.9}$$

where we used the two-dimensional partial integration formula

$$\begin{aligned}&\textstyle\iint_{s\leq t}(h(s) - h(t))^2 d\tilde{J}(s)dJ(t)\\ &= 2\textstyle\iiiint_{s\leq u\leq v\leq t} dh(u)dh(v)d\tilde{J}(s)dJ(t)\\ &= -2\textstyle\iint_{u\leq v}\tilde{J}(s)J(t)dh(u)dh(v).\end{aligned} \tag{3.10}$$

Hence, (3.8) follows from the fact that under $H_0 : \tau_n = 0$

(3.11) $\widehat{\beta}_{n0}(u) = F^{-1}(u) + O_p(n^{-1/2})$ uniformly for u in compact subsets of $(0,1)$

which was proved in Gutenbrunner and Jurečková (1992) under the conditions of the theorem, and contiguity. If $\widehat{\sigma}_n^2(\nu)$ is defined via (3.4), $\widehat{\sigma}_n^2(\nu) \underset{p}{\to} \sigma^2(\nu, F)$ because $\hat{q}_n(u) \underset{p}{\to} (f \circ F^{-1})^{-1}$ uniformly on compact subsets of $(0,1)$ if $s_n^{-1} = o(n^{1/2})$.

We next show that

$$(\mathbf{L}_n /\!\!/ \mathbf{R}_n) \underset{\mathcal{D}}{\to} N(\mathbf{a}, \mathbf{C}) \tag{3.12}$$

holds, with mean $\mathbf{a} = (\mu(\nu, F)\mathbf{Q}_2^{-1}\eta_2) /\!\!/ (\mu_R(b, F)\eta_3)$ and with the block diagonal covariance matrix $\mathbf{C} = (\sigma^2(\nu, F)\mathbf{Q}_2^{-1} \mid \mathbf{0}) /\!\!/ (\mathbf{0} \mid \mathbf{Q}_3)$.

(2.5) and (2.6) together with the orthogonality of $\mathbf{X}_{n2}$ and $\mathbf{X}_{n3}$ under H_0 imply (3.12) with $\mathbf{a} = \mathbf{0}$. Because of (F.1), the law of $\mathbf{V}_n = \Sigma_n \mathbf{U}_n$ under alternatives with (T.1-2) or ($\tilde{\mathrm{T}}$) is contiguous to its law under H_0. Hence, under those alternatives, we still have

$$\begin{aligned} \mathbf{L}_n &= (\mathbf{X}_{n2}'\mathbf{X}_{n2})^{-1}\mathbf{X}_{n2}'\mathbf{\Psi}_{\nu,F}(\mathbf{V}_n) + o_p(1) \\ &= (\mathbf{X}_{n2}'\mathbf{X}_{n2})^{-1}\mathbf{X}_{n2}'\mathbf{\Psi}_{\nu,F}(\Sigma_n \mathbf{U}_n) + o_p(1), \end{aligned} \tag{3.13}$$

and

$$\mathbf{R}_n = \mathbf{X}_{n3}' \mathbf{b} \circ \mathbf{F}(\Sigma_n \mathbf{U}_n) + o_p(1), \tag{3.14}$$

and, according to the central limit theorem, it suffices to investigate the limits of expectation and covariances of the right hand sides of (3.13) and (3.14). We shall use the representations

$$E[\psi_{\nu,F}(\sigma_{ni}U_{ni})] = \int \frac{u - F(F^{-1}(u)/\sigma_{ni})}{f \circ F^{-1}(u)} d\nu(u), \tag{3.15}$$

$$E[b \circ F(\sigma_{ni}U_{ni})] = \int [u - F(F^{-1}(u)/\sigma_{ni})] db(u), \tag{3.16}$$

$$\begin{aligned} &\mathrm{Cov}(\psi_{\nu,F}(\sigma_{ni}U_{ni}), b \circ F(\sigma_{ni}U_{ni})) = \\ &= \iint \Big[F\Big(\frac{F^{-1}(u \wedge v)}{\sigma_{ni}}\Big) - F\Big(\frac{F^{-1}(u)}{\sigma_{ni}}\Big) F\Big(\frac{F^{-1}(v)}{\sigma_{ni}}\Big)\Big] \times \\ &\qquad \times [f \circ F^{-1}(u)]^{-1} d\nu(u) db(v) \end{aligned} \tag{3.17}$$

and similar expressions for the variances of $\psi_{\nu,F}(\sigma_{ni}U_{ni})$ and $b \circ F(\sigma_{ni}U_{ni})$) together with the Taylor expansion

$$F(F^{-1}(u)/\sigma_{ni}) = u - [F^{-1}(u) f \circ F^{-1}(u) + g_{ni}(u)]\tau_{ni} \equiv u - h_{ni}(u)\tau_{ni} \tag{3.18}$$

where, because of the continuity of f, $\max_{1 \le i \le n} \sup_{u \in A} |g_{ni}(u)| \to 0$ as $n \to \infty$ for any compact $A \subset (0,1)$ [resp. $h_{ni}(u) \to h(u) \equiv F^{-1}(u) f \circ F^{-1}(u)$ uniformly on compact subsets].

From these representations we get

$$E\psi_{\nu,F}(\sigma_{ni}U_{ni}) = \left[\int F^{-1}d\nu + \int (g_{ni}/f\mathrm{o}F^{-1})d\nu\right]\tau_{ni}$$
$$= \left[\mu(\nu,F) + c_{ni}\right]\tau_{ni} \tag{3.19}$$

with $\|\mathbf{c}_n\|_\infty = \max_{1\le i\le n}|c_{ni}| \to 0$, implying

$$E[\mathbf{X}_{n2}'\mathbf{\Psi}_{\nu,F}(\mathbf{\Sigma}_n\mathbf{U}_n)] = \mu(\nu,F)\mathbf{X}_{n2}'\tau_n$$
$$+ O(\|\mathbf{c}_n\|_\infty\|\tau_n\|_2\|\mathbf{X}_{n2}\|_2) \to \mu(\nu,F)\eta_2, \tag{3.20}$$

and

$$E[b\mathrm{o}F(\sigma_{ni}U_{ni})] = [\textstyle\int F^{-1}f\mathrm{o}F^{-1}db + c_{ni}^*]\tau_{ni} \quad \text{with } \max_{1\le i\le n}|c_{ni}^*| \to 0 \tag{3.21}$$

implying (by definition of $\mu_R(b,F) = \int F^{-1}f\mathrm{o}F^{-1}db$)

$$E[\mathbf{X}_{n3}'\mathbf{b}\mathrm{o}\mathbf{F}(\mathbf{\Sigma}_n\mathbf{U}_n)] \to \mu_R(b,F)\eta_3. \tag{3.22}$$

(3.17) and (3.18) yield

$$\begin{aligned}
\mathrm{Cov}(\psi_{\nu,F}(\sigma_{ni}U_{ni}), b\mathrm{o}F(\sigma_{ni}U_{ni})) &= \\
&= \textstyle\iint \left[(u\wedge v - uv) - \tau_{ni}\{h_{ni}(u\wedge v) - h_{ni}(u)v - h_{ni}(v)u + \right. \\
&\quad \left. + \tau_{ni}h_{ni}(u)h_{ni}(v)\}\right]\left[f\mathrm{o}F^{-1}(u)\right]^{-1}d\nu(u)db(v) \\
&\equiv \mathrm{Cov}(\psi_{\nu,F}(U_{ni}), b\mathrm{o}F(U_{ni})) - \tau_{ni}K_{ni} \equiv K_1 - \tau_{ni}K_{ni} \quad \text{(say)},
\end{aligned} \tag{3.23}$$

K_{ni} uniformly bounded. With $\mathbf{D}_n = diag(\tau_{ni}), \mathbf{K}_n = diag(K_{ni})$ we therefore have

$$\mathbf{Cov}(\mathbf{X}_{n2}'\mathbf{\Psi}_{\nu,F}(\mathbf{\Sigma}_n\mathbf{U}_n), \mathbf{X}_{n3}'\mathbf{b}\mathrm{o}\mathbf{F}(\mathbf{\Sigma}_n\mathbf{U}_n)) = K_1\mathbf{X}_{n2}'\mathbf{X}_{n3} - \mathbf{X}_{n2}'\mathbf{D}_n\mathbf{K}_n\mathbf{X}_{n3}$$
$$= \mathbf{0} - O(\|\tau_n\|_\infty\|\mathbf{X}_{n2}\|_2\|\mathbf{X}_{n3}\|_2) \to \mathbf{0}. \tag{3.24}$$

Similarly we obtain

$$\max_{1\le i\le n}|\mathrm{Var}\psi_{\nu,F}(\sigma_{ni}U_{ni}) - \sigma^2(\nu,F)| \to 0 \tag{3.25}$$

and

$$\max_{1\le i\le n}|\mathrm{Var}[b\mathrm{o}F(\sigma_{ni}U_{ni})] - 1| \to 0. \tag{3.26}$$

(3.20,22,24-26) together with the central limit theorem imply (3.12) and hence statement (a) of the theorem.

The proof of (b) follows the same lines. In particular, because of the assumed orthogonality (1.5) of the design components, $\int \check{\gamma}_{n2}d\nu$ is asymptotically equivalent to $\mathbf{L}_n$ and $\int \check{\gamma}_{n3}d\nu$ has an analogous representation, replacing $\mathbf{X}_{n2}$ by $\mathbf{X}_{n3}$ in (2.5).

(c) and (d) result by equating $\psi_{\nu,F}$,$b\mathrm{o}F$ and ψ_f^*, which is the same task as in the one or two sample scale problems.

The invariance (e) follows from the equivariance on regression quantiles combined with the fact that ν has total charge zero, and the invariance of regression rank scores (see Koenker and Bassett (1978) and Gutenbrunner and Jurečková (1992)). □

4. Computational aspects

The representation of statistics like $\mathbf{L}_n$ (2.3), $\mathbf{R}_n$ (2.4) or $\widehat{\sigma}_n^2(\nu)$ (3.3) as integrals is most suitable for theoretical investigations. In practice however, these statistics are obtained as finite sums from the solution of (2.1) and (2.2), given by the Koenker and d'Orey (1987,1990) algorithm. This algorithm yields a decomposition

$$(0,1) = \sum_{l=1}^{\widehat{l}} I_l$$

of the unit interval, with intervals I_l on which $\widehat{\beta}_n(\cdot)$ is constant and $\widehat{\mathbf{\Delta}}_n(\cdot)$ is linear, together with the values

$$\widehat{\beta}_n(u) = \widehat{\beta}_n^{(l)}, \quad \widehat{\gamma}_{nk}(u) = \widehat{\gamma}_{nk}^{(l)} \quad (k = 2,3) \quad \text{if} \quad u \in I_l$$

and the values of $\widehat{\mathbf{\Delta}}_n(\cdot)$ at the endpoints of the I_l. From these quantities, the piecewise constant function

$$\dot{\mathbf{\Delta}}_n(u) = -d\widehat{\mathbf{\Delta}}_n(u)/du = \dot{\mathbf{\Delta}}_n^{(l)} \quad \text{if} \quad u \in I_l$$

is easily obtained. Integration by parts yields

$$\int \widehat{\mathbf{\Delta}}_n db = \int b(u)\dot{\mathbf{\Delta}}_n(u)du = \sum_l b_l \dot{\mathbf{\Delta}}_n^{(l)}, \quad b_l = \int_{I_l} b(u)du.$$

For $\mathbf{L}_n$ we have

$$\mathbf{L}_n = \int \widehat{\gamma}_{n2} d\nu = \sum_l \nu(I_l)\widehat{\gamma}_{n2}^{(l)}.$$

Example (heteroscedastic simple linear regression): Within the model

$$Y_i = \beta_0 + \gamma_2 x_i + e^{\tau_i} U_i, \quad U_i \text{ i.i.d.} \sim F$$

where $\sum \tau_i = \sum x_i = 0$, $\sum x_i^2 = 1$ may be assumed, we might be interested in testing

$$H_0 : \sum \tau_i^2 = 0, \qquad \text{vs.} \qquad H_1 : (\sum \tau_i x_i)^2 + (\sum \tau_i x_i^2)^2 > 0.$$

In this case, $p = 0, q = 1, r = 1$, and we orthonormalize according to

$$z_i = (x_i^2 - (\sum x_j^3)x_i - \tfrac{1}{n})/(\sum_k (x_k^2 - (\sum x_j^3)x_k - \tfrac{1}{n})^2)^{1/2}$$

and set $\mathbf{X}_2 = (x_1, \ldots, x_n)'$, $\mathbf{X}_3 = (z_1, \ldots, z_n)'$.

L_n (2.3), R_n (2.4), T_n (3.1) and S_n (3.2) (with $\nu_2 = \nu$) are now given by

$$L_n = \sum_l \nu(I_l)\widehat{\gamma}_{n2}^{(l)}, \quad R_n = \sum_i \sum_l z_i b_l \dot{\Delta}_{ni}^{(l)} \quad \text{and}$$

$$T_n = L_n^2/\widehat{\sigma}_n^2(\nu) + R_n^2, \quad S_n = (\check{L}_{n2}^2 + \check{L}_{n3}^2)/\check{\sigma}_n^2(\nu), \quad \check{L}_{nk} = \sum_l \nu(\check{I}_l)\check{\gamma}_{nk}^{(l)},$$

with the $\check{}$-quantities computed w.r.t. model $Y_i = \check{\beta}_0 + \check{\gamma}_2 x_i + \check{\gamma}_3 x_i^2 + e^{\tau_i} U_i$.
If $\sum_i x_i \tau_i \to \eta_2$, $\sum_i z_i \tau_i \to \eta_3$, T_n resp. S_n according to the theorem are asymptotically χ^2 with 2 degrees of freedom and noncentrality

$$\eta_2^2 \mu^2(\nu, F)/\sigma^2(\nu, F) + \eta_3^2 \mu_R^2(b, F) \quad \text{resp.} \quad (\eta_2^2 + \eta_3^2)\mu^2(\nu, F)/\sigma^2(\nu, F).$$

A relative advantage of T_n over S_n is that T_n uses regression quantiles and rank scores computed within the smaller model involving only $\mathbf{X}_1$, whereas S_n needs the regression quantiles from the enlarged model. The relation between both statistics is somewhat like the one between Lagrange multiplier- and Wald-statistic in the maximum likelihood setting.

5. Generalizations

In this section we outline the possible extension to other types of alternatives than

$$f(x, \tau_{ni}) = \exp(-\tau_{ni}) f(x \exp(-\tau_{ni})),$$

namely Behnen's (1972) nonparametric contiguous alternatives

$$f(x, \tau_{ni}) = f(x)(1 - \tau_{ni} \dot{h}_n \circ F(x)), \tag{5.1}$$

with given functions $\dot{h}_n$ satisifying $\int \dot{h}_n(u)du = 0$ and, in order to guarantee contiguity of the alternatives,

$$(\tilde{F}.1) \qquad \|\dot{h}_n - \dot{h}\|_2 \to 0 \quad \text{for some } \dot{h} \in L_2((0,1)) \quad \text{and} \quad \|\tau_n\|_\infty \|\dot{h}_n\|_\infty^3 \to 0.$$

To obtain positive asymptotic noncentrality under the alternative, we additionally require

$$h(u) < 0 = h(0.5) < h(v) \text{ if } 0 < u < 0.5 < v < 1, \text{ where } h(u) = \int_0^u \dot{h}(v)dv.$$

Then, if condition (F.1) is replaced by ($\tilde{\text{F}}$.1) and ν satisfies (W.1) *and* (W.2), assertions (a) and (b) of the theorem hold with $\mu(\nu, F)$ (2.7) and $\mu_R(b, F)$ (3.5) replaced by

$$\mu(\nu, h, F) = \int (h / f \circ F^{-1}) d\nu \quad \text{and} \quad \mu_R(b, h) = \int h db.$$

The densities (5.1) form a general class of alternatives which are invariant against monotonic transformations and hence naturally appear in the context of ordinal variables.

REFERENCES

[1] Bassett, G. and Koenker R. (1982). An empirical quantile function for linear models with iid errors. *JASA*, **77**, 407-415.

[2] Behnen, K. (1972). A characterization of certain rank-order tests with bounds for the asymptotic relative efficiency. *Ann. Math. Statist.*, **43**, 1839-1851.

[3] Gutenbrunner, C. and Jurečková, J. (1992). Regression rank scores and regression quantiles. *Ann. Statist.*, **20**, 305-330.

[4] Gutenbrunner, C. and Jurečková, J., Koenker, R. and Portnoy, S. (1993). Tests of linear hypotheses based on regression rank scores. *Journ. of Nonpar. Statist.*,**2**, 307–331.

[5] Huber, P. J. (1981). *Robust Statistics*. New York 1981.

[6] Koenker, R. and Bassett, G. (1978). Regression quantiles, *Econometrica*, **46**, 33–50.

[7] Koenker, R. and Bassett, G. (1982). Robust tests for heteroscedasticity based on regression quantiles, *Econometrica*, **50**, 43–61.

[8] Koenker, R. and d'Orey, V. (1987). Computing regression quantiles. *Applied Statistics*,**36**, 383–393.

[9] Koenker, R. and d'Orey, V. (1990). Remark on algorithm 229. Preprint.

[10] Portnoy, S. (1992). A regression quantile based statistic for testing nonstationarity of errors. *Nonparametric Statistics and Related Topics* (ed: A.K. Md.E. Saleh). Amsterdam 1992

[11] Welsh, A.H. (1987). One-step L-estimators for the linear model. *Ann. Statist.*, **15**, 626–641.

PARAMETER ESTIMATION BY PROJECTING ON STRUCTURAL STATISTICAL MODELS[1]

L. Györfi, I. Vajda, E. van der Meulen

Technical University Budapest
Czech Academy of Sciences Prague
Catholic University Leuven

Nonparametric estimates Q_n of distributions from a family $\mathcal{Q}$ based on i.i.d. observations and consistent in a metric on distributions are considered. Point estimates are defined using metric projections of the empirical distributions Q_n on parametric families $\mathcal{P} = \{P_\theta : \theta \in \Theta\}$. The family $\mathcal{Q}$ need not be parametrized by the same parameter as $\mathcal{P}$ or may be nonparametric as well. Hence a very general model of robust statistics is considered. Under weak regularity assumptions concerning $\mathcal{P}$ and $\mathcal{Q}$ it is shown that these estimates strongly consistently estimate parameter values true in a minimum distance sense. If $\mathcal{Q}$ coincides with $\mathcal{P}$ then the parameter values true in the minimum distance sense coincide with those true in the common sense. Special attention is paid to location and scale families $\mathcal{P}$ and $\mathcal{Q}$ where new results concerning consistent estimability are obtained from general theorems.

AMS 1990 subject classification: 72 F 10, 62 F 12, 62 F 35

Key words and phrases: Minimum distance estimates, strong consistency, location and scale.

1 Introduction

Let D be a metric distance on the space $\mathcal{M}$ of all probability measures on $\mathbb{R}$. Consider for every $n \in \mathbb{N}$ a random sample $X_1, \ldots, X_n$ with components i.i.d. according to $Q \in \mathcal{M}$. Suppose that there is given an empirical distribution $Q_n \in \mathcal{M}$ depending on the sample $X_1, \ldots, X_n$ and a subset $\mathcal{Q} \subset \mathcal{M}$ such that $D(Q, Q_n)$, $Q \in \mathcal{Q}$ are measurable in $(X_1, \ldots, X_n)$ and

(1) $$\lim_{n\to\infty} D(Q, Q_n) = 0 \quad \text{a.s.}, \ Q \in \mathcal{Q}.$$

Consider a parametrized subset

$$\mathcal{P} = \{P_\theta : \theta \in \Theta\} \subset \mathcal{M}, \quad \text{with} \quad P_{\theta_1} \neq P_{\theta_2} \text{ for } \theta_1 \neq \theta_2,$$

where Θ is a separable locally compact metric space with a metric d. Let $\mathcal{B}$ be the Borel σ-field of Θ and suppose that $D(P_\theta, Q_n)$ is measurable with respect to both variables θ and $(X_1, \ldots, X_n)$, and that

$$\inf_{\theta \in B} D(P_\theta, Q_n)$$

[1]This work was supported by the scientific exchange program between the Hungarian and Czech Academies of Sciences and the Royal Belgian Academy of Sciences, and by GA CR grant 202/93/0232.

is measurable in $(X_1, \ldots, X_n)$ for every open subset $B \subset \Theta$.

Each D induces a distance d_D on Θ by the formula

(2) $$d_D(\theta_1, \theta_2) = D\left(P_{\theta_1}, P_{\theta_2}\right).$$

As an example consider the induced variational distance

$$d_V(\theta_1, \theta_2) = V\left(P_{\theta_1}, P_{\theta_2}\right),$$

where

$$V(P, Q) = 2 \sup_A (P(A) - Q(A))$$

is the total variation on $\mathcal{M}$, the induced Hellinger distance

$$d_{\sqrt{H}}(\theta_1, \theta_2) = \sqrt{H(P_{\theta_1}, P_{\theta_2})},$$

where

$$H(P, Q) = 1 - \int \sqrt{pq}\, \mathrm{d}\lambda = \frac{1}{2} \int (\sqrt{p} - \sqrt{q})^2\, \mathrm{d}\lambda$$

for Radon–Nikodym densities $p = \mathrm{d}P/\mathrm{d}\lambda$, $q = \mathrm{d}Q/\mathrm{d}\lambda$ w.r.t. some dominating σ-finite $\lambda \gg \{P, Q\}$ is the Hellinger distance on $\mathcal{M}$, and the induced Kolmogorov and Prohorov distances

$$d_K(\theta_1, \theta_2) = K\left(P_{\theta_1}, P_{\theta_2}\right), \; d_\Pi(\theta_1, \theta_2) = \Pi\left(P_{\theta_1}, P_{\theta_2}\right),$$

where $K(P, Q)$ and $\Pi(P, Q)$ are the Kolmogorov and Prohorov distance on $\mathcal{M}$ respectively (cf. Chap. 2 in [5]).

Let us consider two distances δ_1, δ_2 on an arbitrary space $\mathcal{S}$. The distance δ_2 is said topologically stronger than δ_1, in symbols $\delta_1 \preceq \delta_2$, if a subset $S \subset \mathcal{S}$ open in $(\mathcal{S}, \delta_1)$ is open also in $(\mathcal{S}, \delta_2)$ or, equivalently, if the convergence in δ_2 implies the convergence in δ_1.

It is known (cf. (4.13), (4.14) in Chap. 2 of [5]) that

$$\Pi \preceq V \quad \text{and} \quad K \preceq V \qquad \text{on } \mathcal{M}.$$

By Proposition (2.35) in [6], it holds $\sqrt{H} \preceq V$ and $V \preceq \sqrt{H}$, i.e. $\sqrt{H}$ and V are topologically equivalent, in symbols

$$\sqrt{H} \simeq V.$$

Therefore

$$d_\Pi \preceq d_V, \quad d_K \preceq d_V, \quad \text{and} \quad d_V \simeq d_{\sqrt{H}} \quad \text{on} \quad \Theta.$$

But for typical families $\mathcal{P}$ all distances under consideration are topologically equivalent on Θ,

(3) $$d_\Pi \simeq d_K \simeq d_V \simeq d_{\sqrt{H}},$$

in spite the fact of that the equivalences $\Pi \simeq K$ and $K \simeq V$ do not hold on $\mathcal{M}$.

Definition 1. An estimator is a sequence of measurable mappings $\hat{\theta}_n : \mathbb{R}^n \longrightarrow \Theta$. If there exists a sequence $\varepsilon_n \downarrow 0$ such that an estimator $\hat{\theta}_n$ satisfies for all n the relation

$$D\left(P_{\hat{\theta}_n}, Q_n\right) \le \inf_{\theta\in\Theta} D(P_\theta, Q_n) + \varepsilon_n \text{ a.s.}, \quad Q \in \mathcal{Q}, \tag{4}$$

then $\hat{\theta}_n$ is said to be *an asymptotically minimum distance estimator* (AMDE).

In this paper we investigate asymptotic properties of AMDE's for general $\mathcal{P}$, $\mathcal{Q}$ and D. The particular case $\mathcal{Q} \equiv \mathcal{P}$ (i.e. $\mathcal{Q} = \{Q_\theta : \theta \in \Theta\}$ where $Q_\theta = P_\theta$ for all $\theta \in \Theta$) has been considered already by Wolfowitz [11]. The general case $\mathcal{Q} \not\equiv \mathcal{P}$ has been introduced by robust statistics (cf. Huber [5]), together with a motivation for estimators of AMDE-type (cf. Beran [1,2] and Millar [7]). In [4] we investigated the consistency and robustness of estimators generated by the distance $D = V$ for $\mathcal{Q} \neq \mathcal{P}$ under quite strict conditions on $\mathcal{P}$ and $\mathcal{Q}$. Here we restrict ourselves to the strong consistency of AMDE's under quite weak conditions on $\mathcal{P}$, $\mathcal{Q}$ and D. Special attention is paid to statistical models with a group structure on parametrized sets $\mathcal{P}$ (structural statistical models).

2 Basic results

Let us first consider the particular case

$$\mathcal{Q} = \mathcal{P}. \tag{5}$$

Note that (5) is just the set-theoretical equality of $\mathcal{Q}$ and $\mathcal{P}$ which is weaker than the relation $\mathcal{Q} \equiv \mathcal{P}$ considered above. If $\mathcal{Q} = \mathcal{P}$ then one can assume without loss of generality that $\mathcal{Q} = \{Q_\theta : \theta \in \Theta\}$ where, for a bijection $\tau : \Theta \longrightarrow \Theta$, $Q_\theta = P_{\tau(\theta)}$ for all $\theta \in \Theta$. Statistical models of this type can be used for a minimum distance estimation of the parametric function $\tau(\theta)$. The particular case $\mathcal{Q} \equiv \mathcal{P}$ is characterized by the identity mapping τ.

If (5) holds then there is no question about what the strong consistency of an estimator means. An estimator $\hat{\theta}_n$ is considered to be strongly consistent if

$$\lim_{n\to\infty} \hat{\theta}_n = \theta_0 \text{ a.s.}, \qquad Q = P_{\theta_0} \in \mathcal{P}. \tag{6}$$

Remind that this convergence means that $d(\hat{\theta}_n, \theta_0) \to 0$ a.s.

Theorem 1. If (5) holds and

$$d \preceq d_D \tag{7}$$

then every AMDE is strongly consistent.

Proof. By the triangle inequality and symmetry of D, and by (4), it holds a.s. for every $Q = P_{\theta_0} \in \mathcal{P}$ that

$$\begin{aligned} D\left(P_{\theta_0}, P_{\hat{\theta}_n}\right) &\le D\left(P_{\theta_0}, Q_n\right) + D\left(P_{\hat{\theta}_n}, Q_n\right) \\ &\le D\left(P_{\theta_0}, Q_n\right) + D\left(P_{\theta_0}, Q_n\right) + \varepsilon_n. \end{aligned}$$

It follows from here and from (1) that

$$\lim_{n\to\infty} D\left(P_{\theta_0}, P_{\hat{\theta}_n}\right) = 0 \text{ a.s.}, \quad Q = P_{\theta_0} \in \mathcal{P}.$$

Applying (7) one obtains (6) Q. E. D.

Let us now consider the case where (5) does not hold, i.e. where $\mathcal{Q} \neq \mathcal{P}$. If $\mathcal{Q}$ is a non-parametric family then it is not clear what the consistency of point estimators defined by (4) should mean. This problem remains if $\mathcal{Q}$ is a semi-parametric or even parametric, parametrized by the same parameter space Θ figuring in the definition of $\mathcal{P}$, i.e. if

(8) $$\mathcal{Q} = \{Q_\theta : \theta \in \Theta\}.$$

The naive consistency condition

(6*) $$\lim_{n\to\infty} \hat{\theta}_n = \theta_0 \text{ a.s.}, \quad Q = Q_{\theta_0} \in \mathcal{Q},$$

cannot be accepted as a definition as it contradicts in some cases (6). To this end consider (5) with the function $\tau(\theta)$ different from the identity. Then for every $Q \in \mathcal{Q}$ there exists a $\theta_0 \in \Theta$ such that

$$Q = Q_{\theta_0} = P_{\tau(\theta_0)},$$

where $\tau(\theta_0) \neq \theta_0$ for at least one $\theta_0 \in \Theta$.

The last example indicates that the strong consistency of an estimator under consideration means in general the relation

(9) $$\lim_{n\to\infty} \hat{\theta}_n = \theta_Q \text{ a.s.}, \quad Q \in \mathcal{Q},$$

where $\theta_Q \in \Theta$ is a parameter value "true in the family $\mathcal{P}$" under the data distribution Q. The following "realistic" definition of the true parameter is also motivated by this example.

Definition 2. A parameter $\theta_Q \in \Theta$ is said to be *true for a data distribution* $Q \in \mathcal{M}$, if P_{θ_Q} is the unique D-projection of Q on $\mathcal{P}$ in the sense that

(10) $$D\left(P_{\theta_Q}, Q\right) = \inf_{\theta\in\Theta} D(P_\theta, Q).$$

It follows from the reflexivity of D and from the assumption $P_{\theta_1} \neq P_{\theta_2}$ for $\theta_1 \neq \theta_2$ that if (5) holds then the relation

$$\theta_Q = \theta_0$$

is true for every $Q = P_{\theta_0} \in \mathcal{P}$. Therefore in this case the general consistency condition (9) reduces to (6). Note that the last relation is equivalent with

$$\theta_Q = \tau(\theta_0), \qquad Q = Q_{\theta_0} = P_{\tau(\theta_0)} \in \mathcal{Q}.$$

We can summarize that the theory based on definition (9), with true values θ_Q according to Definition 2, is an extension of the natural theory considered in the special case (5) to arbitrary statistical models under consideration.

First we formulate conditions sufficient for the existence of parameters θ_Q true for $Q \in \mathcal{M}$.

Lemma 1. Let $D(P_\theta, Q)$ be lower semicontinuous on Θ and let there exist a compact $B_Q \subset \Theta$ such that $\theta_Q \in B_Q$ and

$$\inf_{\theta \in \Theta - B_Q} D(P_\theta, Q) > \inf_{\theta \in \Theta} D(P_\theta, Q). \tag{11}$$

Then there exists a D-projection P_{θ_Q} of Q on $\mathcal{P}$. If this projection is unique then θ_Q is true for Q.

Proof. It follows from (11) that

$$\inf_{\theta \in B_Q} D(P_\theta, Q) = \inf_{\theta \in \Theta} D(P_\theta, Q),$$

i.e. that every D-projection of Q on $\{P_\theta : \theta \in B_Q\}$ is also a D-projection on $\mathcal{P}$. The existence of a D-projection P_{θ_Q} on $\{P_\theta : \theta \in B_Q\}$ is clear from the assumed lower semicontinuity of $D(P_\theta, Q)$ and from the compactness of B_Q. The last assertion of Lemma 1 is clear from Definition 2.

Remark 1. Note that the obvious inequality

$$|D(P_{\theta_1}, Q) - D(P_{\theta_2}, Q)| \le D(P_{\theta_1}, P_{\theta_2})$$

implies that if $d_D \preceq d$ then $D(P_\theta, Q)$ is continuous on Θ. It is also clear that if $D(\cdot, Q)$ is strictly convex on $\mathcal{M}$ (take e.g. the Hellinger distance $\sqrt{H}$) and the D-projection of Q on $\mathcal{P}$ coincides with the D-projection on the convex hull $\mathcal{C}(\mathcal{P}) \subset \mathcal{M}$, then the D-projection P_{θ_Q} is unique. Thus, for $\mathcal{P}$ and D satisfying the assumption

$$d \simeq d_D \qquad \text{(cf. (7))}$$

and (11), with a guaranteed uniqueness of all D-projections P_{θ_Q}, $Q \in \mathcal{Q}$, the next Theorem is not only an extension but also a sharpening of Theorem 1.

Theorem 2. If the assumptions of Lemma 1 are satisfied for all $Q \in \mathcal{Q}$ and the corresponding D-projections are unique then every AMDE is strongly consistent in the sense of (9).

Proof. By the triangle inequality, the symmetry of D, and (4), it holds for every $Q \in \mathcal{Q}$

$$\begin{aligned} D(P_{\hat\theta_n}, Q) &\le D(P_{\hat\theta_n}, Q_n) + D(Q, Q_n) \\ &\le D(P_{\hat\theta_n}, Q_n) + D(Q, Q_n) + D(P_{\theta_Q}, Q_n) - \inf_{\theta \in \Theta} D(P_\theta, Q_n) \\ &\le \varepsilon_n + D(Q, Q_n) + D(P_{\theta_Q}, Q_n) \\ &\le \varepsilon_n + 2D(Q, Q_n) + D(P_{\theta_Q}, Q). \end{aligned}$$

Hence, by (1) and the assumption that $\varepsilon_n \downarrow 0$ in (4),

$$\limsup_{n \to \infty} D(P_{\hat\theta_n}, Q) \le D(P_{\theta_Q}, Q) \quad \text{a.s.} \tag{12}$$

This together with (11) implies that a. s. all but finitely many elements of the sequence $\hat{\theta}_n$ belong to the compact B_Q. Let $\theta_* \in B_Q$ be an accumulation point of $\hat{\theta}_n$. If $\hat{\theta}_{n_k} \longrightarrow \theta_*$ then, by the lower semicontinuity of $D(P_\theta, Q)$ and (12),

$$D(P_{\theta_*}, Q) \leq \liminf_{k\to\infty} D(P_{\theta_{n_k}}, Q) \leq D(P_{\theta_Q}, Q).$$

It follows from here and Definition 2 that $\theta_* = \theta_Q$ a. s. Hence (9) holds Q. E. D.

Theorem 1 demonstrates that AMDE's exhibit similar properties as MLE's: the estimates are constructed on the basis of data $X_1, \ldots, X_n$ and data generation families $\mathcal{P}$, and these estimates are consistent under weak regularity assumptions concerning $\mathcal{P}$. Since even for non-dominated families $\mathcal{P}$ there exist empirical distributions Q_n and distances D satisfying (1) (and also (7), cf. [3]), the regularity assumptions of Theorem 1 are in fact weaker than the corresponding regularity assumptions for MLE's. Even for dominated families $\mathcal{P}$ there exist examples where both the MLE and AMDE exist, but the MLE is inconsistent. Such an example is given e. g. by the family of discrete probability densities (for $1/2 \leq c < 1$)

$$p_\theta(x) = 1 - c + \frac{c\, 1_{\{\theta\}}(x)}{\lambda(x)}, \quad \theta \in \mathbb{N},$$

concentrated on natural $x \in \mathbb{N}$, with respect to the dominating σ-finite (probability) measure

$$\lambda(x) = \frac{1}{\sqrt{\log(e + x - 1)}} - \frac{1}{\sqrt{\log(e + x)}}, \quad x \in \mathbb{N}.$$

Here (3) holds for the usual Euclidean distance d and (1) holds for the usual empirical distributions Q_n, $\mathcal{Q} \equiv \mathcal{P}$, and all distances Π, K, V, $\sqrt{H}$. At the same time, the MLE is inconsistent (see [9]).

Thus the AMDE's seem to be better adapted to data generating families than the MLE's. Theorem 2 extends the adaptivity of AMDE's to cases where the data generating families $\mathcal{Q}$ differ from the projection family $\mathcal{P}$. It is however to be underlined that here we speak about the first order adaptivity (the convergence itself) and not about the adaptivity in the usual sense (the rate of convergence) considered by Beran [1, 2], Stone [7] or Víšek [10].

3 Fisher consistency

In this section we generalize the simple functional relation between the value θ_0 true in the naive sense considered in (6*) and the value θ_Q true in the sense of Definition 2, observed above in the particular case (5), to arbitrary families $\mathcal{Q}$ parametrized in the sense of (8).

Definition 3. Let $\mathcal{Q}$ be parametrized in the sense of (8) and let the parameter θ_Q true in the sense of Definition 2 exist for every $Q \in \mathcal{Q}$. Then $\tau : \Theta \longmapsto \Theta$ defined by

$$\tau(\theta_0) = \theta_Q \quad \text{for } Q = Q_{\theta_0} \in \mathcal{Q} \tag{13}$$

is called a *τ-mapping* corresponding to given $\mathcal{Q}$ and $\mathcal{P}$. If the τ-mapping is bijective (i. e. a one-to-one mapping onto Θ) and the inverse mapping τ^{-1} is continuous, then $\mathcal{P}$ is said to be *weakly Fisher consistent with* $\mathcal{Q}$. If τ is the identity mapping on Θ then $\mathcal{P}$ is said *strongly Fisher consistent with* $\mathcal{Q}$.

Remark 2. The τ-mapping is nothing but a transformation of parameter values true in the naive sense to values true in the sense of Definition 2. The weak Fisher consistency means that τ exists and that the values true in the naive sense are continuously recoverable from the values true in the sense of Definition 2. The strong Fisher consistency means that the values true in the naive sense coincide with the values true in the sense of Definition 2. If (5) holds and the mapping τ considered there is continuous then τ is the mapping of Definition 3 and $\mathcal{P}$ is weakly Fisher consistent with $\mathcal{Q}$. If $\mathcal{P} \equiv \mathcal{Q}$ then the strong Fisher consistency takes place.

Remark 3. We see that if $\mathcal{Q}$ satisfies (8) and $\mathcal{P}$ is weakly Fisher consistent with $\mathcal{Q}$ then, for every estimator θ_n strongly consistent in the sense of (9),

$$\lim_{n\to\infty} \tau^{-1}(\theta_n) = \theta_0 \quad \text{a.s.}, \quad Q = Q_{\theta_0} \in \mathcal{Q}.$$

If $\mathcal{P}$ is strongly Fisher consistent with $\mathcal{Q}$ then

$$\lim_{n\to\infty} \theta_n = \theta_0 \quad \text{a.s.}, \quad Q = Q_{\theta_0} \in \mathcal{Q}.$$

Thus in this case the consistency of an estimator θ_n in the sense of (9) coincides with the naive consistency (6*).

The applicability of these concepts follows from the results of the next Section.

4 Structural statistical models

Throughout this section we restrict ourselves to distances D on $\mathcal{M}$ such that, for every $P, Q \in \mathcal{M}$,

$$D(P,Q) = \sup_{\mathcal{D}} D(P^{\mathcal{D}}, Q^{\mathcal{D}}), \tag{14}$$

where the supremum extends over all finite measurable decompositions $\mathcal{D}$ of $\mathbb{R}$, and $P^{\mathcal{D}}$, $Q^{\mathcal{D}}$ denote restrictions of P, Q to algebras generated by $\mathcal{D}$. By Theorem 1.30 in [6], (14) holds e.g. for the distances $\sqrt{H}$ and V defined in Section 1. The validity of (14) for K and Π is evident.

We shall consider families $\mathcal{P}$ and $\mathcal{Q}$ *structural with parents* P, Q in the following sense:

(i) There is an associative multiplication $\theta_1 \theta_2$ in Θ such that Θ is a group. The inverse to θ is denoted by θ^{-1} and the neutral element is denoted by e.

(ii) For every $\theta_0 \in \Theta$ the mappings $\theta \longmapsto \theta\,\theta_0$, $\theta \longmapsto \theta_0\,\theta$ are continuous.

(iii) For every $\theta \in \Theta$ there exists a continuous bijection $g_\theta : \mathbb{R} \longmapsto \mathbb{R}$ such that the multiplication $g_{\theta_1} \circ g_{\theta_2}$ defined by

$$(g_{\theta_1} \circ g_{\theta_2})(x) = g_{\theta_1}(g_{\theta_2}(x))$$

makes a group out of $\mathcal{G} = \{g_\theta : \theta \in \Theta\}$.

(iv) $\theta \longmapsto g_\theta$ is an isomorphism between the groups Θ and $\mathcal{G}$. In particular, g_e is the neutral element of $\mathcal{G}$ and $g_\theta^{-1} = g_{\theta^{-1}}$ for all $\theta \in \Theta$.

(v) For every $\theta \in \Theta$ it holds that $P_\theta = P\,g_\theta^{-1}$, $Q_\theta = Q\,g_\theta^{-1}$.

As an example let us consider $\Theta = \mathbb{R} \times \mathbb{R}_+$ and let for every $\theta = (\mu, \sigma) \in \mathbb{R} \times \mathbb{R}_+$

$$g_{\mu,\sigma}(x) = \mu + \sigma x, \quad x \in \mathbb{R}.$$

Then every pair of distributions P, Q defines families $\mathcal{P} = \{P_\theta : \theta \in \Theta\}$, $\mathcal{Q} = \{Q_\theta : \theta \in \Theta\}$ by the formula

$$P_{\mu,\sigma} = P\, g_{\mu,\sigma}^{-1}, \qquad Q_{\mu,\sigma} = Q\, g_{\mu,\sigma}^{-1}.$$

These families are structural under the multiplication in $\mathbb{R} \times \mathbb{R}_+$ defined by

$$(\mu_1, \sigma_1)(\mu_2, \sigma_2) = (\mu_1 + \sigma_1 \mu_2,\ \sigma_1 \sigma_2)$$

with

$$e = (0, 1), \qquad (\mu, \sigma)^{-1} = (-\mu/\sigma,\ 1/\sigma).$$

They are called *location and scale families with parents P, Q*. If the parameter space is reduced to the subgroups

$$\mathbb{R} \times \{1\} \quad \text{or} \quad \{0\} \times \mathbb{R}_+$$

of the group $\mathbb{R} \times \mathbb{R}_+$, which are equivalent to the obvious groups $\mathbb{R}$ or $\mathbb{R}_+$, respectively, then $\mathcal{P}$, $\mathcal{Q}$ are called *location families with parents P, Q* or *scale families with parents P, Q* respectively.

Theorem 3. Let $\mathcal{P}$, $\mathcal{Q}$ be structural families with parents P, Q and there exists θ_Q true for the parent data distribution Q. Then the τ-mapping exists and satisfies the relation

$$\tau(\theta) = \theta\, \tau(e), \quad \theta \in \Theta,$$

where $\tau(e)$ coincides with the parameter $\theta_Q \in \Theta$ true for Q. This means that $\mathcal{P}$ is weakly Fisher consistent with $\mathcal{Q}$.

Proof. Since $g_{\theta_0}^{-1}$ is a continuous bijection $\mathbb{R} \longmapsto \mathbb{R}$ for every $\theta_0 \in \Theta$, (14) remains true with $\mathcal{D}$ replaced by $\mathcal{D}_0 = g_{\theta_0}^{-1}\, \mathcal{D}$.

Let us now consider a fixed $\theta_0 \in \Theta$ and an arbitrary $\theta \in \Theta$. Since for every $A \in \mathcal{D}$ and $A_0 = g_{\theta_0}^{-1}(A)$

$$Q_{\theta_0}(A) = Q\, g_{\theta_0}^{-1}(A) = Q(A_0),$$

and

$$P_\theta(A) = P\, g_\theta^{-1}(A) = P\, g_\theta^{-1} \circ g_{\theta_0}(A_0) = P\left(g_{\theta_0}^{-1} \circ g_\theta\right)^{-1}(A_0) = P\, g_{\theta_0^{-1}\theta}^{-1}(A_0) = P_{\theta_0^{-1}\theta}(A_0),$$

it follows from (14) that

$$D(P_\theta, Q_{\theta_0}) = D(P_{\theta_0^{-1}\theta}, Q) = D(P_{\theta_0^{-1}\theta}, Q_e).$$

Therefore

$$\tau(e) = \theta_0^{-1}\, \tau(\theta_0).$$

Further, it also follows from (14)

$$\tau(e) = \theta_Q,$$

so that

$$\theta_0\, \theta_Q = \tau(\theta_0) \quad \text{Q. E. D.}$$

Corollary. Let $\mathcal{P}$, $\mathcal{Q}$ be location and scale families with parents P, Q, where Q satisfies the assumptions of Lemma 1. Then the τ-mapping exists and satisfies the relation

$$\tau(\mu,\sigma) = (\mu,\sigma)(\mu_Q, \sigma_Q), \quad (\mu,\sigma) \in \mathbb{R} \times \mathbb{R}_+,$$

where $(\mu_Q, \sigma_Q) = \tau(0,1)$. This means that $\mathcal{P}$ is weakly Fisher consistent with $\mathcal{Q}$. The family $\mathcal{P}$ is strongly Fisher consistent with $\mathcal{Q}$ iff

$$(\mu_Q, \sigma_Q) = (0,1).$$

The assumptions of Lemma 1 may not be easy to verify. Next we present more easily verifiable stronger conditions for the particular case of the Hellinger distance. For simplicity we restrict ourselves to the location model. We say that $\phi : \mathbb{R} \longrightarrow [0,\infty)$ is *unimodal* if there is an $x_0 \in \mathbb{R}$ called a *mode* such that ϕ is nonincreasing in the domain $x \geq x_0$ and nondecreasing in the domain $x \leq x_0$. The mode is in general not unique. If an unimodal ϕ is symmetric about $x_0 \in \mathbb{R}$ then x_0 is a mode. This special mode is unique unless ϕ is constant on $\mathbb{R}$.

Lemma 2. Let us consider a location family $\mathcal{P} = \{P_\mu : \mu \in \mathbb{R}\}$. Then for every $Q \in \mathcal{M}$ there exists a *Hellinger projection* P_{μ_Q} on $\mathcal{P}$ which is defined by (19) for $D = \sqrt{H}$.

Proof. If $P_\mu \perp Q$ for all $\mu \in \mathbb{R}$ then $H(P_\mu, Q) = 1$ on $\mathbb{R}$ and every $P_\mu \in \mathcal{P}$ is the Hellinger projection. If P_{μ_0} is nonsingular with Q then there exists $\varepsilon > 0$ such that

$$H(P_{\mu_0}, Q) < 1 - 2\sqrt{\varepsilon}.$$

On the other hand, for every $\varepsilon > 0$ there exists $\mu_\varepsilon > 0$ such that

$$P_{\mu_0}(\mathbb{R} - [-\mu_\varepsilon, \mu_\varepsilon]) < \varepsilon, \quad Q(\mathbb{R} - [-\mu_\varepsilon, \mu_\varepsilon]) < \varepsilon.$$

It holds for every $\mu \in \mathbb{R}$

$$\begin{aligned} H(P_\mu, Q) &\geq 1 - \sqrt{P_\mu(A)\,Q(A)} - \sqrt{P_\mu(\mathbb{R}-A)\,Q(\mathbb{R}-A)} \\ &\geq 1 - \sqrt{Q(A)} - \sqrt{P_\mu(\mathbb{R}-A)}. \end{aligned}$$

In particular, for $A = \mu - \mu_0 + [-\mu_\varepsilon, \mu_\varepsilon]$ it holds

$$\begin{aligned} & P_\mu(\mathbb{R}-A) = P_\mu(\mu - \mu_0 + (\mathbb{R} - [-\mu_\varepsilon, \mu_\varepsilon])) = P_{\mu_0}(\mathbb{R} - [-\mu_\varepsilon, \mu_\varepsilon]) < \varepsilon, \ \mu \in \mathbb{R}, \\ & Q(A) = Q([\mu - \mu_0 - \mu_\varepsilon, \mu - \mu_0 + \mu_\varepsilon] \cap [-\mu_\varepsilon, \mu_\varepsilon]) + \\ + \ & Q(\mu - \mu_0 - \mu_\varepsilon, \mu - \mu_0 + \mu_\varepsilon] \cap (\mathbb{R} - [-\mu_\varepsilon, \mu_\varepsilon])) \\ \leq \ & Q(\mathbb{R} - [-\mu_\varepsilon, \mu_\varepsilon]) < \varepsilon, \quad \mu \notin [\mu_0 - 2\mu_\varepsilon, \mu_0 + 2\mu_\varepsilon]. \end{aligned}$$

Therefore

$$\inf_{\mu \notin [\mu_0 - 2\mu_\varepsilon, \mu_0 + 2\mu_\varepsilon]} H(P_\mu, Q) \geq 1 - 2\sqrt{\varepsilon}$$

while

$$\inf_{\mu \in [\mu_0 - 2\mu_\varepsilon, \mu_0 + 2\mu_\varepsilon]} H(P_\mu, Q) \leq H(P_{\mu_0}, Q) < 1 - 2\sqrt{\varepsilon}.$$

Therefore (11) holds and the desired assertion follows from the argument of Lemma 1.

Theorem 4. Let the parent P of the family of Lemma 2 be absolutely continuous and let its density p be unimodal, symmetric about 0, and decreasing everywhere on $[0, \infty)$. If $Q \in \mathcal{M}$ is not purely atomic and the nonatomic part of Q is absolutely continuous with a density q unimodal and symmetric then the point of symmetry $\mu_0 \in \mathbb{R}$ of q is unique, and the Hellinger projection P_{μ_Q} of Lemma 2 satisfies the relation

$$\mu_Q = \mu_0.$$

Remark 4. It follows from Theorem 4 that μ_0 is true in Hellinger sense for Q. Let us consider a location family $\mathcal{P}$ with parent P and a nonparametric family $\mathcal{Q}$. If the assumptions of Theorem 4 hold for P and all $Q \in \mathcal{Q}$ then the location value true in the Hellinger sense for $Q \in \mathcal{Q}$ is the center of symmetry of the absolutely continuous part of Q. If $\hat{\theta}_n$ is an estimator of location strongly consistent in the sense of (9) then it tends for every $Q \in \mathcal{Q}$ a. s. to this center of symmetry.

Remark 5. The proof of Theorem 4 is based on the fact (cf. e. g. [3]) that if ϕ_1, ϕ_2 are unimodal symmetric about 0 with at least one common point of decrease then

$$\int [\phi_2(x-\theta) - \phi_2(x)]\, \phi_1(x)\, dx < 0, \quad \theta \neq 0.$$

Proof of Theorem 4. We need to prove for every $\mu \neq \mu_0$

$$\int \left[\sqrt{p(x-\mu)} - \sqrt{p(x-\mu_0)}\right] \sqrt{q(x)}\, dx < 0,$$

i. e.

$$\int \left[\sqrt{p(x+\theta)} - \sqrt{p(x)}\right] \sqrt{q_0(x)}\, dx < 0$$

or, equivalently,

$$\int \left[\sqrt{q_0(x-\theta)} - \sqrt{q_0(x)}\right] \sqrt{p(x)}\, dx < 0,$$

where $q_0(x) = q(x+\mu_0)$, $\theta = \mu_0 - \mu$. This however holds by Remark 5.

Acknowledgement. We would like to thank to referee for his suggestions, in particular for simplification of Remark 5.

References

[1] R. Beran (1974): Asymptotically efficient adaptive rank estimates in location models. *Ann. Statist.* 2, 63–74.

[2] Beran R. (1978): An efficient and robust adaptive estimator of location. *Ann. Statist.* 6, 292–313.

[3] Davies L. (1990): The asymptotics of S-estimators in the linear regression model. *Ann. Statist.* 17, 1651–1675.

[4] Györfi L., Vajda I., van der Meulen E. (1992): Family of point estimates yielded by L_1-consistent density estimate. *Statistical Data Analysis* (Ed. Y. Dodge). North Holland, Amsterdam, 415–430.

[5] Huber P. J. (1981): *Robust Statistics*. Wiley, New York.

[6] Liese F., Vajda I. (1987): *Convex Statistical Distances*. Teubner, Leipzig.

[7] P. V. Millar (1981): Robust estimation via minimum distance method. *Zeitschr. Wahrsch. verw. Geb.* 55, 73–89.

[8] C. Stone (1975): Adaptive maximum likelihood estimators of a location parameter. *Ann. Statist.* 3, 267–284.

[9] I. Vajda (1993): Conditions for consistency of MLE's. This volume.

[10] J. Á. Víšek (1992): Adaptive maximum-likelihood-like estimation in linear models, parts 1, 2. *Kybernetika* 28, 357–382, 454–471.

[11] J. Wolfowitz (1957): The minimum distance method. *Ann. Math. Statist.* 28, 75–88.

SOME REMARKS ON THE JOINT ESTIMATION OF THE INDEX AND THE SCALE PARAMETER FOR STABLE PROCESSES

R. HÖPFNER[1] and J.JACOD[2]

[1]Institut für Mathematische Stochastik, Albert-Ludwigs-Universität, Hebelstrasse 27, D-79104 FREIBURG IM BREISGAU.

[2]Laboratoire de Probabilités (CNRS-URA 224), Université P. et M. Curie, 4 Place Jussieu, F-75252 PARIS Cedex.

ABSTRACT We deal with joint estimation of index α and scale parameter ξ in a statistical model where one observes all jumps of a stable increasing process with height not less than Y_n and up to time T_n, or more generally all points of a certain Poisson random measure in a window $[0,T_n]\times[Y_n,\infty)$. We investigate local asymptotic normality of the model and properties of maximum likelihood estimators for different types of asymptotic behaviour of T_n, Y_n as $n\to\infty$.

AMS Subject Classification: 60G55, 62F12.

Key words: point processes, stable processes, local asymptotic normality.

1 - INTRODUCTION

Consider a non-decreasing stable process $X=(X_t)_{t\geq 0}$ without drift, with index $\alpha\in(0,1)$ and scale parameter $\xi>0$, that is whose Laplace transform is:

$$\lambda \longrightarrow E(\exp -\lambda X_t) = \exp -t\xi\Gamma(1-\alpha)\lambda^{\alpha}. \tag{1.1}$$

The question of estimating the pair (α,ξ) has been motivated and studied by Basawa and Brockwell [1], [2].

The laws of X for different values of (α,ξ) are mutually locally singular, so the inference problem is trivial when one observes the path of X over any finite interval. In practice, however, it is usually impossible to observe the path of X, while various partial observation schemes are possible. For example Basawa and Brockwell propose to observe all jumps in a fixed time interval $[0,t]$ and of size bigger than y (the laws of X restricted to the observed σ-field are now equivalent): as $y\to 0$ they proved consistency and asymptotic normality of the MLE $(\hat{\alpha},\hat{\xi})$, with the following limiting distribution if the true value is (α,ξ) and $W\in\mathcal{N}(0,1)$:

$$(t\xi y^{-\alpha})^{1/2}\begin{bmatrix}\hat{\alpha}-\alpha\\ (\hat{\xi}-\xi)/\log y\end{bmatrix} \longrightarrow \begin{bmatrix}\alpha W\\ \xi\alpha W\end{bmatrix}. \tag{1.2}$$

They did not investigate LAN (local asymptotic normality), nor did they prove efficiency for the MLE. They also mention that similar results are true for different observation schemes, where t is still fixed and one observes the biggest jumps of X within $[0,t]$ down to the nth one.

Here we consider more general observation schemes based on the observation of (some) jumps of X, or equivalently of (part of) the random measure μ associated with jumps, that is $\mu = \sum_{s>0,\Delta X_s \neq 0} \delta_{(s,\Delta X_s)}$ (δ_a is the Dirac measure at a). Under (1.1), μ is a Poisson measure with intensity on $\mathbb{R}_+^2$:

$$\alpha\xi\, x^{-(1+\alpha)}\, 1_{(0,\infty)}(t)\, 1_{(0,\infty)}(x)\, dt\, dx. \tag{1.3}$$

Conversely with μ as above, $X_t = \int_0^t\int_0^\infty x\, \mu(ds,dx)$ is a stable process having (1.1). We can even consider Poisson measures having intensity measure (1.3) with $\alpha>0$; we call $(S_i,X_i)_{i\geq1}$ the (random) points in $\mathbb{R}_+^2$ which support the measure μ. Among many possibilities, we study three different observation schemes, which can be described as follows in terms of the asymptotic:

<u>Scheme 1</u>: For each n we fix $y_n>0$, $t_n>0$ and we observe all points (S_i,X_i) inside $[0,t_n]\times[y_n,\infty)$.

<u>Scheme 2</u>: For each n we fix $y_n>0$; we can rearrange the points (S_i,X_i) inside $[0,\infty)\times[y_n,\infty)$ as a sequence (S_i',X_i') with $S_1'<S_2'<\ldots$, and we observe the n points $(S_1',X_1'),\ldots,(S_n',X_n')$.

<u>Scheme 3</u>: For each n we fix $t_n>0$; we can rearrange the points (S_i,X_i) inside $[0,t_n]\times(0,\infty)$ as a sequence (S_i',X_i') with $X_1'>X_2'>\ldots$, and we observe the n points $(S_1',X_1'),\ldots,(S_n',X_n')$.

For Scheme 1, under the parameter value (α_0,ξ_0), the window $(0,t_n]\times[y_n,\infty)$ contains in the mean $\xi_0 t_n y_n^{-\alpha_0}$ points. In order to treat Schemes 1-3 simultaneously, we adopt the <u>convention</u> to write again n for the mean number of observation points in Scheme 1, that is

$$\text{for Scheme 1, under } (\alpha_0,\xi_0), \text{ write } \xi_0 t_n y_n^{-\alpha_0} = n. \tag{1.4}$$

Now about the results. Let $(\hat\alpha_n,\hat\xi_n)$ be the MLE for (α,ξ) at stage n. For all schemes, regardless on the way y_n or t_n behave, $\sqrt{n}(\hat\alpha_n-\alpha)/\alpha$ converges in law to $\mathcal{N}(0,1)$ as $n\to\infty$, if α is the true value. Second, if $y_n\to y\in(0,\infty)$ for Schemes 1 and 2, or $t_n/n\to T\in(0,\infty)$ for Scheme 3, we have the usual LAN property with localizing rate $\sqrt{n}$ for both components, and the MLE is asymptotically regular and efficient for (α,ξ) in the sense of Hájek [4].

Things are more complicated when $y_n \to 0$ or $y_n \to \infty$ for Schemes 1 and 2, or $t_n/n \to 0$ or $t_n/n \to \infty$ for Scheme 3, as is apparent from Basawa and Brockwell's result. The usual localization gives LAN with rate $\sqrt{n}$ for ξ, and a (faster) rate $\sqrt{n}\ |\log y_n|$ or $\sqrt{n}\ |\log t_n/n|$ for α, but the limiting Gaussian model is degenerate. There is also a "non-classical" localization (a non-linear transform) giving LAN with a non-degenerate limit.

Finally we investigate one-dimensional submodels where ξ is a known function of α. The situation is then opposite to the previous one: if $y_n \to y \in (0,\infty)$ or $t_n/n \to T \in (0,\infty)$, it may happen that the MLE is not consistent. If $y_n \to 0$ or $y_n \to \infty$ (resp. $t_n/n \to 0$ or $t_n/n \to \infty$), there is a localization giving LAN with rate $\sqrt{n}\ |\log y_n|$ or $\sqrt{n}\ |\log t_n/n|$ (better than for the complete model), and the MLE is efficient. Further, this rate depends on the observation schemes: it is better to have $|\log y_n|$ or $|\log t_n/n|$ as big as possible, i.e. we should observe either on a very small time interval and down to a very small value of the "size", or the other way around points with very big size only, on a very large time interval.

2 - THE MODEL

1) Let $(\Omega,\mathcal{F})$ be the canonical space of all σ-finite integer-valued measures on $\mathbb{R}^2_+$, with the canonical measure $\mu = \sum_{i\geq 1} \delta_{(S_i,X_i)}$. For $\alpha>0$, $\xi>0$, $P_{\alpha,\xi}$ is the unique probability measure on $(\Omega,\mathcal{F})$ under which μ is a Poisson random measure with intensity given by (1.3). We call $\mathcal{F}^y_t$ the σ-field generated by the restriction of μ to $[y,\infty)\times[0,t]$. If $y>0$ and $t<\infty$, the restrictions of the measures $P_{\alpha,\xi}$ to $\mathcal{F}^y_t$ are all equivalent, and $Z^y_t(\alpha',\xi'|\alpha,\xi) = dP_{\alpha',\xi'}/dP_{\alpha,\xi}|_{\mathcal{F}^y_t}$ is

$$\left.\begin{aligned} &Z^y_t(\alpha',\xi'|\alpha,\xi) = \exp\ \{t(\xi y^{-\alpha} - \xi' y^{-\alpha'}) + N^y_t \log(\tfrac{\xi'\alpha'}{\xi\alpha}) + \overline{N}^y_t(\alpha-\alpha')\}, \\ &\text{where } N^y_t = \sum_{i\geq 1} 1_{\{X_i\geq y, S_i\leq t\}}, \quad \overline{N}^y_t = \sum_{i\geq 1} 1_{\{X_i\geq y, S_i\leq t\}} \log(X_i). \end{aligned}\right\}\tag{2.1}$$

(see e.g. [7], III-5-43). Formula (2.1) remains true if t is replaced by a finite $(\mathcal{F}^y_t)_{t\geq 0}$-stopping time T. Similarly $(\mathcal{F}^{1/z}_t)_{z\geq 0}$ is a filtration, and if $1/Y$ is a finite stopping time relative to this filtration, with $Y>0$, then (2.1) holds with y replaced by Y.

2) In what follows the true value is (α_0,ξ_0). We consider the observation schemes described in the introduction, and introduce some unified notation:

Scheme 1: At stage n we have y_n, t_n related by (1.4), and we observe $\mathcal{F}^{y_n}_{t_n}$. We set $Y_n=y_n$, $T_n=t_n$.

Scheme 2: At stage n we have y_n, and $T_n = \inf(t: N_t^{y_n}=n)$ is an $(\mathcal{F}_t^{y_n})_{t\geq 0}$-stopping time. We observe $\mathcal{F}_{T_n}^{y_n}$, and set $Y_n=y_n$.

Scheme 3: At stage n we have t_n and we observe $\mathcal{F}_{t_n}^{Y_n}$, where $Y_n = \sup(y: N_{t_n}^y=n)$ (so $1/Y_n$ is an $(\mathcal{F}_{t_n}^{1/z})_{z\geq 0}$-stopping time). Set $T_n=t_n$.

In all cases, we set

$$N_n = N_{T_n}^{Y_n}, \qquad \overline{N}_n = \overline{N}_{T_n}^{Y_n}, \qquad L_n = \log Y_n.$$

These are observable, and $N_n=n$ for Schemes 2 and 3, and L_n is deterministic for Schemes 1 and 2. The density of $P_{\alpha,\xi}$ w.r.t. P_{α_0,ξ_0}, in restriction to the observed σ-field, is

$$Z_n(\alpha,\xi) = \exp \{T_n(\xi_0 e^{-\alpha_0 L_n} - \xi e^{-\alpha L_n}) + N_n \log \frac{\alpha\xi}{\alpha_0\xi_0} - \overline{N}_n(\alpha-\alpha_0)\}. \tag{2.2}$$

A simple computation shows that the MLE for the pair (α,ξ) at stage n is

$$\hat{\alpha}_n = N^n/(\overline{N}_n - N_n L_n), \qquad \hat{\xi}_n = (N_n/T_n) \exp \hat{\alpha}_n L_n. \tag{2.3}$$

REMARK 1: When $\alpha<1$, so $X_t = \int_0^t\int_0^\infty x\,\mu(ds,dx)$ is under $P_{\alpha,\xi}$ a stable process with (1.1), then $P_{\alpha,\xi}(X_1>r) \sim \xi r^{-\alpha}$ as $r\to\infty$ (see e.g. Feller [3]). In this case one can view $\hat{\alpha}_n$ and $\hat{\xi}_n$ as continuous-time versions of the estimators introduced by Hill [6] for the index of regular variation α and the thickness ξ of the tail of the law of X_1: see also Hall [5]. ■

3) The asymptotic behaviour of the following variables is essential:

$$U_n = \frac{1}{\sqrt{n}}[N_n + \alpha_0(N_n L_n - \overline{N}_n)], \qquad V_n = (N_n-n)/\sqrt{n}, \qquad W_n = (T_n\xi_0 e^{-\alpha_0 L_n} - n)/\sqrt{n}.$$

LEMMA 1: a) *For Scheme 1 we have* $W_n = 0$ *and* (U_n,V_n) *converges in law under* P_{α_0,ξ_0} *to* $\mathcal{N}(0,I_2)$.

b) *For Schemes 2 and 3 we have* $V_n = 0$ *and* (U_n,W_n) *converges in law under* P_{α_0,ξ_0} *to* $\mathcal{N}(0,I_2)$.

Proof. For Scheme 1 (1.4) yields $W_n=0$, while in Schemes 2 and 3 we have $N_n=n$, hence $V_n=0$. Throughout the proof $P=P_{\alpha_0,\xi_0}$.

a) For Scheme 1, a standard computation on Poisson measures yields

$$\varphi_n(u,v) := \log E[\exp(-uN_n-v\overline{N}_n)] = -n + n\alpha_0 e^{-u-vL_n} \frac{1}{\alpha_0+v}.$$

Therefore (here L_n is deterministic):

$$\log E[e^{-uU_n - vV_n}] = -n + v\sqrt{n} + ne^{-(u+v)/\sqrt{n}} \frac{1}{1-u/\sqrt{n}} \longrightarrow \tfrac{1}{2}(u^2 + v^2).$$

b1) Consider Scheme 2. The points (S_i,X_i) in $[0,\infty)\times[y_n,\infty)$ can be labelled as (S'_i,X'_i) with $0<S'_1<S'_2<\ldots$: the sequences $(S'_i)_{i\geq 1}$ and $(X'_i)_{i\geq 1}$ are independent, and X'_i are i.i.d. with density $x \to \alpha_0 x^{-1-\alpha_0} y_n^{\alpha_0} 1_{(y_n,\infty)}(x)$, and $(S'_i)_{i\geq 1}$ is a Poisson process on $\mathbb{R}_+$ with intensity $\xi_0 y_n^{-\alpha_0}$, and $T_n = S'_n$. Since $N_n=n$ and $Y_n=y_n$, we get that U_n and W_n are independent.

We have $\overline{N}_n = \sum_{1\leq i\leq n} \log X'_i$ and $E(\exp -u \log X'_i) = e^{-uL_n} \frac{\alpha_0}{\alpha_0+u}$, hence

$$\log E(e^{uU_n}) = u\sqrt{n} - n \log(1+u/\sqrt{n}) \longrightarrow u^2/2. \tag{2.4}$$

Next T_n has a gamma distribution with index n and parameter $\xi_0 e^{-\alpha_0 L_n}$, so

$$\log E(e^{-uW_n}) = u\sqrt{n} - n \log(1+u/\sqrt{n}) \longrightarrow u^2/2. \tag{2.5}$$

b2) Consider Scheme 3. The points (S_i,X_i) in $[0,t_n]\times(0,\infty)$ can be labelled as (S'_i,X'_i) with $X'_1>X'_2>\ldots$, and if $Z_i = X_i'^{-\alpha_0}$ the sequence $(Z_i)_{i\geq 1}$ is a Poisson process on $\mathbb{R}_+$ with intensity $\xi_0 t_n$. Further $\overline{N}_n = -\frac{1}{\alpha_0}\sum_{1\leq i\leq n} \log Z_i$ and $U_n = \sqrt{n} + \frac{1}{\sqrt{n}} \sum_{1\leq i\leq n} \log(Z_i/Z_n)$, and $e^{-\alpha_0 L_n} = Z_n$. First Z_n is gamma with index n and parameter $\xi_0 t_n$, hence (2.5) holds. Second the family $(Z_i/Z_n : 1\leq i\leq n-1)$ is independent from Z_n and has the same law as the increasing rearrangement of $n-1$ independent variables, uniform over $(0,1)$. So U_n and W_n are independent and a simple computation shows (2.4). ∎

Note that (2.3) yields

$$\sqrt{n}\,\frac{\hat{\alpha}_n - \alpha_0}{\alpha_0} = \frac{U_n}{1 + (V_n - U_n)/\sqrt{n}} \tag{2.6}$$

and thus, as a first consequence of Lemma 1 we readily obtain the

COROLLARY 1: *For all observation schemes,* $\sqrt{n}(\hat{\alpha}_n - \alpha_0)/\alpha_0$ *converges in law under* P_{α_0,ξ_0} *to* $\mathcal{N}(0,1)$, *where* $\hat{\alpha}_n$ *is the first component of the MLE.*

This is an extension of the asymptotic behaviour of the Hill estimator (see e.g. Theorem 2 of Hall [5]). For studying the second component $\hat{\xi}_n$ of the MLE, we first introduce the following *deterministic* sequence:

$$\ell_n = \alpha_0 L_n \text{ for Scheme 1 and 2,} \qquad \ell_n = \log t_n \xi_0 / n \text{ for Scheme 3.} \tag{2.7}$$

COROLLARY 2: a) *For all observation schemes, the sequence* $(\hat{\xi}_n)$ *is weakly consistent at* ξ_0 *if and only if* $\ell_n/\sqrt{n} \to 0$.

b) *If* $\ell_n/\sqrt{n} \to 0$ *and* $|\ell_n| \to \infty$, *then*

$$\frac{\sqrt{n}}{\xi_0 \ell_n}(\hat{\xi}_n - \xi_0) = \frac{\sqrt{n}}{\alpha_0}(\hat{\alpha}_n - \alpha_0) + o_{P_{\alpha_0,\xi_0}}(1).$$

c) *If* $\ell_n \to \ell \in \mathbb{R}$, *then*

$$\left(\frac{\sqrt{n}}{\alpha_0}(\hat{\alpha}_n - \alpha_0), \frac{\sqrt{n}}{\xi_0}(\hat{\xi}_n - \xi_0)\right) = (U_n, \ell U_n + V_n - W_n) + o_{P_{\alpha_0,\xi_0}}(1),$$

which under P_{α_0,ξ_0} *converges in law to* $\mathcal{N}\left(0, \begin{bmatrix} 1 & \ell \\ \ell & 1+\ell^2 \end{bmatrix}\right)$.

(b) and Corollary 1 give an extension of Basawa and Brockwell's result.

Proof. Since $\ell_n - \alpha_0 L_n = \log(1+W_n/\sqrt{n})$ for Scheme 3, $\ell_n/\sqrt{n} \to 0 \Leftrightarrow L_n/\sqrt{n} \xrightarrow{P} 0$ in all cases by Lemma 1. So the result readily follows from Lemma 1 and Corollary 2, once noticed that by (2.3) we have

$$\frac{\hat{\xi}_n}{\xi_0} = \frac{1+V_n/\sqrt{n}}{1+W_n/\sqrt{n}} \exp\left[\frac{\sqrt{n}}{\alpha_0}(\hat{\alpha}_n - \alpha_0)\frac{L_n \alpha_0}{\sqrt{n}}\right]. \blacksquare$$

3 - LOCALIZATION

1) For proving asymptotic optimality of the MLE we need to localize the model around the true value (α_0, ξ_0). The first localization below is not traditional, in the sense that it is not an affine transformation of the initial coordinates, and gives a sequence of "curved" statistical models, the curve actually depending on n. With the notation (2.7), the parameters (γ, η) of the local model around (α_0, ξ_0) at stage n are given by

$$\alpha = \alpha_0(1 + \gamma/\sqrt{n}), \qquad \xi = \xi_0(1 + \eta/\sqrt{n})e^{\ell_n \gamma/\sqrt{n}} \tag{3.1}$$

REMARK 2: At first glance this localization may seem a bit strange. To get some insight, let us consider only Schemes 1 and 2 and reparametrize the model by (α, β_n), with $\beta_n = \beta_n(\alpha,\xi) := \xi y_n^{-\alpha}$ $(\Leftrightarrow \xi = \beta_n y_n^{\alpha})$.

This reparametrization depending on n is natural, since at stage n one observes in fact, on $[0,t_n]$, the process $X_t^n = \int_0^t \int_{y_n}^{\infty} x\, \mu(ds,dx)$ which, under $P_{\alpha,\xi}$, is a step Markov process with parameter $\beta_n(\alpha,\xi)$ for holding times. The true value of β_n is $\beta_{n,0} = \xi_0 y_n^{-\alpha_0}$, and the local parameters (γ,η) at stage n are given in function of (α,β_n) by

$$\alpha = \alpha_0(1 + \gamma/\sqrt{n}), \qquad \beta_n = \beta_{n,0}(1 + \eta/\sqrt{n}). \tag{3.2}$$

So (3.1) yields a local model in the usual sense for the (α,β_n)-reparametrization. The MLE $\hat{\beta}_n = N_n/T_n$ for β_n is always consistent at (α_0,ξ_0), i.e. $\hat{\beta}_n/\beta_{n,0} \longrightarrow 1$ in P_{α_0,ξ_0}-probability, because $\hat{\beta}_n/\beta_{n,0} = (1+V_n/\sqrt{n})(1+W_n/\sqrt{n})$. ∎

At stage n, the space of local parameters is $\Theta_n = (-\sqrt{n},\infty)\times(-\sqrt{n},\infty)$. Note that $\Theta_n \uparrow \mathbb{R}^2$ as $n\uparrow\infty$. For $n\in\mathbb{N}^*$, we set $P^n_{\gamma,\eta} = P_{\alpha,\xi}$ for (γ,η) related with (α,ξ) by (3.1), and $\mathcal{E}_n = (\Omega,\mathcal{F}^{Y^n}_{T_n},(P^n_{\gamma,\eta})_{(\gamma,\eta)\in\Theta_n})$. Using (2.3) and (2.6), we see that the MLE in the model $\mathcal{E}_n$ is

$$\hat{\gamma}_n = \frac{U_n}{1+(V_n-U_n)/\sqrt{n}}, \qquad \hat{\eta}_n = \begin{cases} \dfrac{V_n-W_n}{1+W_n/\sqrt{n}} & \text{for Schemes 1,2} \\ \sqrt{n}[(1+W_n/\sqrt{n})^{-1-\hat{\gamma}_n/\sqrt{n}} - 1] & \text{for Scheme 3.} \end{cases} \tag{3.3}$$

The limiting model will be the Gaussian shift $\mathcal{E} = (\mathbb{R}^2,\mathcal{R}^2,\mathcal{N}(\begin{bmatrix}\gamma\\ \eta\end{bmatrix},I_2)_{(\gamma,\eta)\in\mathbb{R}^2})$.

THEOREM 1: *For all observation schemes, as* $n\to\infty$ *the sequence* $\mathcal{E}_n$ *weakly converges to* $\mathcal{E}$. *Further the sequence* $(\hat{\gamma}_n,\hat{\eta}_n)$ *converges in law to* $\mathcal{N}(0,I_2)$, *under* P_{α_0,ξ_0} *and is a central sequence for* $\mathcal{E}_n$.

Proof. Set $P=P_{\alpha_0,\xi_0}$, and $\bar{Z}_n(\gamma,\eta) = Z_n(\alpha,\xi)$ with (α,ξ) and (γ,η) related by (3.1). For Schemes 1 and 2 (resp. 3) we deduce from (2.2) and (2.7):

$$\log \bar{Z}_n(\gamma,\eta) = -\eta T_n\xi_0 e^{-\ell_n}/\sqrt{n} + N_n[\log(1+\eta/\sqrt{n})(1+\gamma/\sqrt{n}) + \ell_n\gamma/\sqrt{n}] - \bar{N}_n\alpha_0\gamma/\sqrt{n}$$

$$(\text{resp.} = T_n\xi_0 e^{-\alpha_0 L_n}[1 - (1+\eta/\sqrt{n})e^{\gamma(\ell_n-\alpha_0L_n)/\sqrt{n}}]$$

$$+ N_n[\log(1+\eta/\sqrt{n})(1+\gamma/\sqrt{n}) + \ell_n\gamma/\sqrt{n}] - \bar{N}_n\alpha_0\gamma/\sqrt{n}\ ).$$

Then a simple computation shows that for Scheme 1,

$$\log \bar{Z}_n(\gamma,\eta) = \gamma U_n + \eta V_n - (\gamma^2+\eta^2)/2 + o_P(1), \tag{3.4}$$

while for Schemes 2 and 3, we get

$$\log \bar{Z}_n(\gamma,\eta) = \gamma U_n - \eta W_n - (\gamma^2+\eta^2)/2 + o_P(1). \tag{3.5}$$

(3.4), (3.5), and Lemma 1 prove that $\mathcal{E}_n \longrightarrow \mathcal{E}$ and also that a central sequence is (U_n,V_n) for Scheme 1 and $(U_n,-W_n)$ for Schemes 2 and 3. (3.3) gives $\hat{\gamma}_n = U_n + o_P(1)$, and also $\hat{\eta}_n = V_n + o_P(1)$ for Scheme 1, and $\hat{\eta}_n = -W_n + o_P(1)$ for Scheme 2 or 3. ∎

REMARK 3: a) See LeCam [8] or Strasser [9] for the notion of central sequences. As a consequence of this theorem, $(\hat{\gamma}_n,\hat{\eta}_n)$ is an efficient sequence of estimators for the local parameters (γ,η).

b) For Schemes 1 or 2, by Remark 2 and in particular by (3.2), the estimation sequence $(\hat{\alpha}_n,\hat{\beta}_n)$ for (α,β_n) is regular in the sense of Hájek [4] and efficient at $(\alpha_0,\beta_{n,0})$ (i.e. under P_{α_0,ξ_0}).

c) However, for Schemes 1 or 2, no conclusion can be drawn on optimality of $\hat{\xi}_n$ as an estimator of ξ. By Corollary 2, $\hat{\xi}_n$ need not even converge to ξ.■

2) Suppose now that $\ell_n \to \ell$ with $\ell\in\mathbb{R}$ (recall (2.7)). The second part of (3.1) gives $\xi \sim \xi_0(1 + \eta/\sqrt{n} + \gamma\ell/\sqrt{n})$ and we are led to consider new local coordinates (γ,H) with

$$\alpha = \alpha_0(1 + \gamma/\sqrt{n}), \qquad \xi = \xi_0(1 + H/\sqrt{n}). \tag{3.6}$$

We call $\mathcal{E}'_n$ the local model at stage n, with local parameters (3.6) and $(\hat{\gamma}_n,\hat{H}_n)$ the corresponding MLE. The limiting model is the Gaussian model $\mathcal{E}' = (\mathbb{R}^2,\mathcal{R}^2,\mathcal{N}(I(\ell)\begin{bmatrix}\gamma\\ H\end{bmatrix},I(\ell))_{(\gamma,H)\in\mathbb{R}^2})$, where $I(\ell) = \begin{bmatrix} 1+\ell^2 & -\ell \\ -\ell & 1\end{bmatrix}$.

Since H and $\eta+\gamma\ell$ are "equivalent", an affine transformation readily gives the following corollary (which can of course be proved directly):

COROLLARY 3: *When $\ell_n \to \ell\in\mathbb{R}$, the sequence $\mathcal{E}'_n$ weakly converges to $\mathcal{E}'$. Further, the pair $(\hat{\gamma}_n,\hat{H}_n)$ converges in law to $\mathcal{N}(0,I(\ell)^{-1})$, under P_{α_0,ξ_0}, and is a central sequence for $\mathcal{E}'_n$.*

REMARK 4: The pair $(\hat{\alpha}_n,\hat{\xi}_n)$ is regular and efficient for (α,ξ) at (α_0,ξ_0).■

3) There is also a third localization giving the LAN property. We consider only the case $|\ell_n| \to \infty$. It is motivated by the correct choice of norming constants, taken *separately*, for the α- and ξ-components of the score function at (α_0,ξ_0). Set

$$\alpha = \alpha_0(1 + \Gamma/\ell_n\sqrt{n}), \qquad \xi = \xi_0(1 + H/\sqrt{n}). \tag{3.7}$$

We call $\mathcal{E}''_n$ the local model at stage n with local parameters (3.7), and $(\hat{\Gamma}_n,\hat{H}_n)$ the corresponding MLE. The limiting model is the Gaussian model $\mathcal{E}'' = (\mathbb{R}^2,\mathcal{R}^2,\mathcal{N}(J\begin{bmatrix}\Gamma\\ H\end{bmatrix},J)_{(\Gamma,H)\in\mathbb{R}^2})$, where $J = \begin{bmatrix} 1 & -1\\ -1 & 1\end{bmatrix}$.

THEOREM 2: *If $|\ell_n| \to \infty$, the sequence $\mathcal{E}''_n$ weakly converges to $\mathcal{E}''$ as* $n\to\infty$.

The model $\mathcal{E}''$ is degenerate, with non-invertible Fisher information matrix. The MLE $(\hat{\Gamma}_n,\hat{A}_n)$ do not converge in law here: this is of course in accordance with Corollary 2. The (easy) proof is left to the reader.

4 - ONE-DIMENSIONAL SUBMODELS

For $A\subseteq(0,\infty)$ open and $f: A \to (0,\infty)$ twice continuously differentiable, we look at submodels $\Theta_f = \{(\alpha,f(\alpha)): \alpha\in A\}\subseteq\Theta$. We write $P_\alpha = P_{\alpha,f(\alpha)}$. An example is given by $A=(0,1)$ and $f(\alpha) = 1/\Gamma(1-\alpha)$: this corresponds to the observation of an increasing stable process X with Laplace transform $E(\exp -\lambda X_t) = \exp -t\lambda^\alpha$ (cf. (1.1)). Another example is when $\xi=\xi_0$ is known, so $f(\alpha)=\xi_0$.

The true value of α is $\alpha_0\in A$, and $\xi_0=f(\alpha_0)$. We use the notation ℓ_n of (2.7). The localization for α will be similar to (3.7), except that we do not assume $|\ell_n|\to\infty$, so ℓ_n may vanish. Therefore we set, at stage n:

$$\alpha = \alpha_0(1 + \omega_n\Gamma) \quad \text{with } \omega_n = \omega_n(\alpha_0) := [n(1+\ell_n^2)]^{-1/2}. \tag{4.1}$$

The space of local parameters is $\bar{\Theta}_n = \{\Gamma: \alpha_0(1+\omega_n\Gamma)\in A\}$, so $\bar{\Theta}_n\uparrow\mathbb{R}$. Set $\bar{P}^n_\Gamma = P_\alpha$ for Γ related to α by (4.1), and $\bar{\mathcal{E}}_n = (\Omega,\mathcal{F}^n_{T_n},(\bar{P}^n_\Gamma)_{\Gamma\in\bar{\Theta}_n})$. We consider an MLE $\check{\alpha}_n$ for α at stage n (which is different from $\hat{\alpha}_n$ in (2.3)!, by convention $\check{\alpha}_n=\alpha_1$ with α_1 arbitrary in A if the likelihood does not reach its maximum inside A), and the corresponding MLE for the local model:

$$\check{\Gamma}_n = (\check{\alpha}_n - \alpha_0)/\alpha_0\omega_n \tag{4.2}$$

THEOREM 3: a) *If $\ell_n \to \ell\in\mathbb{R}$, the sequence $\bar{\mathcal{E}}_n$ weakly converges to the model $(\mathbb{R},\mathcal{R},\mathcal{N}(\Gamma,\varphi)_{\Gamma\in\mathbb{R}})$, where*

$$c = \alpha_0\frac{f'(\alpha_0)}{f(\alpha_0)}, \qquad \varphi = [1+(c-\ell)^2]/(1+\ell^2). \tag{4.3}$$

b) *If $|\ell_n| \to +\infty$, the sequence $\bar{\mathcal{E}}_n$ weakly converges to the model $(\mathbb{R},\mathcal{R},\mathcal{N}(\Gamma,1)_{\Gamma\in\mathbb{R}})$, and $\check{\Gamma}_n$ is a central sequence for $\bar{\mathcal{E}}_n$ and converges in law under P_{α_0} to $\mathcal{N}(0,1)$.*

Proof. 1) We prove first the convergence of the models in (a) and (b). We use all notation of the proof of Theorem 1, and $P = P_{\alpha_0}$. Note that

$$\log Z_n(\alpha,f(\alpha)) = nW_n'[1 - \frac{f(\alpha)}{f(\alpha_0)}\, e^{-(\alpha-\alpha_0)L_n}] + N_n \log\frac{f(\alpha)\alpha}{f(\alpha_0)\alpha_0} + N_n(\alpha_0-\alpha). \tag{4.4}$$

Set $\bar{Z}_n(\Gamma) = Z_n(\alpha,f(\alpha))$ with α and Γ related by (4.1). Then

$$\log \bar{Z}_n(\Gamma) = nW_n'[1 - \frac{f(\alpha)}{f(\alpha_0)}\, e^{-\alpha_0\omega_nL_n\Gamma}] + \Gamma\omega_nU_n\sqrt{n}$$
$$+ (n+V_n\sqrt{n})[\log\frac{f(\alpha)\alpha}{f(\alpha_0)\alpha_0} - (1+\alpha_0L_n)\omega_n\Gamma].$$

For Schemes 1 and 2 $\alpha_0 L_n = \ell_n$, while $\alpha_0 L_n = \ell_n - \log(1+W_n/\sqrt{n})$ for Scheme 3. In all cases $\omega_n L_n \sqrt{n}$ remains bounded in probability, so with c as in (4.3):

$$\log \overline{Z}_n(\Gamma) = \Gamma\omega_n\sqrt{n}[U_n + (V_n-W_n)(c-\alpha_0 L_n)] - \Gamma^2\omega_n^2 n[1+(c-\alpha_0 L_n)^2]/2 + o_P(1).$$

Now (a) is obvious, since then $\alpha_0 L_n \to \ell$ in probability (at least). For (b), $\omega_n\sqrt{n} \to 0$ and $|\alpha_0 L_n \omega_n \sqrt{n}| \to 1$, hence if $\overline{V}_n$ and $\overline{W}_n$ equal V_n and W_n multiplied by the sign of ℓ_n, then $\log \overline{Z}_n(\Gamma) = -\Gamma(\overline{V}_n - \overline{W}_n) - \Gamma^2/2 + o_P(1)$: this proves the convergence of $\overline{\mathcal{E}}_n$ and also that $-\overline{V}_n$ for Scheme 1 and $\overline{W}_n$ for Schemes 2 or 3 is a central sequence. So for (b) it remains to prove that

$$\check{\Gamma}_n = -\overline{V}_n + o_P(1) \quad \text{(Scheme 1)}, \qquad \check{\Gamma}_n = \overline{W}_n + o_P(1) \quad \text{(Schemes 2, 3)}. \tag{4.5}$$

2) Now we assume $|\ell_n| \to \infty$. In a first step we prove that

$$P(|\ell_n(\check{\alpha}_n - \alpha_0)| \le \delta\}) \longrightarrow 1 \quad \text{for all } \delta>0. \tag{4.6}$$

Set $V_n(\beta) = \frac{1}{n}\log Z_n(\alpha, f(\alpha))$ and $X(\beta) = f(\alpha)/f(\alpha_0)$ for $\beta = (\alpha-\alpha_0)/\alpha_0$. Since $V_n(0)=0$ and $\check{\alpha}_n$ is the argument of one of the maxima of $V_n(.)$, (4.6) will follow if we prove that for all $\delta>0$, and if $L'_n = \alpha_0 L_n$:

$$\lim_n P(\sup_{\beta: |\beta L'_n| > \delta} V_n(\beta) < 0) = 1. \tag{4.7}$$

If $M_n = N_n/n$ and $W'_n = 1+W_n/\sqrt{n}$, we have by Lemma 1:

$$M_n \xrightarrow{P} 1, \qquad W'_n \xrightarrow{P} 1, \qquad L'_n/\ell_n \xrightarrow{P} 1. \tag{4.8}$$

We may rewrite (4.4) as

$$V_n(\beta) = W'_n[1 - X(\beta)e^{-\beta L'_n}] + M_n[\log X(\beta) + \log(1+\beta) - \beta - \beta L'_n] + \beta U_n/\sqrt{n}. \tag{4.9}$$

Fix $\delta>0$. Since $X(0)=1$ and X is continuous at 0, for $C \in (0,1)$ there is $C'>0$ with $|\beta| \le C' \Rightarrow |X(\beta)-1| \le C$. Hence if $|\beta| \le C'$ and if $u_n = W'_n C + M_n[\log(1+C)(1+C') + C']$, $v_n = W'_n(1-C)$ and $w_n = M_n - U_n/L'_n\sqrt{n}$, we have

$$W'_n > 0 \Rightarrow V_n(\beta) \le u_n + v_n(1-e^{-\beta L'_n}) - w_n \beta L'_n.$$

Now if $a>0$ and $1 < b/a < (e^{\delta}-e^{-\delta})/2\delta$, the function $x \to a(1-e^{-x})-bx$ reaches its maximum on the set $\{x: |x| \ge \delta\}$ at $x=\delta$, for which it equals $a(1-e^{-\delta})-b\delta$. Then on the set $A_n = \{W'_n>0\} \cap \{v_n>0\} \cap \{1<w_n/v_n<(e^{\delta}-e^{-\delta})/2\delta\}$,

$$V_n(\beta) \le u_n + v_n(1-e^{-\delta}) - w_n\delta \quad \text{if } |\beta| \le C', \ |\beta L'_n| \ge \delta.$$

But (4.8) yields $u_n \xrightarrow{P} u := C + C' + \log(1+C)(1+C')$, $v_n \xrightarrow{P} 1-C$ and $w_n \xrightarrow{P} 1$. We choose C, C' small enough so that $u + (1-C)(1-e^{-\delta}) - \delta < 0$ and that $1/(1-C) < (e^{\delta}-e^{-\delta})/2\delta$, so $P(A_n) \to 1$ and finally

$$\lim_n P(\sup_{\beta: |\beta L'_n|>\delta, |\beta|\le C'} V_n(\beta) < 0) = 1. \tag{4.10}$$

Next $-ae^{-v}-bv \le b(-1 - \log a + \log b)$ for all $v\in\mathbb{R}$ if $a,b>0$. Applying this to $v = \beta L'_n$, $a = W'_n X(\beta)$, $b = M_n$, we deduce from (4.9) that on $\{W'_n>0\}$:

$$\begin{aligned} V_n(\beta) &\le W'_n + M_n[-1 - \log W'_n + \log(1+\beta) - \beta + \log M_n] + \beta U_n/\sqrt{n} \\ &\le M_n \log(1+\beta) + A_n\beta + B_n \end{aligned}$$

where $A_n := -M_n + U_n/\sqrt{n} \xrightarrow{P} -1$ and $B_n := W'_n - M_n(1 + \log W'_n + \log M_n) \xrightarrow{P} 0$; since $\sup_{\beta: |\beta|\ge C'}[\log(1+\beta) - \beta] < 0$ when $C'>0$, we deduce

$$\lim_n P(\sup_{\beta: |\beta|\ge C'} V_n(\beta) < 0) = 1. \tag{4.11}$$

Then (4.10) and (4.11) prove (4.8), hence (4.6).

3) Now we prove (4.5). The derivative of $\alpha \to \log Z_n(\alpha,f(\alpha))$ is

$$U_n(\alpha) = nW'_n \frac{f(\alpha)}{f(\alpha_0)}[L_n - \frac{f'(\alpha)}{f(\alpha)}]e^{-(\alpha-\alpha_0)L_n} + N_n[\frac{f'(\alpha)}{f(\alpha)} + \frac{1}{\alpha}] - \overline{N}_n. \tag{4.12}$$

Let B_n be the set where the maximum of $Z_n(.,f(.))$ is reached inside A. On B_n we have $U_n(\check{\alpha}_n)=0$, hence Taylor's expansion yields

$$U_n(\alpha_0) + (\check{\alpha}_n-\alpha_0)U'_n(\beta_n) = 0, \text{ where } \beta_n \text{ is between } \alpha_0 \text{ and } \check{\alpha}_n. \tag{4.13}$$

By (4.12),

$$U'_n(\alpha) = - nW'_n \frac{f(\alpha)}{f(\alpha_0)}[(L_n - \frac{f'(\alpha)}{f(\alpha)})^2 + g(\alpha)]e^{-(\alpha-\alpha_0)L_n} + N_n[g(\alpha) - \alpha^{-2}]$$

where $g = (ff''-f'^2)/f^2$. We know that $\alpha_0 L_n/\ell_n \xrightarrow{P} 1$, and f, $f'f$ and g are locally bounded on A, hence by (4.8), the continuity of f, and $\alpha_0\in A$:

$$\lim_{\delta\to 0} \limsup_n P(\sup_{\alpha: |\ell_n(\alpha-\alpha_0)|\le\delta} |1 + \frac{U'_n(\alpha)}{nL_n^2}| > \eta) = 0,$$

for all $\eta>0$, and it follows by (4.6) that, with β_n as in (4.13):

$$U'_n(\beta_n)/n\ell_n^2 \xrightarrow{P} -1/\alpha_0^2. \tag{4.14}$$

Next (4.3) and (4.12) yield

$$\alpha_0 U_n(\alpha_0) = nW'_n(\alpha_0 L_n-c) + N_n(c+1) - \alpha_0\overline{N}_n = [U_n + (\alpha_0 L_n-c)(W_n - V_n)]\sqrt{n},$$

and thus $U_n(\alpha_0)\omega_n = \overline{W}_n-\overline{V}_n + o_P(1)$. This, (4.14), (4.13) and the fact that $P(B_n) \to 1$ (which follows from (4.6)) readily give (4.5). ∎

REMARK 5: If $\ell_n \to \ell \in \mathbb{R}$ the statement (b) of Theorem 3 may be wrong, and it might even happen that the MLE $\check{\alpha}_n$ are not consistent. However, if $f(\alpha) = \xi_0$ for all α, then whatever the limit ℓ is, $\check{\Gamma}_n$ is a central sequence for $\overline{\mathscr{E}}_n$ and it converges in law to $\mathcal{N}(0,1/\sqrt{\varphi})$. ■

REMARK 6: If one has the choice of the observation schemes, it is suitable to have $\omega_n \longrightarrow 0$ as fast as possible, or equivalently $|\ell_n| \longrightarrow \infty$ as fast as possible. So for Schemes 1 or 2 (resp. Schemes 1 or 3) it is best to take y_n (resp. t_n) as small as possible. This is in contrast with the complete 2-dimensional model, for which the (best) rate of convergence of estimators of α is always $1/\sqrt{n}$, regardless of the scheme. ■

REMARK 7: It is obviously possible to undertake a similar study when α is a known function $g(\xi)$ of ξ, twice continuously differentiable on some open subset B of $(0,\infty)$. We consider two cases:

1) The derivative has $g'(\xi_0)\neq 0$: then in a neighbourhood of ξ_0 there is a bi-continuous one-to-one correspondance between $\alpha=g(\xi)$ and ξ, and the previous results for α-models are immediately transcribed in terms of ξ-models.

2) $g(\xi)=\alpha_0$ is constant on B: then for every fixed n, the model is an exponential family by (2.2). So UMVU estimators for certain functionals of ξ exist: take $(N_n/t_n)e^{\ell_n}$ as estimator for ξ in Scheme 1, $(T_n/n)e^{-\alpha_0 L_n}$ for $1/\xi$ in Schemes 2 or 3 (recall that α_0 is known here). By Lemma 1 these estimators are asymptotically normal with rate $1/\sqrt{n}$, and LAN property will hold for local models $\xi = \xi_0(1+H/\sqrt{n})$ at ξ_0 for all observation schemes. ■

REFERENCES

[1] BASAWA I.V., BROCKWELL P.J. (1978): Inference for gamma and stable processes. *Biometrika* 65, 129-133.

[2] BASAWA I.V., BROCKWELL P.J. (1980): A note on estimation for gamma and stable processes. *Biometrika* 67, 234-236.

[3] FELLER W. (1971): *An introduction to probability theory and its applications*, Vol. 2, J. Wiley and sons: New York.

[4] HÁJEK J. (1970): A characterization of limiting distributions for regular estimates. *Z. Wahr. Th. Verw. Geb.* 14, 323-330.

[5] HALL P. (1982): On some simple estimates of an exponent of regular variation. *J. R. Statist. Soc.* B, 44, 37-42.

[6] HILL B.M. (1975): A simple general approach to inference about the tail of a distribution. *Ann. Math. Statist.* 3, 1163-1174.

[7] JACOD J., SHIRYAEV A.N. (1987): *Limit theorems for stochastic processes.* Springer Verlag: Berlin.

[8] LE CAM L. (1986): *Asymptotic methods in statistical decision theory.* Springer Verlag: Berlin.

[9] STRASSER H. (1985): *Mathematical theory of statistics.* De Gruyter: Berlin.

ASYMPTOTIC BEHAVIOUR OF THE ERROR PROBABILITIES IN THE PSEUDO-LIKELIHOOD RATIO TEST FOR GIBBS-MARKOV DISTRIBUTIONS

Martin Janžura

Institute of Information Theory and Automation

Academy of Sciences of the Czech Republic

Pod vodárenskou věží 4

182 08 Praha 8

CZECH REPUBLIC

Abstract

For testing simple statistical hypotheses given by Gibbs-Markov random fields a Bayes test based on the pseudo-likelihood ratio is constructed in order to deal with numerically available quantities. The asymptotic behaviour of the error probabilities is investigated with the aid of an appropriate version of the large deviations theorem. Under natural assumptions the error probabilities are proved to tend to zero exponentially fast.

Key words: Gibbs distribution, potential, pseudo-likelihood ratio, error probability

AMS classification: 62 M 40

1 Introduction

Let us consider a test of the simple statistical hypothesis

$$H^0 : P = P^0$$

against the alternative

$$H^1 : P = P^1,$$

where P^0, P^1 are probability measures on $(X, \mathcal{B})^T$ where $(X, \mathcal{B})$ is a standard Borel space and $T = Z^d$ is the d-dimensional integer lattice. Such distributions will be called the random fields.

The Bayes test, based on the data $x^V \in X^V$, $V \subset Z^d$ finite, is given by the critical region (the hypothesis is rejected)

$$C_V = \left\{ x \in X^T ; \log \frac{dP_V^1(x_V)}{dP_V^0(x_V)} > \gamma_V \right\}$$

where γ_V is a suitable constant. Under some natural technical conditions this test is proved to be optimal, its asymptotic properties can be studied (cf. Janžura (1984)), but there remains the obvious problem with practical constructing the critical region since it requires to evaluate the log likelihood ratio

$$L_V(x_V) = \log \frac{dP_V^1(x_V)}{dP_V^0(x_V)}$$

which is extremely difficult and mostly impossible even for the basic non-trivial "exponential-like" cases.

Hence, for P^0, P^1 Markovian, i. e. satisfying for every finite $A \subset Z^d$

$$dP^i_{A|A^c}(x_A|x_{A^c}) = dP^i_{A|\partial A}(x_A|x_{\partial A}) \quad \text{a. s. } [P^i]$$

$i = 0, 1,\ \partial A \subset A^c$ finite, we set

$$L_V^A(x_V) = \sum_{t: t+(A\cup\partial A)\subset V} \log \frac{dP^1_{A+t|\partial A+t}(x_{A+t}|x_{\partial A+t})}{dP^0_{A+t|\partial A+t}(x_{A+t}|x_{\partial A+t})}.$$

For fixed finite A we call $L_V^A(x_V)$ the log pseudo-likelihood ratio and propose the test based on the critical region

$$C_V^A = \{x \in X^T;\ L_V^A(x_V) > \gamma_V^A\}$$

again with some appropriate constant γ_V^A. We shall study the asymptotic behaviour for $V \nearrow T$ of the error probabilities for the test. Thanks to the well-known correspondence between the Markov and the Gibbs random fields we shall rather follow the Gibbsian approach. Moreover, we shall study more general setting which will yield a solution of the present problem as a special case in Section 5. As the main source of basic references on Gibbs random fields we follow the comprehensive book of Georgii (1988).

2 Gibbs random fields

Let $\mathcal{P}$ denote the set of all random fields, i. e. the probability measures on $(X, \mathcal{B})^T$. Let us recall that $(X, \mathcal{B})$ is a standard Borel space and $T = Z^d$ is a d-dimensional integer lattice. Let us denote by $k(T)$ the system of all non-void finite subsets of T.

Further we denote $\mathcal{P}_\Theta \subset \mathcal{P}$ the subset of all shift-invariant (stationary) random fields, i. e.

$$P \in \mathcal{P}_\Theta \quad \text{iff} \quad P = P \circ \theta_t^{-1} \quad \text{for every } t \in T,$$

where θ_t is the shift defined by $[\theta_t(x)]_s = x_{t+s}$ for every t, $s \in T$, $x \in X^T$.

For every $S \subset T$ let us denote by $\mathcal{F}_S = \mathrm{Pr}_S^{-1}(\mathcal{B}^S)$ the sub-σ-algebra generated by the projection function $\mathrm{Pr}_S : X^T \to X^S$, by P_S the restriction of a random field

$P \in \mathcal{P}$ to the σ-algebra $\mathcal{F}_S$, and by $\mathcal{L}_S$ the set of all bounded $\mathcal{F}_S$-measurable functions. (Sometimes we shall not distinguish between $x_V \in X^V$ and $\mathrm{Pr}_V^{-1}(x_V) \in \mathcal{F}_V$, as well as between P_S and $\leq_S P$ defined by $\overline{P}_S(F) = P_S(\mathrm{Pr}_S^{-1}(F))$ for every $F \in \mathcal{B}^S$.)

Let

$$\mathcal{L} = \bigcup_{S \in k(T)} \mathcal{L}_S,$$

be the set of all local (cylinder) bounded measurable functions.

Further, let us fix a reference probability measure ω on $(X, \mathcal{B})$.

For every positive integer n we denote by

$$V_n = \{t \in T;\ |t_i| \leq n \text{ for every } i = 1, \dots, d\}$$

the corresponding cube.

Suppose for some $P \in \mathcal{P}_\Theta$ there exist $f \in \mathcal{L}$ and a constant $c(f)$ satisfying

$$(\mathcal{G}\mathrm{I}) \qquad \operatorname*{esssup}_{[\omega^T]} \left| |V_n|^{-1} \left[\log \frac{dP_{V_n}}{d\omega^T_{V_n}} - \sum_{t \in V_n} f \circ \theta_t \right] + c(f) \right| \longrightarrow 0 \quad \text{as } n \to \infty.$$

Then we write $P \in G(f)$ and call P to be the Gibbs random field with respect to the potential $f \in \mathcal{L}$.

The main basic results concerning Gibbs random fields are summarized in the following theorem. We denote by

$$\mathcal{R}(Q|\omega^T) = \lim_{n \to \infty} |V_n|^{-1} \int \log \frac{dQ_{V_n}}{d\omega^T_{V_n}}\, dQ \geq 0$$

the asymptotic I-divergence (asymptotic Kullback–Leibler information, relative entropy rate) which exists for every $Q \in \mathcal{P}_\Theta$ (cf. e. g. Georgii (1988), Theorem 15.12).

Theorem 2.1

For every $f \in \mathcal{L}$ it holds

$$G(f) =$$

$$(\mathcal{G}\mathrm{II}) \qquad = \left\{ P \in \mathcal{P}_\Theta;\ \int f\, dP - \mathcal{R}(P|\omega^T) = \max_{Q \in \mathcal{P}_\Theta} \left[\int f\, dQ - \mathcal{R}(Q|\omega^T) \right] = c(f) \right\}$$

$$(\mathcal{G}\mathrm{III}) \qquad = \left\{ P \in \mathcal{P}_\Theta;\ d\, P_{A|A^c}(\cdot|\cdot) = \prod\nolimits_A^f(\cdot|\cdot) \text{ a. s. } [\omega^A \otimes P_{T \setminus A}] \text{ for every } A \in k(T) \right\}$$

where $\prod_A^f(\cdot|\cdot) = \dfrac{\exp\left\{ \sum_{t:(t+W)\cap A \neq \emptyset} f \circ \theta_t \right\}}{\int \exp\left\{ \sum_{t:(t+W)\cap A \neq \emptyset} f \circ \theta_t \right\} d\omega^A}$ if $f \in \mathcal{L}_W$.

Moreover

$$c(f) = \lim_{n \to \infty} |V_n|^{-1} c_{V_n}(f)$$

where $c_{V_n}(f) = \log \int \exp\left\{ \sum_{t \in V_n} f \circ \theta_t \right\} d\omega^T$ for every positive integer n, and $G(f)$ is a non-void compact (in the topology of local convergence) face in $\mathcal{P}_\Theta$.

Proof:

The implication $(\mathcal{G}\mathrm{I}) \Longrightarrow (\mathcal{G}\mathrm{II})$ is rather straightforward. The equivalence $(\mathcal{G}\mathrm{II}) \Longleftrightarrow (\mathcal{G}\mathrm{III})$ holds thanks to the finite range of the potential $f \in \mathcal{L}_W$ from Theorem 15.39 in Georgii (1988). The remaining implication $(\mathcal{G}\mathrm{III}) \Longrightarrow (\mathcal{G}\mathrm{I})$ follows from the bounds given in Section 15.3 in Georgii (1988) and the existence of limit $c(f) = \lim_{n\to\infty} |V_n|^{-1} c_{V_n}(f)$ by Theorem 15.30 in Georgii (1988). The rest of the proof follows from Theorems 4.23, 5.19, and 15.39 again in Georgii (1988). □

Remark 2.2

Here the definition of Gibbs random fields $(\mathcal{G}\mathrm{I})$ is not the standard one. But we have shown that at least for the finite range potentials it is equivalent with the original one $(\mathcal{G}\mathrm{III})$ which also proves the corresponding Gibbs random fields to be Markovian.

Let us emphasize that we can rather deal with the equality a. s. with respect to $[\omega^T]$ instead of $[\omega^A \times P_{T\setminus A}]$ in $(\mathcal{G}\mathrm{III})$, since by $(\mathcal{G}\mathrm{I})$ for every $B \in k(T)$ we have ω^B equivalent (i. e. mutually absolutely continuous) with every P_B, $P \in G(f)$, $f \in \mathcal{L}$, and we observe $\prod_A^f(\cdot|\cdot) \in \mathcal{L}_{A\cup\partial_f A}$, where $A \cup \partial_f A = \bigcup_{t:(t+W)\cap A\neq\emptyset} (t+W) \in k(T)$ if $f \in \mathcal{L}_W$, $W \in k(T)$.

Moreover, the definition $(\mathcal{G}\mathrm{I})$ is usually more transparent and it also indicates the important fact that Gibbs random fields can be understood as a generalized concept of exponential distributions. □

Remark 2.3

From the definition $(\mathcal{G}\mathrm{I})$ we can also directly conclude that the I-divergence exists with a Gibbs random field as the reference distribution, namely

$$\mathcal{R}(P|Q) = c(g) - \int g\,dP + \mathcal{R}(P|\omega^T) \geq 0$$

holds for $P \in \mathcal{P}_\Theta$, $Q \in G(g)$. If moreover $P \in G(f)$, we obtain

$$\mathcal{R}(P|Q) = c(g) - c(f) + \int f - g\,dP \geq 0.$$

□

3 Equivalence of potentials

There are two kinds of possible non-uniqueness in the definition of Gibbs random fields. The first one is given by the phase transitions, i. e. the case of $|G(f)| > 1$, which is essential and can not be avoided. Nevertheless, the phase transitions can be handled and often cause only technical problems. The second kind of non-uniqueness follows from the equivalence of potentials. We call the potentials $f, g \in \mathcal{L}$ to be equivalent, we write $f \approx g$, if $G(f) = G(g)$. The equivalence of potentials should be and usually can be avoided. Let us introduce a list of equivalent definitions.

Proposition 3.1

Let $f, g, h \in \mathcal{L}$. Then the following statements are equivalent:

a) $G(f) = G(g)$;

b) $G(f) \cap G(g) \neq \emptyset$;

c) $\operatorname*{esssup}_{[\omega^T]} \left| |V_n|^{-1} \sum_{t \in V_n} (f-g) \circ \theta_t - \text{const.} \right| \longrightarrow 0$ as $n \to \infty$;

d) $\int (f-g)\, dQ = \text{const.}$ for every $Q \in \bigcup_{q \in \mathcal{L}} G(q)$;

e) $\mathcal{R}(Q_h^a | Q_h^0)$ and $\mathcal{R}(Q_h^0 | Q_h^a) = 0$ for every $Q_h^0 \in G(h)$, $Q_h^a \in G(h + a(f-g))$, $a \in \mathbb{R}$;

f) $\mathcal{R}(Q_h^a | Q_h^0) = 0$ or $\mathcal{R}(Q_h^0 | Q_h^a) = 0$ for some $Q_h^0 \in G(h)$, $Q_h^a \in G(h + a(f-g))$, $a \neq 0$.

Proof:

a) $\Longrightarrow$ b) is obvious. Let us denote

$$\beta_n^{P,f} = \operatorname*{esssup}_{[\omega^T]} \left| |V_n|^{-1} \left[\log \frac{dP_{V_n}}{d\omega_{V_n}^T} - \sum_{t \in V_n} f \circ \theta_t \right] + c(f) \right|$$

and

$$\beta_n^{f,g} = \operatorname*{esssup}_{[\omega^T]} \left| |V_n|^{-1} \sum_{t \in V_n} (f-g) \circ \theta_t - c(f) + c(g) \right|.$$

Let $P \in G(f) \cap G(g)$. Then

$$\beta_n^{f,g} \leq \beta_n^{P,f} + \beta_n^{P,g} \longrightarrow 0 \quad \text{as } n \to \infty,$$

and therefore b) $\Longrightarrow$ c).

Since $Q_V \ll \omega_V^T$ for every $V \in k(T)$ and $Q \in \bigcup_{a \in \mathcal{L}} G(q)$ we obtain

$$\left| \int (f-g)\, dQ - c(f) + c(g) \right| \leq \beta_n^{f,g} \longrightarrow 0 \quad \text{as } n \to \infty.$$

Thus c) $\Longrightarrow$ d).

Further, under d) we have

$$0 \leq \mathcal{R}(Q_h^a | Q_h^0) + \mathcal{R}(Q_h^0 | Q_h^a) = a \int (f-g)\, (d\, Q_h^a - d\, Q_h^0) = 0.$$

Therefore d) $\Longrightarrow$ e), and e) $\Longrightarrow$ f) is obvious.

From f) by Theorem 2.1 and Remark 2.3 we obtain $G(h) \cap G(h+a(f-g)) \neq \emptyset$ with some $a \neq 0$, and therefore $\beta_n^{f,g} \longrightarrow 0$ as $n \to \infty$ with $c(g) = c(f) - \frac{c(h+a(f-g))-c(h)}{a}$. If $P \in G(f)$ we obtain $\beta_n^{P,g} \leq \beta_n^{P,f} + \beta_n^{f,g} \longrightarrow 0$ as $n \to \infty$ and therefore $P \in G(g)$ with the symmetric role of f and g.

Thus f) $\Longrightarrow$ a) and the proof is completed. □

In the above Proposition 3.1 we do not deal with the local characteristics $\Pi_A^f(\cdot|\cdot)$ for every $A \in k(T)$ and $f \in \mathcal{L}$, as defined by ($\mathcal{G}$III) in Theorem 2.1.

From ($\mathcal{G}$III) and Remark 2.2 we can easily conclude that

$$f \approx g \quad \text{iff} \quad \mathcal{E}_A^{f,g} : \left[\Pi_A^f(\cdot|\cdot) = \Pi_A^g(\cdot|\cdot) \text{ a. s. } [\omega^T]\right] \quad \text{holds for every } A \in k(T).$$

Further, let us denote by

$$\mathcal{H}_A(P|Q) = \int \log \frac{d\,P_{A|A^c}(\cdot|\cdot)}{d\,Q_{A|A^c}(\cdot|\cdot)}\, dP \geq 0,$$

whenever the expression makes sense, the corresponding conditional I-divergence.

Proposition 3.2

i) For $f, g \in \mathcal{L}$ and every $A \in k(T)$ we have $\mathcal{E}_A^{f,g}$ iff $\mathcal{E}_{\{0\}}^{f,g}$.

ii) Moreover, the following statements are equivalent

a) $f \approx g$;

b) $\mathcal{E}_A^{f,g}$ holds for some $A \in k(T)$;

c) $\mathcal{H}_A(P|Q) = 0$ for some $A \in k(T)$, $P \in G(f)$, $Q \in G(g)$.

Proof:

For $B \subset A$ we can easily verify

$$\Pi_A^f(x_A|x_{T\setminus A}) = \Pi_B^f(x_B|x_{T\setminus B}) \cdot \int \Pi_A^f(y_B, x_{A\setminus B}|x_{T\setminus A})\, d\omega^B(y_B)$$

which yields $\mathcal{E}_A^{f,g} \Longrightarrow \mathcal{E}_B^{f,g}$ for every $B \subset A$, including one-point sets $\{t\}$, $t \in A$. From the definition of $\Pi_A^f(\cdot|\cdot)$ we can directly deduce its translation invariance, i. e. $\Pi_A^f(\cdot|\cdot) = \Pi_{A-B}^f(\cdot|\cdot) \circ \theta_s$ for every $s \in T$. Thus $\mathcal{E}_{\{t\}}^{f,g} \Longrightarrow \mathcal{E}_{\{0\}}^{f,g}$.

The last implication holds also in the opposite way, i. e.

$$\mathcal{E}_{\{0\}}^{f,g} \Longrightarrow \mathcal{E}_{\{t\}}^{f,g}, \quad t \in T.$$

Again, it can be easily verify

$$\Pi^f_{A\cup B}(x_{A\cup B}|x_{T\setminus(A\cup B)}) = \Pi^f_A(x_A|x_{T\setminus A})\cdot\left(\int \frac{\Pi^f_A(y_A|x_{T\setminus A})}{\Pi^f_B(x_B|y_A, x_{T\setminus(A\cup B)})}\,d\omega^A(y_A)\right)$$

which yields

$$\left[\mathcal{E}^{f,g}_A \text{ and } \mathcal{E}^{f,g}_B\right] \Longrightarrow \mathcal{E}^{f,g}_{A\cup B},$$

and by induction we finally obtain $\mathcal{E}^{f,g}_{\{0\}} \Longrightarrow \mathcal{E}^{f,g}_A$ for every $A \in k(T)$. Thus the proof of i) is completed. In ii) the equivalence a) $\Longleftrightarrow$ b) follows directly from i) and the preceding considerations. The equivalence b) $\Longleftrightarrow$ c), i.e. $\mathcal{E}^{f,g}_A \Longleftrightarrow \mathcal{H}_A(P|Q) = 0$, follows from the well-known properties of I-divergences, namely $\mathcal{H}_A(P|Q) = 0$ iff $d\,P_{A|A^c} = d\,Q_{A|A^c}$ a. s. $[\omega^A \times P_{T\setminus A}]$, but this is for $P \in G(f)$ and $Q \in G(g)$ equivalent by Theorem 2.1 and Remark 2.2 to $\mathcal{E}^{f,g}_A$. □

Now, let us define function $\Phi_{g,f} : R \to R$ by the following formula

$$\Phi_{g,f} : \gamma \longmapsto a : \int f\,dQ^a_g = \gamma \quad \text{for some} \quad Q^a_g \in G(g + a\,f),$$

where $g,\ f \in \mathcal{L},\ f \not\approx 0$.

Proposition 3.3

The function $\Phi_{g,f}$ is well defined on an open interval, continuous and non-decreasing.

Proof:

We have $0 \le \mathcal{R}(Q^a_g|Q^b_g) + \mathcal{R}(Q^b_g|Q^a_g) = (b-a)\,\left[\int f\,dQ^b_g - \int f\,dQ^a_g\right]$. If $\int f\,dQ^b_g = \int f\,dQ^a_g$ for $b \neq a$ we would obtain $f \approx 0$ by Proposition 3.1 which contradicts the assumption. Thus the function $\Phi_{g,f}$ is well defined.

Moreover the above inequality can be rewritten as

$$0 \le [\Phi_{g,f}(\gamma) - \Phi_{g,f}(\delta)]\cdot(\gamma - \delta)$$

which proves the monotoneity.

Now, let us denote $\delta^+_g = \sup_{a\in R}\min_{P\in G(g+a\,f)} \int f\,dP$ and $\delta^-_g = \inf_{a\in R}\max_{P\in G(g+a\,f)} \int f\,dP$. Obviously $-\|f\| \le \delta^-_g < \delta^+_g \le \|f\|$. For $\gamma \in (\delta^-_g, \delta^+_g)$ there must exist $a(\gamma) \in R$ with $\gamma \le \int f\,dQ^a_g$ for every $Q^a_g \in G(g + a\,f)$, $a > a(\gamma)$, and $\int f\,dQ^a_g \le \gamma$ for every $Q^a_g \in G(g + a\,f)$, $a < a(\gamma)$.

Let us take $a^+_n \to a(\gamma)_+$ resp. $a^-_n \to a(\gamma)_-$ as $n \to \infty$, and $Q^+_n \in G(g + a^+_n f)$ resp. $Q^-_n \in G(g + a^-_n f)$ for every n. With the aid of Proposition 15.14 in Georgii (1988) we can prove the existence of cluster points Q^+ and Q^- of the sequences $\{Q^+_n\}$ and $\{Q^-_n\}$, respectively, satisfying $Q^+,\ Q^- \in G(g + a(\gamma)\,f)$. Since $\int f\,dQ^- \le \gamma \le \int f\,dQ^+$

and $G(g+a(\gamma)f)$ is a convex set we obtain $\Phi_{g,f}(\gamma) = a(\gamma)$. Thus $\Phi_{g,f}$ is defined on the whole interval (δ_g^-, δ_g^+).

Finally, let $a^0 = \Phi_{g,f}(\gamma^0)$ and $\varepsilon > 0$. We set

$$\delta = \min_{P\in G(g+(a^0-\varepsilon)f)\cup G(g+(a^0+\varepsilon)f)} \left| \int f\,dP - \gamma^0 \right| > 0$$

to obtain: $|\gamma - \gamma^0| < \delta$ implies $|\Phi_{g,f}(\gamma) - a^0| < \varepsilon$, and the continuity is proved as well.
□

Remark 3.4

The above proposition illustrates quite well the problems caused by the phase transitions. With no phase transitions we would deal with a strictly increasing continuous function $a \longmapsto \int f\,dQ_g^a$, and the same properties would follow for its inverse $\Phi_{g,f}$. In general, unfortunately, we have only a multi-function, and therefore the inverse may not be strictly increasing. Nevertheless, the above proved properties of the function $\Phi_{g,f}$ are sufficient for our purposes. Note that $\Phi_{g,f}$ is constant only inside the intervals of the type $\left[\min_{P\in G(g+af)} \int f\,dP,\ \max_{P\in G(g+ag)} \int f\,dP\right]$, being equal to this particular $a \in R$, while for different intervals it obviously differs. □

4 Large derivations theorem

For a fixed $f \in \mathcal{L}$, $f \not\approx 0$, let us denote

$$C_V^{\gamma_V} = \left\{ x \in X^T;\ \sum_{t\in V} f \circ \theta_t(x) > \gamma_V \right\}.$$

Theorem 4.1

Let $|V_n|^{-1}\gamma_{V_n} \longrightarrow \int f\,dQ_g^{a^0}$ as $n \to \infty$ for some $Q_g^{a^0} \in G(g+a^0 f)$, $a^0 > 0$, $g \in \mathcal{L}$. Then

$$|V_n|^{-1} \log Q_g^0 \left[C_{V_n}^{\gamma_{V_n}}\right] \longrightarrow -\mathcal{R}(Q_g^{a^0} | Q_g^0) < 0 \quad \text{as } n \to \infty$$

for every $Q_g^0 \in G(g)$.

Proof:

The upper bound is known to be easy. We can directly observe

$$|V_n|^{-1} \log Q_g^0 \left[C_{V_n}^{\gamma_{V_n}}\right] \le |V_n|^{-1} \left\{ a\,\gamma_{V_n} + c_{V_n}(g+af) - c_{V_n}(g) + \delta(|V_n|) \right\} \longrightarrow$$
$$\longrightarrow \mathcal{R}(Q_g^{a^0} | Q_g^a) - \mathcal{R}(Q_g^{a^0} | Q_g) \quad \text{as } n \to \infty, \text{ for every } a \ge 0,$$

$Q_g^a \in G(g+af)$, with minimum for $a = a^0$.

The lover bound, which is more complicated, can be found e.g. in Janžura (1987) or in Föllmer and Orey (1988). □

Theorem 4.2

Let $g,\ h \in \mathcal{L}$, $|V_n|^{-1}\gamma_{V_n} \longrightarrow \gamma^0 \in (\gamma_g^-, \gamma_h^+)$ as $n \to \infty$, where

$$gamma_g^- = \max_{P \in G(g)} \int f\,dP \quad \text{and} \quad \gamma_h^+ = \min_{P \in G(h)} \int f\,dP.$$

Then there exist $a^0 > 0 > b^0$ and $Q_g^{a^0} \in G(g + a^0 f)$, $Q_h^{b^0} \in G(h + b^0 f)$ satisfying $\gamma^0 = \int f\,dQ_g^{a^0} = \int f\,dQ_h^{b^0}$.

Moreover

$$|V_n|^{-1} \log Q_g^0\left(C_{V_n}^{\gamma V_n}\right) \longrightarrow -\mathcal{R}(Q_g^{a^0}|Q_g^0) < 0 \quad \text{and}$$
$$|V_n|^{-1} \log Q_h^0\left(\left[C_{V_n}^{\gamma V_n}\right]^c\right) \longrightarrow -\mathcal{R}(Q_h^{b^0}|Q_h^0) < 0 \quad \text{as } n \to \infty$$

for every $Q_g^0 \in G(g)$, $Q_h^0 \in G(h)$.

Proof:

It is sufficient to prove the existence of a^0, b^0 with the claimed properties. The rest of the proof then follows from a double application of the preceding theorem. (Note that the strict inequality in the definition of $C_V^{\gamma V}$ is not substantial.)

Thus, we have to prove that the open interval (γ_g^-, γ_h^+) is contained in the domains (δ_g^-, δ_g^+) and (δ_h^-, δ_h^+) of both $\Phi_{g,f}$ and $\Phi_{h,g}$ (cf. Section 3), respectively, since then $\Phi_{g,f}(\gamma_g^-, \gamma_g^+) \subset (0, \infty)$ resp. $\Phi_{h,f}(\gamma_g^-, \gamma_h^+) \subset (-\infty, 0)$ follow from the monotoneity properties (cf. Proposition 3.3 and Remark 3.4).

But for $a - b > 0$ we have

$$0 \leq \frac{\mathcal{R}(Q_g^a|Q_h^b) + \mathcal{R}(Q_h^b|Q_g^a)}{a-b} = \int f\,dQ_g^a - \int f\,dQ_h^b + \frac{\int (g-h)\,(dQ_g^a - dQ_h^b)}{a-b}$$

for every $Q_g^a \in G(g + af)$, $Q_h^b \in G(h + bf)$.

By $a \to \infty$ we obtain $\delta_g^+ \geq \int f\,dQ_h^b$ for every $b \in R$, $Q_h^b \in G(h + bf)$, and therefore especially $\delta_g^+ \geq \gamma_h^+$. Similarly, by $b \to -\infty$ we obtain $\delta_h^- \leq \int f\,dQ_g^a$, and finally $\delta_h^- \leq \gamma_g^-$.

Since the remaining inequalities are obvious, the proof is completed. □

Remark 4.3

Let us mention that $\gamma_g^- < \gamma_h^+$ implies $g \not\approx h$ as well as $f \not\approx 0$. In general, for $g \approx h$ we could not obtain both the error probabilities tending to zero exponentially fast. □

Remark 4.4

For $f \approx \varepsilon(h - g)$ with $g \not\approx h$ and $\varepsilon > 0$ we obtain a test equivalent to the standard likelihood ratio test. Since $h \approx g + \varepsilon^{-1} f$ we now obtain

$$\int f\,dQ_g^0 < \int f\,dQ_h^0 = \int f\,dQ_g^{\varepsilon^{-1}}$$

for every $Q_g^0 \in G(g)$ and $Q_h^0 = Q_g^{\varepsilon^{-1}} \in G(h)$, and therefore

$$\gamma_g^- < \gamma_h^+.$$

Moreover, for $\gamma^0 \in (\gamma_g^-, \gamma_h^+)$ we obtain $\varepsilon\, a^0 = 1 + \varepsilon\, b^0 \in (0,1)$. □

Remark 4.5

The result of Theorem 4.2 is very natural. Forgetting about the phase transitions, we deal with a pair of random fields Q_g^0, Q_h^0 with $\int f\, dQ_g^0 < \int f\, dQ_h^0$. We construct best rejecting P^0 whenever $|V|^{-1} \sum_{t \in V} f \circ \theta_t(x) > \gamma$ where $\int f\, dQ_g^0 < \gamma < \int f\, dQ_h^0$. Then both the error probabilities tend to zero exponentially fast. □

Proposition 4.6

Let a^0, b^0 be from Theorem 4.2, $g \not\approx h$. We set $\lambda = \frac{-b^0}{a^0 - b^0} \in (0,1)$. Then

$$\lambda\, \mathcal{R}(Q_g^{a^0} | Q_g^0) + (1-\lambda)\, \mathcal{R}(Q_h^{b^0} | Q_h^0) \leq \lambda\, \mathcal{R}(Q_\mu | Q_g^0) + (1-\lambda)\, \mathcal{R}(Q_\mu | Q_h^0)$$

for every $Q_\mu \in G(\mu g + (1-\mu) h)$, $\mu \in [0,1]$, with the equality iff $f \approx \varepsilon(h-g)$ for some $\varepsilon > 0$ and $\mu = \lambda$.

Proof:

Under the assumptions we can verified that

$$\lambda\, \mathcal{R}(Q_g^{a^0} | Q_h^0) + (1-\lambda)\, \mathcal{R}(Q_h^{b^0} | Q_h^0) =$$
$$= \lambda\, \mathcal{R}(Q_\mu | Q_g^0) + (1-\lambda)\, \mathcal{R}(Q_\mu | Q_h^0) - \mathcal{R}(Q_\mu | Q_\lambda) - \lambda\, \mathcal{R}(Q_\lambda | Q_g^{a^0}) - (1-\lambda)\, \mathcal{R}(Q_\lambda | Q_h^{b^0})$$

which proves the inequality. And the equality holds whenever the superfluous terms vanish which occurs just under the additional conditions. □

Corollary 4.7

At least one rate for the standard likelihood ratio test is greater to compare with the corresponding rate of the general test based on the statistics f.

Proof:

For the likelihood ratio test the rates are given by $\mathcal{R}(Q_{\mu^0} | Q_g^0)$ and $\mathcal{R}(Q_{\mu^0} | Q_h^0)$, respectively, with some $\mu^0 \in (0,1)$, and therefore the statement follows from the preceding proposition. □

5 Application to pseudo-likelihood ratio test

Now, let us return back to the idea of pseudo-likelihood ratio test as introduced in Section 1. We suppose $P^0 = Q_g^0 \in G(g)$, $P^1 = Q_h^0 \in G(h)$ with g, $h \in \mathcal{L}$, $g \not\approx h$. Let us set

$$f_A = \log \frac{\Pi_A^h(\cdot|\cdot)}{\Pi_A^g(\cdot|\cdot)}$$

with some fixed $A \in k(T)$.

Proposition 5.1

For the above defined $f_A \in \mathcal{L}$ it holds

$$f_A \not\approx 0 \quad \text{and} \quad \gamma_g^- < 0 < \gamma_h^+ \quad \text{where} \quad \gamma_g^- = \max_{P \in G(g)} \int f_A \, dP, \quad \gamma_h^+ = \min_{P \in G(h)} \int f_A \, dP.$$

Proof:

For every $P \in G(h)$ we have $\int f_A \, dP = \mathcal{H}_A(P|Q_g^0) > 0$ by Proposition 3.2 ii) since $h \not\approx g$, and therefore also $\gamma_h^+ > 0$ since $G(h)$ is compact. Similarly, for every $P \in G(g)$ we have $\int f_A \, dP = -\mathcal{H}_A(P|Q_h^0) < 0$ by the same arguments, and therefore $\gamma_g^- < 0$. We conclude $f_A \not\approx 0$ obviously by Proposition 3.1 d). □

Corollary 5.2

If $|V_n|^{-1}\gamma_{V_n}^A \longrightarrow \gamma^A \in (\gamma_g^-, \gamma_h^+)$, then both the error probabilities of the pseudo-likelihood test given by the critical region

$$C_{V_n}^{\gamma_{V_n}, A} = \left\{ x \in X^T;\ \sum_{t \in V_n} f_A \circ \theta_t(x) > \gamma_{V_n}^A \right\}$$

tend to zero exponentially fast with the rates $\mathcal{R}(Q_g^{a^0}|Q_g^0) > 0$ and $\mathcal{R}(Q_a^{b^0}|Q_h^0) > 0$, respectively, where $\gamma^A = \int f_A \, dQ_g^{a^0} = \int f_A \, dQ_h^{b^0}$, $Q_g^{a^0} \in G(g + a^0 f_A)$, $Q_h^{b^0} \in G(h + b^0 f_A)$, $a^0 > 0 > b^0$.

Proof: The statement follows directly from Proposition 5.1 and Theorem 4.2. □

Remark 5.3

The main advantage of the choice f_A for the proper statistics consists in the following fact. We can be sure without any additional calculations or estimates that for $\gamma^A = 0$ the assumptions are satisfied, the critical region $C_{V_n}^{0,A}$ can be constructed without any problem, and we shall obtain a good enough test.

We could also consider a more wide range since

$$\gamma_h^+ \geq \min_{x_{T\setminus A} \in X^{T\setminus A}} \int \log \frac{\Pi_A^h(y_A|x_{T\setminus A})}{\Pi_A^h(y_A|x_{T\setminus A})} \Pi_A^h(y_A|x_{T\setminus A})\, d\omega^A(y_A) \geq 0$$

and analogically for γ_g^-, which can be sometimes calculated and can offer a non-trivial extension of the one-point "sure" range $\{0\}$.

With a general statistics $f \in \mathcal{L}$ there are always troubles since there is not direct way available for calculating $\int f\, dP$ with $P \in G(g) \cup G(h)$. Therefore, in general, we do not know the "admissible range" (γ_g^-, γ_h^+) and the construction of the critical region fails.

6 Concluding remark

The main result of the present paper, which is contained in Corollary 5.2 and which gives a solution to the problem stated in the Introduction, follows from two results as indicated in its proof. While the importance of the "large derivations type" Theorem 4.2 is not surprising, we should mention the verification of the general assumptions for the case of the pseudo-likelihood ratio test in Proposition 5.1. The proof of its statement is deeply based on the results concerning the phenomenon of the equivalence of potentials, the detailed study of which in Section 3 is thus fully justified. Avoiding problems connected with this equivalence is necessary for meaningful results, but, as demonstrated here, it may be rather intricate to guarantee so.

REFERENCES

Föllmer, H. and Orey, S. (1988) Large derivations for the empirical field of a Gibbs measure. Ann. Probab. 16 (961–977).

Georgii, H.-D. (1988) Gibbs Measures and Phase Transitions. de Gruyter, Berlin.

Janžura, M. (1984) Discrimination rate of Gibbs random fields for unfitted decision procedure. Proc. 3rd Prague Symp. on Asymp. Stat. (eds Mandl, P., Hušková, M.). SNTL Praha, 303–308.

Janžura, M. (1987) Large deviations theorem for Gibbs random fields. Proc. of the 5th Pannonian Symp. on Math. Stat. Visegrad 1985 (eds. Grossmann, W., Mogyorodi, J., Vincze, I., Wertz, W.). Akademiai Kiadí, Budapest, 97–112.

DETECTION OF CHANGE IN VARIANCE

JARUŠKOVÁ Daniela, ANTOCH Jaromír
Czech Technical University, Prague
Charles University, Prague

ABSTRACT. Overview of methods for the detection of change in variance based on the maximum likelihood and quasibayesian approaches is given. Paper provides critical values for maximum of log-likelihood ratios calculated from Bonferroni and Worsley inequalities. The same critical are also estimated using a Monte Carlo simulation study.

1990 MATHEMATICS SUBJECT CLASSIFICATION: 62 F 03, 62 F 15

KEY WORDS AND PHRASES: Change point problem, change in variance, maximum likelihood, quasibayesian, Monte Carlo

1. INTRODUCTION

The detection of a change in the variance is very important problem e.g. in the analysis of the hydrological series. The deforestation of the area and the loss of capacity for water-retention of the soil may cause the increase of variance of the water discharges. On the other side building of reservoirs of different sizes usually has the opposite effect in decreasing the variance.

In our contribution we suppose that the observations $Y_1, \dots, Y_n$ are independent normally distributed with a known mean, which can be considered to be zero, and unknown variances. The problem of the detection of change in variance can be formulated as following testing problem, i.e.

$$\begin{aligned} H_0 &: \ Y_1, \dots, Y_n \sim N(0, \sigma^2), \\ A &: \ \text{there exists} \quad k \in \{1, \dots, n-1\} \quad \text{such that} \\ & \quad Y_1, \dots, Y_k \sim N(0, \sigma_1^2) \quad \text{and} \quad Y_{k+1}, \dots, Y_n \sim N(0, \sigma_2^2), \end{aligned}$$

where σ^2, σ_1^2 and σ_2^2, $\sigma_1^2 \neq \sigma_2^2$, are unknown variances.

If we introduce random variables $X_i = Y_i^2$, then the variables X_i, $i = 1, \dots, n$, follow gamma distribution with the shape parameter $\alpha = 1/2$ and the scale parameter $\beta = 2\sigma^2$, what means that our problem forms a special case of the detection of change in the scale parameter of random variables distributed according to the gamma law with a known and constant shape parameter.

For the detection of change in variance two different approaches, i.e. the maximum likelihood and the quasibayesian one, can be used.

2. MAXIMUM LIKELIHOOD APPROACH

First we shall follow the maximum likelihood approach. Supposing the time of the change in variance known and being k, then we can use for testing the null hypothesis the statistic Z_k equal to twice of the logarithm of likelihood ratio. It has for k fixed the form

$$Z_k = k \log \frac{\overline{x}_n}{\overline{x}_k} + (n-k) \log \frac{\overline{x}_n}{\overline{x}_k^*} = \tag{2.1}$$
$$= -k \log \left(\frac{S_k}{S_n} \cdot \frac{n}{k} \right) - (n-k) \log \left(\left(1 - \frac{S_k}{S_n} \right) \cdot \frac{n}{n-k} \right),$$

where $S_k = \sum_{i=1}^k X_i$, $S_k^* = \sum_{i=k+1}^n X_i$, $\overline{x}_k = S_k/k$ and $\overline{x}_k^* = S_k^*/(n-k)$. If we are looking for the distribution of the random variable Z_k, it is useful to notice that the event $\{Z_k < t\}$ is equivalent to the event $\{a_k(t) < S_k/S_n < b_k(t)\}$, where $a_k(t)$ and $b_k(t)$ are the solutions of the equation

$$-k \log \left(x \frac{n}{k} \right) - (n-k) \log \left((1-x) \frac{n}{n-k} \right) = t,$$

see Worsley (1986). Therefore $P(Z_k < t) = P(a_k(t) < S_k/S_n < b_k(t))$, where the random variable S_k/S_n is distributed according to the beta distribution with the parameters $k/2$ and $(n-k)/2$, respectively the random variable $(n-k)S_k/kS_k^*$ has F-distribution with k and $n-k$ degrees of freedom.

If the time k where the change occurs is not known we can use for testing the null hypothesis H_0 the statistic $\max_{1 \le k < n} Z_k$. Obviously it holds

$$P\left(\max_{1 \le k < n} Z_k < t \right) = P\left(a_1(t) < \frac{S_1}{S_n} < b_1(t), \dots, a_{n-1}(t) < \frac{S_{n-1}}{S_n} < b_{n-1}(t) \right),$$

where the density function f of the vector $\left(\frac{S_1}{S_n}, \dots, \frac{S_{n-1}}{S_n} \right)$ is

$$f(s_1, \dots, s_{n-1}) = \frac{const}{\sqrt{s_1}} \prod_{i=2}^{n-2} \frac{1}{\sqrt{s_i - s_{i-1}}} \frac{1}{\sqrt{1 - s_{n-1}}}, \quad 0 < s_1 < \dots < s_{n-1} < 1.$$

However, it seems to be impossible to use the numerical integration of f to find the distribution and the critical values of $\max_{1 \le k < n} Z_k$, see Worsley (1986). The problems arise from the fact that the density function f is unbounded and the intervals $(a_1(t), b_1(t)), \dots, (a_{n-1}(t), b_{n-1}(t))$ are not identical, so that the control of the numerical errors is difficult.

To find the upper bound of p-value one can use either the Bonferroni inequality

$$P\left(\max_{1 \le k < n} Z_k > t \right) \le \sum_{k=1}^{n-1} P(Z_k > t), \tag{2.2}$$

or the more accurate Worsley inequality

$$(2.3) \qquad P\left(\max_{1\leq k<n} Z_k > t\right) \leq \sum_{k=1}^{n-1} P(Z_k > t) - \sum_{k=1}^{n-2} P(Z_k > t, Z_{k+1} > t).$$

If we find t such that for the given significance level α the equality $\sum_{k=1}^{n-1} P(Z_k > t) = \alpha$, respectively $\sum_{k=1}^{n-1} P(Z_k > t) - \sum_{k=1}^{n-2} P(Z_k > t, Z_{k+1} > t) = \alpha$ holds then this value can be used as a α% critical value for the corresponding conservative test. We tried to find these critical values for $\alpha = 0.05$ and $n = 10, 20, \ldots, 100$, using the program Mathematica combining both the analytical and numerical approach. Unfortunately, we were able to find these (conservative) critical values using the Worsley inequality only for $n \leq 70$. For $n \geq 80$ we met the problems with the numerical non-stability.

Gombay and Horváth (1990) studied the asymptotic behavior of $\max_{1\leq k<n} Z_k$ and showed that

$$P\left(\max_{1\leq k<n} (2\log\log n) Z_k \leq \left(x + 2\log\log n + \frac{1}{2}\log\log\log n - \frac{1}{2}\log\pi\right)^2\right) \longrightarrow$$

$$(2.4) \qquad \longrightarrow \exp\left\{-2e^{-x}\right\} \qquad \text{as} \qquad n\to\infty.$$

The asymptotic distribution can be used for calculating critical values if the number of observations n is large.

To compare the critical values obtained from the preceding approaches we also prepared a simulation study. 100 000 realizations of $\max_{1\leq k<n} Z_k$ were simulated and corresponding empirical distribution function calculated. We repeated this procedure ten times and we estimated the α% critical value by the average of ten obtained sample α% critical values. Tables 1. – 3. provide 10%, 5% and 1% critical values calculated from all described procedures. The values are precise at least up to two decimal places. It is apparent that the convergence to the limit distribution is rather slow.

n	t_S	t_B	t_W	t_A
10	6.56	7.30	6.71	9.35
20	7.25	8.45	7.53	9.69
30	7.58	9.11	7.97	9.89
40	7.81	9.57	8.26	10.02
50	7.97	9.94	8.48	10.12
60	8.09	10.24	8.66	10.20
70	8.20	10.50	8.81	10.26
80	8.29	10.72	-	10.32
90	8.35	10.92	-	10.36
100	8.42	11.10	-	10.40

Table 1. 10% critical values of $max_{1\leq k<n} Z_k$ obtained from the simulation study (t_S), the Bonferroni inequality (t_B), the Worsley inequality (t_W) and the asymptotic distribution (t_A).

n	t_S	t_B	t_W	t_A
10	8.06	8.66	8.16	13.06
20	8.78	9.79	8.98	12.95
30	9.12	10.44	9.41	12.99
40	9.34	10.90	9.70	13.04
50	9.50	11.27	9.91	13.08
60	9.64	11.57	10.09	13.12
70	9.75	11.82	10.24	13.15
80	9.85	12.05	–	13.18
90	9.89	12.24	–	13.21
100	9.96	12.42	–	13.23

Table 2. 5% critical values of $max_{1\leq k<n} Z_k$ obtained from the simulation study (t_S), the Bonferroni inequality (t_B), the Worsley inequality (t_W) and the asymptotic distribution (t_A).

n	t_S	t_B	t_W	t_A
10	11.46	11.83	11.50	23.78
20	12.18	12.93	12.31	22.09
30	12.57	13.56	12.73	21.59
40	12.80	14.02	13.01	21.34
50	12.94	14.37	13.24	21.20
60	13.08	14.67	13.40	21.10
70	13.20	14.92	13.54	21.02
80	13.31	15.14	–	20.97
90	13.36	15.33	–	20.92
100	13.41	15.51	–	20.89

Table 3. 1% critical values of $max_{1\leq k<n} Z_k$ obtained from the simulation study (t_S), the Bonferroni inequality (t_B), the Worsley inequality (t_W) and the asymptotic distribution (t_A).

It is well-known that the test based on $max_{1\leq k<n} Z_k$ is more sensitive to change at the very beginning or at the very end of the series. Unfortunately, this is accompanied by the relatively small power if the change occurs in the middle of the series. For this case Deshayes and Picard (1986) suggested to use rather the maximum of weighted log-likelihood ratios $\max_{1\leq k<n} Z_k \cdot k(n-k)/n^2$. To avoid the problem with the behavior of maximum of log-likelihood ratios at the beginning and at the end of the series the statistic $\max_{\lambda_1 n\leq k\leq \lambda_2 n} Z_k$, $\lambda_1, \lambda_2 \in (0,1)$, can be used. The critical values for the both modifications can be calculated from corresponding limiting distributions, i. e.

(2.5)
$$\max_{1\le k<n} \frac{k(n-k)}{n^2} Z_k \xrightarrow{\mathcal{D}} \sup_{0\le s\le 1} B^2(s)$$

resp.

(2.6)
$$\max_{\lambda_1 n\le k\le\lambda_2 n} Z_k \xrightarrow{\mathcal{D}} \sup_{\lambda_1\le s\le\lambda_2} \frac{B^2(s)}{s(1-s)},$$

where $B(s)$ denotes Brownian bridge. The critical values of (2.5) were tabulated by Kiefer (1959), while those of (2.6) by De Long (1981). Moreover, the approximation for the tail behavior of the distribution of (2.6) is given in James et al. (1987). Another modifications assume $\lambda_1 \to 0_+$ and $\lambda_2 \to 1_-$ as $n \to \infty$.

3. QUASIBAYESIAN APPROACH

Following the quasibayesian approach introduced by Chernoff and Zacks and developed by Kander and Zacks (1966), we can use for testing the null hypothesis against one-sided alternative the statistic

$$T_1 = \frac{1}{n}\sum_{i=1}^{n-1}\left(\frac{1}{\sqrt{n}}\sum_{j=i+1}^{n}\frac{X_i-\overline{X}}{\sqrt{2}\,\overline{X}}\right).$$

In the case of the two-sided alternative the statistic

$$T_2 = \frac{1}{n}\sum_{i=1}^{n-1}\left(\frac{1}{\sqrt{n}}\sum_{j=i+1}^{n}\frac{X_i-\overline{X}}{\sqrt{2}\,\overline{X}}\right)^2$$

introduced by Mac Neil (1974) is preferable. The statistic T_1 converges for $n \to \infty$ to the statistic $\int_0^1 B(t)\,dt$ distributed according to the normal distribution $N(0, 1/12)$. The statistic T_2 converges for $n \to \infty$ to the statistic $\int_0^1 B^2(t)\,dt$. The distribution of $\int_0^1 B^2(t)\,dt$ was calculated by Kiefer (1959).

REFERENCES

[1] De Long D.M., *Crossing probabilities for a square root boundary by Bessel process*, Communications Statistics - Theory Methodology **A 10** (1981), 2197-2213.

[2] Deshayes J., Picard D., *Off-line statistical analysis of change-point models*, Detection of abrupt changes in signals and dynamics systems, Lecture Notes in Control and Information Sciences, vol. 77, 1986, pp. 103-168.

[3] Gombay E., Horváth L., *Asymptotic distribution of maximum likelihood tests for change in the mean*, Biometrika **77** (1990), 411-414.

[4] Hsu D. A., *Tests for variance shift at unknown time point*, Applied Statistics **26** (1977), 279,-284.

[5] James B., James K.L., Siegmund D., *Testing for a change-point*, Biometrika **74** (1987), 71-83.

[6] Kander Z., Zacks S., *Test procedures for possible changes in parameters of statistical distributions occurring at unknown time point*, Annals of Mathematical Statistics **35** (1966), 1196-1210.

[7] Kiefer J., *K-sample analogues of the Kolmogorov-Smirnov and Cramér-von Mises tests*, Annals of Mathematical Statistics **30** (1959), 420,- 447.
[8] Mac Neil I. B., *Tests for change of parameter at unknown times and distributions of some related functionals on Brownian motion*, Annals of Statistics **2** (1974), 950-962.
[9] Worsley K. J., *Confidence regions and tests for a change-point in a sequence of exponential family random variables*, Biometrika **73** (1986), 91-104.

Czech Technical University, Department of Mathematics, Thákurova 5, CZ-166 29 Praha 6, Czech Republic; tel: (+42 2) 332 43 83

Charles University, Department of Probability and Mathematical Statistics, Sokolovská 83, CZ-186 00 Praha 8 - Karlín, Czech Republic; tel: (+42 2) 231 78 62; fax: (+42 2) 664 150 95; telex: 121 673 mff

E-mail: jarus@csearn.bitnet, antoch@cspguk11.bitnet

SHRINKAGE OF MAXIMUM LIKELIHOOD ESTIMATOR OF MULTIVARIATE LOCATION

Jana Jurečková[1]
Charles University, Prague

Xavier Milhaud
Université des Sciences et Techniques du Languedoc, Montpellier

Abstract

In the multivariate model with independent components and symmetric marginal distributions, we construct correction terms which improve the asymptotic quadratic risk of MLE and/or other asymptotically efficient estimators. The correction terms have the form of formal posterior expectations of the average log-likelihood ratio, calculated with respect to differentiable superharmonic prior densities.

Keywords : Maximum likelihood estimator, Bayes estimator, superharmonic function.

AMS 1991 Subject Classification : 62H12, 62F15.

1 Introduction

Consider the random vector $\underset{\sim}{X} = (X_1, \ldots, X_p)'$ with the density

$$\prod_{i=1}^{p} f(x_i - \Theta_i), \underset{\sim}{x} = (x_1, \ldots, x_p)' \in R^p$$

$$\underset{\sim}{\Theta} = (\Theta_1, \ldots, \Theta_p)' \in R^p \tag{1.1}$$

where f is an univariate density (with respect to the Lebesgue measure), $0 < f(x) < \infty$, $x \in R^1$. Denote

$$\rho(x) = -ln\ f(x), \quad x \in R^1 \tag{1.2}$$

1 Work supported by Ministère de la Recherche et de la Technologie, France, décision 90R0433

and assume that ρ is differentiable with derivative $\psi(x) = \frac{d\rho(x)}{dx}$. We shall assume throughout that $f(x)$ is symmetric around 0, i.e. $f(x) = f(-x)$ (hence $\rho(-x) = \rho(x)$) for $x \in R^1$ and that $\psi = \rho'$ is continuous and that the distribution has finite Fisher's information, i.e. we assume that

$$0 < I(f) = \int_{R^1} \psi^2(x)\, e^{-\rho(x)}\, dx = \int_{R^1} \psi'(x)\, e^{-\rho(x)} dx \ < \infty \tag{1.3}$$

Let $\underset{\sim}{X}^{(1)}, \underset{\sim}{X}^{(2)}, \ldots, \underset{\sim}{X}^{(n)}$ be a random sample of size n from the distribution with density (1.1), where

$$\underset{\sim}{X}^{(j)} = (X_{ij}, \ldots, X_{pj})', \ \ j = 1, \ldots, n \tag{1.4}$$

We shall consider the problem of estimating the parameter $\underset{\sim}{\Theta}$. If $\hat{\underset{\sim}{\Theta}} = (\hat{\Theta}_1, \ldots, \hat{\Theta}_p)'$ is the maximum likelihood estimator (MLE) of $\underset{\sim}{\Theta}$, then, due to the independence of the components of $\underset{\sim}{X}$, $\hat{\Theta}_i$ coincides with the MLE of Θ_i based on the observations $(X_{i1}, \ldots, X_{in})$. Actually, $\hat{\underset{\sim}{\Theta}}$ is a solution of the minimisation

$$\sum_{i=1}^{p} \sum_{j=1}^{n} \rho\,(X_{ij} - t_i) := min \tag{1.5}$$

with respect to $\underset{\sim}{t} = (t_1, \ldots, t_p)' \in R^p$ and hence also a solution (or one of the solutions) of the system of equations

$$\sum_{j=1}^{n} \psi\,(X_{ij} - t_i) = 0, \ \ i = 1, \ldots, p\,. \tag{1.6}$$

If $E\hat{\Theta}_i^2 < \infty$, then, due to the symmetry of ρ, $\hat{\Theta}_i$ is an unbiased estimator of Θ_i, equivariant with respect to the translation of $(X_{i1}, \ldots, X_{in})$, i.e.,

$$\hat{\Theta}_i(X_{i1} + C, \ldots, X_{in} + C) = \hat{\Theta}_i(X_{i1}, \ldots, X_{in}) + C, \ \ C \in R^1\,. \tag{1.7}$$

Specifically, $\hat{\underset{\sim}{\Theta}} = \underset{\sim}{X}^{(1)}$ if we have only one observation $\underset{\sim}{X}^{(1)}$. This situation was studied by STEIN (1981) under the condition that $\underset{\sim}{X}^{(1)}$ has the p-dimensional normal distribution with the expectation $\underset{\sim}{\Theta} = (\Theta_1, \ldots, \Theta_p)'$ and with the identity covariance matrix. In the case $p \geq 3$, he found a large class of estimators of $\underset{\sim}{\Theta}$ dominating $\hat{\underset{\sim}{\Theta}} = \underset{\sim}{X}^{(1)}$ with respect to the quadratic risk. This class also covers the formal Bayes estimators of $\underset{\sim}{\Theta}$ generated by superharmonic prior densities.

JUREČKOVÁ and SALEH (1990) showed that, in the same model, the quadratic risk of the estimator $\hat{\underset{\sim}{\Theta}} = \underset{\sim}{X}^{(1)}$ could be uniformly improved not only in the case of normal distribution, but also in the case of normal distribution contamined by the exponential one. An analogous result could be proved also for a larger class of distributions with densities of the type $\prod_{i=1}^{p} f(x_i - \Theta_i) = exp\,\{-\sum_{i=1}^{p} \rho(x_i - \Theta_i)\}$. The correcting factor typically is the bayesian aposterior expectation of the vector

$$(I(f))^{-1}\left(\psi\left(X_{11}-Z_1\right),\ \psi\left(X_{21}-Z_2\right),\ldots,\psi\left(X_{p1}-Z_p\right)\right)'$$

given $\underset{\sim}{X}^{(1)} = (X_{11},\ldots,X_{p1})'$, where $\underset{\sim}{Z} = (Z_1,\ldots,Z_p)'$ is supposed to have a prior distribution Π with a superharmonic density π; let us denote this aposterior expectation as $E^{\underset{\sim}{X}^{(1)}}\{\frac{1}{I}\underset{\sim}{\psi}(\underset{\sim}{X}^{(1)}-\underset{\sim}{Z})\}$. Moreover, the same authors also proved that

$$E_{\underset{\sim}{\Theta}}\left\|\frac{1}{I}\underset{\sim}{\psi}\left(\underset{\sim}{X}^{(1)}-\underset{\sim}{\Theta}\right)-E^{\underset{\sim}{X}^{(1)}}\left\{\frac{1}{I}\underset{\sim}{\psi}\left(\underset{\sim}{X}^{(1)}-\underset{\sim}{Z}\right)\right\}\right\|^2 \leq p/I(f) \qquad (1.8)$$

where $||.||$ stands for the euclidean distance in R^p. (1.8) could be interpreted in such a way as that the risk of the "pseudoestimator" $\underset{\sim}{\Theta}+\frac{1}{I}\underset{\sim}{\psi}(\underset{\sim}{X}^{(1)}-\underset{\sim}{\Theta})-E^{\underset{\sim}{X}^{(1)}}\{\frac{1}{I}\underset{\sim}{\psi}(\underset{\sim}{X}^{(1)}-\underset{\sim}{Z})\}$ never exceeds the Cramer-Rao lower bound.

In the present paper, we shall show that (1.8) could be extended to the sample of n observations for a large class of distributions $f(x)=e^{-\rho(x)}$. More precisely, we shall show that

$$E_{\underset{\sim}{\Theta}}\sum_{i=1}^{p}\left\{\frac{1}{nI}\sum_{j=1}^{n}\psi\left(X_{ij}-\Theta_i\right)-E^{\underset{\sim}{X}}\left(\frac{1}{nI}\sum_{j=1}^{n}\psi\left(X_{ij}-Z_i\right)\right)\right\}^2 \leq \frac{p}{nI(f)} \qquad (1.9)$$

for all $\underset{\sim}{\Theta}\in R^p$, where $E^{\underset{\sim}{X}}(.)$ stands for the aposterior expectation, given the sample $\underset{\sim}{X}=(\underset{\sim}{X}^{(1)},\ldots,\underset{\sim}{X}^{(n)})$; (1.9) holds provided the prior distribution Π of $\underset{\sim}{Z}=(Z_1,\ldots,Z_p)$ has a superharmonic density π. For moderate and large sample sizes, it has an impact on the risk of MLE $\hat{\underset{\sim}{\Theta}}_n$ of $\underset{\sim}{\Theta}$ (and as well on the risk of other asymptotically efficient estimators).

Actually, we have under general conditions,

$$\hat{\Theta}_i-\Theta_i=\frac{1}{nI(f)}\sum_{j=1}^{n}\psi\left(X_{ij}-\Theta_i\right)+R_{ni} \qquad (1.10)$$

where nR_{ni} has a nondegenerate asymptotic (nonnormal) distribution (see, e.g., JUREČKOVÁ (1985) for the proof), $i=1,\ldots,p$. Hence, we see from (1.9) that

$$\left(E^{\underset{\sim}{X}}\left\{\frac{1}{nI}\sum_{j=1}^{n}\psi\left(X_{ij}-Z_i\right)\right\},\ i=1,\ldots,p\right) \qquad (1.11)$$

improves the risk of an approximation of $\hat{\underset{\sim}{\Theta}}$. We shall show that (1.11) also improves the asymptotic risk of $\hat{\underset{\sim}{\Theta}}$ for a class of symmetric distributions with a smooth likelihood function.

The paper is organized in the following way : Because the correcting factor (1.11) is a posterior expectation corresponding to the superharmonic prior density, in Section 2 we summarize the basic characteristics and propositions concerning the superharmonic functions. The regularity conditions and the main results are formulated in Section 3 and the proofs are postponed to Section 4.

2 Superharmonic Functions

A function $u : D \to (-\infty, \infty)$, where D is an open subset of R^p, is called superharmonic if

1. u is lower semicontinuous;
2. u is not identically $+\infty$ in any open connected component of D;
3. for any ball $B(\underset{\sim}{\xi}, \delta), \bar{B} \subset D$ with center $\underset{\sim}{\xi}$ and radius δ, the value $u(\underset{\sim}{\xi})$ is greater than or equal to the average of u over the boundary $\partial B(\xi, \delta)$ of $B(\underset{\sim}{\xi}, \delta)$; more precisely,

$$u(\underset{\sim}{\xi}) \geq \left[l_{p-1}\left(\partial B(\underset{\sim}{\xi}, \delta)\right)\right]^{-1} \int_{\partial B(\underset{\sim}{\xi},\delta)} u dl_{p-1} = \frac{1}{\pi_p \delta^{p-1}} \int_{\partial B(\underset{\sim}{\xi},\delta)} u \, d\, l_{p-1} \tag{2.1}$$

where l_{p-1} is the Lebesgue measure and $\pi_p = 2\pi^{\frac{p}{2}}/\Gamma(\frac{p}{2})$.

A detailed exposition of the superharmonic functions could be found in BLUMENTAL and GETOOR (1968), CHUNG (1982) and DOOB (1984), among others.

We shall only restrict ourselves to some basic properties which we shall need in the context of the prior densities, namely :

(i) A function u, twice differentiable on D, is superharmonic if and only if $\Delta u \leq 0$, where $\Delta u = \sum_{i=1}^{p} \frac{\partial^2 u}{\partial t_i^2}$ is the Laplacian of u.

(ii) Let μ be a finite measure on the Borel subsets of $R^p (p \geq 3)$. Then the function

$$u(\underset{\sim}{\xi}) = \int_{R^p} \frac{1}{\|\underset{\sim}{\xi} - \underset{\sim}{t}\|^{p-2}} \mu(d\underset{\sim}{t}), \quad \underset{\sim}{\xi} \in R^p \tag{2.2}$$

is superharmonic in R^p; moreover, if the measure μ satisfies $\mu(d\underset{\sim}{t}) = h(\underset{\sim}{t})\, d\underset{\sim}{t}, \ \underset{\sim}{t} \in R^p$, where h is a positive function with a compact support in R^p, then

$$u(\underset{\sim}{\xi}) \to 0 \quad as \|\underset{\sim}{\xi}\| \to \infty \,. \tag{2.3}$$

In this way, we could obtain a class of superharmonic functions which are twice differentiable and satisfy (2.3). Such functions, in the role of prior densities, will form the shrinkage factors in the next section.

3 Assumptions and main results

Let $\underset{\sim}{X} = (X_1, \ldots, X_p)'$ be a random vector with density (1.1). Our first result will be proved under the following regularity conditions :

(A.1) $0 < f(x) < \infty$ and $f(x) = f(-x)$, $x \in R^1$

(A.2) The function ρ defined in (1.2) is absolutely continuous with continuous derivative $\psi = \rho'$ and $0 < I(f) = \int \psi^2(x) e^{-\rho(x)} dx < \infty$.

(A.3) The integral $\int f(x-\Theta)dx$ is twice differentiable in Θ under the integral sign.

(A.3*) ψ is a continuous function which has a bounded derivative in $(-c,c)$ and which is constant outside $(-c,c)$, where $c>0$ is a fixed number.

Consider the problem of simultaneous estimator of the vector parameter $\underset{\sim}{\Theta} = (\Theta_1,\ldots,\Theta_p)'$ in (1.1). In the following theorem, we shall first improve the quadratic risk of the "pseudoestimator".

$$(\Theta_i + (nI(f))^{-1}\sum_{j=1}^{n}\psi(X_{ij}-\Theta_i) : i=1,\ldots,p) \tag{3.1}$$

with the aid of its posterior conditional expectation corresponding to the prior measure Π on the Borel subsets of R^p. We shall assume that Π has a density π with respect to the Lebesgue measure and π satisfies the following conditions :

(P.1) $\pi(z_1,\ldots,z_p)$ is bounded and twice continuously differentiable superharmonic function.

(P.2) $\lim_{\|\underset{\sim}{z}\|\to\infty} \pi(z_1,\ldots,z_p) = 0$

Remark : The existence of Π satisfying (P.1) and (P.2) follows from Section 2.

Let us now consider the parameter $\underset{\sim}{\Theta}$ in (1.1) as the random vector $\underset{\sim}{Z}$ with distribution Π. Then the conditional distribution of $\underset{\sim}{X}$ given $\underset{\sim}{Z}=\underset{\sim}{\Theta}$ has the density (1.1) and the conditional density of the sample $\mathbf{X} = (\underset{\sim}{X}^{(1)},\ldots,\underset{\sim}{X}^{(n)})$ given $\underset{\sim}{Z}=\underset{\sim}{\Theta}$ is

$$\prod_{i=1}^{p}\prod_{j=1}^{n} f(X_{ij}-\Theta_i) = exp\left\{-\sum_{i=1}^{p}\sum_{j=1}^{n}\rho(X_{ij}-\Theta_i)\right\} \tag{3.2}$$

The unconditional density of $\mathbf{X}$ with respect to the Lebesgue measure on R^{pn} is then equal to

$$g(\mathbf{X}) = \int_{R^p} exp\left\{-\sum_{i=1}^{p}\sum_{j=1}^{n}\rho(X_{ij}-z_i)\right\}\pi(\underset{\sim}{z})d\underset{\sim}{z} \tag{3.3}$$

We shall show that adding the vector

$$(nI(f))^{-1}\sum_{j=1}^{n}\frac{\partial ln\, g(\mathbf{X})}{\partial X_{ij}} \quad : i=1,\ldots,p) \tag{3.4}$$

to (3.1) improves its quadratic risk for all $\underset{\sim}{\Theta}$.

Theorem 3.1 : Assume that the density $f(x)=e^{-\rho(x)}$ satisfies either (A.1), (A.2) and (A.3) or (A.1), (A.2) and (A.3*) and that the prior density π satisfies (P.1) and (P.2). Then there exists $K_{\underset{\sim}{\Theta}}>0$ such that

$$E_{\underset{\sim}{\Theta}}\sum_{i=1}^{p}\left\{(nI(f))^{-1}\sum_{j=1}^{n}\left[\psi(X_{ij}-\Theta_i)+\frac{\partial ln\, g(\mathbf{X})}{\partial X_{ij}}\right]\right\}^2$$

$$= \frac{p}{nI(f)} - \frac{K_{\underset{\sim}{\Theta}}}{n^2} \leq E_{\underset{\sim}{\Theta}} \sum_{i=1}^{n} \left\{ (nI(f))^{-1} \sum_{j=1}^{n} \psi(X_{ij} - \Theta_i) \right\}^2 \tag{3.5}$$

for any $\underset{\sim}{\Theta} \in R^p$

The correction term $(nI(f))^{-1} \sum_{j=1}^{n} \frac{\partial ln\, g(\mathbf{X})}{\partial X_{ij}}$ could be also rewritten in the form

$$\frac{1}{nI} \sum_{j=1}^{n} \frac{\partial log\, g(\mathbf{X})}{\partial X_{ij}} = -E^{\mathbf{X}} \left(\frac{1}{nI} \sum_{j=1}^{n} \psi(X_{ij} - Z_i) \right) \tag{3.6}$$

of the posterior expectation of $\frac{1}{nI} \sum_{j=1}^{n} \psi(X_{ij} - Z_i), i = 1, \ldots, p$; we used the notation $I = I(f)$.

Hence, the quadratic risk of the "pseudoestimator" $(\Theta_i + (nI)^{-1} \sum \psi(X_{ij} - \Theta_i) : i = 1, \ldots, p)$ is for all $\underset{\sim}{\Theta} \in R^p$ improved by its aposterior expectation with respect to a superharmonic prior density. On the other hand, this "pseudoestimator" aproximates the MLE $\hat{\underset{\sim}{\Theta}}_n$ [and other asymptotically efficient estimators] as $n \to \infty$ see (1.10).

Hence, there is a natural question whether (3.6) analogously improves the quadratic risk of $\hat{\underset{\sim}{\Theta}}_n$, at least for moderate and large sample sizes. The following theorem gives a positive reply to this question for a class of symmetric unimodal densities satisfying some regularity conditions; namely, the conditions (A.1), (A.2) and the conditions which follow :

(A.4) ρ has 5 continuous derivatives, $\rho' = \psi = \psi^{(0)}$ and $\psi^{(j)}, j = 1, 2, 3, 4$ and $|\psi^{(4)}(x + h) - \psi^{(4)}(x)| < C(1 + |x|^d + |h|^d)|h|$ for some $C > 0$ and $d > 0$.

(A.5) $\int |\psi^{(j)}(x)|^{r_1}\, exp\{-\rho(x)\}dx < \infty$ for some $r_1 \geq 8$ and for $j = 0, \ldots, 4$.

(A.6) There exists $r_2 \geq 8$ such that

$$\int |x|^{dr_2} exp\{-\rho(x)\}dx = K_d < \infty$$

with $d > 0$ from condition (A.4).

Theorem 3.2 : Assume that $f(x) = e^{-\rho(x)}$ is a symmetric density satisfying the conditions (A.1), (A.2), (A.4) - (A.6). Let π be the prior density on R^p satisfying (P.1) and (P.2). Let $\hat{\underset{\sim}{\Theta}}_n$ be the MLE of $\underset{\sim}{\Theta}$ based on n independent observations from distribution (1.1). Then, as $n \to \infty$,

$$\sum_{i=1}^{p} E_\Theta \left\{ \hat{\Theta}_i - \Theta_i + (nI(f))^{-1} \sum_{j=1}^{n} \frac{\partial ln\, g(\mathbf{X})}{\partial X_{ij}} \right\}^2$$
$$= \frac{p}{nI(f)} - \frac{K_{\underset{\sim}{\Theta}}}{n^2} + o(n^{-2}) \tag{3.7}$$

for all $\underset{\sim}{\Theta} \in R^p$, where $g(\underset{\sim}{\mathbf{X}})$ is defined in (3.3) and $K_{\underset{\sim}{\Theta}}$ is given in Theorem 3.1.

4 Proofs of Theorems 3.1 and 3.2

Proof of Theorem 3.1 : For the sake of simplicity, denote

$$A(\mathbf{X}, \underset{\sim}{z}) = exp\left\{ -\sum_{i=1}^{p} \sum_{j=1}^{n} \rho(X_{ij} - z_i) \right\} \tag{4.1}$$

Then we could write

$$\begin{aligned}
\sum_{j=1}^{n} \frac{\partial lng(\mathbf{X})}{\partial X_{ij}} &= -(\int_{R^p} A(\mathbf{X}, \underset{\sim}{z})\, \pi(\underset{\sim}{z})\, d\underset{\sim}{z})^{-1} \int_{R^p} \sum_{j=1}^{n} \psi(X_{ij} - z_i) A(\mathbf{X}, \underset{\sim}{z})\, \pi(\underset{\sim}{z}) d\underset{\sim}{z} \\
&= (\int_{R^p} A(\mathbf{X}, \underset{\sim}{z})\, \pi(\underset{\sim}{z})\, d\underset{\sim}{z})^{-1} \int_{R^p} A(\mathbf{X}, \underset{\sim}{z})\, \nabla_i\, \pi(\underset{\sim}{z})\, d\underset{\sim}{z} \\
&= C_i(\mathbf{X})/D(\mathbf{X}) \quad \text{(say)}.
\end{aligned} \tag{4.2}$$

Rewrite the risk on the left-hand side of (3.5) in the form

$$\frac{p}{nI} + \frac{1}{(nI)^2} \sum_{i=1}^{p} E_{\underset{\sim}{\Theta}} \left(\sum_{j=1}^{n} \frac{\partial lng(\mathbf{X})}{\partial X_{ij}} \right)^2$$
$$+ \frac{2}{(nI)^2} \sum_{i=1}^{p} E_{\underset{\sim}{\Theta}} \left\{ \sum_{j=1}^{n} \psi(X_{ij} - \Theta_i) \sum_{k=1}^{n} \frac{\partial lng(\mathbf{X})}{\partial X_{ik}} \right\} \tag{4.3}$$

and consider each term separately. The middle term could be rewritten, using the same arguments as in (4.2), as

$$\frac{1}{(nI)^2} E_{\underset{\sim}{\Theta}} \left\{ (\int_{R^p} A(\mathbf{X}, \underset{\sim}{z})\, \pi(\underset{\sim}{z}) d\underset{\sim}{z})^{-2} \sum_{i=1}^{p} (\int_{R^p} A(\mathbf{X}, \underset{\sim}{z})\, \underset{i}{\nabla}\, \pi(\underset{\sim}{z}) d\underset{\sim}{z})^2 \right\}. \tag{4.4}$$

The last term in (4.3) could be written as

$$\begin{aligned}
&\frac{2}{(nI)^2} \sum_{i=1}^{p} \sum_{j=1}^{n} \sum_{k=1}^{n} E_{\underset{\sim}{\Theta}} \left\{ \frac{\partial lng(\mathbf{X})}{\partial X_{ik}} \psi(X_{ij} - \Theta_i) \right\} \\
&= \frac{2}{(nI)^2} \sum_{i=1}^{p} \sum_{j=1}^{n} \sum_{k=1}^{n} E_{\underset{\sim}{\Theta}} \frac{\partial^2 lng(\mathbf{X})}{\partial X_{ij} \partial X_{ik}} \\
&= \frac{2}{(nI)^2} E_{\underset{\sim}{\Theta}} \sum_{i=1}^{p} \sum_{k=1}^{n} \left\{ \frac{1}{D} \frac{\partial C_i}{\partial X_{ik}} - \frac{C_i}{D^2} \frac{\partial D}{\partial X_{ik}} \right\}
\end{aligned} \tag{4.5}$$

with C_i and D from (4.2), $i = 1, \ldots, p$; hence, it could be further rewritten as

$$\frac{2}{(nI)^2} \sum_{i=1}^{p} E_{\underset{\sim}{\Theta}} \left\{ ((\int_{R^p} A(\mathbf{X}, \underset{\sim}{z})\, \pi(\underset{\sim}{z})\, d\underset{\sim}{z})^{-1} \int_{R^p} A(\mathbf{X}, z)\, \nabla_i^2\, \pi(\underset{\sim}{z})\, d\underset{\sim}{z}) \right.$$

$$- \left(\int_{R^p} A(\mathbf{X}, \underset{\sim}{z}) \, \pi(\underset{\sim}{z}) \, d\underset{\sim}{z} \right)^{-2} \left(\int_{R^p} A(\mathbf{X}, \underset{\sim}{z}) \, \nabla_i \, \pi(\underset{\sim}{z}) \, d\underset{\sim}{z} \right)^2 \Bigg\} . \tag{4.6}$$

Combining (4.3), (4.4) and (4.6), we finally obtain the risk (3.5) in the form

$$\sum_{i=1}^{p} E_{\underset{\sim}{\Theta}} \left\{ \frac{1}{nI} \sum_{j=1}^{n} \left[\psi\,(X_{ij} - \Theta_i) + \frac{\partial ln g(\mathbf{X})}{\partial X_{ij}} \right] \right\}^2 = \frac{p}{nI(f)} - \frac{K_{\underset{\sim}{\Theta}}}{n^2}$$

where

$$K_{\underset{\sim}{\Theta}} = - \frac{2}{I^2} E_{\underset{\sim}{\Theta}} \left\{ \left(\int_{R^p} A(\mathbf{X}, \underset{\sim}{z}) \, \pi(\underset{\sim}{z}) \, d\underset{\sim}{z} \right)^{-1} \int_{R^p} A(\mathbf{X}, z) \, \Delta \, \pi(\underset{\sim}{z}) \, d\underset{\sim}{z} \right\}$$

$$+ \frac{1}{I^2} E_{\underset{\sim}{\Theta}} \left\{ \left(\int_{R^p} A(\mathbf{X}, \underset{\sim}{z}) \, \pi(\underset{\sim}{z}) \, d\underset{\sim}{z} \right)^{-1} \sum_{i=1}^{p} \left(\int_{R^p} A(\mathbf{X}, \underset{\sim}{z}) \, \nabla_i \, \pi(\underset{\sim}{z}) \, d\underset{\sim}{z} \right)^2 \right\} . \tag{4.7}$$

The second term in (4.7) is positive and the first term is nonnegative, because $\Delta\pi \leq 0$ due to (P.1). □

Proof of Theorem 3.2 We shall use the following notation :

$$S_i^{(\ell+1)}(\Theta_i + t_i) = (-1)^{\,\ell} \sum_{j=1}^{n} \psi^{(\ell)}(X_{ij} - (\Theta_i + t_i)) \tag{4.8}$$

$\ell = 0, \ldots, 4$, $i = 1, \ldots, p$. For the sake of brevity, we shall denote

$$S_i^{(\ell)} = S_i^{(\ell)}(\Theta_i) \, , \; \hat{S}_i^{(\ell)} = S_i^{(\ell)}(\hat{\Theta}_i)$$

$\ell = 0, \ldots, 4$ and $i = 1, \ldots, p$, with $\hat{\Theta}_i$ and Θ_i being the MLE and the true parameter value, respectively. Moreover, let us write the correction term in (3.7) in the following form :

$$(nI)^{-1} \sum_{i=1}^{p} \sum_{j=1}^{n} \frac{\partial ln \, g(\mathbf{X})}{\partial X_{ij}} = \sum_{i=1}^{p} B_i = B \tag{4.9}$$

where

$$B_i = E^{\mathbf{X}} \{ (nI)^{-1} S_i^{(1)}(Z_i) \}, \quad i = 1, \ldots, p \tag{4.10}$$

Then the risk in (3.7) could be rewritten in the following way :

$$R = \sum_{i=1}^{p} E_{\underset{\sim}{\Theta}} (\hat{\Theta}_i - \Theta_i + B_i)^2 = R_1 + R_2 \tag{4.11}$$

where

$$R_1 = \sum_{i=1}^{p} E_{\underset{\sim}{\Theta}} (\hat{\Theta}_i - \Theta_i)^2 \tag{4.12}$$

$$R_2 = \sum_{i=1}^{p} E_{\underset{\sim}{\Theta}} [B_i (B_i + 2(\hat{\Theta}_i - \Theta_i))] \tag{4.13}$$

Let us treat the behavior of R_2. Following Ibragimov and Hasminskii (1981) (shortly [IH]), consider the likelihood ratio process

$$Z_n(t) = \prod_{i=1}^{p} Z_{ni}(t_i) \tag{4.14}$$

with

$$Z_{ni}(t_i) = exp\{-\sum_{j=1}^{n}[\rho(X_{ij} - (\Theta_i + n^{-1/2}t_i)) - \rho(X_{ij} - \Theta_i)]\} \tag{4.15}$$

$i = 1, \ldots, p$.

First, notice that the conditions N.1 - N.4 in Section III.1 of [IH] could be verified for our location model; hence, Theorem III.1.1 of [IH] implies the convergence of moments of MLE to the corresponding moments of the asymptotic normal distribution. Moreover, the following lemma could be proved along the lines of the proof of Lemma 1.5.2 in [IH].

Lemma 4.1 Under the conditions of theorem 3.2 :

i) For all $N > 0$ and all positive integer k, there exists $n_0 = n_0(N, k)$ such that the sequence of random variables

$$\left\{\left(\frac{\int |t_i|^k Z_n(t)\pi(n^{-1/2}t)dt}{\int Z_n(t)\pi(n^{-1/2}t)dt}\right)^N : n = n_0, n_0 + 1, \ldots\right\} \tag{4.16}$$

is uniformly integrable

ii) For all $N > 0$, there exist $n_0 = n_0(N), C_N > 0$ such that for all $T > 0$

$$P\left\{\int_{||t||>T} Z_n(t)\pi(n^{-1/2}t)dt > T^{-1}\right\} < C_N T^{-N}.$$

Moreover, we shall later use the following lemma which easily follows from Holder's inequality :

Lemma 4.2 Let $\{X_n\}_{n=1}^{\infty}$ and $\{A_n\}_{n=1}^{\infty}$ be sequences of random variables and of random events, respectively, defined on the same probability space, such that $\{|X_n|^k\}$ are uniformly integrable for a fixed positive integer k and $P(A_n^c) = O(n^{-q})$ as $n \to \infty$ for some $q > 0$. Then

$$E(|X_n|I[A_n^c] = o(n^{-q(1-k^{-1})}) \quad n \to \infty. \tag{4.17}$$

Let us fix two positive numbers α, N, (α sufficiently small and N sufficiently large; without loss of generality put $\alpha = 0.1, N = 100$). Put $A_n = A_{1n} \cap A_{2n} \cap A_{3n} \cap A_{4n} \cap A_{5n}$ where

$$A_{1n} = \bigcap_{i=1}^{p}\{|\hat{\Theta}_i - \Theta_i| < n^{\alpha-1/2}\} \tag{4.18}$$

$$A_{2n} = \bigcap_{i=1}^{p} A_{2ni} \quad A_{2ni} = \{n^{-1}\sum_{j=1}^{n}|X_{ij} - \Theta_i|^d < 2K_d\} \tag{4.19}$$

$i = 1, \ldots, p$, where d and K_d are constants from the condition (A.6);

$$A_{3n} = \bigcap_{i=1}^{p} \bigcap_{\ell=1}^{5} A_{3ni}^{(\ell)}, \quad \text{where}$$

$$A_{3ni}^{(\ell)} = \{n^{-1/2}(S_i^{(\ell)} - ES_i^{(\ell)}) < \Lambda(\log n)^{1/2}\} \tag{4.20}$$

$i = 1, \ldots, p$, $\ell = 1, 2, 3, 4, 5$ and $\Lambda > 0$ is the constant whose existence is guaranteed by the Von Bahr (1965) moment inequality for the 8th moment of a sum;

$$A_{4n} = \bigcap_{i=1}^{p} \left\{ \int_{|t_i| > 2n^{\alpha}} |t_i|^k P(dt/\mathbf{X}) < n^{-\alpha N} \right\} \tag{4.21}$$

where we denote

$$P(dt/\mathbf{X}) = Z_n(t)\pi(n^{-1/2}t)dt \Big/ \int Z_n(u)\pi(n^{-1/2}u)du \tag{4.22}$$

and

$$A_{5n} = \left\{ \int_{||t|| > n^{\alpha}} Z_n(t)\pi(t/\sqrt{n})dt < n^{-\alpha} \right\} \tag{4.23}$$

Then, by Theorem III.1.1 in [IH] we obtain

$$P(A_{1n}^c) = o(n^{-M}) \tag{4.24}$$

for every integer M as $n \to \infty$. Moreover, by (A.6) and by the Markov inequality,

$$P(A_{2n}^c) = 0(n^{-4}) \tag{4.25}$$

while

$$P(A_{3n}^c) = o(n^{-3}(\log n)^{-4}) \tag{4.26}$$

by the von Bahr moment inequality; finally, by an adaptation of Lemma 1.5.2. in [IH] we obtain

$$P(A_{4n}^c) = o(n^{-\alpha N}) \tag{4.27}$$

and

$$P(A_{5n}^c) = o(n^{-N}) \quad \text{as } n \to \infty \tag{4.28}$$

(4.23) - (4.28) then imply

$$P(A_n^c) = o(n^{-3}(\log n)^{-4}). \tag{4.29}$$

Lemma 4.3 Under the conditions of theorem 3.2,

$$\hat{\Theta}_i - \Theta_i = (\hat{\Theta}_i - \Theta_i)I[A_n^c] + \left\{ \frac{S_i^{(1)}}{nI} + u_{ni} + v_{ni} \right\} I[A_n] \tag{4.30}$$

where

$$u_{ni} = \frac{S_i^{(1)}}{nI^2} \left[\frac{S_i^{(2)}}{n} + I \right], \quad i = 1, \ldots, n \tag{4.31}$$

and

$$|v_{ni}| \leq c\, n^{-3/2}(\log n)^4 \quad \text{for some} \quad c > 0 \tag{4.32}$$

Proof of Lemma 4.3. It follows from Theorem 5 in Chibisov (1973) that for $\omega \in A_n$

$$|\hat{\Theta}_i - \Theta_i - \frac{S_i^{(1)}}{nI}\left[1 + \left[\frac{S_i^{(2)}}{n} + I\right]\right]| \leq c\, n^{-3/2}(\log n)^4 = o(n^{-1}) \tag{4.33}$$

for some $c > 0$. □

Using Lemma 4.3, we can decompose R_2 in (4.13) as

$$R_2 = R_{21} + R_{22} + R_{23} \tag{4.34}$$

where
$$R_{21} = \sum_{i=1}^{p} E_{\underset{\sim}{\Theta}}\{B_i[B_i - 2(S_i^{(1)}/nI)]\} \tag{4.35}$$

$$R_{22} = -2\sum_{i=1}^{p} E_{\underset{\sim}{\Theta}}\{B_i[(\hat{\Theta}_i - \Theta_i) - (S_i^{(1)}/nI)].I[A_n^c]\} \tag{4.36}$$

$$R_{23} = -2\sum_{i=1}^{p} E_{\underset{\sim}{\Theta}}\{B_i[(\hat{\Theta}_i - \Theta_i) - (S_i^{(1)}/nI)].I[A_n]\} \tag{4.37}$$

Then, it follows from the proof of Theorem 3.1 that

$$R_{21} = -K_{\underset{\sim}{\Theta}}/n^2 \ , \ \ K_{\underset{\sim}{\Theta}} > 0 \tag{4.38}$$

Hence it remains to estimate the terms R_{22} and R_{23}. We shall first prove

Lemma 4.4 Under the assumptions of Theorem 3.2,

$$R_{22} = o(n^{-2}). \tag{4.39}$$

Proof of Lemma 4.4 Regarding condition (A.4), we get by the Taylor expansion of B_i in (4.10)

$$|B_i I[A_n^c]| \leq \left\{\sum_{\ell=1}^{5} |S_i^{(\ell)}/nI| \int_{R^p} |n^{-1/2}t_i|^{\ell-1} P(dt/\mathbf{X})\right.$$

$$\left. + k\int_{R^p}\{1 + n^{-1}\sum_{j=1}^{n}|X_{ij} - \Theta_i|^d + |n^{-1/2}t_i|^d\}|n^{-1/2}t_i|^5 P(dt/\mathbf{X})\right\} I[A_n^c] \tag{4.40}$$

By (A.5), (A.6) and (4.29), the absolute moments of B_i up to the order 8 are bounded uniformly in n; hence we obtain (4.39) from (4.40) and from Lemma 4.1 and 4.2. □

Let us consider R_{23}; put

$$\tau_i = t_i - n^{1/2}(\hat{\Theta}_i - \Theta_i), \quad \text{hence} \quad \Theta_i + n^{-1/2}t_i = \hat{\Theta}_i + n^{-1/2}\tau_i \tag{4.41}$$

$$\hat{Z}_{ni}(\tau_i) = Z_{ni}(\tau_i + n^{1/2}(\hat{\Theta}_i - \Theta_i)), i = 1, \dots, p \tag{4.42}$$

$$\hat{Z}_n(\tau) = \prod_{i=1}^{p} \hat{Z}_{ni}(\tau_i) = \prod_{i=1}^{p} exp\, S_i(\hat{\Theta}_i + n^{-1/2}\tau_i) \tag{4.43}$$

and

$$\hat{P}(d\tau/\mathbf{X}) = \hat{Z}_n(\tau)\pi(\hat{\Theta} + n^{-1/2}\tau)d\tau / \int \hat{Z}_n(\sigma)\pi(\hat{\Theta} + n^{-1/2}\sigma)d\sigma \tag{4.44}$$

Moreover, we shall use the following decomposition :

$$\hat{P}(d\tau/\mathbf{X}) = \hat{P}^{*}(d\tau/\mathbf{X}) + \hat{P}^{**}(d\tau/\mathbf{X}) \tag{4.45}$$

where
$$\hat{P}^{*}(d\tau/\mathbf{X}) = \prod_{i=1}^{p} I[\{0 < |\tau_i| < 2n^{\alpha}\}]\hat{P}(d\tau/\mathbf{X}) \tag{4.46}$$

and
$$\hat{P}^{**}(d\tau/\mathbf{X}) = I[\bigcup_{i=1}^{p}\{|\tau_i| \geq 2n^{\alpha}\}]\hat{P}(d\tau/\mathbf{X}) \tag{4.47}$$

Consider the behavior of the likelihood process in the neighborhood of $\hat{\underset{\sim}{\Theta}}$. For $\eta = (\eta_1, \ldots, \eta_p)' \in [-2n^{\alpha}, 2n^{\alpha}]^p$ write

$$S_i(\hat{\Theta}_i + n^{-1/2}\eta_i) = \hat{S}_i + n^{-1/2}\eta_i\hat{S}_i^{(1)} - \frac{1}{2}I\eta_i^2 + r_{ni} \tag{4.48}$$

with
$$r_{ni} = \eta_i^2 \int_0^1 [n^{-1}S_i^{(2)}(\hat{\Theta}_i + vn^{-1/2}\eta_i) + I](1-v)dv \tag{4.49}$$

Regarding (4.41), we have

$$\begin{aligned} Z_{ni}(n^{-1/2}t_i) &= exp\,(\hat{S}_i + \hat{S}_i^{(1)}.n^{-1/2}\tau_i).exp\{-\frac{1}{2}I\tau_i^2\}.exp\{r_{ni}\} \\ &= exp\,\{\hat{S}_i\}.exp\{-\frac{1}{2}I\tau_i^2\}.exp\{r_{ni}\} \end{aligned} \tag{4.50}$$

Concerning the third factor in (4.50), we shall prove

Lemma 4.5 If $|\eta_i| < 2n^{\alpha}$ for $n \geq n_0$, then $I(A_n)r_{ni} \to 0$ as $n \to \infty$.

Proof of Lemma 4.5. Expanding r_{ni} around Θ_i, we obtain

$$r_{ni} = \eta_i^2 \int_0^1 (\sum_{\ell=1}^{5} r_{ni}^{(\ell)}(v)(1-v)dv \tag{4.51}$$

with
$$r_{ni}^{(1)}(v) = \frac{1}{2}(n^{-1}S_i^{(2)} + I)$$

$$r_{ni}^{(2)}(v) = \frac{1}{2}(\hat{\Theta}_i - \Theta_i + n^{-1/2}v\eta_i)S_i^{(3)}/n$$

$$r_{ni}^{(3)}(v) = \frac{1}{4}(\hat{\Theta}_i - \Theta_i + n^{-1/2}v\eta_i)^2 S_i^{(4)}/n$$

$$r_{ni}^{(4)}(v) = \frac{1}{12}(\hat{\Theta}_i - \Theta_i + n^{-1/2}v\eta_i)^3 S_i^{(5)}/(6n)$$

$$r_{ni}^{(5)}(v) = K(\hat{\Theta}_i - \Theta_i + n^{-1/2}v\eta_i)^4\gamma_{ni} \tag{4.52}$$

where γ_{ni} is a random variable, $|\gamma_{ni}| \leq 1$ and $K > 0$ is an appropriate constant. If $|\eta_i| < 2n^{\alpha}$ the last two terms in (4.52) are $o(n^{-3/2})$. Moreover $r_{ni}^{(\ell)}I[A_n], l = 1,2,3$ is of order $O(n^{-1/2}log^{1/2}n), O(n^{\alpha-1}log^{1/2}n), O(n^{2\alpha-1})$ respectively. From (4.51) and $\alpha = 0,1$ we obtain the lemma. □

Lemma 4.6 For $\eta_i \in [-2n^{\alpha}, 2n^{\alpha}]$ and for $\omega \in A_n$,

$$\hat{Z}_{ni}(\eta_i) = exp\{-1/2I\eta_i^2\}P_{ni}(\eta_i) \tag{4.53}$$

where $$P_{ni}(\eta_i) = 1 + 1/2\eta_i^2(n^{-1}\hat{S}_i^{(2)} + I) + (2!n^{3/2})^{-1}\eta_i^3\hat{S}_i^{(3)}$$

$$+\eta_i^4\{(4!n^2)^{-1}\hat{S}_i^{(4)} + \frac{1}{8}(n^{-1}\hat{S}^{(2)} + I)^2\} + n^{-1.3}|\eta_i|^5 C_i\gamma_{ni}^* \tag{4.54}$$

where C_i is an appropriate constant and γ_{ni}^* is a random variable bounded by 1.

Proof of Lemma 4.6. Lemma follows by the standard Taylor expansion of $e^{r_{ni}}$ around $\hat{\Theta}_i$. □

Similarly, if we take an expansion around $\hat{\underset{\sim}{\Theta}}$ for $|\eta_i| < 2n^\alpha$ and use the fact that $\hat{S}_i^{(1)} = 0$, we obtain

$$(nI)^{-1}S_i^{(1)}(\hat{\Theta}_i + n^{-1/2}\eta_i) = n^{-3/2}I^{-1}\eta_i\hat{S}^{(2)} + (2n^2I)^{-1}\eta_i^2\hat{S}^{(3)}$$

$$+C_2n^{-3/2}\gamma_{ni}^{**}|\eta_i|^3 \tag{4.55}$$

and
$$\pi(\hat{\underset{\sim}{\Theta}} + n^{-1/2}\eta) = \hat{\pi}\{1 + \sum_{i=1}^{p}(\hat{\pi}_i'/\hat{\pi})n^{-1/2}\eta_i$$

$$+\frac{1}{2}\sum_{i=1}^{p}\sum_{k=1}^{p}(\hat{\pi}_{ij}''/\hat{\pi})n^{-1}\eta_i\eta_k + n^{-3/2}\pi^*(\eta)||\eta||^3\} \tag{4.56}$$

where C_2 is an appropriate constant, γ_{ni}^{**} is a random variable bounded by 1 and π^* is a bounded function of η.

Finally, utilizing (4.41) - (4.47), we may write $B_i = B_i^* + B_i^{**}$ with

$$B_i^* = \int_{R^p}(nI)^{-1}S_i^{(1)}(\hat{\Theta}_i + n^{-1/2}\tau_i)\hat{P}^*(d\tau/\mathbf{X}) \tag{4.57}$$

and
$$B_i^{**} = \int_{R^p}(nI)^{-1}S_i^{(1)}(\hat{\Theta}_i + n^{-1/2}\tau_i)\hat{P}^{**}(d\tau/\mathbf{X}), i = 1, \ldots, p. \tag{4.58}$$

Let us first study the behavior of B_i^*. We may write

$$B_i^* = N_i/D_n \tag{4.59}$$

where
$$N_i = \int_{R^p}[S^{(1)}(\hat{\Theta}_i + n^{-1/2}\eta_i)/(nI)]\prod_{k=1}^{P}(exp\{-\frac{1}{2}I\eta_k^2\}P_{nk}(\eta_k).$$

$$.I[|\eta_k| \leq 2n^\alpha])\pi(\hat{\underset{\sim}{\Theta}} + n^{-1/2}\eta)d\eta \tag{4.60}$$

and
$$D_n = \int_{R^p} Z_n(t)\pi(\hat{\underset{\sim}{\Theta}} + n^{-1/2}t)dt \tag{4.61}$$

Lemma 4.7 For $\omega \in A_n$, it holds

$$N_i = (\hat{\pi}_i'/\hat{\pi})\frac{\hat{S}_i^{(2)}}{n^2I} + o(n^{-1}), i = 1, \ldots, p \tag{4.62}$$

Proof of Lemma 4.7 It follows from Lemma 4.6 and from the representations (4.55) and (4.56). □

Concerning D_n, we can prove the following approximation of the denominator of B_i^* :

Lemma 4.8 For $\omega \in A_n$, $n = 1, 2, \ldots$

$$D_n^{-1} = 1 + o(1) \quad \text{as} \quad n \to \infty \tag{4.63}$$

Proof of Lemma 4.8. We take the decomposition : $D_n = D_{1n} + D_{2n}$

where
$$D_{1n} = \int_{R^p} \hat{Z}_n(\tau) I[||\tau|| < 2n^\alpha] \pi(\underset{\sim}{\hat{\Theta}} + n^{-1/2}\tau) d\tau$$

$$D_{2,n} = \int_{R^p} \hat{Z}_n(\tau) I[||\tau|| > 2n^\alpha] \pi(\underset{\sim}{\hat{\Theta}} + n^{-1/2}\tau) d\tau$$

For $\omega \in A_n$, it follows from 4.23 that $D_{2n} = o(1)$. The same argument as in Lemma 4.7 gives $D_{1,n} = 1 + o(1)$. □

Lemma 4.9 Denote

$$B_{i\nu}^{**} = \int_{R^p} (nI)^{-1} S_i^{(1)}(\hat{\Theta}_i + n^{-1/2}\tau_i) I[|\tau_\nu| > 2n^\alpha] \hat{P}(d\tau/\mathbf{X}), i, \nu = 1, \ldots, p \tag{4.64}$$

Then, given $N > 0$, there exists $c_N > 0$ such that

$$P\{I[A_n]|B_{i\nu}^{**}| > n^{-\alpha N}\} < c_N n^{-\alpha N} \tag{4.65}$$

Proof of Lemma 4.9 Regarding (4.55) - (4.59), the probability in (4.65) is dominated by

$$P\{I[A_n] \int_{R^p} I[|t_\nu| > n^\alpha](nI)^{-1}|S_i^{(1)}(\Theta_i + n^{-1/2}t_i)| P(dt/\mathbf{X}) > n^{-\alpha N}\}$$

Now, $(nI)^{-1}|S_i^{(1)}(\Theta_i + n^{-1/2}t_i)|$ is majorized by a polynomial in $|t_i|$ on A_n; moreover, we have $1 < (|t_\nu|/n^\alpha)^N$ on the integration domain; hence, (4.65) follows from Lemma 4.1 (i). □

Finally, we are in a position to prove

Lemma 4.10 Under the conditions of Theorem 3.2

$$R_{23} = o(n^{-2}) \tag{4.66}$$

Proof of Lemma 4.10 First, notice that by Lemma 4.3

$$R_{23} = -2 \sum_{i=1}^{p} E_\Theta\{B_i(u_{ni} + v_{ni}) I[A_n]\} \tag{4.67}$$

Moreover, by (4.50), (4.44), (4.45), (4.46), Lemma 4.1 and Lemma 4.5,

$$R_{23} = R_{23}^* + R_{23}^{**}$$

where R_{23}^* and R_{23}^{**} are defined by (4.35) with B_i replaced by B_i^* and B_i^{**}, respectively. Then

$$|R_{23}^{**}| \leq \sum_{i=1}^{p}\sum_{\nu=1}^{p}[E_{\underset{\sim}{\Theta}}\{|B_{i\nu}^{**}(u_{ni}+v_{ni})|I[A_n]\}] \tag{4.68}$$

and

$$E_{\underset{\sim}{\Theta}}\{|B_{i\nu}^{**}(u_{ni}+v_{ni})|I[A_n]\}$$

$$\leq E_{\Theta}\{|B_{i\nu}^{**}(u_{ni}+v_{ni})|I[A_n]\}\}.I[I[A_n]B_{i\nu}^{**} \leq n^{-\alpha N}]\} + P_{\Theta}\{I[A_n]B_{i\nu}^{**} > n^{-\alpha N}\}$$

$$\leq o(n^{-2}) + c_N\, n^{-\alpha N} = o(n^{-2}) \tag{4.69}$$

as $n \to \infty$ for sufficiently large $N(\alpha N = 10)$.

Finally, let us write

$$R_{23}^{*} = \sum_{i=1}^{p} R_{23i}^{*} \tag{4.70}$$

where by Lemma 4.3

$$R_{23i}^{*} = E\left\{B_i^{*}\left[\frac{S_i^{(1)}}{nI^2}\left[\frac{S_i^{(2)}}{n}+I\right]+v_{ni}\right]\right\} \tag{4.71}$$

and, by Lemma 4.7 and 4.8,

$$B_i^{*} = \left\{\frac{\hat{S}_i^{(2)}}{n^2 I}\,\frac{\hat{\pi}_i'}{\hat{\pi}} + o(n^{-1})\right\}.I[A_n] \tag{4.72}$$

By expansion of $\hat{S}_i^{(2)}$ and of $\hat{\pi}_i/\hat{\pi}$ in neighborhood of Θ we obtain

$$B_i^{*} = \left\{\frac{S_i^{(2)}}{n^2 I}\,\frac{\pi_i'}{\pi} + o(n^{-1})\right\} I[A_n]$$

hence

$$R_{23}^{*} = o(n^{-2})$$

□

Acknowledgement. The authors would like to thank Prof. M. HUŠKOVÁ for interesting discussions of the subject, which in turn led to an improvement of Theorem 3.2.

BIBLIOGRAPHY

von BAHR B. (1967). On the central limit theorem in R_k. *Arkiv för matematik* 7, 61-69.

BLUMENTHAL, R.M. and GETOOR, R.K. (1968). Markov Processes and Potential Theory. *Academic Press*, New York.

CHIBISOV, D.M. (1973). An asymptotic expansion for a class of estimators containing maximum likelihood estimators. *Theor. Probab. and Applications* 18, 295-303.

CHUNG, KAI LAI (1982). Lectures from Markov Processes to Brownian Motion. *Springer Verlag*, New York.

DOOB, J.L. (1984). Classical Potential Theory and Its Probabilistic Counterpart. *Springer Verlag*, New York.

HUDSON, H.M. (1978). A natural identity for exponential families with applications in multiparameter estimation. *Ann. Statist.*, 6, 473-484.

IBRAGIMOV, I.A. and HASMINSKII, R.Z. (1981). Statistical Estimation : Asymptotic Theory. *Springer Verlag*, New York.

JUREČKOVÁ, J. (1985). Representation of M-estimators with the second-order asymptotic distribution. *Statistics & Decisions*, 3, 263-276.

JUREČKOVÁ, J. and SALEH, A.K.MD. E. (1990). Robustified versions of Stein's multivariate location estimation. *Statist. & Probab. Letters*, 9, 375-380.

STEIN, C.M. (1981). Estimation of the mean of a multivariate normal distribution. *Ann. Statist.*, 9, 1135-1151.

Charles University
Department of Statistics
Sokolovska 83, 18600 Prague 8
Czech Republic

Université Montpellier II
Département des Sciences Mathématiques
Place E. Bataillon, 34095 Montpellier
France

ALMOST SURE INVARIANCE PRINCIPLES FOR V- AND U-STATISTICS BASED ON WEAKLY DEPENDENT RANDOM VARIABLES

Shuya Kanagawa
Department of Mathematics, Faculty of Liberal Arts & Education, Yamanashi University, 4-4-37, Takeda, Kofu 400, JAPAN

Ken-ichi Yoshihara
Department of Mathematics, Faculty of Engineering, Yokohama National University, 156, Tokiwadai, Hodogaya, Yokohama 240, JAPAN

Almost sure invariance principles for V- and U-statistics for some degenerate kernel functions with degree $m \geq 2$ are established for strictly stationary sequences of random variables satisfying some mixing conditions. The error term obtained is $o\left(t^{m/2-\gamma}\right)$ which implies Donsker's and Strassen's invariance principles. The basic method of the proof is to expand the kernel function in its Fourier series and apply almost sure invariance principles for partial sums of Banach space valued random variables due to Dehling (1983).

Key words: U-statistics; V-statistics; almost sure invariance principle; mixing random variables; Banach space valued random variables
AMS 1980 classification: 60F15; 60E20

1. Introduction and results

Let $h: \mathbf{R}^m \to \mathbf{R}$ be a symmetric function and $\left\{\xi_j,\ j \geq 1\right\}$ be a strictly stationary sequence of random variables with a probability distribution μ. For $n \geq 2$, V-statistic $V_n(h)$ and U-statistic $U_n(h)$ with degree $m \geq 2\,(m \leq n)$ are defined as

$$V_n(h) := \sum_{1 \leq i_1, \cdots, i_m \leq n} h\left(\xi_{i_1}, \cdots, \xi_{i_m}\right),\quad U_n(h) := \sum_{\substack{1 \leq i_1, \cdots, i_m \leq n \\ i_j \neq i_k \text{ for } j \neq k}} h\left(\xi_{i_1}, \cdots, \xi_{i_m}\right),$$

respectively. Furthermore assume that h is a degenerate and square integrable kernel, i.e. for all $i=1,\cdots,m$

$$\int h(x_1,\cdots,x_m)\mu(dx_i)=0 \quad\text{and}\quad \int\cdots\int h^2(x_1,\cdots,x_m)\prod_{k=1}^{m}\mu(dx_k)<\infty.$$

Under some assumptions Harel and Puri (1989) (Corollary 3.1) obtained that

$$n^{1/2}\left|n^{-m}V_n(h)-\frac{1}{m!}\binom{n}{m}^{-1}U_n(h)\right|\to 0 \quad\text{in probability as } n\to\infty.$$

Let $L^2\left(\mathbf{R}^m,\prod_{k=1}^{m}\mu(dx_k)\right)$ be a Hilbert space of square integrable functions with respect to $\prod_{k=1}^{m}\mu(dx_k)$ and $\{g_i, i\geq 0\}$ an orthogonal basis of $L^2(\mathbf{R}^1,\mu(dx))$ with $g_0\equiv 1$. We can expand the kernel function $h\in L^2\left(\mathbf{R}^m,\prod_{k=1}^{m}\mu(dx_k)\right)$ in its Fourier series,

$$h(x_1,\cdots,x_m)=\sum_{0\leq i_1,\cdots,i_m<\infty}\lambda_{i_1\cdots i_m}\prod_{j=1}^{m}g_{i_j}(x_j),$$

where

$$\lambda_{i_1\cdots i_m}:=\int\cdots\int h(x_1,\cdots,x_m)\prod_{j=1}^{m}g_{i_j}(x_j)\prod_{k=1}^{m}\mu(dx_k).$$

Let $\{V(t), t\geq 0\}$ and $\{U(t), t\geq 0\}$ be random elements of $D[0,\infty)$ defined by

$$V(t):=\sum_{0\leq i_1,\cdots,i_m<\infty}\lambda_{i_1\cdots i_m}\sigma_{i_1}W_{i_1}(t)\cdots\sigma_{i_m}W_{i_m}(t),$$

$$U(t):=\sum_{\substack{0\leq i_1,\cdots,i_m<\infty\\ i_j\neq i_k \text{ for } j\neq k}}\lambda_{i_1\cdots i_m}\sigma_{i_1}W_{i_1}(t)\cdots\sigma_{i_m}W_{i_m}(t),$$

respectively, where $\{\{W_j(t), t\geq 0\}, j\geq 0\}$ are Wiener processes satisfying

$$E\left[W_i(t)W_j(s)\right]=\lim_{n\to\infty}\frac{1}{n\sigma_i\sigma_j}\sum_{k=1}^{[nt]}\sum_{\ell=1}^{[ns]}E\left[g_i(\xi_k)g_j(\xi_\ell)\right],$$

$$\sigma_j^2:=1+2\sum_{i=1}^{\infty}E\left[g_j(\xi_1)g_j(\xi_{i+1})\right]>0.$$

In the case where $\{\xi_j\}$ are independent random variables Dehling, Denker and Philipp (1984) and Dehling (1989) obtained Strassen's law of the iterated logarithm for $U_n(h)$:

almost surely $\{(2n \log\log n)^{-m/2} U_{[nt]}(h),\ 0 \le t \le 1\}_{n \ge 3}$ is relatively compact and the set of its limit points coincides with some compact subset G in $C[0,1]$.

Kanagawa and Yoshihara (1993) extended their results to the almost sure invariance principle under mixing conditions, using Hoeffding's decomposition scheme (Hoeffding, 1948) for the kernel function h, although the U-statisics which we treated have degree 2:

without changing the distributions of $\{U_{[t]}(h),\ t \ge 0\}$ and $\{U(t),\ t \ge 0\}$ we can redefine them on a richer probability space such that for some $\gamma > 0$,

$$\left|U_{[t]}(h) - U(t)\right| = o\left(t^{1-\gamma}\right) \quad \text{a.s.} \quad \text{as } t \to \infty.$$

In this paper, using the Fourier expansion of the kernel function h, we extend the above result to the case where the degree is $m \ge 2$. Let $\mathcal{M}_a^b$ denote the σ-algebra generated by $\xi_a, \cdots, \xi_b$, $-\infty \le a \le b \le \infty$. We say that $\{\xi_j\}$ is ϕ-mixing if as $n \to \infty$

$$\phi(n) := \sup_{A \in \mathcal{M}_{-\infty}^0,\, B \in \mathcal{M}_n^\infty} \left|\frac{P(AB) - P(A)P(B)}{P(A)}\right| \downarrow 0.$$

Furthermore $\{\xi_j\}$ is said to be absolutely regular if as $n \to \infty$

$$\beta(n) := E\left(\sup_{A \in \mathcal{M}_n^\infty} \left|P\left(A \middle| \mathcal{M}_{-\infty}^0\right) - P(A)\right|\right) \downarrow 0$$

and $\{\xi_j\}$ is said to be strong mixing if as $n \to \infty$

$$\alpha(n) := \sup_{A \in \mathcal{M}_{-\infty}^0,\, B \in \mathcal{M}_n^\infty} \left|P(AB) - P(A)P(B)\right| \downarrow 0.$$

It is easy to see that $\alpha(n) \le \beta(n) \le \phi(n)$ for each n (e.g. Yoshihara, 1993). For a technical reason we first treat V-statistics. The result for U-statistics is given in Corollary to Theorems 1 and 2.

Theorem 1. *Assume that for some* $s>0$

(1.1) $$\sum_{N+1\le i_1,\cdots,i_m<\infty}\left|\lambda_{i_1\cdots i_m}\right|=O\left(N^{-s}\right)\quad \text{as } n\to\infty.$$

Moreover suppose that $\{\xi_j\}$ *is a strictly stationary sequence of random variables with zero mean and for some* $\delta>0$

(1.2) $$\sup_{k\ge 0} \boldsymbol{E}\left(\left|g_k(\xi_1)\right|^{2m+\delta}\right)<\infty.$$

Assume that $\{\xi_j\}$ *satisfies the* ϕ*-mixing condition with*

(1.3) $$\phi(n)=O\left(n^{-2(1+\varepsilon)}\right)\quad as\ n\to\infty\ for\ some\ \varepsilon>0,$$

or the absolutely regular condition with

(1.4) $$\beta(n)=O\left(n^{-4(1+\varepsilon)}\right)\quad as\ n\to\infty\ for\ some\ \varepsilon>0.$$

Then, without changing the distributions of $\{V_{[t]}(h),\, t\ge 0\}$ *and* $\{V(t),\, t\ge 0\}$, *we can redefine them on a richer probability space such that for some* $\gamma>0$

(1.5) $$\left|V_{[t]}(h)-V(t)\right|=o\left(t^{m/2-\gamma}\right)\quad a.s.\quad as\ t\to\infty.$$

Theorem 2. *Let* $\{\xi_j\}$ *be a strictly stationary strong mixing sequence of random variables with*

(1.6) $$\alpha(n)=O\left(n^{-4(1+\varepsilon)}\right)\quad as\ n\to\infty\ for\ some\ \varepsilon>0.$$

Furthermore suppose (1.1) and (1.2). Then, without changing the distributions of $\{V_{[t]}(h),\, t\ge 0\}$ *and* $\{V(t),\, t\ge 0\}$, *we can redefine them on a richer probability space such that*

(1.7) $$\left|V_{[t]}(h)-V(t)\right|=o\left((t\log\log t)^{m/2}\right)\quad a.s.\quad as\ t\to\infty.$$

By a slight modification of the above theorems we can easily obtain the next corollary.

Corollary 1. *Under the same assumptions as in Theorem 1, without changing the distributions of* $\{U_{[t]}(h),\, t\ge 0\}$ *and* $\{U(t),\, t\ge 0\}$, *we can redefine them on a richer probability space such that for some* $\gamma>0$

$$\left|U_{[t]}(h)-U(t)\right|=o\left(t^{m/2-\gamma}\right)\quad a.s.\quad as\ t\to\infty. \tag{1.8}$$

Moreover, under the same assumptions as in Theorem 2, we can replace (1.7) to

$$\left|U_{[t]}(h)-U(t)\right|=o\left((t\log\log t)^{m/2}\right)\quad a.s.\quad as\ t\to\infty. \tag{1.9}$$

2. Proofs of Theorems 1 and 2

The main technique of the proofs is to estimate the V-statistics $V_{[t]}(h)$ by partial sums of Banach space valued random variables as following. Let X be a separable Banach space equipped with the norm $\|\cdot\|$ defined as

$$X:=\left\{x=(x_1,x_2,\cdots)\in\mathbf{R}^\infty:\ \sum_{0\le i_1,\cdots,i_m<\infty}\left|\lambda_{i_1\cdots i_m}\right|\left|x_{i_1}\right|^m<\infty\right\},$$

$$\|x\|:=\left\{\sum_{0\le i_1,\cdots,i_m<\infty}\left|\lambda_{i_1\cdots i_m}\right|\left|x_{i_1}\right|^m\right\}^{1/m}$$

Since (1.1) and (1.2) imply $\boldsymbol{E}\left(\left\{\sum_{0\le i_1,\cdots,i_m<\infty}\left|\lambda_{i_1\cdots i_m}\right|\left|g_{i_1}(\xi_k)\right|^m\right\}^{1/m}\right)<\infty$, we can define X-valued random variables $G_k:=\left(g_0(\xi_k),\, g_1(\xi_k),\cdots\right)$ for each $k\ge 1$ and $S(t):=\sum_{k=1}^{[t]}G_k$. Furthermore denote $W(t)$ a X-valued Wiener process definded by $\{W(t),\, t\ge 0\}:=\left\{\left(\sigma_0W_0(t),\sigma_1W_1(t),\cdots\right),\, t\ge 0\right\}$. Since $\lambda_{i_1,\cdots,i_m}$ is symmetric in its m arguments, we can observe by Hölder's ineqality

$$\left|V_{[t]}(h)-V(t)\right| \tag{2.1}$$

$$=\left|\sum_{1\le j_1,\cdots,j_m\le n}\ \sum_{0\le i_1,\cdots,i_m<\infty}\lambda_{i_1\cdots i_m}\prod_{k=1}^{m}g_{j_k}\left(\xi_{j_k}\right)-\sum_{0\le i_1,\cdots,i_m<\infty}\lambda_{i_1\cdots i_m}\sigma_{i_1}W_{i_1}(t)\cdots\sigma_{i_m}W_{i_m}(t)\right|$$

$$= \left| \sum_{0 \le i_1, \cdots, i_m < \infty} \lambda_{i_1 \cdots i_m} \sum_{k=1}^{[t]} g_{i_1}(\xi_k) \cdots \sum_{k=1}^{[t]} g_{i_m}(\xi_k) - \sum_{0 \le i_1, \cdots, i_m < \infty} \lambda_{i_1 \cdots i_m} \sigma_{i_1} W_{i_1}(t) \cdots \sigma_{i_m} W_{i_m}(t) \right|$$

$$\le \left| \sum_{0 \le i_1, \cdots, i_m < \infty} \lambda_{i_1 \cdots i_m} \left(\sum_{k=1}^{[t]} g_{i_1}(\xi_k) - \sigma_{i_1} W_{i_1}(t) \right) \sum_{k=1}^{[t]} g_{i_2}(\xi_k) \cdots \sum_{k=1}^{[t]} g_{i_m}(\xi_k) \right|$$

$$+ \left| \sum_{0 \le i_1, \cdots, i_m < \infty} \lambda_{i_1 \cdots i_m} \sigma_{i_1} W_{i_1}(t) \left(\sum_{k=1}^{[t]} g_{i_2}(\xi_k) - \sigma_{i_2} W_{i_2}(t) \right) \sum_{k=1}^{[t]} g_{i_3}(\xi_k) \cdots \sum_{k=1}^{[t]} g_{i_m}(\xi_k) \right|$$

$$+ \cdots + \left| \sum_{0 \le i_1, \cdots, i_m < \infty} \lambda_{i_1 \cdots i_m} \sigma_{i_1} W_{i_1}(t) \cdots \sigma_{i_{m-1}} W_{i_{m-1}}(t) \left(\sum_{k=1}^{[t]} g_{i_m}(\xi_k) - \sigma_{i_m} W_{i_m}(t) \right) \right|$$

$$\le \|S(t) - W(t)\| \, \|S(t)\|^{m-1} + \|W(t)\| \, \|S(t) - W(t)\| \, \|S(t)\|^{m-2} + \cdots + \|W(t)\|^{m-1} \|S(t) - W(t)\|.$$

We estimate the right hand side of (2.1) by the next almost sure invariance principles for partial sums of X-valued random variables due to Dehling (1983).

Theorem A. *(Dehling, 1983) Let $\{\eta_j, j \ge 1\}$ be a weakly stationary sequence of random variables with values in the separable Banach space X with a norm $\|\cdot\|$ such that $E[\eta_j] = 0$ for each $j \ge 1$ and $\sup_{j \ge 1} E\left[\|\eta_j\|^{2+\delta}\right] < \infty$ for some $\delta > 0$. Suppose that $\{\eta_j, j \ge 1\}$ is either ϕ-mixing with rate*

$$\phi(n) = O\left(n^{-2(1+\varepsilon)}\right) \quad as \ n \to \infty \ for\ some\ \varepsilon > 0$$

or absolutely regular with rate

$$\beta(n) = O\left(n^{-(1+\varepsilon)(1+2/\delta)}\right) \quad as \ n \to \infty \ for\ some\ \varepsilon > 0 \ and \ 0 < \delta \le 2/3.$$

Let P_N be a sequence of bounded operators on X with N-dimensional range. Assume that

$$\text{(2.2)} \qquad E\left[\left\| n^{-1/2} \sum_{j=a+1}^{a+n} (\eta_j - P_N \eta_j) \right\|^2\right] = O(N^{-u}) \quad as \ N \to \infty \ for\ some\ u > 0$$

uniformly in $a \ge 0$ and $n \ge 1$, and

(2.3) $$\sup_{\|x\|=1} \|P_N x\| = O(N^r) \quad as \quad N \to \infty \quad \textit{for some} \quad r \geq 0.$$

Then we can redefine the sequence $\{\eta_j, j \geq 1\}$ *on a new probability space together with Brownian motion* $W(t)$ *with covariance structure* T *such that*

(2.4) $$\left\| \sum_{j=1}^{[t]} \eta_j - W(t) \right\| = O\left(t^{1/2-\gamma}\right) \quad as \quad t \to \infty \quad \textit{for some} \quad \gamma > 0,$$

where T *is defined by the following absolutely convergent series,*

$$T(f,g) := E\left[f(\eta_1)g(\eta_1)\right] + \sum_{k \geq 2} E\left[f(\eta_1)g(\eta_k)\right] + \sum_{k \geq 2} E\left[f(\eta_k)g(\eta_1)\right], \quad f,g \in X^*,$$

where X^* *is the dual space of* X.

For the strong mixing sequence Strassen's type almost sure invariance principle with error term $O\left((t \log\log t)^{1/2}\right)$ is obtained.

Theorem B. *(Dehling, 1983) Let* $\{\eta_j, j \geq 1\}$ *be a weakly stationary strong mixing sequence of random variables with values in the separable Banach space* X *with norm* $\|\cdot\|$ *such that* $E[\eta_j] = 0$ *for each* $j \geq 1$ *and* $\sup_{j \geq 1} E\left[\|\eta_j\|^{2+\delta}\right] < \infty$ *for some* $0 < \delta \leq 2/3$ *and suppose that the mixing rate is*

$$\alpha(n) = O\left(n^{-(2+\varepsilon)(1+2/\delta)}\right) \quad as \quad n \to \infty \quad \textit{for some} \quad \varepsilon > 0.$$

Let P_N *be a sequence of bounded operators on* X *with N-dimensional range satisfying (2.3) and*

(2.5) $$E\left[\left\| n^{-1/2} \sum_{j=1}^{n} (\eta_j - P_N \eta_j) \right\|^2\right] = O\left((\log N)^{-3\rho}\right) \quad as \quad N \to \infty \quad \textit{for some} \quad \rho > 1.$$

Then we can redefine the sequence $\{\eta_j, j \geq 1\}$ *on a new probability space together with Brownian motion* $W(t)$ *with covariance structure* T *such that*

(2.6) $$\left\| \sum_{j=1}^{[t]} \eta_j - W(t) \right\| = O\left((t \log\log t)^{1/2}\right) \quad \text{a.s.} \quad as \quad t \to \infty.$$

To apply Dehling's results we check (2.2) and (2.3).

Lemma 1. *Let P_N be a sequence of bounded operators on X defined by $P_N x := (x_1, x_2, \cdots, x_N, 0, \cdots)$ for any $x = (x_1, x_2, \cdots) \in X$ and $N \geq 1$. Under the assumptions in Theorem 1 or 2, $\{G_k, k \geq 1\}$ satisfies (2.2) or (2.3), respectively.*

Proof. We first show (2.2). By Hölder's inequality

$$(2.7) \quad E\left\| n^{-1/2} \sum_{j=a+1}^{a+n} \left(G_j - P_N G_j\right)\right\|^2$$

$$= n^{-1} E\left(\left\{ \sum_{N+1 \leq i_1, \cdots, i_m < \infty} \left|\lambda_{i_1 \cdots i_m}\right| \left|\sum_{j=a+1}^{a+n} G_j - P_N G_j\right|^m \right\}^{2/m}\right)$$

$$\leq n^{-1} \left\{ \sum_{N+1 \leq i_1, \cdots, i_m < \infty} \left|\lambda_{i_1 \cdots i_m}\right| \right\}^{2/m} \sup_{i_1 \geq 0} E\left\{ \left|\sum_{j=a+1}^{a+n} g_{i_1}\left(\xi_j\right)\right|^m \right\}^{2/m}$$

Using Theorem 1 in Yokoyama (1980), we obtain from the stationarity of $\{\xi_j\}$ and (1.2) that

$$(2.8) \qquad n^{-1} \sup_{i_1 \geq 0} E\left\{ \left|\sum_{j=a+1}^{a+n} g_{i_1}\left(\xi_j\right)\right|^m \right\}^{2/m} = O(1) \quad \text{as } n \to \infty.$$

Combining (1.1), (2.7) and (2.8) we obtained the desired result. Furthermore (2.3) can be shown easily from the definition of $\|\cdot\|$. ❑

We now turn to the proof of the theorem. Recall that $S(t) := \sum_{k=1}^{[t]} G_k$. Since it is easy to show the assumptions except for (2.2) and (2.3), using Lemma 1 we may apply Theorem A to redefine $\{S(t), t \geq 0\}$ and $\{W(t), t \geq 0\}$ on a new probability space such that

$$(2.9) \qquad \|S(t) - W(t)\| = O\left(t^{1-\gamma}\right) \quad \text{a.s. as } t \to \infty \text{ for some } \gamma > 0.$$

On the other hand, by the law of the iterated logarithm for $\{S(t), t \geq 0\}$ and $\{W(t), t \geq 0\}$ (see e.g. Dehling, 1983) we have

(2.10) $$\| S(t) \| = O\left((t \log\log t)^{1/2}\right) \quad \text{a.s. as } t \to \infty,$$

(2.11) $$\| W(t) \| = O\left((t \log\log t)^{1/2}\right) \quad \text{a.s. as } t \to \infty.$$

Hence from (2.1) and (2.9)-(2.11) we conclude the proof of Theorem 1. Furthermore we can also prove Theorem 2 by the same way as Theorem 1. ❑

Acknowledgement. The authors wish to thank Professor M. Hitsuda of Kumamoto University for his valuable comments.

REFERENCES

H. Dehling (1983), Limit theorems for sums of weakly dependent Banach space valued random variables, Wahrscheinlichkeitstheorie verw. Gebiete **63**, 393-432.

H. Dehling, M. Denker and W. Philipp (1984), Invariance principles for von Mises and U-statistics, Z. Wahrscheinlichkeitstheorie verw. Gebiete **67**, 139-167.

M. Harel and M. L. Puri (1989), Limiting behaviour of U-statistics, V-statistics, and one sample rank order statistics for non stationary absolutely regular processes, J. Multivariate Anal. **30**, 181-204.

W. Hoeffding (1948), A non-parametric test of independence, Ann. Math. Statist. **19** , 546-557.

S. Kanagawa and K. Yoshihara (1993), The almost sure invariance principles of degenerate U-statistics of degree two for stationary random variables, to appear in Stochastic Processes Applications.

R. Yokoyama (1980), Moment bounds for stationary mixing sequences, Z. Wahrscheinlichkeitstheorie verw. Gebiete **52**, 45-57.

K. Yoshihara, *Weakly Dependent Stochastic Sequences and Their Applications, Vol. 2, Asymptotic Statistics based on Weakly Dependent Date* (Sanseido Co. Ltd., Tokyo, 1993).

ON STABILITY IN TWO–STAGE STOCHASTIC NONLINEAR PROGRAMMING

Vlasta Kaňková
Institute of Information Theory and Automation
Academy of Sciences of the Czech Republic
P. O. Box 18, CZ–182 08 Prague 8
Czech Republic

Two–stage stochastic programming problems are very often assigned to practical optimization problems with random elements. Especially, these models are employed if the basic solution should be determined without knowing the random parameter realization and if the obtained effect can be corrected by a new optimization problem (called the recourse problem) depending on the random elements realization.
It is well-known that then the total problem depends on the random elements only through the corresponding probability measure. Consequently, the probability measure can be treated as a parameter in such problems and it is surely reasonable to study the stability with respect to it. The aim of this paper is to study the stability of two–stage nonlinear programming problem with respect to the distribution function. Of course, the linear case is also included in our consideration.

Key words: Two–stage stochastic programming problems, stability, stability region, distribution functions space, Kolmogorov metric.
AMS classification: 90C15

1. Introduction

Let $(\Omega,\ S,\ P)$ be a probability space,
$\xi = \xi(\omega) = [\xi_1(\omega), \ldots, \xi_s(\omega)]$ be an s–dimensional random vector defined on $(\Omega,\ S,\ P)$, $F(z)$ and $Z = Z_F \subset E_s$ denote the distribution function of the random vector $\xi(\omega)$ and the support of the probability measure corresponding to the distribution function $F(z)$, respectively.

Let, moreover, $f_i(u),\ i = 0, 1, 2, \ldots, l$ and $h_i(x,\ z),\ i = 0, 1, 2, \ldots, l$ be real–valued, continuous functions defined on E_r and $E_n \times E_s$, respectively,
$X \subset E_n$, $Z \subset E_s$, $U \subset E_r$ be nonempty sets.
(E_n, $n \geq 1$ denotes an n–dimensional Euclidean space.)

If the mapping $\mathcal{K}(x,\ z)$ and the function $g(x,\ z)$ are defined by

$$\mathcal{K}(x,\ z) = \{u \in U : f_i(u) \leq h_i(x, z),\ i = 1, 2, \ldots, l\},\quad x \in E_n,\ z \in E_s, \tag{1}$$

$$g(x,\, z) = \sup\{\, f_0(u)\,|u \in \mathcal{K}(x,\, z)\,\},\;\; x \in E_n,\; z \in E_s, \tag{2}$$

then we can introduce two-stage stochastic nonlinear programming problem as the problem: Find

$$\sup_X \mathsf{E}_F\{g(x,\, \xi(\omega)) + h_0(x, \xi(\omega))\} = \varphi(F). \tag{3}$$

E_F denotes the operator of the mathematical expectation considered with respect to the distribution function $F(z)$.

Remarks. 1. The problem defined by (1), (2) and (3) can be assigned, for example, to an original optimization problem with random elements in the form: Find

$$\max\{h_0(x,\, \xi(\omega))|\, x \in X : g_i(x,\, \xi(\omega)) \le 0,\, i = 1, 2, \dots, l\},$$

where $g_i(x,\, z)$, $i = 1, 2, \dots, l$ are real-valued, continuous functions defined on $E_n \times E_s$. Evidently, then $h_i(x,\, z) = -g_i(x,\, z)$, $i = 1, 2, \dots, l$ and the inner problem given by (1) and (2) can express either an additional profit corresponding to the unexhausted constraints or it can correspond to a penalty.

2. In general, it may happen that some symbols in (1), (2) and (3) are not well defined. However, this situation cannot appear under the assumptions considered in this paper.

3. Of course, it is possible to introduce more general problems of two-stage stochastic programming (see, for example, Kaňková (1978), Kaňková (1980 a) or Kaňková (1980 b)).

The aim of the paper is to study the stability of the problem given by the relations (1), (2) and (3) in the distribution functions space. In detail, the aim of the paper will be, first, for $\delta > 0$ to determine a neighbourhood $\mathcal{F}(F,\, \delta)$ of the theoretical distribution function $F(\cdot)$ and real-valued functions $K_1(\cdot)$, $K_2(\cdot)$ defined on E_1 such that

$$G \in \mathcal{F}(F,\, \delta) \implies |\varphi(F) - \varphi(G)| \le K_1(\delta), \tag{4}$$

$$G \in \mathcal{F}(F,\, \delta) \implies \|x(F) - x(G)\|^2 \le K_2(\delta),$$

where

$$x(F) = \arg\max\{\mathsf{E}_F[g(x,\, \xi(\omega)) + h_0(x,\, \xi(\omega))]\,|x \in X\}.$$

($\|\cdot\|$ denotes the Euclidean norm in E_n.)

Further, we shall introduce two real-valued functions $M_1(\cdot)$, $M_2(\cdot)$ defined on E_1 such that

$$|\varphi(F) - \varphi(G)| \le M_1(\sup\, |F(z) - G(z)|), \tag{5}$$

$$\|x(F) - x(G)\|^2 \le M_2(\sup\, |F(z) - G(z)|).$$

Using this result, we shall express the corresponding upper bound with the help of the Kolmogorov metric. Of course, this will be possible only under some additional assumptions.

Remark. It is evident that there is a slight inaccuracy in the definition of the symbol $\varphi(G)$. The exact definition of $\varphi(G)$ should be

$$\varphi(G) = \sup_X \mathsf{E}_G\{g(x, \xi^G(\omega)) + h_0(x, \xi^G(\omega))\},$$

where $\xi^G(\omega)$ is a random vector defined on (Ω, S, P) with the corresponding distribution function $G(z)$. Similar situation exists in the case of $x(G)$.

2. Some Auxiliary Definitions

The class of strongly concave (convex) functions is rather important for concave (convex) programming problems. We shall recall here the corresponding definition.

Definition 1. Let $h(x)$ be a real-valued function defined on a convex set $K \subset E_n$. $h(x)$ is a strongly concave function with a parameter $\varrho > 0$ if

$$h(\lambda x^1 + (1-\lambda)x^2) \geq \lambda h(x^1) + (1-\lambda)h(x^2) + \lambda(1-\lambda)\varrho\|x^1 - x^2\|^2$$

for every $x^1, x^2 \in K$, $\lambda \in \langle 0, 1\rangle$.

Furthermore, we shall define two functions $F_\delta(z)$, $F^\delta(z)$ for $\delta > 0$, $z \in E_s$ by

$$F_\delta(z) = F(z) - \frac{1}{2^{s+2}} \inf_{z' \in B(z, 2\sqrt{s}\delta)} P_F\{Z[z', \frac{\delta}{2}]\}, \tag{6}$$

$$F^\delta(z) = F(z) + \frac{1}{2^{s+2}} \inf_{z' \in B(z, 2\sqrt{s}\delta)} P_F\{Z[z', \frac{\delta}{2}]\},$$

where

$$Z[z, \delta] = \{z' \in E_s : z'_i \in \langle z_i, z_i+\delta), i = 1, 2, \ldots, s, z = (z_1, \ldots, z_s), z' = (z'_1, \ldots, z'_s)\},$$

$B(z, \delta)$ denotes the δ-neigbourhood of the point $z \in Z$ and $P_F(\cdot)$ the probability measure corresponding to the distribution function $F(\cdot)$.

To the inner problem (2) we can assign a parametric optimization problem: Find

$$\bar{g}(y) = \sup\{f_0(u) \mid u \in \bar{\mathcal{K}}(y)\}, \quad y \in E_l, \tag{7}$$

where

$$\bar{\mathcal{K}}(y) = \{u \in U : f_i(u) \leq y_i, i = 1, 2 \ldots, l\}, \quad y = (y_1, \ldots, y_l) \in E_l. \tag{8}$$

Evidently, then

$$\mathcal{K}(x, z) = \bar{\mathcal{K}}(h(x, z)), \quad g(x, z) = \bar{g}(h(x, z))$$

for $x \in X$, $z \in Z$, $h(x, z) = (h_1(x, z), h_2(x, z), \ldots, h_l(x, z))$.

The sets $Y \in E_l$, $Y(\varepsilon)$ for $\varepsilon > 0$ will be defined by the relations

$$Y = \{y \in E_l : y = (y_1, \ldots, y_l),\ y_i = h_i(x, z),\ i = 1, 2, \ldots, l \text{ for some } x \in X, z \in Z\}, \tag{9}$$

$$Y(\varepsilon) = \text{conv}\{y \in E_l : y = y_1 + y_2,\ y_1 \in Y,\ y_2 \in B(\varepsilon)\}, \tag{10}$$

where $B(\varepsilon) = B(0, \varepsilon)$ denotes the ε-neighbourhood of $0 \in E_l$.

$E_r^+ = \{x \in E_r : x = (x_1, \ldots, x_r),\ x_i \geq 0,\ i = 1, 2, \ldots, r\}$.

3. Main Results

The aim of this section is to present the assumptions under which the relations (4) and (5) are valid. Let $\varepsilon > 0$, $\delta > 0$ be arbitrary, given. We make the following assumptions.

i. a. $h_i(x, z)$, $i = 0, 1, 2, \ldots, l$ are uniformly continuous functions on $X \times Z$,

b. $f_0(u)$ is a Lipschitz function on U with the Lipschitz constant L,

c. for every $x \in X$, $h_i(x, z)$, $i = 0, 1, 2, \ldots, l$ are Lipschitz functions on Z with the Lipschitz constants L_i not depending on $x \in X$,

ii. a. $f_i(u)$, $i = 1, 2, \ldots, l$ are linear functions, U is a convex set, without loss generality, we can consider in this case the constraints in (1) to be in the form of equalities,

b. for $y \in Y$, $\bar{\mathcal{K}}(y)$ is a nonempty, compact set,

c. the matrix A of the type $(l \times r)$, $l \leq r$ fulfils the relation

$$\bar{\mathcal{K}}(y) = \{u \in U : Au = y\},\ y \in Y$$

and, moreover, all its submatrices of the types $(l \times l)$, $A(1), \ldots, A(m)$ are nonsingular,

d. the real-valued constant C is defined by the relation $C = l \max_{i,j\,k} a_{i,j}(k)$, where for every $k \in \{1, 2, \ldots, m\}$, $a_{i,j}(k)$, $i, j = 1, 2, \ldots, l$, denote the elements of the inverse matrix to $A(k)$,

ii'. a. U is a convex, compact set, $M_1 = \sup_{u^1, u^2 \in U} \|u^1 - u^2\|$,

b. $f_i(u)$, $i = 1, 2, \ldots, l$ are convex functions on U,

c. $\bar{\mathcal{K}}(y) \neq \emptyset$ for $y \in Y(\varepsilon)$,

d. the real-valued constant C fulfils the relation $C = \frac{M_1}{\varepsilon_0}$ for an $\varepsilon_0 \in (0, \varepsilon)$,

ii". $f_i(u)$, $i = 1, 2, \ldots, l$ are real-valued, continuous functions on U such that

a. $f_i(0) = 0$, $i = 1, 2, \ldots, l$, $0 \in E_r$,

b. there exists a constant $\gamma_1 > 0$ such that

$$f_i(u) - f_i(u') \geq \gamma_1 \sum_{j=1}^{r} (u_j - u'_j),\ i = 1, 2, \ldots, l$$

for every $u = (u_1, \ldots, u_r)$, $u' = (u'_1, \ldots, u'_r)$, $u \geq u'$ componentwise, $u', u \in E_r^+$,

c. $h_i(x, z) \geq 0$ for $i = 1, 2, \ldots, l$, $x \in X$, $z \in Z$,

iii. a. $F_\delta(z)$, $F^\delta(z)$ are defined by the relations (6),

b. $G(z)$ is an arbitrary s-dimensional distribution function fulfilling the relation

$$G(z) \in \langle F_\delta(z),\, F^\delta(z)\rangle \text{ for } z \in E_s,$$

c. there exist the finite $\mathsf{E}_F\xi(\omega)$, $\mathsf{E}_G\xi(\omega)$,

iv. a. X is a convex set,

b. for every $z \in Z$, $h_0(x, z)$ is a strongly concave function on X with a parameter $\varrho > 0$

c. $f_0(u)$ is a concave function on U,

d. for every $z \in Z$, $h_i(x, z)$, $i = 1, 2, \ldots, l$ are concave functions on X.

Now we can already introduce the first assertion of this paper.

Theorem 1. Let $\varepsilon > 0$, $\delta > 0$ be arbitrary. Let, moreover, $X \subset E_n$ be a compact set. If either the assumptions (i), (ii), (iii) or the assumptions (i), (ii'), (iii) (for these ε, δ) are fulfilled, then

$$|\varphi(F) - \varphi(G)| \leq 4\sqrt{s}\delta\,[L_0 + LC\sum_{i=1}^{l} L_i]. \tag{11}$$

If, moreover, the assumption (iv) is fulfilled, then also

$$\|x(F) - x(G)\|^2 \leq \frac{16}{\varrho}\sqrt{s}\delta\,[L_0 + LC\sum_{i=1}^{l} L_i]. \tag{12}$$

The assumption on the convexity of the functions $f_i(\cdot)$, $i = 1, \ldots, l$ is essential in Theorem 1. Naturally, the question arises if it is possible to replace this assumption by another one. The next Theorem answers this question.

Theorem 2. Let $\delta > 0$ be arbitrary. Let, moreover, $U = E_r^+$, $X \subset E_n$ be a compact set. If the assumptions (i), (ii"), (iii) (for this δ) are fulfilled, then

$$|\varphi(F) - \varphi(G)| \leq 4\sqrt{s}\delta\,[L_0 + L\frac{r}{\gamma_1}\sum_{i=1}^{l} L_i]. \tag{13}$$

If, moreover, the assumption (iv) is fulfilled and if $f_i(u)$, $i = 1, 2, \ldots, l$ are convex functions on U, then also

$$\|x(F) - x(G)\|^2 \leq \frac{16}{\varrho}\sqrt{s}\delta\,[L_0 + L\frac{r}{\gamma_1}\sum_{i=1}^{l} L_i]. \tag{14}$$

Up to now we have dealt with the case of general type of the distribution function $F(z)$. In what follows, we shall try to present some stronger assertions for the continuous case. To this end we shall assume:

v. $\xi(\omega)$ fulfils the following conditions:

a. the probability measure $P_F(\cdot)$ of the random vector $\xi(\omega)$ is absolutely continuous with respect to the Lebesgue measure in E_s. We denote by $\bar{f}(z)$, $z \in E_s$ the probability density corresponding to the distribution function $F(z)$ of the random vector $\xi(\omega)$,

b. there exist real numbers c_j, $j = 1, 2 \ldots, s$ such that

$$P_F\{\omega : \xi(\omega) \in \prod_{j=1}^{s} \langle 0, c_j \rangle\} = 1,$$

c. there exists a constant $\vartheta_1 > 0$ such that $\vartheta_1 < \bar{f}(z)$ for every $z \in \prod_{j=1}^{s} \langle 0, c_j \rangle$.

For this special case we can introduce the next two assertions.

Theorem 3. Let $\varepsilon > 0$ be arbitrary and let $X \subset E_n$ be a compact set. If

1. either the assumptions (i), (ii), (v) or the assumptions (i), (ii'), (v) (for this ε) are fulfilled,

2. $G(z)$ is an arbitrary distribution function with the support $\prod_{j=1}^{s} \langle 0, c_j \rangle$ such that

$$\sup |F(z) - G(z)| \leq \frac{\vartheta_1}{2} (\frac{\min c_j}{2})^s,$$

then

$$|\varphi(F) - \varphi(G)| \leq 16\sqrt{s}\,[L_0 + LC \sum_{i=1}^{l} L_i] (\frac{2 \sup |F(z) - G(z)|}{\vartheta_1})^{\frac{1}{s}}. \tag{15}$$

If, moreover, the assumption (iv) is fulfilled, then also

$$\|x(F) - x(G)\|^2 \leq \frac{64}{\varrho}\sqrt{s}\,[L_0 + LC \sum_{i=1}^{l} L_i] (\frac{2 \sup |F(z) - G(z)|}{\vartheta_1})^{\frac{1}{s}}. \tag{16}$$

Theorem 4. Let $U = E_r^+$, $X \subset E_n$ be a compact set. If

1. the assumptions (i), (ii"), (v) are fulfilled,

2. $G(z)$ is an arbitrary distribution function with the support $\prod_{j=1}^{s} \langle 0, c_j \rangle$ such that

$$\sup |F(z) - G(z)| \leq \frac{1}{2}\vartheta_1 (\frac{\min_j c_j}{2})^s,$$

then

$$|\varphi(F) - \varphi(G)| \leq 16\sqrt{s}\,[L_0 + L\frac{r}{\gamma_1} \sum_{i=1}^{l} L_i] (2\frac{\sup |F(z) - G(z)|}{\vartheta_1})^{\frac{1}{s}}. \tag{17}$$

If, moreover, the assumption (iv) is fulfilled and if $f_i(u)$, $i = 1, 2, \ldots, l$ are convex functions on U, then also

$$\|x(F) - x(G)\|^2 \leq \frac{64}{\varrho}\sqrt{s}\,[L_0 + L\frac{r}{\gamma_1} \sum_{i=1}^{l} L_i](\frac{2 \sup |F(z) - G(z)|}{\vartheta_1})^{\frac{1}{s}}. \tag{18}$$

4. Appendix

The aim of this section is to give the proofs of the assertions presented in the former parts of the paper. These proofs can be divided into two groups. First, we shall deal with the case of Theorem 1 and Theorem 2. To this end, we shall recall the results of Kaňková (1992 b), (Theorem 1).

Theorem 5. Let $\delta > 0$ be arbitrary, $X \subset E_n$ be a compact set. If

1. a. $\bar{g}(x, z)$ is a uniformly continuous function on $X \times Z$,
   b. for every $x \in X$, $\bar{g}(x, z)$ is a Lipschitz function of $z \in Z$ with the Lipschitz constant $\bar{L}$ not depending on $x \in X$,
2. a. $F_\delta(\cdot)$, $F^\delta(\cdot)$ are defined by the relation (6),
   b. $G(z)$ is an arbitrary s–dimensional distribution function fulfilling the assumption (iii.b),
   c. for every $x \in X$ there exist the finite $\mathsf{E}_F\bar{g}(x, \xi(\omega))$, $\mathsf{E}_G\bar{g}(x, \xi(\omega))$,

then

$$|\sup_X \mathsf{E}_F\, \bar{g}(x, \xi(\omega)) - \sup_X \mathsf{E}_G\, \bar{g}(x, \xi(\omega))| \leq 4\sqrt{s}\delta\bar{L}. \tag{19}$$

If, moreover,

3. a. X is a convex set,
   b. for every $z \in E_s$ $\bar{g}(x, z)$ is a strongly concave function on X with a parameter $\varrho > 0$,

then also

$$\|\bar{x}(F) - \bar{x}(G)\|^2 \leq \frac{16}{\varrho}\delta\bar{L}\sqrt{s}, \tag{20}$$

where
$\bar{x}(F) = \arg\max\{\mathsf{E}_F\, \bar{g}(x, \xi(\omega)) | x \in X\}$, $\bar{x}(G) = \arg\max\{\mathsf{E}_G\, \bar{g}(x, \xi(\omega)) | x \in X\}$.

Evidently, the assertion of Theorem 1 will be proven if we verify the asumptions of Theorem 5. To this end, first, it follows from the theory of linear programming that if the assumptions (ii) hold, then

$$\Delta[\bar{\mathcal{K}}(y), \bar{\mathcal{K}}(y')] \leq C\|y - y'\| \tag{21}$$

for every $y, y' \in Y$ and the corresponding value of C.
($\Delta[\cdot, \cdot] = \Delta_r[\cdot, \cdot]$ denotes the Hausdorff distance in the space of nonempty subsets of E_r, see e. g. Kaňková (1980 a).)

If we consider the assumptions (ii'), then we can employ a modified version of the assertion proven in Kaňková (1980 b), (Lemma 1). (See also Kaňková (1988).)

Lemma 1. Let $\varepsilon > 0$ be arbitrary given. Let, moreover, U be a compact, convex set, $Y \subset E_l$ be a nonempty set. If $f_i(u)$, $i = 1, 2, \ldots, l$ are convex functions on U and if $\bar{\mathcal{K}}(y) \neq \emptyset$ for $y \in Y(\varepsilon)$, then

$$\Delta[\bar{\mathcal{K}}(y), \bar{\mathcal{K}}(y')] \leq C\|y - y'\| \tag{22}$$

for every $y, y' \in Y$, $C = \frac{M_1}{\varepsilon_0}$, $M_1 = \sup_{u^1, u^2 \in U} \|u^1 - u^2\|$, $0 < \varepsilon_0 < \varepsilon$.

Lemma 2. If

1. for $y \in Y$, $\bar{\mathcal{K}}(y) \neq \emptyset$ is a compact set,
2. $f_0(u)$ is a Lipschitz function on U with the Lipschitz constant L,
3. there exists a constant C such that

$$\Delta[\bar{\mathcal{K}}(y), \bar{\mathcal{K}}(y')] \leq C\|y - y'\| \text{ for every } y, y' \in E_l,$$

then $\bar{g}(y)$ is a Lipschitz function on Y with the Lipschitz constant LC.

The assertion of Lemma 2 can be proven by the method of the proof of Lemma 2 in Kaňková (1978).

Since it is easy to see that fulfilling of the assumptions 1 and 2 of Theorem 5 follows from Lemma 1, Lemma 2 and the assumptions of Theorem 1, we can see that the relation (11) holds. It remains to prove the relation (12). To this end, first, we shall recall the next Lemma (cf. Lemma 1 of Kaňková (1978)).

Lemma 3. Let X, U be convex, nonempty sets, $\mathcal{K}(x, z) \neq \emptyset$ for $x \in X, z \in Z$ be a compact set. If either the assumptions (ii.a), (iv.c), (iv.d) or the assumptions (ii'.b), (iv.c), (iv.d) are fulfilled, then for every $z \in Z$, $g(x, z)$ is a concave function on X.

Consequently, according to the assumption (iv.b), the objective function in (3) is a strongly concave function on X with the parameter ϱ. Evidently, the relation (12) follows immediately from the next auxiliary assertion, proven in Kaňková (1993).

Lemma 4. Let $X \subset E_n$ be a nonempty, compact, convex set. Let, further, $h(x)$ be a strongly concave with a parameter $\varrho' > 0$, continuous, real–valued function defined on X. If $x_0 \in X$ is defined by the relation

$$x_0 = \arg\max_{x \in X} h(x),$$

then

$$\|x - x_0\|^2 \leq \frac{2}{\varrho'} |h(x_0) - h(x)|$$

for every $x \in X$.

We have completed the proof of Theorem 1. It follows from this proof that the Lipschitz property of the mapping $\bar{\mathcal{K}}(\cdot)$ is essential for the Lipschitz property of $\bar{g}(y)$ and consequently also for the Lipschitz property of $g(x, z)$. Further, we shall try to find a constant C fulfilling the relation (21) under the assumption (ii"). To this end we shall prove the next auxiliary assertion.

Lemma 5. Let $U = E_r^+$. If the assumption (ii") is fulfilled, then

$$\Delta[\bar{\mathcal{K}}(y), \bar{\mathcal{K}}(y')] \leq \frac{r}{\gamma_1}\|y - y'\| \text{ for every } y, y' \in E_l^+. \tag{23}$$

Proof. Evidently, if we define the sets $\bar{\mathcal{K}}_i(y_i)$, $y_i \in E_1^+$, $i = 1, 2, \ldots, l$ by

$$\bar{\mathcal{K}}_i(y_i) = \{u \in E_r^+ : f_i(u) \leq y_i\},\ i = 1, 2, \ldots, l,$$

then for $y_i \in E_1^+$, $i = 1, 2, \ldots, l$, $\bar{\mathcal{K}}_i(y_i)$ are nonempty, compact sets and, moreover,

$$\bar{\mathcal{K}}(y) = \bigcap_{i=1}^{l} \bar{\mathcal{K}}_i(y_i) \neq \emptyset \text{ for } y = (y_1, \ldots y_l). \tag{24}$$

The proof of the assertion (23) will be divided into two parts. First we shall prove, for every $i = 1, 2, \ldots, l$, the validity of the implication

$$y_i, y_i' \in E_1^+, y_i < y_i' \Longrightarrow \Delta[\bar{\mathcal{K}}_i(y_i), \bar{\mathcal{K}}_i(y_i')] \leq \frac{|y_i - y_i'|}{\gamma_1}\sqrt{r}. \tag{25}$$

To this end, let $i \in \{1, 2 \ldots, l\}$ be arbitrary given. Since, evidently $\bar{\mathcal{K}}_i(y_i) \subset \bar{\mathcal{K}}_i(y_i')$, it is necessary only to prove the inequality

$$\sup_{u' \in \bar{\mathcal{K}}_i(y_i')} \inf_{u \in \bar{\mathcal{K}}_i(y_i)} \|u - u'\| \leq \frac{|y_i - y_i'|}{\gamma_1}\sqrt{r}. \tag{26}$$

However to prove (26) it is surely sufficient for an arbitrary $u' = (u_1', \ldots, u_r') \in \bar{\mathcal{K}}_i(y_i')$ to find $u^* = u^*(u')$ such that

$$\|u' - u^*\| \leq \frac{|y_i - y_i'|}{\gamma_1}\sqrt{r} \text{ and simultaneously } u^* \in \bar{\mathcal{K}}_i(y_i). \tag{27}$$

It is easy to see that if $u' \in \bar{\mathcal{K}}_i(y_i)$, then we can set $u' = u^*$. Consequently, it remains to consider the case $u' \notin \bar{\mathcal{K}}_i(y_i)$. If we define in this case the point $u^0 = (u_1^0, \ldots, u_r^0)$ by $u_j^0 = u_j' - \frac{|y_i - y_i'|}{\gamma_1}$, $j = 1, 2, \ldots, r$, then two different cases can happen

a. there exists $\nu \in \{1, 2, \ldots, r\}$ such that $u_\nu^0 \geq 0$,

b. $u_j^0 < 0$ for all $j = 1, 2, \ldots, r$.

Let us consider, first, the case a. In this case we can define the point $u^* = (u_1^*, u_2^*, \ldots, u_r^*)$ by the following way $u_\nu^* = u_\nu^0$, $u_j^* = u_j'$ for $j \neq \nu$. It follows from the assumptions that

$$f_i(u^*) < f_i(u') \text{ and simultanously } |f_i(u^*) - f_i(u')| \geq |y_i - y_i'|.$$

Consequently also $u^* \in \bar{\mathcal{K}}_i(y_i)$. Since $\|u^* - u'\| = \frac{|y_i - y_i'|}{\gamma_1}$ we have find u^* fulfilling the relation (27), in the case a. It remains to consider the case b. However, as it is

$\|u'\| \leq \frac{|y_i - y_i'|}{\gamma_1}\sqrt{r}$ in this case, we can see that the existence of the point u^* follows from the assumptions too. Since $i \in \{1, 2, \ldots, r\}$ was arbitrary we have proven the validity of the implication (25).

Employing the implication (25) we shall prove the assertion of Lemma 5. It means we shall prove the validity of the relation (23). To this end, let $y, y' \in E_l^+$, $y = (y_1, \ldots, y_l)$, $y' = (y_1', \ldots, y_l')$ be arbitrary given. We shall define $\underline{y} = (\underline{y}_1, \ldots, \underline{y}_l)$, $\overline{y} = (\overline{y}_1, \ldots, \overline{y}_l)$ by the relations

$$\underline{y}_j = \min(y_j, y_j'),\ \overline{y}_j = \max(y_j, y_j'),\ j = 1, 2, \ldots, l. \tag{28}$$

Evidently

$$\underline{y}, \overline{y} \in E_l^+,\ \|\underline{y} - \overline{y}\| = \|y - y'\| \tag{29}$$

and simultanously

$$\bar{\mathcal{K}}(\underline{y}) \subset \bar{\mathcal{K}}(y) \subset \bar{\mathcal{K}}(\overline{y}),\ \bar{\mathcal{K}}(\underline{y}) \subset \bar{\mathcal{K}}(y') \subset \bar{\mathcal{K}}(\overline{y}).$$

Since it follows from the properties of the Hausdorff distance that

$$\Delta[\bar{\mathcal{K}}(y), \bar{\mathcal{K}}(y')] \leq \Delta[\bar{\mathcal{K}}(\underline{y}), \bar{\mathcal{K}}(\overline{y})], \tag{30}$$

the assertion of Lemma 5 will be proven if we verify the inequality

$$\Delta[\bar{\mathcal{K}}(\underline{y}), \bar{\mathcal{K}}(\overline{y})] \leq \frac{\|\underline{y} - \overline{y}\|}{\gamma_1} r. \tag{31}$$

Since, furthermore, it follows from the definition of the mapping $\bar{\mathcal{K}}(\cdot)$ that $\bar{\mathcal{K}}(\underline{y}) \subset \bar{\mathcal{K}}(\overline{y})$, we can see that to prove (31) it is sufficient to prove the implication

$$u \in \bar{\mathcal{K}}(\overline{y}) \Longrightarrow \text{ there exists } \underline{u} \in \bar{\mathcal{K}}(\underline{y}) \text{ such that } \|u - \underline{u}\| \leq \frac{\|\underline{y} - \overline{y}\|}{\gamma_1} r. \tag{32}$$

So let $u \in \bar{\mathcal{K}}(\overline{y})$, $u = (u_1, u_2, \ldots, u_r)$ be arbitrary. According to the implication (25) there exist $u^i \in \bar{\mathcal{K}}_i(\underline{y}_i)$, $i = 1, 2, \ldots, l$ such that

$$\|u - u^i\| \leq \frac{|\overline{y}_i - \underline{y}_i|}{\gamma_1}\sqrt{r},\ u^i = (u_1^i, \ldots, u_r^i).$$

Consequently, we can define the point $\underline{u} = (\underline{u}_1, \ldots, \underline{u}_r)$ by the relation

$$\underline{u}_j = \min_i(u_j^i),\ j = 1, 2, \ldots, r.$$

Evidently, it follows from the relation (24) that $\underline{u} \in \bar{\mathcal{K}}(\underline{y})$ and, moreover,

$$\|u - \underline{u}\| \leq \frac{\|\underline{y} - \overline{y}\|}{\gamma_1} r.$$

According to (29) we have verified the validity of the relation (23) and consequently also the relation (21) for $C = \frac{r}{\gamma_1}$. □

Now it is easy to see that the assertions of Theorem 2 follow from Lemma 2, Lemma 3, Lemma 4 and Lemma 5 by the same way as the assertions of Theorem 1 followed from the theory of the linear programming, Lemma 1, Lemma 2, Lemma 3 and Lemma 4. Consequently, it remains to deal with Theorem 3 and Theorem 4. To this end we shall introduce the modified result of Kaňková (1992 a).

Theorem 6. Let $\varepsilon > 0$ be arbitrary, $X \subset E_n$ be a compact set. If

1. for every $x \in X$, $\bar{g}(x, z)$ is a Lipschitz function of $z \in Z$ with the Lipschitz constant $\bar{L}$ not depending on $x \in X$,
2. the assumption (v) is fulfilled,
3. $2\sqrt[s]{\frac{2\varepsilon}{\vartheta_1}} \le \min_j c_j$,
4. $G(z)$ is an arbitrary distribution function with the support $\prod_{j=1}^{s} \langle 0, c_j \rangle$ such that its Kolmogorov distance from $F(z)$ is nongreater than ε, i.e.

$$\sup |F(z) - G(z)| \le \varepsilon,$$

then

$$|\sup_X \mathsf{E}_F \bar{g}(x, \xi(\omega)) - \sup_X \mathsf{E}_G \bar{g}(x, \xi(\omega))| \le 16\sqrt{s}\, \bar{L}\, (\frac{2\varepsilon}{\vartheta_1})^{\frac{1}{s}}.$$

We can obtain the assertion of Theorem 6 repeating the proof of Lemma 6 and Theorem 3, Kaňková (1992 a) (where we set $X_F(\alpha) = X$). Utilizing, further, this assertion and repeating the proof of Theorem 1 and Theorem 2 we obtain the assertion of Theorem 3 and Theorem 4.

5. Conclusion

Empirical estimates, parameters estimates, approximative method of solution as well as many numerical method of solution are based on the possibility to replace theoretical distribution function by some approximating one, Dupačová (1987), Dupačová (1990), Kaňková (1980 b), Vogel (1992). Consequently, the stability problem considered with respect to the probability measure is surely a very serious problem of stochastic programming. This problem has already been discussed many times in the literature. However, the corresponding discrepancy is mostly expressed by a rather complicated measure of distance in the space of probability measures, see e.g. Römisch and Wakolbinger (1987), Römisch and Schulz (1991). The exception is Gröwe and Römisch (1990), where a special case of the objective is considered.

The presented paper tries to consider the stability of two-stage nonlinear stochastic programming problems with respect to the distribution function and consequently to the Kolmogorov metric. The corresponding discrepancy is expressed by a rather simple distance. The basis of this work are the former results of Kaňková (1992 a) and Kaňková (1992 b).The considered model does not include only a usual case of linear and convex constraints, however, it also admits some type of general left-side of constraints.

Of course, the results of this paper can be further applied to estimate problems, contamination case or to the problems of approximation as well as generalized to multistage case. However, this all problems doesn't belong to the framework of this paper.

Acknowledgement. This work was supported by the Grant Agency of the Czech Republic under Grant No. 402/93/0631 and by the Grant Agency of the Academy of Sciences of the Czech Republic under Grant No. 27 561.

REFERENCES

P. Billingsley (1977): Convergence of Probability Measures. J. Wiley, New York.

J. Dupačová (1987): Stochastic programming with incomplete information: A survey of results on postoptimization and sensitivity analysis. Optimization *18*, 507–532.

J. Dupačová (1990): Stability and sensitivity analysis for stochastic programming. Annals of Operations Research *2*, 115–142.

N. Gröwe and W. Römisch (1990): A stochastic programming model for optimal power dispatch: Stability and numerical treatment. Humboldt Univ. Berlin., Sekt. Mathem. Preprint 266.

P. Kall (1976): Stochastic Linear Programming. Springer–Verlag, Berlin - Heildelberg - New York.

V. Kaňková (1978): Differentiability of the optimalized function in a two–stage stochastic nonlinear programming problem (in Czech). Ekonomicko–matematický obzor *14*, 322–330.

V. Kaňková (1980 a): Optimization problem with parameter and its application to the problems of two–stage stochastic nonlinear programming. Kybernetika *16*, 411–425.

V. Kaňková (1980 b): Approximative solution of problems of two–stage stochastic nonlinear programming (in Czech). Ekonomicko–matematicky obzor *16*, 64–76.

V. Kaňková (1988): A note on the differentiability in two–stage stochastic nonlinear programming problems. Kybernetika *24*, 207–215.

V. Kaňková (1992 a): A note on the stability in stochastic programming problems. In: Trans. of the Eleventh Prague Conference 1990 (Academia, Prague, 1992), Volume B, 51–59.

V. Kaňková (1992 b): A note on distribution sensitivity in stochastic programming. Submitted for Trans. of the IV International Conference on Stochastic Programming - Udine, Italy 1992.

V. Kaňková (1993): Stability in stochastic programming - the case of unknown location parameter. Kybernetika *29*, 80–101.

W. Römisch and A. Wakolbinger (1987): Obtaining convergence for approximations in stochastic programming. In: Parametric Optimization and Related Topics (J. Guddat, ed.), Akademie Verlag, Berlin, 1987, 327–343.

W. Römisch and R. Schultz (1991): Distribution sensitivity in stochastic programming. Mathematical Programming *50*, 197–226.

S. Vogel (1992): On stability in multiobjective programming - a stochastic approach. Mathematical Programming *56*, 91–119.

SPREAD INEQUALITY AND EFFICIENCY OF FIRST AND SECOND ORDER

CHRIS A.J. KLAASSEN AND SHANTI A. VENETIAAN

University of Amsterdam

ABSTRACT. The asymptotic performance of location estimators is considered as the sample size tends to infinity. Expanding the lower bound of the spread inequality - a finite sample inequality for the distribution of estimators - we obtain an asymptotic inequality which is sharp to second order. As a by-product this yields a simple proof of the well-known result that first order efficiency implies second order efficiency.

1. Introduction.

In this paper we consider the spread inequality for location estimators. This is a finite sample inequality for the quantiles of the distribution of estimators; see Klaassen (1984). Expanding the bound on these quantiles to the order $n^{-\frac{1}{2}}$ we obtain an inequality for the $n^{-\frac{1}{2}}$-term of Edgeworth expansions of efficient estimators. In fact, this inequality shows that the spread inequality is sharp to order $n^{-\frac{1}{2}}$ and it illustrates the well-known result " first order efficiency implies second order efficiency ". This phenomenon was discovered by Pfanzagl (1979). He gives results for the power functions of test statistics. He also proves that the result holds in estimation problems. For particular testing problems this property was noticed before by Pfanzagl (1973), (1975), Chibisov (1974) and Bickel and Van Zwet (1978).

Bickel, Chibisov and Van Zwet (1981) present an intuitive explanation of the phenomenon. They prove a simple theorem for testing problems under weak assumptions. Similar results are presented in sections 6.1, 6.4 and 8.7 of Pfanzagl and Wefelmeyer(1985) and in Section 1.6 of Akahira (1986).

Pfanzagl (1979) and Bickel, Chibisov and Van Zwet (1981) consider the topic within a quite general framework. We will discuss estimation of a one-dimensional location parameter by a translation equivariant estimator. This enables us to state the result in a straightforward way.

In Section 2 we present the announced results and also sketch the basic idea of the proof. In Section 3 we derive an asymptotic expansion for the lower bound provided by the spread inequality and prove the result.

1991 *Mathematics Subject Classification.* Primary 62F12; secondary 62E20, 62F10.

Key words and phrases. First order efficiency, second order efficiency, spread inequality, location estimators.

2. First order efficiency implies second order efficiency.

Let $X_1, ..., X_n$ be i.i.d. random variables with common density $f(. - \theta)$. $T_n = t_n(X_1, ..., X_n)$ is a translation equivariant estimator of the location parameter $\theta \in \mathbf{R}$. Let

$$G_n(x) = P_\theta(\sqrt{nI(f)}(T_n - \theta) \leq x) \tag{2.1}$$

be the distribution function of T_n normed via the Fisher information $I(f) = \int (f'/f)^2 f < \infty$. We define the distribution function H_n of the score function of the observations by

$$H_n(z) = P_\theta(\frac{1}{\sqrt{nI(f)}} \sum_{i=1}^{n} -\frac{f'}{f}(X_i - \theta) \leq z). \tag{2.2}$$

Note that a distribution function K_n whose score function has a distribution described by H_n, may be defined via its quantile function

$$K_n^{-1}(u) = \int_{\frac{1}{2}}^{u} \frac{1}{\int_s^1 H_n^{-1}(t)dt} ds. \tag{2.3}$$

The fact that T_n cannot estimate θ arbitrarily accurately may be expressed now by the statement that the quantiles of G_n are at least as far apart as those of K_n, i.e.

$$G_n^{-1}(v) - G_n^{-1}(u) \geq K_n^{-1}(v) - K_n^{-1}(u), 0 < u < v < 1. \tag{2.4}$$

The proof of this spread inequality may be found in Klaassen (1984).

In Theorem 3.1 of Section 3 below conditions are given under which an Edgeworth expansion for K_n or, more precisely, a Cornish-Fisher expansion for K_n^{-1} holds. Under those conditions

$$\begin{aligned} K_n^{-1}(v) - K_n^{-1}(u) &= \Phi^{-1}(v) - \Phi^{-1}(u) \\ &\quad - \frac{\kappa_3}{12\sqrt{n}}\{(\Phi^{-1}(v))^2 - (\Phi^{-1}(u))^2\} + o(\frac{1}{\sqrt{n}}), \end{aligned} \tag{2.5}$$

where Φ is the standard normal distribution function; in the sequel ϕ will denote its density. In particular, (2.4) and (2.5) yield

$$\liminf_{n \to \infty} (G_n^{-1}(v) - G_n^{-1}(u)) \geq \Phi^{-1}(v) - \Phi^{-1}(u), 0 < u < v < 1,$$

and T_n is first order efficient if $G_n^{-1}(u) \to \Phi^{-1}(u)$ or $G_n \xrightarrow{w} \Phi$, as $n \to \infty$. Often, G_n admits an Edgeworth expansion and G_n^{-1} a Cornish-Fisher expansion which, if T_n is first order efficient, equals

$$G_n^{-1}(u) = \Phi^{-1}(u) - \frac{1}{\sqrt{n}} p(\Phi^{-1}(u)) + o(\frac{1}{\sqrt{n}}) \tag{2.6}$$

for some function $p(.)$. Combining (2.4) through (2.6) we see that

$$\begin{aligned} 0 &\leq \lim_{n\to\infty} \sqrt{n}\{G_n^{-1}(v) - G_n^{-1}(u) - (K_n^{-1}(v) - K_n^{-1}(u))\} \\ &= p(\Phi^{-1}(u)) - p(\Phi^{-1}(v)) + \frac{\kappa_3}{12}\{(\Phi^{-1}(v))^2 - (\Phi^{-1}(u))^2\}, 0 < u < v < 1, \end{aligned} \tag{2.7}$$

and hence

$$p(y) - p(x) \leq \frac{\kappa_3}{12}(y^2 - x^2), \quad x < y. \tag{2.8}$$

If $p(.)$ is differentiable this inequality is equivalent to

$$p'(x) \leq \frac{\kappa_3}{6}x, \quad x \in \mathbb{R}. \tag{2.9}$$

Assuming that the first two terms of the Hoeffding decomposition (cf. Van Zwet (1984), Section 2) of $\sqrt{nI(f)}(T_n - \theta)$ under θ may be used to obtain a stochastic expansion of order $o(n^{-\frac{1}{2}})$ as in (10.2.2′) of Pfanzagl and Wefelmeyer (1985), and assuming regularity conditions satisfied, one arrives at (2.6) respectively (2.14) with $p(.)$ an even quadratic polynomial; cf. Proposition 10.3.1 with $f_1 = g_1 = 0$ of Pfanzagl and Wefelmeyer (1985). So, often $p(.)$ is an even function and (2.9) yields

$$p(x) = \frac{\kappa_3}{12}x^2 + c, \quad \text{for some} \quad c \in \mathbb{R}, \tag{2.10}$$

since the function $\kappa_3 x/6 - p'(x)$ is odd and nonnegative and hence vanishes ((2.10) may be derived from (2.8) directly without assuming differentiability of $p(.)$). If T_n is asymptotically median unbiased of order $o(n^{-\frac{1}{2}})$ (cf. Definition 8.2.14, p.254, of Pfanzagl and Wefelmeyer (1985)), then (2.6) implies $c = p(0) = 0$. All in all we have shown that, if T_n is efficient to first order, if T_n is asymptotically median unbiased to order $o(n^{-\frac{1}{2}})$ and if it has an Edgeworth (Cornish-Fisher) expansion (2.6) with $p(.)$ an even function, then under the regularity conditions of Theorem 2.1 below, $p(.)$ is determined completely by (2.10) with $c = 0$. This is a result of the type "first order efficiency implies second order efficiency".

Assuming a more general stochastic expansion for the estimator, Pfanzagl and Wefelmeyer (1985) arrive at $p(x) = ax^2 + bx + c$, a quadratic polynomial with possibly a linear term; cf. their Sections 10.2, 10.3 and 11.1. Then (2.9) becomes

$$2(a - \frac{\kappa_3}{12})x + b \leq 0, \quad x \in \mathbb{R},$$

and hence $a = \kappa_3/12$, $b \leq 0$. Consequently, we obtain

$$p(x) = \frac{\kappa_3}{12}x^2 + bx + c, \quad b \leq 0. \tag{2.11}$$

Again, if T_n is asymptotically median unbiased of order $o(n^{-\frac{1}{2}})$, then c vanishes. If $l(.)$ is a symmetric, differentiable loss function with continuous bounded nonnegative derivative on $[0,\infty)$, and if $\sqrt{n}|G_n^{-1}(u) - \Phi^{-1}(u)|$ is bounded uniformly in n by a (0,1)-integrable function, then (2.6) and (2.11) yield

$$\begin{aligned} E_\theta l(\sqrt{nI(f)}(T_n - \theta)) &= E_\Phi l(Z) - \frac{b}{\sqrt{n}}E_\Phi\{l(Z)(Z^2 - 1)\} + o(\frac{1}{\sqrt{n}}) \\ &\geq E_\Phi l(Z) + o(\frac{1}{\sqrt{n}}); \end{aligned} \tag{2.12}$$

see also Pfanzagl (1985), p.781-782.

A precise formulation of our main result runs as follows.

Theorem 2.1. *$X_1, ..., X_n$ are i.i.d. with density $f(. - \theta)$. Let $I(f) = \int (f'/f)^2 f$ and $\kappa_3 = (I(f))^{-\frac{3}{2}} \int (-f'/f)^3 f$ be well-defined and finite. Assume that Cramér's condition*

$$\limsup_{|t|\to\infty} \left| \int \exp(-it\frac{f'}{f}) f \right| < 1 \tag{2.13}$$

is satisfied. Let $T_n = t_n(X_1, ..., X_n)$ be an efficient translation equivariant estimator of θ. Consequently, G_n defined by (2.1) converges weakly to Φ. If G_n admits an Edgeworth expansion

$$G_n(x) = \Phi(x) + \frac{1}{\sqrt{n}} p(x)\phi(x) + o(\frac{1}{\sqrt{n}}), \quad x \in \mathbf{R}, \tag{2.14}$$

with $p(.)$ an arbitrary function, then (2.8) holds. This implies (2.9) if $p(.)$ is differentiable, (2.10) if $p(.)$ is an even function, and (2.11) if $p(.)$ is a quadratic polynomial.

3. Proof of Theorem 2.1.

Because $\kappa_3 < \infty$ and (2.13) implies that $-f'/f(X_1)$ does not have a lattice distribution under f, H_n admits the Edgeworth expansion to order $o(n^{-\frac{1}{2}})$ (cf. Feller (1971), Theorem XVI.4.1, p. 539)

$$\tilde{H}_n(x) = \Phi(x) - \frac{\kappa_3}{6\sqrt{n}}(x^2 - 1)\phi(x). \tag{3.1}$$

We introduce

$$\bar{H}_n(x) = \begin{cases} 0 & \text{for } x \in (-\infty, a_n) \\ \tilde{H}_n(x) & \text{for } x \in (a_n, b_n) \\ 1 & \text{otherwise,} \end{cases} \tag{3.2}$$

with $a_n = -c \log n$ and $b_n = c \log n, c > 0$. An expansion for the inverse of H_n is

$$H_n^{*-1}(t) = \Phi^{-1}(t) + \frac{\kappa_3}{6\sqrt{n}}[(\Phi^{-1}(t))^2 - 1], \tag{3.3}$$

as we will show by proving the following result.

Lemma 3.1. *Uniformly in s,*

$$\left| \int_s^1 H_n^{*-1}(t)dt - \int_s^1 H_n^{-1}(t)dt \right| = o(\frac{1}{\sqrt{n}}). \tag{3.4}$$

Proof. We bound the left hand side of (3.4) via

$$
\begin{aligned}
(3.5)\qquad & |\int_s^1 H_n^{*-1}(t)dt - \int_s^1 H_n^{-1}(t)dt| \\
& \leq |\int_s^1 (H_n^{*-1}(t) - \bar{H}_n^{-1}(t))dt| + |\int_s^1 (\bar{H}_n^{-1}(t) - H_n^{-1}(t))dt|.
\end{aligned}
$$

Note that $\bar{H}_n$ is a monotone increasing function and so $\bar{H}_n^{-1}$ exists. This follows from the fact that $\tilde{H}_n$ is monotone increasing on $[a_n, b_n]$ for n large enough. To find out where the extreme values of $\tilde{H}_n$ appear we look at $\tilde{H}_n'(x) = \phi(x)(1+n^{-\frac{1}{2}}p_1(x))$, with p_1 a polynomial of degree 3. This means that the extreme values closest to 0 appear at values x of the order $\mathcal{O}(n^{\frac{1}{6}})$. We will first consider the second term at the right hand side of (3.5);

$$
\begin{aligned}
& |\int_s^1 (\bar{H}_n^{-1}(t) - H_n^{-1}(t))dt| \leq \int_{-\infty}^{\infty} |\bar{H}_n(x) - H_n(x)|dx \\
(3.6)\qquad & \leq \int_{-\infty}^{\infty} |\tilde{H}_n(x) - H_n(x)|dx + \int_{-\infty}^{\infty} |\tilde{H}_n(x) - \bar{H}_n(x)|dx \\
& = \int_{-\infty}^{\infty} |\tilde{H}_n(x) - H_n(x)|dx + \int_{-\infty}^{a_n} |\tilde{H}_n(x)|dx + \int_{b_n}^{\infty} |\tilde{H}_n(x) - 1|dx.
\end{aligned}
$$

For the first term at the right hand side of (3.6) we apply theorem VI.3.5 (A), p.169, from Petrov (1975) with $r = 3$ and $p = 1$ and we see

$$
(3.7)\qquad \int_{-\infty}^{\infty} |\tilde{H}_n(x) - H_n(x)|dx = o(\frac{1}{\sqrt{n}}).
$$

The second and third term of (3.6) can be treated as follows.

$$
\begin{aligned}
(3.8)\qquad & \int_{b_n}^{\infty} |1 - \tilde{H}_n(x)|dx \leq \int_{b_n}^{\infty} |1 - \Phi(x)|dx + \int_{b_n}^{\infty} q_1(x)\phi(x)dx \\
& = \mathcal{O}(\int_{b_n}^{\infty} \frac{\phi(x)}{x}dx) + [q_2(x)\phi(x)]_{b_n}^{\infty} = o(\frac{1}{\sqrt{n}}),
\end{aligned}
$$

where q_1 and q_2 are polynomials and a is a positive constant. Combining (3.6) through (3.8) we see that the second term at the right hand side of (3.5) is of the order $o(n^{-\frac{1}{2}})$.

To study the first term at the right hand side of (3.5) we note that for n large enough the following holds

$$
\begin{aligned}
(3.9)\qquad & |\int_s^1 (H_n^{*-1}(t) - \bar{H}_n^{-1}(t))dt| \leq \int_{\tilde{H}_n(a_n)}^{\tilde{H}_n(b_n)} |H_n^{*-1}(t) - \bar{H}_n^{-1}(t)|dt \\
& + \int_{\tilde{H}_n(b_n)}^{1} |H_n^{*-1}(t) - b_n|dt.
\end{aligned}
$$

First note that

$$\tilde{H}_n(b_n) = \Phi(b_n) - \frac{\kappa_3}{6\sqrt{n}}(b_n^2 - 1)\phi(b_n).$$

Consequently, we have

$$\tilde{H}_n(b_n) \geq \Phi(b_n/2)$$

for n sufficiently large and

$$\begin{aligned} \int_{\tilde{H}_n(b_n)}^{1} |H_n^{*-1}(t) - b_n| dt &= \mathcal{O}\Big(\int_{b_n/2}^{\infty} |y + \frac{\kappa_3}{6\sqrt{n}}(y^2-1) - b_n|\phi(y)dy\Big) \\ &= \mathcal{O}\Big(\int_{b_n/2}^{\infty} (y^2-1)\phi(y)dy\Big) = o(\frac{1}{\sqrt{n}}). \end{aligned} \tag{3.10}$$

Hence, the second term at the right hand side of (3.9) is of the order $o(n^{-\frac{1}{2}})$.

Finally, we consider the first term at the right hand side of (3.9) and note

$$\int_{\tilde{H}_n(a_n)}^{\tilde{H}_n(b_n)} |H_n^{*-1}(t) - \tilde{H}_n^{-1}(t)| dt = \int_{a_n}^{b_n} |H_n^{*-1}(\tilde{H}_n(x)) - x| d\tilde{H}_n(x). \tag{3.11}$$

Uniformly on (a_n, b_n),

$$\begin{aligned} \Phi^{-1}(\tilde{H}_n(x)) &= x - \frac{\kappa_3}{6\sqrt{n}}(x^2-1) + \mathcal{O}(\frac{q(x)}{n}) \\ &= x - \frac{\kappa_3}{6\sqrt{n}}(x^2-1) + o(\frac{1}{\sqrt{n}}) \end{aligned}$$

holds with $q(.)$ a polynomial. Consequently, we have

$$\sup_{a_n < x < b_n} |H_n^{*-1}(\tilde{H}_n(x)) - x| = o(\frac{1}{\sqrt{n}}),$$

which results in

$$\int_{a_n}^{b_n} |H_n^{*-1}(\tilde{H}_n(x)) - x| d\tilde{H}_n(x) = o(\frac{1}{\sqrt{n}}). \tag{3.12}$$

Combining (3.11) and (3.12) we see that the first term at the right hand side of (3.9) is of the order $o(n^{-\frac{1}{2}})$. □

Lemma 3.1 will be used to prove a Cornish-Fisher expansion for K_n^{-1}.

Theorem 3.1. *Under the assumptions of Theorem 2.1 and with K_n^{-1} as in (2.3), the expansion*

$$K_n^{*-1}(u) = \Phi^{-1}(u) - \frac{\kappa_3}{12\sqrt{n}}(\Phi^{-1}(u))^2 \tag{3.13}$$

satisfies

(3.14) $$|K_n^{*-1}(u) - K_n^{-1}(u)| = o(\frac{1}{\sqrt{n}}), o < u < 1.$$

Proof. First note

$$|K_n^{*-1}(v) - K_n^{*-1}(u) - \int_u^v \frac{1}{\int_s^1 H_n^{-1}(t)dt} ds|$$

(3.15) $$\leq |K_n^{*-1}(v) - K_n^{*-1}(u) - \int_u^v \frac{1}{\int_s^1 H_n^{*-1}(t)dt} ds|$$

$$+ |\int_u^v \frac{1}{\int_s^1 H_n^{*-1}(t)dt} ds - \int_u^v \frac{1}{\int_s^1 H_n^{-1}(t)dt} ds|.$$

We will study the first term of the right hand side of (3.15). We are done if we can show that

(3.16) $$\sup_{u<s<v} |\frac{1}{\phi(\Phi^{-1}(s))}(1 - \frac{\kappa_3}{6\sqrt{n}}\Phi^{-1}(s)) - \frac{1}{\int_s^1 H_n^{*-1}(t)dt}| = o(\frac{1}{\sqrt{n}}), \quad 0 < u < v < 1.$$

Note that

$$\int_s^1 H_n^{*-1}(t)dt = \phi(\Phi^{-1}(s))(1 + \frac{\kappa_3}{6\sqrt{n}}\Phi^{-1}(s)).$$

By straightforward calculation and the fact that s is bounded away from 0 and 1, we get (3.16).

For the second term at the right hand side of (3.15) we apply lemma 3.1. The fact that the lemma holds uniformly in s, and for $u < s < v$, $\int_s^1 H_n^{-1}(t)dt$ stays bounded away from 0, and the fact that the integration interval is bounded give

$$\int_u^v \frac{1}{\int_s^1 H_n^{*-1}(t)dt} ds = \int_u^v \frac{1}{\int_s^1 H_n^{-1}(t)dt + R_n(s)} ds$$

$$= \int_u^v \frac{1}{\int_s^1 H_n^{-1}(t)dt}(1 + \tilde{R}_n(s))ds = \int_u^v \frac{1}{\int_s^1 H_n^{-1}(t)dt} ds(1 + o(\frac{1}{\sqrt{n}})),$$

where $\sup_{0\leq s\leq 1} |R_n(s)| = o(\frac{1}{\sqrt{n}})$ and $\sup_{0\leq s\leq 1} |\tilde{R}_n(s)| = o(\frac{1}{\sqrt{n}})$. □

With Theorem 3.1 we obtain (2.5). To complete the proof of Theorem 2.1 by the arguments given in Section 2 (cf. (2.6) through (2.12)), we note that by standard techniques (2.6) may be obtained from (2.14).

Acknowledgements. We would like to thank Wolfgang Wefelmeyer for a useful discussion and reference.

REFERENCES

Akahira, M., *The Structure of Asymptotic Deficiency of Estimators*, Queen's Papers in Pure and Applied Mathematics **75** (1986), Queen's University Press, Kingston.

Bickel, P.J., D.M. Chibisov and W.R. van Zwet, *On efficiency of first and second order*, Internat. Statist. Rev. **49** (1981), 169-175.

Bickel, P.J. and W.R. van Zwet, *Asymptotic expansions for the power of distributionfree tests in the two-sample problem*, Ann. Statist. **6** (1978), 937-1004.

Chibisov, D.M., *Asymptotic expansions for some asymptotically optimal tests*, Proc. Prague Symp. on Asymptotic Statistics (J. Hájek, ed.), vol. 2, 1974, pp. 37-68.

Feller, W., *An Introduction to Probability Theory and Its Applications*, vol. II, John Wiley and Sons, Inc., New York, 1971.

Klaassen, C.A.J., *Location estimators and spread*, Ann. Statist. **12** (1984), 311-321.

Petrov, V.V., *Sums of Independent Random Variables*, Springer Verlag, Berlin, 1975.

Pfanzagl, J., *Asymptotic expansions related to minimum contrast estimators*, Ann. Statist. **1** (1973), 993-1026.

Pfanzagl, J., *On asymptotically complete classes*, Statistical Inference and Related Topics, (M.L. Puri, ed.), vol. **2**, Academic Press, New York, 1975, pp. 1-43.

Pfanzagl, J., *First order efficiency implies second order efficiency*, Contributions to Statistics [J.Hajek Memorial Volume] (J.Jurecková, ed.), Academia Prague, 1979, pp. 167-196.

Pfanzagl, J., *On second-order asymptotics for differentiable functionals*, Proceedings of the Berkeley Conference in honor of Jerzy Neyman and Jack Kiefer (L.M. Le Cam and R.A. Olshen, eds.), vol. II, Wadsworth, Monterey, 1985, pp. 769-787.

Pfanzagl, J. and W. Wefelmeyer, *Asymptotic Expansions for General Statistical Models*, Lecture Notes in Statistics 31, Springer-Verlag, New York, 1985.

Van Zwet, W.R., *A Berry-Esseen bound for symmetric statistics*, Z. Wahrsch. Verw. Gebiete **66** (1984), 425-440.

DEPARTMENT OF MATHEMATICS, PLANTAGE MUIDERGRACHT 24, 1018 TV AMSTERDAM, NETHERLANDS.

Confidence Intervals for Regression Quantiles

Roger Koenker
University of Illinois at Champaign-Urbana

ABSTRACT. Several methods to construct confidence intervals for regression quantile estimators (Koenker and Bassett (1978)) are reviewed. Direct estimation of the asymptotic covariance matrix requires an estimate of the reciprocal of the error density (sparsity function) at the quantile of interest; some recent work on bandwidth selection for this problem will be discussed. Several versions of the bootstrap for quantile regression will be described as well as a recent proposal by Parzen, Wei, and Ying (1992) for resampling from the (approximately pivotal) estimating equation. Finally, we will describe a new approach based on inversion of a rank test suggested by Gutenbrunner, Jurečková, Koenker, and Portnoy (1993) and introduced in Hušková(1994). The latter approach has several advantages: it may be computed relatively efficiently, it is consistent under certain heteroskedastic conditions and it circumvents any explicit estimation of the sparsity function. A small monte-carlo experiment is employed to compare the competing methods.

AMS 1980 Subject Classification: 62G10, 62J10.

KEYWORDS: regression quantiles, confidence intervals, rank tests, bootstrapping, sparsity estimation.

1. INTRODUCTION

Quantile regression, as introduced in Koenker and Bassett (1978), is gradually developing into a comprehensive approach to the statistical analysis of linear and non-linear response models. By supplementing the exclusive focus of least-squares-based methods on the estimation of conditional mean functions with a general technique for estimating conditional quantile functions, it has expanded the flexibility of both parametric and non-parametric statistical methods.

There is already a well-developed theory of asymptotic inference for quantile regression and related L-statistics based on the Bahadur representation of the regression quantile process. See for example Jurečková(1984) and Koenker and Portnoy (1987). However, when inference on discrete quantiles is desired, one is faced with a rather bewildering array of methods based on direct estimation of the asymptotic covariance matrix, an approach which involves estimation of the reciprocal of the error density at the quantile of interest, or some form of the bootstrap. Versions of both approaches are available in existing statistical packages. Recently, however, several alternative approaches to inference have emerged. The objective of this paper is to compare the various methods and offer some guidance from a practical perspective on which methods seem most promising.

Consider the linear model for the τth conditional quantile function of a response variable, Y, given covariates $x \in \mathbf{R}^p$,

$$Q_Y(\tau|x) = x'\beta(\tau)$$

It will be assumed throughout that the first coordinate of x is identically one. We will focus on constructing a confidence interval for the pth coordinate of the parameter vector, β_p. A point estimate $\hat{\beta}(\tau)$, of the parameter $\beta(\tau)$, is obtained by solving

$$\min_{b \in \mathbf{R}^p} \sum_{i=1}^{n} (y_i - x_i' b)$$

where $\rho_\tau(u) = u(\tau - I(u < 0))$. In the simplest case where

$$y_i = x_i'\beta + u_i$$

and the $\{u_i\}$ are iid F with $f = F'$ and $f(F^{-1}(\tau)) > 0$ in a neighborhood of τ. Under mild design conditions we have (Koenker and Bassett(1978))

$$\sqrt{n}(\hat{\beta}(\tau) - \beta(\tau)) \rightsquigarrow \mathfrak{N}(0, \omega^2(\tau, F)D^{-1})$$

where $\beta(\tau) = \beta + F^{-1}(\tau)e_1$, $e_1 = (1, 0, ..., 0)'$, $\omega^2(\tau, F) = \tau(1-\tau)/f^2(F^{-1}(\tau))$ and $D = \lim\ n^{-1} \sum x_i x_i'$.

It is natural to begin the discussion with direct estimation of the covariance matrix of regression quantiles under iid error conditions. It is an somewhat unhappy fact of life that the asymptotic precision of quantile estimates in general, and quantile regression estimates in particular (Koenker and Bassett(1978)) depend upon the reciprocal of a density function evaluated at the quantile of interest – a quantity Tukey has termed the "sparsity function" and Parzen calls the quantile-density function. It is perfectly natural that the precision of quantile estimates should depend on this quantity since it reflects the density of observations near the quantile of interest. Thus, to estimate the precision of the τth quantile regression estimate directly, the nuisance quantity

$$s(\tau) = [f(F^{-1}(\tau))]^{-1}$$

must be estimated and therefore we must venture into the realm of smoothing. In fact, it may be possible to pull oneself out of this swamp by the bootstraps, or other statistical necromancy, but we defer the exploration of these strategies and explore the direct approach in next section.

2. Sparsity Estimation

Fortunately, there is a large literature on estimating $s(\tau)$ including Siddiqui (1960), Bofinger (1975), Sheather and Maritz (1983), Welsh (1986) and Hall and Sheather (1988). Siddiqui's idea is simplest and has received the most attention in the literature so we will focus on it. Differentiating the identity, $F(F^{-1}(t)) = t$ we have

$$\frac{d}{dt}F^{-1}(t) = s(t)$$

so it is natural to estimate $s(t)$ by using a simple difference quotient of the empirical quantile function, i.e.,

$$\hat{s}_n(t) = [\hat{F}_n^{-1}(t+h_n) \ - \ \hat{F}_n^{-1}(t-h_n)]/2h_n$$

where $\hat{F}^{-1}(\cdot)$ is an estimate of F^{-1} and h_n is a bandwidth which tends to zero as $n \to \infty$. Bofinger (1975) showed that

$$h_n = n^{-1/5}\left(\frac{4.5s^2(t)}{(s''(t))^2}\right)^{1/5}$$

was the optimal (minimum mean squared error) choice of h_n under mild regularity conditions on F. This result follows from standard density estimation asymptotics.

Of course, if we knew $s(t)$ and $s''(t)$ we wouldn't need h_n, but fortunately $s(t)/s''(t)$ is not very sensitive to F so little is lost if we choose h_n for some typical distributional shape - say, the Gaussian. Sheather and Maritz(1983) discuss preliminary estimation of s and s'' as a means of estimating a plug-in h_n. In general,

$$\frac{s(t)}{s''(t)} = \frac{f^2}{2(f'/f)^2 + [(f'/f)^2 - f''/f]}$$

and, for example, if f is Gaussian, $(f'/f)(F^{-1}(t)) = \Phi^{-1}(t)$ and the term in square brackets is 1 , so the optimal bandwidth becomes,

$$h_n = n^{-1/5}\left(\frac{4.5\phi^4(\Phi^{-1}(t))}{(2\Phi^{-1}(t)^2+1)^2}\right)^{1/5}.$$

In Figure 1, I have plotted this bandwidth, in the solid lines, as a function of n for three distinct quantiles $t = .50, 75$, and $.95$. For symmetric F the h_n's at t and $1-t$ are obviously the same.

The rule suggested above which is based upon standard density estimation asymptotics has been recently questioned by Hall and Sheather (1988). Based on Edgeworth expansions for studentized quantiles, they suggest

$$h_n = n^{-1/3} z_\alpha^{2/3}[1.5s(t)/s''(t)]^{1/3}$$

where z_α satisfies $\Phi(z_\alpha) = 1 - \alpha/2$ for the construction of $1-\alpha$ confidence intervals. This bandwidth rule is illustrated in Figure 1 by the dotted curves. It gives somewhat narrower bandwidths for modest to large n. Since the Hall and Sheather rule is explicitly designed for confidence interval construction for the quantiles, rather than simply optimizing mse-performance for the sparsity estimate itself, it seems reasonable to use it for inference. Since the optimal constant in the Hall-Sheather expression depends on the same sparsity functional as the Bofinger bandwidth, the same argument suggests that it may not be unreasonable to use the normal model version. Note that $s(t)/s''(t)$ is location-scale invariant so only the *shape* of the distribution influences this constant.

Having chosen a bandwidth h_n the next question is: how should we compute $\hat{F}^{-1}$? The simplest approach is to use the residuals from the quantile regression fit. Let $r_i : i = 1,...,n$ be these residuals, and $r_{(i)} : i = 1,...,n$ be the corresponding order statistics. Then define the usual empirical quantile function, $\hat{F}^{-1}(t) = r_{(j)}$ for $t \in$

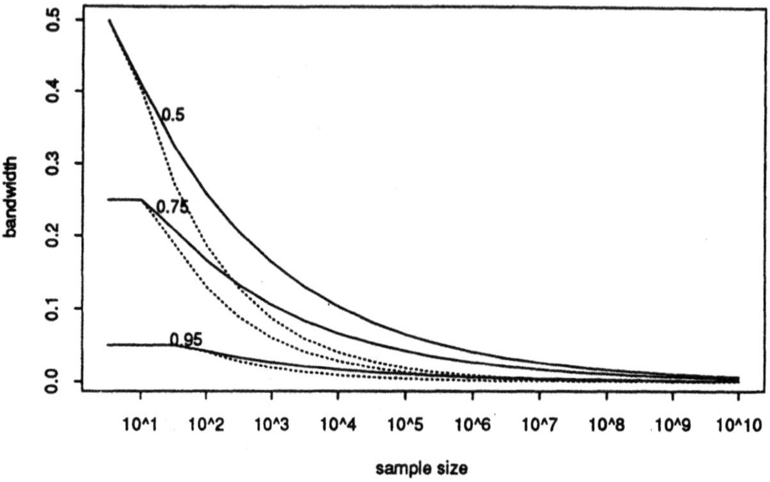

FIGURE 1. Siddiqui Bandwidths for Gaussian Sparsity Estimation

$[(j-1)/n,\ j/n\ )$. Alternatively, one may wish to interpolate to get a piecewise linear version

$$\tilde{F}^{-1}(t) = \begin{cases} r_{(1)} & if\ t \in [0, 1/2n) \\ \lambda r_{(j+1)} + (1-\lambda) r_{(j)} & if\ t \in [(2j-1)/2n, (2j+1)/2n)\ j = 1, ..., n-1 \\ r_{(n)} & if\ t \in [(2n-1)/2n, 1] \end{cases}$$

where $\lambda = tn - j + 1/2$. Alternative schemes are obviously possible. A possible pitfall of the residual-based estimates of F^{-1} is that if the number of parameters estimated, say p, is large relative to n, then since there must be p residuals equal to zero at the fitted quantile we must make sure that the bandwidth is large enough to avoid the zero residuals. The simplest approach here seems to be to ignore the zero residuals in the construction of $\hat{F}^{-1}$ and $\tilde{F}^{-1}$ and treat the effective sample size as $n - p$.

An alternative, perhaps less obvious, approach to obtain $\hat{F}^{-1}$ is to employ the empirical quantile function suggested in Bassett and Koenker (1982). In effect this amounts to using $\hat{F}^{-1} = \bar{x}'\hat{\beta}(t)$ where $\hat{\beta}(\cdot)$ is the usual regression quantile process. Like the EQF based on residuals, this is a piecewise constant function, but now the jumps are no longer equally spaced on $[0, 1]$. Nevertheless, the same ideas still apply and either the piecewise constant form of the function or the linear interpolant can be used. See Bassett and Koenker(1989) for a detailed treatment of the (strong) consistency of this method.

Finally, we should address the question: what happens if $t \pm h_n$ falls outside $[0,\ 1]$? This can easily happen when n is small. Obviously, some *ad hoc* adjustment is needed in this case, with perhaps a warning to users that the plausibility of the asymptotic theory is strained in such situations. In Figure 2.1 the flattening out of the bandwidth functions for small n reflects a simple rule of this sort.

3. Inversion of Rank Tests

In Gutenbrunner, Jurečková, Koenker and Portnoy (1993) we have developed a new approach to rank-based inference for the linear regression model. The classical theory of rank tests as exposited in the monograph of Hájek and Šidák (1967) begins with the so-called rankscore functions,

$$\hat{a}_{ni}(t) = \begin{cases} 1 & \text{if } t \leq (R_i - 1)/n \\ R_i - tn & \text{if } (R_i - 1)/n < t \leq R_i/n \\ 0 & \text{if } R_i/n < t \end{cases}$$

where R_i is the rank of the ith observation, Y_i, in the sample $\{Y_1, \dots, Y_n\}$. Integrating $\hat{a}_{ni}(t)$ with respect to various score generating functions φ yields vectors of rank-like statistics which may be used for constructing tests. For example, integrating with respect to Lebesgue measure yields the Wilcoxon scores,

$$b_i = \int_0^1 \hat{a}_{ni}(t)dt = (R_i - 1/2)/n \qquad i = 1, \dots, n,$$

while using $\varphi(t) = \text{sgn}(t - 1/2)$ yields the sign scores, $b_i = \hat{a}_{ni}(1/2)$.

How can this idea be extended to regression when, under the null, a nuisance regression parameter is present? This question was answered by Gutenbrunner and Jurečková(1992) who observed that the Hájek-Šidák rankscores may be viewed as a special case of a more general formulation for the linear model in which the functions $\hat{a}_{ni}(t)$ are defined in terms of the linear program

$$\max\{y'a | X'a = (1-t)X'1, \ a \in [0,1]^n\} \tag{3.1}$$

This problem is formally dual to the linear program defining the regression quantiles. Algorithmic details are given in Koenker and d'Orey (1993). Tests of the hypothesis $\beta_2 = 0 \in \mathbf{R}^q$ in the model $y = X_1\beta_1 + X_2\beta_2 + u$ based on the regression rankscore process may be constructed by first computing $\{\hat{a}_{ni}(t)\}$ at the restricted model,

$$y = X_1\beta_1 + u$$

computing the n-vector b with elements $b_i = \int \hat{a}_{ni}(t)d\varphi(t)$, forming the q-vector, $S_n = n^{-1/2}X_2'b$, and noting that, under the null $S_n \rightsquigarrow \mathfrak{N}(0, A^2(\varphi)Q)$ where $A^2(\varphi) = \int_0^1 \varphi^2(t)dt$, $Q = lim_{n\to\infty}Q_n$, $Q_n = (X_2 - \hat{X}_2)'(X_2 - \hat{X}_2)/n$ and $\hat{X}_2 = X_1(X_1'X_1)^{-1}X_1'X_2$. So the test statistic $T_n = S_n'Q^{-1}S_n/A^2(\varphi)$ has an asymptotic χ_q^2 null distribution, and noncentral χ_q^2 distribution under appropriate contiguous alternatives.

In the special case that X_1 is simply a column vector of ones this reduces to the original formulation of Hájek and Šidák. When $\varphi(t) = \text{sgn}(t - 1/2)$ it specializes to the score-test proposed for ℓ_1-regression in Koenker and Bassett (1982). An important feature of these rank tests is that they require no estimation of nuisance parameters, since the functional $A(\varphi)$ depends only on the score function and not on the (unknown) distribution of the vector u. This is familiar from the theory of elementary rank tests, but stands in sharp contrast with other methods of testing in the linear model where, typically, some estimation of a scale parameter, e.g., σ^2, is required to compute the test statistic.

This raises the question: could we invert a rank test of this form to provide a method of estimating a confidence interval for quantile regression, thus circumventing the problem of estimating $s(t)$. Hušková(1994) considers this problem in considerable generality establishing the validity of sequential fixed-width confidence intervals for general score functions φ. Unfortunately, for general score functions these intervals are difficult to compute. However, in the case of a fixed quantile one particularly natural choice of φ yields extremely tractable computations and we will focus on this case.

Specializing to the scalar β_2 case and using the τ-quantile score function

$$\varphi_\tau(t) = \tau - I(t < \tau)$$

and proceeding as above, we find that

$$\hat{b}_{ni} = -\int_0^1 \varphi_\tau(t) d\hat{a}_{ni}(t) = \hat{a}_{ni}(\tau) - (1-\tau) \tag{3.2}$$

with

$$\bar{\varphi} = \int_0^1 \varphi_\tau(t) dt = 0$$

$$A^2(\varphi_\tau) = \int_0^1 (\varphi_\tau(t) - \bar{\varphi})^2 dt = \tau(1-\tau).$$

Thus, a test of the hypothesis $H_o : \beta_2 = \xi$ may be based on $\hat{a}_n$ from solving,

$$\max\{(y - x_2\xi)'a | X_1'a = (1-\tau)X_1'1, a \in [0,1]^n\} \tag{3.3}$$

and the fact that

$$S_n(\xi) = n^{-1/2} x_2' \hat{b}_n(\xi) \rightsquigarrow \mathfrak{N}(0, A^2(\varphi_\tau) q_n^2) \tag{3.4}$$

under H_o; where $q_n^2 = n^{-1} x_2'(I - X_1(X_1'X_1)^{-1}X_1')x_2$. That is we may compute $T_n(\xi) = S_n(\xi)/(A(\varphi_\tau)q_n)$ and reject H_o if $|T_n(\xi)| > \Phi^{-1}(1-\alpha/2)$. This takes us back to the linear program (3.3) which may now be viewed as a one parameter parametric linear programming problem in ξ. In ξ the dual vector $\hat{a}_n(\xi)$ is piecewise constant; ξ may be altered without compromising the optimality of $\hat{a}_n(\xi)$ as long as the sign of the residuals in the primal quantile regression problem do not change. When ξ gets to such a boundary the solution does change, but may be restored by taking one simplex pivot. The process may continue in this way until $T_n(\xi)$ exceeds the specified critical value. Since $T_n(\xi)$ is piecewise constant we interpolate in ξ to obtain the desired level for the confidence interval. See Beran and Hall (1993) for a detailed analysis of the effect of interpolation like this in the case of confidence intervals for ordinary quantiles. This interval, unlike the Wald type sparsity intervals, is not symmetric; but it *is* centered on the point estimate $\hat{\beta}_2(\tau)$ in the sense that $T_n(\hat{\beta}_2(\tau)) = 0$. This follows immediately from the constraint $X'\hat{a} = (1-\tau)X'1$ in the full problem.

The primary virtue of this approach is that it inherits the scale invariance of the test statistic T_n and therefore circumvents the problem of estimating the sparsity function. Implemented in S, using an adaptation of the algorithm described in Koenker and d'Orey (1993), it has essentially the same computational efficiency as the sparsity methods. More computationally intensive resampling methods offer an alternative route which we explore in the next section.

4. Resampling Methods

There has been considerable recent interest in resampling methods for estimating confidence intervals for quantile-type estimators. However, despite the fact that confidence intervals for quantiles was one of the earliest success stories for the bootstrap (in contrast to the delete-1 jackknife which fails in this case) recent results have been considerably more guarded in their enthusiasm. Hall and Martin (1989) conclude:

> It emerges from these results that the standard bootstrap techniques perform poorly in constructing confidence intervals for quantiles... The percentile method does no more than reproduce a much older method with poor coverage accuracy at a fixed level: bias corrections fail for the same reason; bootstrap iteration fails to improve the order of coverage accuracy; and percentile-t is hardly an efficacious alternative because of non-availability of suitable variance estimates.

Nevertheless, there has been considerable recent interest, particularly among econometricians, in using the bootstrap to compute standard errors in quantile regression applications. See Buchinsky(1994), Hahn(1993) and Fitzenberger(1993) for examples.

There are several possible implementations of the bootstrap for quantile regression applications. As in other regression applications we have a choice between the residual bootstrap and the xy-pairs bootstrap. The former resamples with replacement from the residual vector and adds this to the fitted vector $X\hat{\beta}_n(\tau)$ and reestimates, in so doing it assumes that the error process is iid. The latter resamples xy pairs, and therefore is able to accomodate some forms of heteroskedasticity. As in the sparsity estimation approaches we may consider replacing the residual EQF by the EQF obtained directly from the the regression quantile process, but this maintains the iid error assumption. More interesting is the possibility of resampling directly from the full regression quantile process which we will call the Heqf bootstrap. By this I mean for each bootstrap realization of n observations we draw n p-vectors from the estimated process $\hat{\beta}_n(t)$. There are, say, J distinct such realizations

$$\hat{\beta}_n(t) = \hat{\beta}_n(t_j) \quad \text{for} \quad t_j \leq t < t_{j+1}$$

$j = 1, ..., J$ and each is drawn with probability $\pi_j = t_{j+1} - t_j$. For each design row x_i we associate the bootstrapped y observation which is the inner product of that design row and the corresponding ith draw from the regression quantile process. This procedure has the virtue that it is again capable of accommodating certain forms of heteroskedastic regression models, in particular those with linear conditional quantile functions.

Finally, we will describe a new resampling method due to Parzen, Wei and Ying (1993) which is quite distinct from the bootstrap. It arises from the observation that the function

$$S(b) = n^{-1/2} \sum_{i=1}^{n} x_i(\tau - I(y_i \leq x_i'b)) \tag{4.1}$$

which is the estimating equation for the τth regression quantile is a pivotal quantity for the true τth quantile regression parameter $b = \beta_\tau$. That is, its distribution may be generated exactly by a random vector U which is a weighted sum of independent, recentered Bernoulli variables which play the role of the indicator function. They show further that for large n the distribution of $\hat{\beta}_n(\tau) - \beta_\tau$ can be approximated by the conditional distribution of $\hat{\beta}_U - \hat{\beta}_n(\tau)$, where $\hat{\beta}_U$ solves an augmented quantile regression problem with $n+1$ observations and $x_{n+1} = -n^{1/2}u/\tau$ and y_{n+1} is sufficiently large that $I(y_{n+1} \le x'_{n+1}b)$ is always zero. This is essentially the same as solving $S(b) = -u$ for given realization of u. This approach, by exploiting the asymptotically pivotal role of the quantile regression "gradient condition", also achieves a robustness to certain heteroskedastic quantile regression models. In practice, one might be able to exploit the fact that the solution to the augmented problems is close to the original one, since they differ by only one observation, but we have not tried to do this in our simulation experiments which are reported in the next section.

5. Monte-Carlo Comparison of Methods

In this final section we report on a small monte-carlo experiment to compare the performance of the methods described above. We focus primarily on the computationally less demanding sparsity estimation and inverted rankscore methods, but some results are reported for three of the resampling methods. Preliminary results indicated that the Hall and Sheather bandwidths performed considerably better than the Bofinger choice so we have restricted our reported results mainly to this form of sparsity estimation. We also focus exclusively on the problem of confidence intervals for the median regression parameters, partly because this is the most common practical problem, and also because it restricts the amount of computation and reporting required. In subsequent work, it is hoped to provide a much more exhaustive monte-carlo experiment.

We considered first an iid error model in which both x's and y's were generated from the Student t distribution. The degrees of freedom parameter varies over the set $\{1, 3, 8\}$ for both x's and y's. The first column of the design matrix is ones, all other entries are iid draws from the specified t distribution. For each cell of the experiment the design matrix is drawn once, and 1000 replications of the response vector, y, are associated with this fixed design matrix. Throughout, we have studied only the sample size $n = 50$.

All of the computations were carried out in the 'S' language of Becker, Chambers, and Wilks(1988) on a Sun workstation. Further details on the experiment are available from the author on request, since space limitations dictate a rather abbreviated treatment here.

In Table 1 we report observed monte-carlo coverage frequencies for nine situations and three non-resampling methods. Confidence intervals are computed for all three slope coefficients for each situation so in each cell we report the number of times the interval covers the true parameter (zero in all cases) in 3000 trials. Throughout the experiment the nominal size is .10. In these iid error situations we see that the size of the rank inversion method is quite accurate throughout, as is the Hall-Sheather sparsity estimate. However the Bofinger results are considerably less satisfactory.

Generally, the rank-inversion intervals are shorter than the sparsity intervals except for the anomalous cases of Cauchy design.

TABLE 1. Confidence Interval Performance – IID Errors

	coverage			length		
	dfy= 1	dfy= 3	dfy= 8	dfy= 1	dfy= 3	dfy= 8
dfx= 1						
rank-inverse	0.892	0.923	0.922	0.320	0.392	0.359
sparsity-HS	0.893	0.907	0.909	0.240	0.142	0.079
sparsity-BS	0.932	0.927	0.931	0.288	0.153	0.083
dfx= 3						
rank-inverse	0.875	0.904	0.890	0.625	0.504	0.501
sparsity-HS	0.923	0.911	0.923	0.614	0.505	0.544
sparsity-BS	0.954	0.932	0.937	0.736	0.544	0.577
dfx= 8						
rank-inverse	0.887	0.885	0.884	0.791	0.617	0.585
sparsity-HS	0.941	0.920	0.919	0.921	0.683	0.640
sparsity-BS	0.968	0.948	0.935	1.107	0.737	0.680

To compare the performance of the resampling methods we report in Table 2 results for 3 iid error situations and 5 methods. Since the resampling methods are quite slow, 500 resamples are done for each of them, we restrict attention to only the diagonal cases of the previous table with the degrees of freedom parameter for x's and y's equal. We are primarily interested in resampling as a means of acheiving consistent confidence intervals in heteroskedastic situations so we restrict attention to the Parzen-Wei-Ying (PWY) approach, the heteroskedastic empirical quantile function bootstrap (Heqf), and the xy-pairs bootstrap. It can be seen from the table that again the rank-inversion method is quite reliable in terms of size, and also performs well with respect to length. The PWY resampling method has empirical size less than half the nominal 10 percent, while the xy-bootstrap is also undersized. The Heqf-bootstrap is accurately sized except for the Cauchy situation. It is obviously difficult to compare the lengths acheived by various methods, given the discrepancies in size, however the rank inversion approach seems to perform reasonably well in this respect.

TABLE 2. Confidence Interval Performance – IID Errors

	coverage			length		
	df= 1	df= 3	df= 8	df= 1	df= 3	df= 8
rank-inverse	0.900	0.893	0.879	0.335	0.427	0.558
sparsity-HS	0.872	0.922	0.915	0.217	0.455	0.613
PWY	0.961	0.957	0.957	0.411	0.520	0.680
Heqf-BS	0.802	0.881	0.895	0.220	0.380	0.512
XY-BS	0.929	0.948	0.945	0.331	0.486	0.640

A more challenging problem for estimation of confidence intervals for quantile regression problems involves heteroskedastic situations. We consider a simple case which bears a close resemblance to the previous iid error situations. Again, we generate 3 columns of the design matrix X as iid draws, this time from the lognormal distribution. The response vectors are then drawn from a Student t distribution with location 0 and scale given by $\sigma_i = \sum_{i=1}^{4} x_i/5$. For all i, $x_{1i} = 1$. Again the design is fixed for a given configuration, and hence scale is fixed. In this model all the conditional quantile functions are linear, so the Heqf-bootstrap is applicable, however the simple sparsity estimation approach is obviously not consistent under these conditions.

TABLE 3. Confidence Interval Performance – Heteroskastic Errors

	coverage			length		
	df= 1	df= 3	df= 8	df= 1	df= 3	df= 8
rank-inverse	0.887	0.902	0.878	1.196	0.793	0.621
sparsity-HS	0.763	0.717	0.656	0.702	0.552	0.357
PWY	0.953	0.950	0.946	1.557	0.907	0.715
Heqf-BS	0.754	0.813	0.804	0.971	0.655	0.486
XY-BS	0.907	0.911	0.897	1.332	0.799	0.612

Again the rank-inversion approach seems to perform well. As expected the sparsity approach fails miserably. The Parzen-Wei-Ying resampler is again substantially undersized – a rather puzzling result. The xy bootstrap also performs very well, but the Heqf version of the bootstrap has very poor coverage frequencies suggesting that this approach is probably not reliable. Since the rank-inversion method is on the order of 10 times faster than any of the bootstrap methods even for moderate sized problems it appears to have a substantial advantage.

References

Bassett, G.W. and Koenker, R.W. (1982). An empirical quantile function for linear models with iid errors. *Journal of the American Statistical Association.* **77**, 407-415.

Bassett, G.W. and Koenker, R.W. (1986). Strong consistency of regression quantiles and related empirical processes *Econometric Theory*, **2**, 191-201.

Beran, R. and Hall, P. (1993). Interpolated nonparametric prediction intervals and confidence intervals, J. Royal Stat. Soc. (B), **55**, 643-652.

Bofinger, E. (1975). Estimation of a density function using order statistics, *Australian J. of Statistics*, **17**, 1-7.

Buchinsky, M. (1991) The Theory and Practice of Quantile Regression, Ph.d. Thesis, Department of Economics, Harvard University.

Gutenbrunner, C. and Jurečková, J. (1992) Regression quantile and regression rank score process in the linear model and derived statistics, *Ann. Statist.* **20**,

305-330.

Gutenbrunner, C., Jurečková, J., Koenker, R. and Portnoy, S. (1993). Tests of linear hypotheses based on regression rank scores, *J. of Nonparametric Statistics*, **2**, 307-331.

Hahn, J. (1991) Bootstrapping Quantile Regression Models, preprint, Department of Economics, Harvard University.

Hušková, M. (1994) Some sequential procedures based on regression rank scores, *J. of Nonparametric Statistics*, forthcoming.

Fitzenberger, B. (1993) The moving blocks bootstrap and robust inference in linear regression, Ph.d. Thesis, Department of Economics, Stanford University.

Hájek, J. and Šidák, Z. (1967). *Theory of Rank Tests*, Academia, Prague.

Hall, P. and Martin, M.A. (1989) A note on the accuracy of bootstrap percentile method confidence intervals for a quantile. *Stat. and Prob. Letters.* **8**, 197-200.

Hall, P. and Sheather, S.J. (1988). On the distribution of a studentised quantile. *J. R. Statist. Soc.*B, **50**, 381-391.

Jurečková, J. (1984). Regression quantiles and trimmed least squares estimator under a general design, *Kybernetika*, **20**, 345-357.

Koenker, R. and Bassett, G. (1978). Regression quantiles. *Econometrica*, **46**, 33-50.

Koenker, R. and Bassett, G (1982). Tests of linear hypotheses and l_1 estimation, *Econometrica*, **50**, 1577-1584.

Koenker, R. and d'Orey, V. (1993) A Remark on Computing Regression Quantiles, *Applied Statistics*, **36**, 383-393.

Koenker, R.W. and Portnoy, S. (1987). L-Estimation for the Linear Model., *Journal of the American Statistical Association*, **82**, 851-857.

Parzen, M. I. Wei, L.J., and Ying, Z. (1992) A resampling method based on pivotal estimating functions, preprint.

Siddiqui, M.M. (1960). Distribution of quantiles in samples from a bivariate population. *J. Res. Nat. Bur. Standards*, **64B**, 145-150.

Sheather, S.J. and J.S. Maritz (1983) An estimate of the asymptotic standard error of the sample median, *Australian J. of Statistics*, **25**, 109-122.

Welsh, A.H. (1988). Asymptotically efficient estimation of the sparsity function at a point. *Stat and Prob Letters*, **6**, 427-432.

SPATIAL QUANTILES AND THEIR BAHADUR-KIEFER REPRESENTATIONS

by V. Koltchinskii[1]
University of Giessen, 35392, Giessen, Germany

One possible extension of quantile function to multivariate case is considered. Bahadur-Kiefer type approximation for the corresponding empirical quantile process is obtained.

Key words and phrases: spatial quantiles, Bahadur-Kiefer representation.
AMS 1980 subject classifications: 60F05, 62E20.

1. Introduction. Let P be a probability in $\mathbf{R}^1$, and let F_P be its distribution function. A p–th quantile of P is any value $x \in \mathbf{R}^1$ such that $F_P(x-) \leq p \leq F_P(x)$. If $p = 1/2$, a p–th quantile is called a median of P. The inverse function of F_P, defined by

$$F_P^{-1}(u) := \inf\{x : F_P(x) \geq u\}, u \in [0,1],$$

is called the quantile function of P.

It is well known that medians of P could be also defined as the values of $s \in \mathbf{R}^1$ which minimize the functional

$$M_P(s) := \int_{\mathbf{R}^1} (|s-x| - |x|)P(dx).$$

Considering along with M_P the functional $M_{P,t}, t \in \mathbf{R}^1$, defined by

$$M_{P,t}(s) := M_P(s) - ts, s \in \mathbf{R}^1$$

it's easy to check that the values of $s \in \mathbf{R}^1$ which minimize $M_{P,t}$ for $t = 2p-1$ are exactly the p–th quantiles of the distribution P. More specifically, let

$$G_P(s) := \int_{\mathbf{R}^1} \operatorname{sign}(s-x)P(dx) = F_P(s) + F_P(s-) - 1, s \in \mathbf{R}^1.$$

If $s \in \mathbf{R}^1$ isn't an atom of P, then M_P is continuously differentiable at the point s, and $M_P'(s) = G_P(s)$. For any $t \in (-1,1)$ $G_P^{-1}(t)$ minimizes the functional $M_{P,t}$. If P is nonatomic, then $G_P(s) = 2F_P(s)-1, s \in \mathbf{R}^1$, and $G_P^{-1}(t) = F_P^{-1}((1+t)/2), t \in (-1,1)$, so, in fact, one could consider G_P^{-1} as a reparametrisation of the quantile function: instead of the initial parameter p (related to the probability) we consider now the quantile function as a function of the new parameter t (the value of the derivative of M_P).

The functional M_P is only one particular example of the functionals determining M–estimators of location of the distribution P. Considering another M– estimators of location one could relate to them appropriate "quantile" functions. The idea could be used

[1] Supported by the Alexander von Humboldt Foundation while on leave from Kiev University

to extend quantiles to multivariate case (see, J. Eihnmahl and Mason (1992) for another approaches and further references).

In this paper we are going to consider only one example of such extension.

2. $r-$ quantiles in multivariate case. Let P be a Borel probability in $\mathbf{R}^d$. For $r \geq 1$ we define the functional

$$M_P^{[r]}(s) := \int_{\mathbf{R}^d} (|s-x|^r - |x|^r) P(dx), s \in \mathbf{R}^d,$$

where $|.|$ denotes the Euclidean norm in $\mathbf{R}^d$. It's easy to check that $M_P^{[r]}$ is well defined if

$$\int_{\mathbf{R}^d} |x|^{r-1} P(dx) < \infty. \tag{2.1}$$

We define also

$$G_P^{[r]}(s) := r \int_{\{x \neq s\}} (s-x)|s-x|^{r-2} P(dx), s \in \mathbf{R}^d.$$

Under condition (2.1) $G_P^{[r]}$ is a map from $\mathbf{R}^d$ into $\mathbf{R}^d$.

In what follows, we'll omit "r" in the notations $M_P^{[r]}, G_P^{[r]}$ and so on, and write simply M_P, G_P and so on, if it is clear which value of r we mean.

It's easy to show that in the case $r > 1$ for all P, satisfying (2.1), and in the case $r = 1$ for all nonatomic P, M_P is continuously differentiable in $\mathbf{R}^d$, and $(M_P)'(s) = G_P(s), s \in \mathbf{R}^d$.

We'd like also to mention the following useful asymptotic properties of M_P and G_P. For any Borel probability P in $\mathbf{R}^d$, satisfying (2.1),

$$M_P(s) = |s|^r + o(|s|^r), s \to \infty, \tag{2.2}$$

$$G_P(s) = r|s|^{r-2}s + o(|s|^{r-1}), s \to \infty. \tag{2.3}$$

Consider along with the functional M_P the functional $M_{P,t}, t \in \mathbf{R}^d$, defined by

$$M_{P,t}(s) = M_P(s) - (t,s), s \in \mathbf{R}^d.$$

2.1. Theorem. Let $r > 1$. For any Borel probability P in $\mathbf{R}^d$, satisfying (2.1), G_P is a homeomorphism from $\mathbf{R}^d$ onto $\mathbf{R}^d$. For any $t \in \mathbf{R}^d$ the unique minimal point of the functional $M_{P,t}$ in $\mathbf{R}^d$ is $G_P^{-1}(t)$.

Proof. Note that (2.2) implies for any $t \in \mathbf{R}^d$

$$M_{P,t}(s) \to \infty, s \to \infty.$$

Since $M_{P,t}$ is continuous in $\mathbf{R}^d$, this implies that it has a minimal point in $\mathbf{R}^d$. Let us prove that the minimal point is unique. Clearly, $M_{P,t}$ is convex in $\mathbf{R}^d$. Moreover, denote

$$g(s;x) := |s-x|^r - |x|^r - (t,s),$$

so, that

$$M_{P,t}(s) = \int_{\mathbf{R}^d} g(s;x)P(dx).$$

The elementary inequality

$$\frac{|y_1+y_2|^r}{2} < \frac{|y_1|^r + |y_2|^r}{2}, y_1 \neq y_2, y_1, y_2 \in \mathbf{R}^d,$$

implies

$$g(\frac{s_1+s_2}{2};x) < \frac{g(s_1;x)+g(s_2;x)}{2}, s_1 \neq s_2, s_1, s_2 \in \mathbf{R}^d, x \in \mathbf{R}^d.$$

Therefore

$$M_{P,t}(\frac{s_1+s_2}{2}) < \frac{M_{P,t}(s_1)+M_{P,t}(s_2)}{2}, s_1 \neq s_2. \tag{2.4}$$

Suppose that s_1, s_2 are both minimal points of $M_{P,t}$, $s_1 \neq s_2$, and $M_{P,t}(s_1) = M_{P,t}(s_2) = a$. Then

$$M_{P,t}(\frac{s_1+s_2}{2}) < a,$$

so, s_1, s_2 couldn't be minimal points. This contradiction shows that the minimal point is unique. Let us denote it s_0. Then $M'_{P,t}(s_0) = 0$, which implies $G_P(s_0) = t$.

Now we show that s_0 is the unique solution of the equation $G_P(s) = t$. Let s_1 be another solution, and let $s_1 \neq s_0$. Take $v := \frac{(s_1-s_0)}{|s_1-s_0|}, \rho := |s_1 - s_0|$. Consider the function

$$\varphi(\lambda) := M_{P,t}(s_0 + \lambda v), \lambda \in \mathbf{R}^1.$$

Clearly, φ is convex in $\mathbf{R}^1$. We have $\varphi'(0) = (G_P(s_0) - t, v) = 0$ and since φ is convex, φ' is nondecreasing function in $\mathbf{R}^1$. Since $G_P(s_1) = t$,

$$\varphi'(\rho) = (G_P(s_1) - t, v) = 0.$$

This implies that φ' is identically 0 in $[0,\rho]$, so, φ is a constant in this interval. But condition (2.4) implies

$$\varphi(\frac{\rho}{2}) < \frac{\varphi(0)+\varphi(\rho)}{2},$$

so, φ couldn't be a constant. Thus we have proved that for any $t \in \mathbf{R}^d$ the equation $G_P(s) = t$ has the unique solution in $\mathbf{R}^d$, so, the map G_P is one-to-one from $\mathbf{R}^d$ onto $\mathbf{R}^d$. Moreover, $G_P^{-1}(t)$ is the unique minimal point of the functional $M_{P,t}$.

It remains to show that G_P is a homeomorphism. Define the map i from the unit open ball $B_d := \{s \in \mathbf{R}^d : |s| < 1\}$ onto $\mathbf{R}^d$ by

$$i(s) = \frac{s}{1-|s|}, s \in B_d.$$

Clearly, i is a homeomorphism of B_d onto $\mathbf{R}^d$. The inverse i^{-1},

$$i^{-1}(t) = \frac{t}{1+|t|}, t \in \mathbf{R}^d$$

defines a compactification of $\mathbf{R}^d$, since it is a homeomorphism of $\mathbf{R}^d$ onto a dense subset of the compact $\bar{B}_d$ (the closure of B_d). In the usual way we can define now $\bar{\mathbf{R}}^d \supset \mathbf{R}^d$ such that i has an extension to a homeomorphism of $\bar{B}_d$ onto $\bar{\mathbf{R}}^d$ with the inverse i^{-1}. For any $v \in \mathbf{R}^d, |v| = 1$

$$G_P(\rho v) \to i(v), \rho \to \infty \text{ in } \bar{\mathbf{R}}^d.$$

It means that the one-to-one map $i^{-1} \circ G_P \circ i$ from B_d onto B_d can be extended to the one-to-one map from $\bar{B}_d$ to $\bar{B}_d$, and moreover, $i^{-1} \circ G_P \circ i$ is continuous on $\bar{B}_d$. Since $\bar{B}_d$ is a compact, we can conclude that $i^{-1} \circ G_P \circ i$ is a homeomorphism from $\bar{B}_d$ onto $\bar{B}_d$, which, of course, implies that G_P is a homeomorphism from $\mathbf{R}^d$ onto $\mathbf{R}^d$. ◇

The case $r = 1$ is more complicated since the kernel of the transformation $P \mapsto G_P$ is not continuous. This case was studied by Dudley and Koltchinskii (1992).

Let $B(s;\delta)$ denote the open ball with center s and radius $\delta > 0$, and let $\bar{B}(s;\delta)$ be its closure. We keep the notations B_d and $\bar{B}_d$ for the open and the closed unit balls with center 0 respectively.

Let D_P denote the set of all atoms of the probability P. Denote $p(s) := P(\{s\}), s \in D_P$. For $s \in D_P$ the ball $\bar{B}(G_P(s);p(s))$ will be called the "hole" of the atom s. It was shown by Dudley and Koltchinskii (1992) that all "holes" are mutually disjoint and all of them belongs to $\bar{B}_d$.

Define

$$U_P := B_d \setminus \bigcup_{s \in D_P} \bar{B}(G_P(s);p(s)).$$

So, to get the set U_P one should take B_d and exclude off it the "holes" of all atoms of P. Define also

$$V_P := U_P \bigcup G_P(D_P),$$

$$U_P^\circ := B_d \setminus \overline{\bigcup_{s \in D_P} B(G_P(s);p(s))}.$$

The following version of theorem 2.1 was proved by Dudley and Koltchinskii (1992).

2.2. Theorem. Let $r = 1$. If P is not concentrated in a line, then G_P is a one-to-one map from $\mathbf{R}^d$ onto V_P, and it is a homeomorphism of $\mathbf{R}^d \setminus \overline{D_P}$ and U_P°. In particular, if P is nonatomic, then G_P is a homemorphism of $\mathbf{R}^d$ and B_d. For any $t \in B_d$, the unique minimum of the functional $M_{P,t}$ is at the point $G_P^{-1}(t)$, if $t \in U_P$ (or, t doesn't belong to any "hole"), or, at the atom $s \in D_P$, if t belongs to the "hole" of this atom.

Thus, for $r > 1$ the inverse G_P^{-1} is well defined on $\mathbf{R}^d$. For $r = 1$ the domain of G_P^{-1} is the set V_P, but we'll extend it to the whole unit ball B_d, just setting $G_P^{-1}(t) = s$ for any t from the "hole" $\bar{B}(G_P(s); p(s))$ of an atom $s \in D_P$. The definition is correct since all "holes" are mutually disjoint.

The inverse $(G_P^{[r]})^{-1}$ will be called the *r−quantile function* of the distribution P.

Let us mention briefly some other properties of the maps $G_P^{[r]}, r \geq 1$.

1. If $r \geq 1$, $r \neq 2k, k = 1, 2, \ldots$, and if P, Q are Borel probabilities in $\mathbf{R}^d$ such that

$$G_P(s) = G_Q(s), s \in \mathbf{R}^d$$

$$\int_{\mathbf{R}^d} |x|^r P(dx) = \int_{\mathbf{R}^d} |x|^r Q(dx) < \infty,$$

then $P = Q$.

The proof is based on the uniqueness results for the solution of the inverse problem for the potentials of measures in $\mathbf{R}^d$ (see, Stephenson (1977)).

2. If $r = 1$, then $G_P^{-1}(0)$ coincides with the Haldane's median of P, (see, Haldane (1948)) and if $r > 1$, then it coincides with the L_r−mean of P.

3. If P is invariant with respect to the group of all rotations around a point $s_0 \in \mathbf{R}^d$, then there exists a strictly increasing and continuous function φ on $(0, \infty)$ such that

$$G_P(s) = \frac{s - s_0}{|s - s_0|}\varphi(|s - s_0|), s \neq s_0,$$

$$G_P(s_0) = 0.$$

In the case $r = 1$ φ is a distribution function on $(0, \infty)$, and $\varphi(0+) = 0$ if and only if P is nonatomic (so, it doesn't have an atom at the point s_0.) The inverse map is given by

$$G_P^{-1}(t) = G_P^{-1}(0) + \varphi^{-1}(|t|)\frac{t}{|t|}, \tag{2.5}$$

for any t such that $p(s_0) < |t| < 1$, and $G_P^{-1}(0) = s_0$.

In the case $r > 1$ we have

$$\varphi(\lambda) = r\lambda^{r-1} + o(\lambda^{r-1}), \lambda \to \infty,$$

and $\varphi(0+) = 0$. The inverse map is given by (2.5) for any $t \neq 0$ and $G_P^{-1}(0) = s_0$.

4. If $P_\theta, \theta \in \mathbf{R}^d$ denotes the θ−shift of the distribution P (so, P_θ is the distribution of $X + \theta$, given X has the distribution P), then for any $r \geq 1$

$$G_{P_\theta}(s) = G_P(s - \theta), s \in \mathbf{R}^d,$$

$$G_{P_\theta}^{-1}(t) = \theta + G_P^{-1}(t),$$

$t \in \mathbf{R}^d$ for $r > 1$, and $t \in B_d$ for $r = 1$.

3. Bahadur-Kiefer representations. Let $(X_1, ..., X_n)$ be a sample from a distribution P. Let P_n be the empirical measure based on this sample. We are interested in the Bahadur-Kiefer type approximation of the r-quantile process $n^{1/2}(G_{P_n}^{-1} - G_P^{-1})$.

For $K \subset \mathbf{R}^d$ and $x : K \mapsto \mathbf{R}^d$ we denote

$$\|x\|_K := \sup_{t \in K} |x(t)|.$$

The inverse of a linear operator A from $\mathbf{R}^d$ onto $\mathbf{R}^d$ will be denoted $\mathrm{inv}(A)$.

We take and fix $r \geq 1$. Suppose that G_P is continuously differentiable in an open set $U \in \mathbf{R}^d$ with the nondegenerate derivative $G'_P(s), s \in U$, and, moreover, suppose that G'_P is locally Lipschitz in U, so, that for any compact $K \subset U$ there exists a constant $L > 0$ such that

$$\|G'_P(s) - G'_P(t)\| \leq L|s - t|, s, t \in K$$

($\|\cdot\|$ is the operator norm).

Let $V := G_P(U)$, and for $K \subset U$ let

$$\Delta_n(K) := \|G_{P_n}^{-1} - G_P^{-1} + \mathrm{inv}(G'_P \circ G_P^{-1})(G_{P_n} - G_P) \circ G_P^{-1}\|_K.$$

The following results were proved by Koltchinskii (1993).

3.1. Theorem. Let $r = 1$. Suppose that P has a bounded density. For any compact $K \subset V := G_P(U)$ there exists a constant $C > 0$ such that

$$Pr^*(\{\Delta_n(K) \geq \delta_n u\}) \leq C \exp\{-\frac{u}{C}\}, u > 0,$$

where

$$\delta_n := \begin{cases} n^{-1} \log^{1/2} n & \text{for } d > 2 \\ n^{-1} \log n & \text{for } d = 2 \\ n^{-3/4} \log^{1/2} n & \text{for } d = 1. \end{cases}$$

Note that Niemiro (1992) proved that in the case $d > 1$ for any $\varepsilon > 0$ a.s.

$$|G_{P_n}^{-1}(0) - G_P^{-1}(0) + \mathrm{inv}(G'_P \circ G_P^{-1}(0))(G_{P_n} - G_P) \circ G_P^{-1}(0)| = o(n^{-1+\varepsilon}), n \to \infty.$$

3.2. Theorem. Let $r > 1$. Suppose that P has a bounded density with a compact support. For any compact $K \subset V := G_P(U)$ there exists a constant $C > 0$ such that

$$Pr^*(\{\Delta_n(K) \geq \delta_n u\}) \leq C \exp\{-\frac{u}{C}\}, u > 0,$$

where

$$\delta_n := \begin{cases} n^{-r/2-1/4} \log^{1/2} n & \text{for } d = 1, r \in (1, 3/2) \\ n^{-1} \log n & \text{for } d = 1, r = 3/2 \\ n^{-1} \log^{1/2} n & \text{for } d = 1, r > 3/2, \text{ or for } d > 1. \end{cases}$$

The proofs are based on the inequalities for functional inverses obtained by Koltchinskii and Dudley (1992) as well as on some exponential estimates for empirical processes indexed by Vapnik-Chervonenkis classes of functions.

REFERENCES

Dudley, R.M. and Koltchinskii, V. (1992) The spatial quantiles (preprint).

Einmahl, John H.J. and Mason, D. (1992) Generalized quantile process. *Ann. Statist.*, 20, 1062-1078.

Haldane, J.B.S. (1948) Note on the median of a multivariate distribution. *Biometrika* 35, 414-415.

Koltchinskii, V. (1993) Bahadur-Kiefer approximation for spatial quantiles (preprint)

Koltchinskii, V. and Dudley, R.M. (1992) Differentiability of inverse operators and asymptotic properties of inverse functions (preprint).

Niemiro, W. (1992) Asymptotics for M−estimators defined by convex minimization. *Ann. Statist.* 20, 1514-1533.

Stephenson, K. (1977) Certain integral equalities, which imply equimeasurability of functions. *Canad. J. Math.* 29, 827-844.

ASYMPTOTIC EXPANSIONS OF ERROR PROBABILITIES FOR TESTS

Harald Luschgy
University of Münster

Abstract

Expansions of the second kind error of Neyman-Pearson tests are derived. No special properties of the observation process are required but the log-likelihood ratio process must satisfy a central limit condition.

Key words: Neyman-Pearson tests, second order expansions, martingales, diffusions, processes with independent increments, autoregressive processes.
AMS 1980 subject classifications: 62 F 05, 62 M 02, 62 M 07

1 Introduction

Let P_0 and P_1 be probability measures on a measurable space $(\Omega, \mathcal{F})$ equipped with a filtration $\mathbb{F} = (\mathcal{F}_t)_{t\geq 0}$. Let $\beta_t(\alpha) = E_{P_1}(1 - \varphi_{\alpha,t})$ denote the second kind error of the Neyman-Pearson test $\varphi_{\alpha,t}$ for testing $\{P_0|\mathcal{F}_t\}$ against $\{P_1|\mathcal{F}_t\}$ at level α. For a second order analysis of the error probability we need a central limit condition for the log-likelihood ratio process

$$Z_t = Z_t(P_0, P_1) = \log \frac{dP_0}{dP_1}|\mathcal{F}_t.$$

Proposition 1 *(Luschgy, 1991) Let $\alpha_t \to \alpha$, $\alpha \in (0,1)$, and let $r_i : \mathbb{R}_+ \to \mathbb{R}_+$, $i = 1,2$, denote rate functions satisfying $\lim_{t\to\infty} r_i(t) = \infty$ and $\lim_{t\to\infty} \frac{r_1(t)}{r_2(t)} = \infty$. Assume there are $H, K \in (0,\infty)$ sucht that*

$$\mathcal{L}(r_2(t)^{-1}(Z_t - K\, r_1(t))|P_0) \xrightarrow{\mathcal{D}} N(0, H). \tag{1}$$

Then

$$\log \beta_t(\alpha_t) = -K r_1(t) - H^{1/2}\Phi^{-1}(\alpha) r_2(t) + o(r_2(t)) \text{ as } t \to \infty. \tag{2}$$

In this paper we derive solutions of the central limit problem for Z for some classes of models. Essentially the same central limit condition as in (1) has been used by Mykland (1992) to prove asymptotic expansions of the distributions of martingale estimators.

2 Processes with Continuous Sample Paths

Let $\mathbb{F}$ be the right continuous filtration generated by a process $X = (X_t)_{t\geq 0}$ with continuous sample paths starting at $X_0 = x$ such that under P_i, the observed process X is a semimartingale. Assume that $P_0|\mathcal{F}_t$ and $P_1|\mathcal{F}_t$ are mutually absolutely continuous for every t and $\mathcal{F} = \sigma(X_t : t \geq 0)$. Then under some additional conditions Z admits the representation

$$Z_t = M_t + \frac{1}{2} < M >_t, \tag{3}$$

where M is a continuous local martingale with $M_0 = 0$ under P_0 (see Jacod and Shiryayev, 1987, Chap. III.5). The idea of proving (1) is to derive in a first step a representation of the form

$$Z_t - Kr_1(t) = N_t + o_{P_0}(r_2(t)), \tag{4}$$

where N is a continuous local martingale under P_0 and then in a second step to apply the CLT for martingales to get

$$\mathcal{L}(r_2(t)^{-1}N_t|P_0) \xrightarrow{\mathcal{D}} N(0, H). \tag{5}$$

The main tool for (4) is Ito's formula. Write $K_t(P_0, P_1)$ for the Kullback-Leibler information of $P_0|\mathcal{F}_t$ with respect to $P_1|\mathcal{F}_t$. If $E_{P_0} < M >_t < \infty$, then

$$K_t(P_0, P_1) = E_{P_0} Z_t = \frac{1}{2} E_{P_0} < M >_t < \infty.$$

Of course, $Kr_1(t)$ is related to $K_t(P_0, P_1)$.

In this setting it can be seen that necessary conditions for the central limit condition (1) (with arbitrary limiting distribution) to hold are

$$\frac{< M >_t}{r_1(t)} \to 2K \text{ in } P_0-\text{probability}, \ (\frac{M_t}{r_1(t)^{1/2}}, \frac{\frac{1}{2} < M >_t - Kr_1(t)}{r_3(t)}) = O_{P_0}(1),$$

where $r_3 = r_2$ in case $r_1(t)^{1/2}/r_2(t) \to a \in [0, \infty)$ and $r_3 = r_1^{1/2}$ in case $r_1(t)^{1/2}/r_2(t) \to \infty$. In particular, $< M >$ must satisfy a central limit condition.

2.1 Diffusions

Let X denote the coordinate process on $\Omega = C(\mathbb{R}_+, \mathbb{R})$. Under P_i, suppose X satisfies the stochastic differential equation

$$dX_t = a_i(X_t)dt + b(X_t)dW_t(a_i), \ X_0 = x \text{ a.s.}, \tag{6}$$

where $W(a_i)$ is a standard Wiener process, a_i and b are continuous, $b > 0$. Then $P_0|\mathcal{F}_t$ and $P_1|\mathcal{F}_t$ are mutually absolutely continuous and (3) holds with

$$\begin{aligned} M_t &= \int_0^t \frac{a_0(X_s) - a_1(X_s)}{b(X_s)} dW_s(a_0) \text{ and} \\ < M >_t &= \int_0^t \frac{(a_0(X_s) - a_1(X_s))^2}{b^2(X_s)} ds. \end{aligned} \tag{7}$$

Consider the generator of the Markov process X under P_0,

$$Gg = a_0 g' + \frac{1}{2} b^2 g''$$

and the G-harmonic function

$$f(x) = \int_0^x \exp(-2\int_0^z \frac{a_0(y)}{b^2(y)} dy) dz.$$

Assume $f(x) \to \pm\infty$ as $x \to \pm\infty$ and

$$D := \int \frac{1}{b^2(y) f'(y)} dy < \infty.$$

Then X is ergodic under P_0 with invariant distribution

$$\pi(dx) = D^{-1} \frac{1}{b^2(x) f'(x)} dx$$

(see Skorokhod, 1989, Theorem I.16). Let

$$K = \frac{1}{2} \int \frac{(a_0 - a_1)^2}{b^2} d\pi$$

and assume $0 < K < \infty$. Let g be a solution of

$$Gg = \frac{1}{2} \frac{(a_0 - a_1)^2}{b^2} - K.$$

Then by Ito's formula

$$\begin{aligned} g(X_t) &= g(x) + \int_0^t Gg(X_s) ds + \int_0^t g'(X_s) b(X_s) dW_s(a_0) \\ &= g(x) + \frac{1}{2} < M >_t - Kt + \int_0^t g'(X_s) b(X_s) dW_s(a_0). \end{aligned}$$

Therefore,

$$\begin{aligned} \text{(8)} \qquad Z_t - Kt &= N_t + o_{P_0}(t^{1/2}), \text{ where} \\ N_t &= \int_0^t [\frac{a_0(X_s) - a_1(X_s)}{b(X_s)} - g'(X_s) b(X_s)] dW_s(a_0). \end{aligned}$$

Let

$$H = \int (\frac{a_0 - a_1}{b} - g'b)^2 d\pi$$

and assume $H < \infty$.

Proposition 2 $\mathcal{L}(t^{-1/2}(Z_t - Kt)|P_0) \xrightarrow{\mathcal{D}} N(0, H)$.

Proof: By the ergodic theorem

$$t^{-1} < N >_t \to H \quad P_0 - \text{a.s.}$$

The assertion follows from (8) and the CLT for martingales. □

2.2 Composite Hypotheses

We indicate how to use a central limit condition for the log-likelihood ratio process to construct asymptotically optimal tests for composite hypotheses. Consider the following model. Let X denote the coordinate process on $\Omega = C(\mathbb{R}_+, \mathbb{R}^2)$. Under P_τ, $\tau \in \mathbb{R}^2$, let the data satisfy the stochastic differential equation

$$(9) \qquad dX_t = -\tau_1 X_t dt + \tau_2 D X_t dt + dW_t(\tau),\ X_0 = 0,$$

where $D = \begin{pmatrix} 0 & -1 \\ 1 & 0 \end{pmatrix}$ and $W(\tau)$ is a 2-dimensional standard Wiener process. Then

$$(10) \qquad \begin{aligned} Z_t(\tau,\xi) &= \log \frac{dP_\tau}{dP_\xi}|\mathcal{F}_t = M_t + \frac{1}{2} < M >_t, \text{ where} \\ M_t &= \int_0^t [A(\tau-\xi)X_s]^T dW_s(\tau), \\ < M >_t &= |\tau - \xi|^2 \int_0^t |X_s|^2 ds, \\ A(\tau) &= \begin{pmatrix} -\tau_1 & -\tau_2 \\ \tau_2 & -\tau_1 \end{pmatrix}. \end{aligned}$$

If $\tau_1 > 0$, one obtains along the lines of Section 2.1

$$(11) \qquad \mathcal{L}(t^{-1/2}(Z_t(\tau,\xi) - Kt)|P_\tau) \xrightarrow{\mathcal{D}} N(0,H),$$

where $K = K(\tau,\xi) = \frac{|\tau-\xi|^2}{2\tau_1}$, $H = H(\tau,\xi) = \frac{\xi_1|\tau-\xi|^2}{\tau_1^2} + \frac{|\tau-\xi|^4}{4\tau_1^3}$.

Now fix $\tau_2, \xi_2 \in \mathbb{R}$, $\tau_2 \neq \xi_2$, and consider the problem of testing the composite null hypothesis $\{P_{(\vartheta,\tau_2)} : \vartheta \in \Theta\}$ against the compositive alternative $\{P_{(\vartheta,\xi_2)} : \vartheta \in \Theta\}$ with $\Theta = [\varepsilon, \infty)$, $\varepsilon > 0$. By Proposition 1, any test φ_t of asymptotic level α, $\alpha \in (0,1)$, satisfies

$$(12) \qquad \begin{aligned} \liminf_{t\to\infty} t^{-1} \log E_{(\vartheta,\xi_2)}(1-\varphi_t) &\geq -\inf_{\delta\in\Theta} K((\delta,\tau_2),(\vartheta,\xi_2)) \\ &= -(\sqrt{(\tau_2-\xi_2)^2+\vartheta^2} - \vartheta) =: -\bar{K}(\vartheta) \end{aligned}$$

for every $\vartheta \in \Theta$. A globally (Hodges-Lehmann) optimal test, that is, a test ψ_t such that the limit $t^{-1}\log E_{(\vartheta,\xi_2)}(1-\psi_t)$ exists and equals $-\bar{K}(\vartheta)$ for every $\vartheta \in \Theta$ can be constructed as follows. In view of (11), consider

$$\varphi_t(\vartheta) = \begin{cases} 1 & \\ & Z_t((\vartheta,\tau_2),(\vartheta,\xi_2)) \begin{matrix} \leq \\ \\ > \end{matrix} K((\vartheta,\tau_2),(\vartheta,\xi_2))t + H((\vartheta,\tau_2),(\vartheta,\xi_2))^{1/2}t^{1/2}\Phi^{-1}(\alpha) \\ 0 & \end{cases}.$$

Replace ϑ in Z and the first order term K by the efficient and strongly consistent estimator

$$\hat{\vartheta}_t = t / \int_0^t |X_s|^2 ds$$

and adjust by replacing the second order term H by $2K((\varepsilon,\tau_2),(\varepsilon,\xi_2))$. We arrive at a test which is of asymptotic level α and globally optimal. For details see Luschgy and Rukhin (1993).

3 Processes with Poisson Jump Measure

Under P_i, let the observed process X be a $\mathbb{R}^d$-valued semimartingale starting at $X_0 = x$ with respect to the right continuous filtration $\mathbb{F}$ generated by X. Assume $\mathcal{F} = \sigma(X_t : t \geq 0)$. Write $\mu = \mu^X$ for the random measure on $\mathbb{R}_+ \times \mathbb{R}^d$ associated with the jumps of X by

$$\mu = \sum_{t \geq 0} \varepsilon_{(t,\Delta X_t)} 1_{\{\Delta X_t \neq 0\}}$$

and ν_i for its predictable compensator under P_i. Assume μ is a Poisson measure under P_i, i.e. ν_i is deterministic and $\nu_i(\{t\} \times \mathbb{R}^d) = 0$ for every t. Assume further that $P_0|\mathcal{F}_t$ and $P_1|\mathcal{F}_t$ are mutually absolutely continuous for every t. Then by Girsanov's theorem

(13) ν_0 and ν_1 are mutually absolutely continuous and
$$\int_0^t \int (Y^{1/2} - 1)^2 d\nu_1 = \int_0^t \int ((\frac{1}{Y})^{1/2} - 1)^2 d\nu_0 < \infty$$
for every t, where $Y = d\nu_0 / d\nu_1$.

Under some additional conditions we have

$$(14) \quad Z_t = M_t + \frac{1}{2} < M >_t + \int_0^t \int (1 - \frac{1}{Y}) d(\mu - \nu_0) + \int_0^t \int (\log Y - 1 + \frac{1}{Y}) d\mu,$$

where M is a continuous local martingale under P_0 with $M_0 = 0$ (see Jacod and Shiryayev, 1987, Chap. III.5). Define the Kullback-Leibler information for the σ-finite measures $\nu_i | \mathcal{B}(\mathbb{R}_+) \otimes \mathcal{B}(\mathbb{R}^d) \cap [0,t] \times \mathbb{R}^d$ by

$$\begin{aligned} K_t(\nu_0, \nu_1) &= \int_0^t \int (Y \log Y - Y + 1) d\nu_1 \\ &= \int_0^t \int (\log Y - 1 + \frac{1}{Y}) d\nu_0 \end{aligned}$$

and assume

(A.1) $K_t(\nu_0, \nu_1) < \infty$ for every t.

Then from (14)

$$(15) \quad \begin{aligned} Z_t &= \tilde{N}_t + \frac{1}{2} < M >_t + K_t(\nu_0, \nu_1), \quad \text{where} \\ \tilde{N}_t &= M_t + \int_0^t \int \log Y \, d(\mu - \nu_0). \end{aligned}$$

Assume further

(A.2) $\int_0^t \int (\log Y)^2 d\nu_0 < \infty$ for every t.

Then the process $\tilde{N}$ is a local square integrable martingale under P_0 and

$$(16) \quad < \tilde{N} >_t = < M >_t + \int_0^t \int (\log Y)^2 d\nu_0.$$

If $E_{P_0} < M >_t < \infty$, we get

$$K_t(P_0, P_1) = \frac{1}{2} E_{P_0} < M >_t + K_t(\nu_0, \nu_1) < \infty.$$

3.1 Processes with Independent Increments

Let X denote the coordinate process on $\Omega = D(\mathbb{R}_+, \mathbb{R}^d)$. Under P_i, let X be a process with independent increments with respect to $\mathbb{F}$ with $X_0 = 0$ a.s. and

$$\begin{aligned} E_{P_i} \exp(iu^T X_t) &= \exp\{iu^T B_i(t) - \frac{1}{2} u^T C(t) u \\ &+ \int_0^t \int (e^{iu^T x} - 1 - iu^T h(x)) \nu_i(ds, dx)\}, \; u \in \mathbb{R}^d, \end{aligned}$$

where $h : \mathbb{R}^d \to \mathbb{R}^d$ denotes a truncation function, B_i and C are (deterministic) continuous functions (componentwise) of finite variation over finite intervals with values in $\mathbb{R}^d$ and the set of symmetric nonnegative definite $d \times d$ matrices, respectively, and $B_i(0) = C(0) = 0$. Choose a continuous increasing function F with $F(0) = 0$ such that

$$C(t) = \int_0^t c(s) dF(s).$$

Assume (13) and the existence of a Borel function $b : \mathbb{R}_+ \to \mathbb{R}^d$ such that

$$\begin{aligned} B_0(t) &= B_1(t) + \int_0^t c(s) b(s) dF(s) + \int_0^t \int (Y - 1) h \, d\nu_1, \\ & \int_0^t |c(s)^{1/2} b(s)|^2 dF(s) < \infty \text{ for every } t. \end{aligned}$$

Then $P_0 | \mathcal{F}_t$ and $P_1 | \mathcal{F}_t$ are mutually absolutely continuous for every t and Z admits the representation (14) with

$$(17) \qquad M_t = \int_0^t b(s) dX_s^c \text{ and } < M >_t = \int_0^t |c(s)^{1/2} b(s)|^2 dF(s),$$

where X^c denotes the continuous martingale part of X under P_0 (see Jacod and Shiryayev, 1987, III.5.35, IV.4.32). Here $< M >$ is deterministic and M is a continuous Gaussian martingale under P_0. Choose

$$r_1(t) = \frac{1}{2} < M >_t + K_t(\nu_0, \nu_1) = K_t(P_0, P_1)$$

and

$$r_2(t) = (< M >_t + \int_0^t \int (\log Y)^2 d\nu_0)^{1/2} = (\mathrm{Var}_{P_0} Z_t)^{1/2}$$

and assume $r_i(t) \to \infty$, $\frac{r_1(t)}{r_2(t)} \to \infty$.

Proposition 3 *Assume*

$$(18) \qquad r_2(t)^{-2} \int_0^t \int (\log Y)^2 \, 1_{\{|\log Y| > r_2(t)\varepsilon\}} d\nu_0 \to 0 \textit{ for every } \varepsilon > 0.$$

Then

$$\mathcal{L}(r_2(t)^{-1}(Z_t - r_1(t)) | P_0) \xrightarrow{\mathcal{D}} N(0, 1).$$

Proof: From (15)

$$Z_t - r_1(t) = \tilde{N}_t.$$

The assertion follows from the CLT for martingales. □

Condition (18) is satisfied, for instance, if $\nu_i(ds, dx) = ds \otimes L_i(dx)$, $i = 0, 1$.

3.2 Random Observation Periods

Assume $\nu_i(ds, dx) = ds \otimes L_i(dx)$. Then the density Y does not depend on t and

$$K_t(\nu_0, \nu_1) = t\, K(L_0, L_1).$$

Assume $K(L_0, L_1) > 0$. Consider the $\mathbb{F}$-stopping time

$$\eta_t = \inf\{s \geq 0 : \frac{1}{2} < M >_s + s\, K(L_0, L_1) > t\}. \tag{19}$$

Observation of the process X up to time η_t corresponds to $\mathcal{F}_{\eta_t}$. From (15)

$$\log \frac{dP_0}{dP_1} | \mathcal{F}_{\eta_t} = Z_{\eta_t} = N_t + t, \tag{20}$$

where $N_t = \tilde{N}_{\eta_t}$.

The process N is a local square integrable martingale under P_0 with respect to the filtration $\mathbb{F}_\eta = (\mathcal{F}_{\eta_t})_{t \geq 0}$.

Proposition 4 *Assume $\frac{\eta_t}{r_2(t)^2} \to \gamma$ in P_0-probability, $\frac{t}{r_2(t)^2} \to \delta$, where $\gamma, \delta \in [0, \infty)$ and $\frac{t}{r_2(t)} \to \infty$. Let $H = 2\delta + \gamma[\int (\log Y)^2 dL_0 - 2K(L_0, L_1)]$. Then*

$$\mathcal{L}(r_2(t)^{-1}(Z_{\eta_t} - t) | P_0) \xrightarrow{\mathcal{D}} N(0, H).$$

Proof: From (16)

$$\begin{aligned} < \tilde{N} >_t &= 2[\frac{1}{2} < M >_t + t\, K(L_0, L_1)] \\ &\quad + t[\int (\log Y)^2 dL_0 - 2K(L_0, L_1)]. \end{aligned}$$

Therefore,

$$< N >_t = < \tilde{N} >_{\eta_t} = 2t + \eta_t[\int (\log Y)^2 dL_0 - 2K(L_0, L_1)].$$

This implies

$$\frac{< N >_t}{r_2(t)^2} \to H \text{ in } P_0 - \text{probability}.$$

Furthermore,

$$r_2(t)^{-2} \eta_t \int (\log Y)^2\, 1_{\{|\log Y| > r_2(t)\epsilon\}} dL_0 \to 0 \text{ in } P_0 - \text{probability}.$$

The assertion now follows from (20) and the CLT for martingales. □

Proposition 1 applies to the filtration $\mathbb{F}_\eta$. The local theory of semimartingale models based on stopping times has been investigated by Luschgy (1993).

3.3 Autoregressive Processes

Let X denote the coordinate process on $\Omega = D(\mathbb{R}_+, \mathbb{R})$. Under P_i, suppose X satisfies the stochastic differential equation

$$dX_t = -\vartheta_i X_t dt + dV_t(\vartheta_i),\ X_0 = 0 \text{ a.s.,} \tag{21}$$

where $\vartheta_i \in \mathbb{R}$, $V(\vartheta_i)$ is a process with independent stationary increments, $V_0(\vartheta_i) = 0$ a.s. and

$$\begin{aligned} E_{P_i} \exp(iuV_t(\vartheta_i)) &= \exp t\{iub_i - \frac{1}{2}cu^2 \\ &\quad + \int (e^{iux} - 1 - iuh(x))dL_i(x)\}, \end{aligned}$$

$b_i \in \mathbb{R}$, $c > 0$. For simplicity assume $c = 1$. Then $\nu_i(ds, dx) = ds \otimes L_i(dx)$ and under the condition (13) $P_0|\mathcal{F}_t$ and $P_1|\mathcal{F}_t$ are mutually absolutely continuous for every t and Z admits the representation (14) with

$$\begin{aligned} M_t &= \int_0^t [(\vartheta_1 - \vartheta_0)X_{s-} + a]dX_s^c, \\ < M >_t &= \int_0^t [(\vartheta_1 - \vartheta_0)X_s + a]^2 ds, \end{aligned} \tag{22}$$

where X^c is the continuous martingale part of X under P_0 and

$$a = b_0 - b_1 - \int (Y-1)h\, dL_1.$$

If $\vartheta_0 = \vartheta_1$, then $< M >$ is deterministic and the central limit problem for Z can be solved as in Section 3.1. So assume $\vartheta_0 \neq \vartheta_1$. In case $\vartheta_0 > 0$ and in case $\vartheta_0 = 0$, $E_{P_0} V_1(\vartheta_0) \neq 0$, the central limit problem for Z has been solved in Luschgy (1992) using Markov methods similar to Section 2.1. The case $\vartheta_0 < 0$ is treated based on random stopping.

Proposition 5 *Let η_t be as in (19). Assume $\vartheta_0 < 0$ and $\int x^2 dL_0(x) < \infty$. Then*

$$\mathcal{L}(t^{-1/2}(Z_{\eta_t} - t)|P_0) \xrightarrow{\mathcal{D}} N(0, 2).$$

Proof: Since $\int x^2 dL_0(x) < \infty$, the martingale convergence theorem gives

$$e^{\vartheta_0 t} X_t \to \int_0^\infty e^{\vartheta_0 s} dV_s(\vartheta_0)\ \ P_0 - \text{a.s.}$$

This implies

$$e^{2\vartheta_0 t}[\frac{1}{2} < M >_t + tK(L_0, L_1)] \to -\frac{(\vartheta_0 - \vartheta_1)^2}{4\vartheta_0}(\int_0^\infty e^{\vartheta_0 s} dV_s(\vartheta_0))^2 =: U\ \ P_0 - \text{a.s.}$$

and, therefore,

$$e^{2\vartheta_0 \eta_t} t \to U\ \ P_0 - \text{a.s.}$$

One obtains

$$\frac{\eta_t}{t} \to 0\ \ P_0 - \text{a.s.}$$

and the assertion follows from Proposition 4. □

References

[1] Jacod, J. and Shiryayev, A. N. (1987): Limit Theorems for Stochastic Processes. Springer-Verlag, Berlin, Heidelberg, New York

[2] Luschgy, H. (1991): Second order behavior of Neyman-Pearson tests for stochastic processes. Statistics and Decisions (to appear)

[3] Luschgy, H. (1992): Asymptotic behavior of Neyman-Pearson tests for autoregressive processes. Preprints "Angewandte Mathematik und Informatik", Univ. Münster

[4] Luschgy, H. (1993): Local asymptotic quadraticity of stochastic process models based on stopping times. Preprings "Angewandte Mathematik und Informatik", Univ. Münster

[5] Luschgy, H. and Rukhin, A. L. (1993): Asymptotic properties of tests for a class of diffusion processes: optimality and adaptation. Math. Methods of Statistics 2, 42 - 51

[6] Mykland, P. A. (1992): Asymptotic expansions and bootstrapping distributions for dependent variables: a martingale approach. Ann. Statist. 20, 623 - 654

[7] Skorokhod, A. V. (1989): Asymptotic Methods in the Theory of Stochastic Differential Equations. Amer. Math. Soc., Providence, R.I.

Universität Münster
Institut für Mathematische Statistik
Einsteinstr. 62
D-48149 Münster

AN APPLICATION OF COMPLEX COMMUTATION FUNCTIONS

Petr Mandl
Charles University, Prague

Abstract

In the paper commutation functions of the form $D_x = \ell_x \exp(-x(\rho + i\lambda))$ are introduced in connection with a model of social pension plans.

Key words: commutation functions, pension plans, spectral decomposition

AMS Subject Classification: 62 P 05

1 Introduction

The members of a *social pension plan* form an open risk community. To model its evolution it is necessary to make demographic hypotheses, which refer mostly to the group of actives. The other groups (retirement and disability pensioners, recipients of survivors' pensions) are created using the service table and other actuarial tables. The present note aims to contribute to the analysis of pension plans under the assumption of relative stability presented in the monographs by P. Thullen (1977, 1982). Randomness is introduced into the model of the stream of new entrants into the active group. This makes it possible to analyze the sensitivity of the financial management methods to deviations from the projected demographic evolution.

Two methods of operating the plan are considered:

I) the defrayal of current pension cost by current contributions,

II) the defrayal of the values of new pensions by current contributions.

2 Pension plans in relatively stable state

To make the presentation simple, we consider following Thullen (1982) a plan with only one entry age x_0 and a retirement age u. Although continuous models are more transparent, we deal with a discrete model having direct connection to numerical computations.

For a *deterministic model* of the pension plan the assumption of relative stability can be introduced by stating that the number of entrants into the active group in year t grows geometrically as

$$e(x_0)\, e^{\rho t}.$$

Provided that this has been so for a sufficiently long time the distribution of the ages in the active group becomes stable and is

$$L^{aa}(x,t) = e(x_0)\, e^{\rho(t-x+x_0)} \frac{\ell^{aa}_x}{\ell^{aa}_{x_0}} = c_0\, e^{\rho t} D_x^{aa(\rho)}. \tag{1}$$

In the commutation function above the force of interest is identified with the growth rate ρ. The total size of the active group is obtained by summation,

$$(2) \qquad \Lambda^{aa}(t) = c_0\, e^{\rho t} \sum_{x_0}^{u-1} D_x^{aa(\rho)} = c_0\, e^{\rho t} N^{aa(\rho)}_{x_0:\overline{u-x_0|}}.$$

From (1) using the actuarial tables similar formulae are derived for the numbers of retirement and disability pensioners

$$(3) \qquad \Lambda^{(a)}(t) = c_0\, e^{\rho t}\, D_u^{aa(\rho)}\, \ddot{a}_u^{(\rho)}, \qquad \Lambda^{i}(t) = c_0\, e^{\rho t}\, N_{x_0}^{ai(\rho)},$$

and for the recipients of widows' and orphans' pensions

$$(4) \qquad \Lambda^{w}(t) = c_0\, e^{\rho t}\, N_{x_0}^{aw(\rho)}, \qquad \Lambda^{k}(t) = c_0\, e^{\rho t}\, N_{x_0}^{ak(\rho)}.$$

Consider the pension plan, in which the retirement and disability pensions amount to the portion γ of the last salary, while the widows' and the orphans' pensions are the portions γ^w and γ^k of the original recipient's pension. One has to introduce a salary scale $\{s_x\}$ into the model. This is performed by modifying the service table $\{\ell_x^{aa}\}$ of the actives into $\{\ell_x^{aa} s_x\}$. The quantity $\Lambda^{aa}(t)$ is then proportional to the total amount of insured salaries. Further let δ be the actual force of interest and σ be the salaries growth rate.

Assume first I), i.e. *the direct defrayal of the pension cost* from the contributions. Then (2) multiplied by the premium rate $\Pi^{\mathrm{I}}_{x_0}$ has to cover the benefits, whose total amount is obtained from (3) and (4). Consequently,

$$(5) \qquad N^{aa(\rho)}_{x_0:\overline{u-x_0|}}\, \Pi^{\mathrm{I}}_{x_0} = \gamma\left(D_u^{aa(\rho)}\, \ddot{a}_u^{(\rho)} + N_{x_0}^{ai(\rho)} + \gamma^w\, N_{x_0}^{aw(\rho)} + \gamma^k\, N_{x_0}^{ak(\rho)}\right).$$

The first term in the bracket $D_u^{aa}\, \ddot{a}_u$ has two components: the demographic one D_u^{aa} and the annuity one $\ddot{a}_u$. In the case of II), *the coverage of new pensions* from the contributions, one has to replace in the annuity component the demographic rate ρ by the financial rate $\delta - \sigma$. Further, $N_{x_0}^{ai(\rho)}$ should be replaced by

$$N_{x_0}^{ai(\rho,\delta-\sigma)} = \sum_{x_0}^{u-1} D_x^{ai(\rho,\delta-\sigma)},$$

where

$$D_x^{ai(\rho,\delta-\sigma)} = e^{-(x+1/2)\rho}\, \ell_x^{aa}\, {}^{*}i_x\, \ddot{a}_{x+1/2}^{(\delta-\sigma)}.$$

Similarly, for the annuity components of $N_{x_0}^{aw}$ and of $N_{x_0}^{ak}$ the force of interest $\delta - \sigma$ is to be used. The commutation functions with two interest rates are called *mixed commutation functions.* Using them one can write the equation for the premium rate under II) as

$$N^{aa(\rho)}_{x_0:\overline{u-x_0|}}\, \Pi^{\mathrm{II}}_{x_0} = \gamma\left(D_u^{aa(\rho)}\, \ddot{a}_u^{(\delta-\sigma)} + N_{x_0}^{ai(\rho,\delta-\sigma)} + \gamma^w\, N_{x_0}^{aw(\rho,\delta-\sigma)} + \gamma^k\, N_{x_0}^{ak(\rho,\delta-\sigma)}\right).$$

The replacement of ρ by $\delta-\sigma$ in (5) yields the equation for the so called *generation premium rate*, balancing the expected contributions and benefits within each generation.

3 Random stream of entrants

We introduce *random fluctuations* into the demographic part of the model by assuming that the number of entrants into the group of actives in year t equals

$$e(x_0)\, e^{\rho t}(1+U_t), \tag{6}$$

where $\{U_t\}$ is a (wide sense) stationary random sequence with zero expectation. Its standard deviation

$$s = \sqrt{E\,U_t^2}$$

thus coincides with the variation coefficient of the number of entrants. We do not take the insurance technical risk into account supposing that it is eliminated to a sufficient extent by the law of large numbers.

The simplest instance of a stationary random sequence is the first order autoregressive sequence satisfying

$$U_t = r\,U_{t-1} + s\sqrt{(1-r^2)}\,Y_t, \tag{7}$$

where $\{Y_t\}$ is a sequence of uncorrelated random variables, $E\,Y_t = 0$, $E\,Y_t^2 = 1$. r is the one step correlation in $\{U_t\}$.

We shall employ *the spectral representation* of $\{U_t\}$ (see e.g. Rosenblatt (1974))

$$U_t = \int_{-\pi}^{\pi} e^{it\lambda}\, dZ_\lambda, \quad t = 0, \pm 1, \dots . \tag{8}$$

In (8) $\{Z_\lambda\}$ is a random process with orthogonal increments. It holds

$$E\,|Z_\lambda - Z_\mu|^2 = F(\lambda) - F(\mu), \quad \lambda > \mu,$$

where $F(\lambda)$, $\lambda \in [-\pi, \pi]$, is the spectral distribution function of $\{U_t\}$. If $F(\lambda)$ is absolutely continuous, its derivative $f(\lambda)$ is called spectral density.

If (7) holds, then

$$f(\lambda) = \frac{s^2}{2\pi} \frac{1-r^2}{1-2r\cos\lambda + r^2}\,.$$

In the subsequent propositions we assume that (6) did hold for a sufficiently long time or that the pension plan is in a state compatible with this assumption.

Proposition 1. Let A_t and B_t be respectively the total amount of the contributions and of the costs in year t. Then

$$A_t = \Pi_{x_0} c_0\, e^{\rho t} \left(N^{aa(\rho)}_{x_0:\overline{u-x_0|}} + \int_{-\pi}^{\pi} e^{i\lambda(t+x_0)} N^{aa(\rho+i\lambda)}_{x_0:\overline{u-x_0|}}\, dZ_\lambda \right). \tag{9}$$

Under I)

$$\begin{aligned} B_t \;=\; & c_0\, e^{\rho t} \Big(\Pi^{\mathrm{I}}_{x_0} N^{aa(\rho)}_{x_0:\overline{u-x_0|}} + \gamma \int_{-\pi}^{\pi} e^{i\lambda(t+x_0)} \big(D^{aa(\rho+i\lambda)}_u \ddot{a}_u^{(\rho+i\lambda)} + \\ & + N^{ai(\rho+i\lambda,\rho+i\lambda)}_{x_0} + \gamma^w N^{aw(\rho+i\lambda,\rho+i\lambda)}_{x_0} + \gamma^k N^{ak(\rho+i\lambda,\rho+i\lambda)}_{x_0} \big)\, dZ_\lambda \Big), \end{aligned} \tag{10}$$

under II)

$$(11) \qquad B_t = c_0\, c^{\rho t} \left(\Pi^{\rm II}_{x_0}\, N^{aa(\rho)}_{x_0:\overline{u-x_0|}} + \gamma \int_{-\pi}^{\pi} e^{i\lambda(t+x_0)} \left(D^{aa(\rho+i\lambda)}_u \ddot{a}^{(\delta-\sigma)}_u + \right.\right.$$
$$\left.\left. + N^{ai(\rho+i\lambda,\delta-\sigma)}_{x_0} + \gamma^w\, N^{aw(\rho+i\lambda,\delta-\sigma)}_{x_0} + \gamma^k\, N^{ak(\rho+i\lambda,\delta-\sigma)}_{x_0} \right) dZ_\lambda \right).$$

(6) together with the spectral decomposition (8) is the motive to introduce complex valued mixed commutation functions.

To present the method of proof of Proposition 1 let us verify (9). From (1), (6) and (8) one obtains

$$L^{aa}(x,t) = e(x_0)\, e^{\rho(t-x+x_0)} \left(1 + \int_{-\pi}^{\pi} e^{i\lambda(t-x+x_0)} dZ_\lambda \right) \ell^{aa}_x \,/\, \ell^{aa}_{x_0} =$$
$$= c_0\, e^{\rho t} \left(D^{aa(\rho)}_x + \int_{-\pi}^{\pi} e^{i\lambda(t+x_0)} D^{aa(\rho+i\lambda)}_x dZ_\lambda \right).$$

Summation over $x = x_0, \ldots, u-1$ and the multiplication by the premium rate yield (9).

To measure the discrepancy between A_t and B_t we introduce *the normed deviation*

$$\Delta = \sqrt{E(A_t - B_t)^2} \,/\, E\, A_t.$$

Note that $E\, A_t = E\, B_t$.

Proposition 2.

$$\Delta = \sqrt{\left(2 \int_0^{\pi} \left| \Pi_{x_0}\, N^{aa}_{x_0:\overline{u-x_0|}} - \gamma \left(D^{aa}_u\, \ddot{a}_u + N^{ai}_{x_0} + \gamma^w\, N^{aw}_{x_0} + \gamma^k\, N^{ak}_{x_0} \right) \right|^2 dF(\lambda) \right)} \,/$$
$$/\, \Pi_{x_0}\, N^{aa}_{x_0:\overline{u-x_0|}}\,,$$

where the interest rates are as in (10) for I) and as in (11) for II).

Proposition 2 follows from Proposition 1 and from the fact that for any function $\varphi(\lambda)$

$$E \int_{-\pi}^{\pi} \varphi(\lambda)\, dZ_\lambda = 0, \qquad E \left| \int_{-\pi}^{\pi} \varphi(\lambda)\, dZ_\lambda \right|^2 = \int_{-\pi}^{\pi} |\varphi(\lambda)|^2\, dF(\lambda)$$

provided that the last integral is finite (see e. g. Rosenblatt (1974)).

4 Numerical illustration

We present some values of Δ for a pension plan providing retirement and disability benefits, $\gamma = 0.5$, ($\gamma^w = 0$, $\gamma^k = 0$), $x_0 = 20$, $u = 65$. (7) is assumed to hold with $s = 0.1$. The EVK - 1960 (B) males table as published in Thullen (1982) was used in the computations, $s_x \equiv 1$.

	I) $\rho = 0$ $\Pi^{\mathrm{I}}_{x_0} = 0.161$	II) $\rho = 0,\ e^{\delta-\sigma} = 1.03$ $\Pi^{\mathrm{II}}_{x_0} = 0.123$	I) $e^{\rho} = 1.03$ $\Pi^{\mathrm{I}}_{x_0} = 0.063$
$r = 0$	0.086	0.084	0.090
0.2	0.106	0.103	0.110
0.4	0.132	0.129	0.137
0.6	0.172	0.168	0.180
0.8	0.257	0.250	0.267

Table of Δ

It is seen that besides s the correlation in $\{U_t\}$ is here the basic factor influencing the fluctuations of $A_t - B_t$.

REFERENCES

N. L. Bowers, Jr. et al. (1986): Actuarial Mathematics. Society of Actuaries, Itasca.

M. Rosenblatt (1974): Random Processes, Springer–Verlag, New York.

P. Thullen (1977): Mathematische Methoden der Sozialen Sicherheit, Verlag Versicherungswirtschaft, Karlsruhe.

P. Thullen (1982): Die Mathematik der sozialen Rentenversicherung unter dynamischen Bedingungen, Verlag Versicherungswirtschaft, Karlsruhe.

Charles University, Department of Probability and Mathematical Statistics,
Sokolovská 83, 186 00 Praha 8, Czech Republic

ONE-SIDED DEVIATIONS OF A RANDOM WALK WITHOUT MOMENT ASSUMPTIONS [1]

Alexander Martikainen
Sankt-Petersburg University
Dept. Mathematics and Mechanics [2]

Abstract

Let $\{S_n\}$ be a random walk with independent increments. We state necessary and sufficient conditions under which the relation $P(S_n - median(S_n) > \varepsilon b_n) = o(d_n)$ holds for any (for some) $\varepsilon > 0$ in the following cases: $b_n = n^{1/t}, 1 < t < 2, d_n = n^{-d}, d > 0$, or $b_n = n^{1/2}(\log n)^s, s \geq 0.5$, d_n is the same as in the previous case, or $b_n = n^{1/2}(\log\log n)^s, s \geq 0.5, d_n = (\log n)^{-d}$. Based on these particular cases we describe a method, which indeed allows us to derive such necessary and sufficient conditions for more general classes of b_n and d_n.

Key words and expressions: strong law of large numbers, law of the iterated logarithm, rates of convergence, large deviations, moderate deviations, one-sided

AMS 1991 Subject Classification. Primary - 60F15.

1. Introduction

Let $X, X_1, X_2, \ldots$ be a sequence of independent identically distributed nondegenerate random variables, $S_n = X_1 + \ldots + X_n, F(x) = P(X < x)$. The goal of the paper is to estimate the rate of decrease probabilities $P(S_n - a_n > b_n)$ for a given norming sequence $\{b_n\}$ and an appropriate centering sequence $\{a_n\}$. We put further $a_n = medianS_n$ or $a_n = n\int_{-b_n}^{b_n} x\,dF(x)$, because these universal centering sequences do not require any moment assumptions. Unlike the papers on large or moderate deviations, we deal with distributions, which can have infinite variance or even infinite mean and try to state conditions, for the relation

$$P(S_n - a_n > \varepsilon b_n) = o(d_n). \tag{1}$$

to hold for any (some) $\varepsilon > 0$. Here d_n is a given sequence of positive numbers, $d_n \searrow 0$.

Two-sided version of (1) with $|S_n - a_n|$ instead of $S_n - a_n$ were investigated completely in numerous papers starting with P.Hsu and H.Robbins (1947), D.Brillinger

---

[1] This paper was written while the author was at Philipps Universität Marburg supported by Alexander von Humboldt Foundation (Bonn, Fed.Rep.Germany).

[2] Current address: Uni Marburg, FB Mathematik, Hans-Meerwein-Str., Postfach Lahnberge 35032, Marburg/Lahn, Germany

(1962), L.Baum and M.Katz (1965) (see V.Petrov (1972) for bibliography). Necessary and sufficient conditions are derived for wide classes of b_n and d_n in the two-sided problem.

But they are not derived for the one-sided problem with a few exceptions. V.Petrov and I.Shirokova (1973) stated nesessary and sufficient conditions for the case $a_n = EX, b_n = n, d_n = \gamma^n, 0 < \gamma < 1$, some results of this type are stated by author in (1982) and without proofs in (1992). A series of sufficient conditions for (1) are derived by several authors. We omit detailed bibliography in this short paper.

The effective method describing here is to write the sum S_n as the sum of two ones

$$S_n = R_n + T_n, \tag{2}$$

where $R_n = X_1 I(X_1 < 0) + ... + X_n I(X_n < 0), T_n = X_1 I(X_1 > 0) + ... + X_n I(X_n > 0)$. The rate of decrease the $P(T_n - \alpha_n > \varepsilon b_n)$ can be found from the above mentioned two-sided results. In Section 2 we state (in Theorem 1) the estimates from below and from above for $P(R_n - \beta_n > \varepsilon b_n)$ and in Sections 3 and 4 we find necessary and sufficient conditions for (1) with $b_n = n^{1/t}, 1 < t < 2, d_n = M(n)n^{-d}, d \geq 0$, in Theorem 2 and 3 ($M(n)$ is a positive, nonincreasing, slowly-varying at infinity sequence), $b_n = n^{1/2}(\log n)^s, s \geq 0.5, d_n = n^{-d}, d > 0$, in Theorems 4 and 5, $b_n = n^{1/2}(\log\log n)^s, s \geq 0.5, d_n = (\log n)^{-d}, d > 1$, in Theorems 6 and 7. In Theorems 2 - 7 we put $a_n = median S_n$, but if the relation (1) holds for some $\varepsilon > 0$ in one of these results, then it holds for the same ε also with $a_n = n \int_{-b_n}^{b_n} x\, dF(x)$.

2. Random variables bounded above. Define the functions

$$D_u(x) = \frac{1}{x} \int_{-x}^{x} y^2\, dF(y) + \int_{u > |y| \geq x} |y|\, dF(y), \qquad u \geq x > 0. \tag{3}$$

It is easy to verify that

$$D_u(x) = \int_{-u}^{u} |y| (\frac{|y|}{x} \wedge 1)\, dF(y), \qquad u \geq x > 0,$$

so that D_u is positive, continuous and nonincreasing on $(0,u]$, $xD_u(x)$ is nondecreasing on the same interval. Moreover, D_u strictly decreases on $[x_F, u]$ where x_F is the infimum of the support of the distribution of $|X|$.

Let $\{b_n\}$ be a nondecreasing sequence of positive numbers. Fix $\tau > 0$. Define $L_n = L_n(\tau)$ as the largest solution in $(0,b_n]$ of the equation

$$nD_{b_n}(L_n) = \tau b_n \tag{4}$$

if (4) has a solution in $(0,b_n]$. Put $L_n = 0$ otherwise (we also put further $e^{-a/L_n} = 0$ if $a > 0$ and $L_n = 0$).

Functions $D_u(x)$ were introduced by the author in (1993). The quantities L_n were defined in this paper in a similar way. The following theorem is a reformulation of lemmas 6 and 11 of A.Martikainen (1993) for our situation.

Put $E(x) = \int_{-x}^{x} y\, dF(y)$ for $x > 0$.

THEOREM 1 *Assume that* $P(X < C) = 1$ *for some constant* $C, EX = 0$ *if* $EX^2 < \infty, b_n \nearrow \infty, b_n/n \to 0$ *and*

$$(S_n - nE(b_n))/b_n \xrightarrow{P} 0. \tag{5}$$

Then

$$P(S_n - nE(b_n) > \varepsilon b_n) > e^{-\gamma b_n/L_n}$$

for any $\gamma > 0.817\tau, \tau > 1.583\varepsilon, \varepsilon > 0$, *all sufficienlty large (s.l.)* n, L_n *and* b_n/L_n. *Furthermore,*

$$P(S_n - nE(b_n) > \varepsilon b_n) \leq e^{-(\varepsilon-\tau)b_n/L_n}$$

for any positive ε, τ *and all s.l.* n.

Now we can show how this general result leads to necessary and sufficient conditions for (1) when X is bounded above.

COROLLARY 1 *Assume that* $P(X < C) = 1$ *for some constant* C, $EX = 0$ *if* $EX^2 < \infty$, $b_n \nearrow \infty, b_n/n \to 0, d_n \searrow 0$ *and the condition (5) holds. Then*

$$P(S_n - nE(b_n) > \varepsilon b_n) < d_n \quad \text{for any } \varepsilon > 0, \text{ all s.l.} n, \tag{6}$$

if and only if

$$nD_{b_n}(b_n/\log(1/d_n)) = o(b_n) \tag{7}$$

The relation (6) holds for some $\varepsilon > 0$ *(instead of any* $\varepsilon > 0$*) and for all s.l.* n, *if and only if (7) holds with* O *instead of* o.

Note that, for any positive b, $\log(1/d_n)$ can be changed in (7) by $b\log(1/d_n)$, because, for any $u > 0$, $D_u(x)$ does not increase and $xD_u(x)$ does not decrease. It means that, if (6) holds in Corollary 1, then, for any positive b, it holds with d_n^b instead of d_n. Therefore, the probability in (6) can not decrease as $(\log n)^{-d'}$ if $d_n = (\log n)^{-d}$, or as $n^{-d'}$ if $d_n = n^{-d}$. But it can not decrease as e^{-dn}, as it follows from Corollary 2 below. This observation can explain also why the probability of a deviation of the sum R_n defined in (2) has a negligible part in the right-side part of the inequality

$$P(S_n - nE(b_n) > \varepsilon b_n) \tag{8}$$

$$\leq P(R_n - n\int_{-b_n}^{0} x\, dF(x) > \varepsilon b_n/2) + P(T_n - n\int_{0}^{b_n} x\, dF(x) > \varepsilon b_n/2),$$

while we study (1) with $d_n = (\log n)^{-d}$ and $d_n = n^{-d}$.

Proof of Corollary 1. Since $(\int_{-z}^{z} x\, dF(x))^2 = o(\int_{-z}^{z} x^2\, dF(x))$ as $z \to \infty$, we have, by (5) and the weak law of large numbers, $n\int_{b_n}^{-b_n} x^2\, dF(x) = o(b_n^2)$, and therefore, for any $\delta \in (0, 1]$,

$$\begin{aligned} nD_{b_n}(\delta b_n) &= \frac{n}{\delta b_n}\int_{-\delta b_n}^{\delta b_n} x^2\, dF(x) + n\int_{\delta b_n \leq |x| < b_n} |x|\, dF(x) \\ &= o(b_n) + b_n nP(|X| \geq \delta b_n) = o(b_n). \end{aligned}$$

Thus, $L_n < \delta b_n$ for all s.l. n, i.e. $L_n = o(b_n)$.

The conditions $b_n/n \to 0, b_n \to \infty$ imply $nD_{b_n}(L)/b_n \geq \frac{n}{L}\int_{-L}^{L} x^2\,dF(x) \to \infty$ for all s.l. L. Therefore, L_n are positive for all s.l. n and $L_n \to \infty$. By Theorem 1 with $\gamma = \tau = 2\varepsilon$, the relation (6) implies

$$\exp(\tau b_n/L_n(\tau)) < d_n \quad \text{for any } \tau > 0, \text{ all s.l. } n. \tag{9}$$

By the same theorem with $\varepsilon = 2\tau$, (9) implies (6), so that these relations are equivalent. It is clear that (9) is equivalent to

$\tau b_n/L_n(\tau) > \log(1/d_n)$ for any $\tau > 0$. The latter one is equivalent to the condition $nD_{b_n}(\tau b_n/\log(1/d_n)) < \tau b_n$ for any $\tau > 0$. Since $D_{b_n}(x)$ is a nonicreasing function of x and $xD_{b_n}(x)$ does not decrease, this condition is equivalent to (7).

The proof of the second statement of Corollary 1 is the same.

COROLLARY 2 *Assume that $P(X < C) = 1$ for some constant $C, EX = 0$ if $EX^2 < \infty, b_n \nearrow \infty, b_n/n \to 0$ and the condition (5) holds. If the relation*

$$P(S_n - nE(b_n) > b_n) < e^{-\alpha b_n}$$

holds with some $\alpha > 0$ for infinitely many n, then the distribution of X is degenerate.

Proof. As in the proof of Corollary 1 we have $L_n = o(b_n)$ and $L_n \to \infty$ for any $\tau > 0$. By Theorem 1 with $\gamma = \tau = 2, \varepsilon = 1$, we obtain inequality $e^{-2b_n/L_n(2)} < e^{-\alpha b_n}$ for infinitely many n. Therefore, $L_n(2) < 2/\alpha$ that contradicts to $L_n(2) \to \infty$.

3. Case of $b_n = n^{1/t}, d_n = n^{-d}, 1 < t < 2, d > 0$. We do not assume further that X is bounded above (with the exception of Lemma 3). First we define the functions $D_u^-(x)$ in the same way as $D_u(x)$ but using $XI(X < 0)$ instead of X:

$$D_u^-(x) = \frac{1}{x}\int_{-x}^{0} y^2\,dF(y) + \int_{-u}^{-x} |y|\,dF(y), \qquad u \geq x > 0.$$

These new functions have evidently the same properties as were pointed above for $D_u(x)$. Let $med\, S_n$ denote the least median of random variable S_n.

THEOREM 2 *Let $M(n)$ be a positive, nonincreasing, slowly-varying at infinity sequence, $1 < t < 2, d > 0$. Then*

$$P(S_n - med\, S_n > \varepsilon n^{1/t}) = o(M(n)n^{-d}) \quad \textit{for any } \varepsilon > 0, \tag{10}$$

if and only if the following conditions are satisfied:

$$P(X < -n^{1/t}) = o(n^{-1}), \tag{11}$$

$$P(X > n^{1/t}) = o(M(n)n^{-d-1}) \tag{12}$$

and

$$D_n^-(n/\log n) = o(n^{1-t}). \tag{13}$$

The relation (10) holds for some $\varepsilon > 0$ (instead of any $\varepsilon > 0$), if and only if the conditions (11) and (12) are satisfied and the condition (13) holds with O instead of o.

The case $d = 0$ will be studied in Theorem 3.

COROLLARY 3 *Let $1 < t < 2, d > 0$. If the conditions (12) and*

$$nP(X < -n^{1/t}) = o((\log n)^{1-t}) \tag{14}$$

are satisfied then (10) holds. If o is replaced for O in (12) or (14), then (10) can fail.

This corollary demonstrates a series of distributions, for which (10) holds while $P(|S_n - med\, S_n| > \varepsilon n^{1/t}) \neq o(n^{-d})$. By the result of C.Heyde and V.Rohatgi (1967), we have this effect, if (12) and (14) hold, but $nP(X < -n^{1/t}) \neq o(n^{-d})$.

The case $0 < t \leq 1$ was investigated by N.Amosova (1978) under additional assumptions.

The proofs of these results involve Corollary 1 and the following three lemmas.

LEMMA 1 *Let $b_n \nearrow \infty, \sup b_{2n}/b_n < \infty$. If $P(S_n - med\, S_n > \varepsilon b_n) \to 0$ for any (some) $\varepsilon > 0$, then (5) holds (the sequence $(S_n - med\, S_n)/b_n$ is stochastically bounded).*

This well-known result follows, for example, from the Esseen's estimate (1966) for the Levy's concentration function $Q(X, \lambda) = \sup_x P(x \leq X < x + \lambda)$:

$$Q(S_r, \lambda) \leq A(rP(|X^s| > \lambda))^{-1/2}$$

for any positive $\lambda, r = 1, 2, \ldots$ Here A is an absolute constant, X^s is the symmetrized random variable. Replacing in this estimate S_r for S_{nr}, X for S_n and λ for εb_{nr}, we get

$$Q(S_{nr}, \varepsilon b_{nr}) \leq A(rP(|S_n^s| > \varepsilon b_{nr}))^{-1/2}.$$

By the definition of the median, $\liminf Q(S_{nr}, \varepsilon b_{nr}) \geq 0.5$ for any $\varepsilon > 0$ and $r = 1, 2, \ldots$ Therefore, $\limsup P(|S_n^s| > \varepsilon b_n r) \leq A^2/2r$. Since $\sup b_{nr}/b_n < \infty$ for any $r > 1$, and positive ε can be chosen arbitrarily small, we obtain $S_n^s/b_n \xrightarrow{P} 0$. Applying the symmetrization inequalities completes the proof.

The proof of the second statement of Lemma 1 is the same.

LEMMA 2 *Let $b_n \nearrow \infty$ and $d_n \searrow 0$. If*
$P(S_n - med\, S_n > \varepsilon b_n) = o(d_n)$ for any (for some) $\varepsilon > 0$,
then, for any (for some) $\varepsilon > 0$,

$$nP(X > \varepsilon b_n) = o(d_n). \tag{15}$$

One can find this result, for example, in the paper of the author (1982). In fact, the idea (without the formulation) was used by P. Erdős (1949), C.Heyde and V.Rohatgi (1967) and in many other papers.

It is easy to check that (15) holds for each $\varepsilon > 0$ if it holds for some $\varepsilon > 0$ and $\liminf b_{2n}/b_n > 1, d_n \searrow 0, \liminf d_{n+1}/d_n > 0$. These conditions are satisfied in all results of this and following sections.

LEMMA 3 *Let X be a nonpositive random variable, $C \geq 0$ be a constant and the functions $D_u(x)$ and $D'_u(x)$ are defined by (3) for the random variables X and $X' = X + C$, respectively. Then*

$$D'_u(x) \leq (1+o(1))(D_u(x) + uP(X < -u)) + \frac{C^2}{x} \quad \text{and}$$

$$D'_u(x) \geq (1+o(1))D_u(x) + \frac{2CEX}{x},$$

as $x \to \infty$ and $u > x + C$.

Proof. As usually, put $F(x) = P(X < x)$. Then, for $u > x > C$,

$$D'_u(x) = \frac{1}{x}\int_{-x-C}^{0} (y+C)^2\, dF(y) + \int_{-u-C}^{-x-C} |y+C|\, dF(y) \tag{16}$$

$$= \frac{1}{x}\int_{-x-C}^{0} y^2\, dF(y) + \frac{2C}{x}\int_{-x-C}^{0} y\, dF(y) + \frac{C^2}{x}\int_{-x-C}^{0} dF(y) +$$

$$\int_{-u-C}^{-x-C} |y|\, dF(y) - C\int_{-u-C}^{-x-C} dF(y).$$

Therefore,

$$D'_u(x) \leq \frac{1}{x}\int_{-x-C}^{0} y^2\, dF(y) + \frac{C^2}{x} + \int_{-u-C}^{-x-C} |y|\, dF(y)$$

$$\leq \frac{(1+o(1))}{x+C}\int_{-x-C}^{0} y^2\, dF(y) + \int_{-u}^{-x-C} |y|\, dF(y) + \frac{C^2}{x} + (u+C)P(X < -u)$$

$$= (1+o(1))D_u(x+C) + \frac{C^2}{x} + (u+C)P(X < -u),$$

as $x \to \infty, u > x + C$. It gives the first inequality of Lemma 3, because $D_u(x+C) \leq D_u(x)$.

The second inequality becomes trivial when $EX = -\infty$. So, assume that EX is finite. By (16),

$$D'_u(x) \geq \frac{1}{x}\int_{-x-C}^{0} y^2\, dF(y) + \frac{2CEX}{x} + \int_{-u-C}^{-x-C} |y|\, dF(y) - C\int_{-u-C}^{-x-C} dF(y)$$

$$\geq \frac{1}{x+C}\int_{-x-C}^{0} y^2\, dF(y) + \frac{2CEX}{x} + (1+o(1))\int_{-u-C}^{-x-C} |y|\, dF(y)$$

$$= (1+o(1))D_{u+C}(x+C) + \frac{2CEX}{x} \geq (1+o(1))D_u(x+C) + \frac{2CEX}{x}.$$

as $x \to \infty, u > x + C$. The second inequality of Lemma 3 follows now from the inequality $(x+C)D_u(x+C) \geq xD_u(x)$.

Proof of Theorem 2. Suppose that (10) holds. By Lemma 1, $(S_n - nE(n^{1/t}))/n^{1/t} \xrightarrow{P} 0$ and therefore (11) holds. Lemma 2 implies (12). Let $X_n^+ = X_n I(X_n \geq 0), X_n^- = X_n I(X_n < 0), X^- = X_1^-, T_n = X_1^+ + ... + X_n^+, R_n = X_1^- + ... + X_n^-, E^+(x) = \int_0^x y\,dF(y)$ and $E^-(x) = \int_{-x}^0 y\,dF(y)$ for $x > 0$. By the result of C.Heyde and V.Rohatgi (1967), (12) yields the relation

$$P(|T_n - nE^+(n^{1/t})| > \varepsilon n^{1/t}) = o(M(n)n^{-d}) \quad \text{for any } \varepsilon > 0. \tag{17}$$

Since $med\,S_n$ can be replaced in (10) for $nE(n^{1/t})$, we have

$$P(R_n - nE^-(n^{1/t}) > \varepsilon n^{1/t}) = o(M(n)n^{-d}) \quad \text{for any } \varepsilon > 0. \tag{18}$$

Note that EX^- is finite (it follows from (11)). Applying Corollary 1 to the sequence of independent identically distributed random variables $\{X_n^- - EX_n^-\}$ yields the condition $nD'_{n^{1/t}}(n^{1/t}/\log n) = o(n^{1/t})$, where $D'_u(x)$ is constructed for $X^- - EX^-$ in the same way as $D_u(x)$ is constructed for X. Putting here n instead of $n^{1/t}$ and taking into account the monotonocity of functions $D'_u(x)$ and $xD'_u(x)$, we obtain $D'_n(n/\log n) = o(n^{1-t})$. By (11), $P(X < -n) = o(n^{-t})$ and therefore, by Lemma 3, (13) holds.

Now suppose that the conditions (11), (12) and (13) hold. Then $nP(|X| > n^{1/t}) \to 0$ and (5) holds by the weak law of large numbers. The result of C.Heyde and V.Rohatgi gives (17). Corollary 1 proves (18). The inequality (8) completes the proof of the first statement of Theorem 2. We omit the proof of the last statement which is essentially the same.

Proof of Corollary 3. A simple calculation proves that (14) implies (13). If (14) holds with O instead of o, put $P(X < -x) = x^{-t}(\log x)^{1-t}$ for all s.l. x. Then (13) fails and, by Theorem 2, (10) can not hold. If o is changed in (12) by O, put $P(X > x) = M(x^t)x^{-t(1+d)}$ for all s.l. x (here $M(x)$ is a nonincreasing continuation of $M(n)$ on $x : x > 0$). By Lemma 2, (10) can not hold.

Now we state the result for the case $d = 0$. The main difference between the cases $d > 0$ and $d = 0$ is that in the first case the argument of $D^-_{n^{1/t}}$ in (13) does not depend on d, while it depends on $M(x)$ in the second case.

THEOREM 3 *Let $M(x)$ be a positive, nonincreasing, slowly-varying at infinity function, $1 < t < 2$. Then*

$$P(S_n - med\,S_n > \varepsilon n^{1/t}) = o(M(n)) \quad \textit{for any } \varepsilon > 0, \tag{19}$$

if and only if the conditions (11),

$$P(X > n^{1/t}) = o(M(n)/n) \tag{20}$$

and

$$nD^-_{b_n}(b_n/\log(1/M(n))) = o(n^{1/t}) \tag{21}$$

are satisfied.

The relation (19) holds for some $\varepsilon > 0$ (instead of any $\varepsilon > 0$), if and only if the conditions (11) and (20) are satisfied and the condition (21) holds with O instead of o.

Proof is essentially the same as for Theorem 2.

4. Case of $b_n = n^{1/2}(\log n)^s$ and $b_n = n^{1/2}(\log\log n)^s, s \geq 0.5$. In this paper we do not state general results, applicable to general classes of b_n and d_n. The purpose of this section is to illustrate the possibilities of the method described in the previous section to state necessary and sufficient conditions for (1) with norming sequences which are more useful in the law of the iterated logarithm than in the strong law of large numbers.

THEOREM 4 *Let $s > 0.5, d > 0$. The relation*

$$P(S_n - med\, S_n > \varepsilon n^{1/2}(\log n)^s) = o(n^{-d}) \tag{22}$$

holds for any $\varepsilon > 0$, if and only if the following conditions are satisfied:

$$P(X < -n^{1/2}(\log n)^s) = o(1/n), \quad P(X > n^{1/2}(\log n)^s) = o(n^{-d-1}) \tag{23}$$

and

$$D_n^-(n/\log n) = o((\log n)^{2s}/n). \tag{24}$$

The relation (22) holds for some $\varepsilon > 0$, if and only if the conditions (23) are satisfied and the condition (24) holds with O instead of o.

The case $s = 0.5$ has some peculiarities.

THEOREM 5 *Let $d > 0$. The relation*

$$nP(S_n - med\, S_n > \varepsilon(n\log n)^{1/2}) = o(n^{-d}), \tag{25}$$

holds for some $\varepsilon > 0$ if and only if the following conditions are satisfied:

$$P(X < -(n\log n)^{1/2}) = o(1/n), \quad P(X > (n\log n)^{1/2}) = o(n^{-1-d}),$$

$$EX^2 < \infty \quad \text{and} \quad \int_{-n}^{-n/\log n} |x|\, dF(x) = O((\log n)/n).$$

If the relation (25) holds for any $\varepsilon > 0$, then the distribution of X is degenerate.

It is possible to prove that the least ε for which (25) holds, exists and is equal to $(2d\, var\, X)^{1/2}$. It is easy to prove also that (22) can not hold for a nondegenegate distribution if $s < 0.5$.

LEMMA 4 *Let $s > 0.5, d > 0$. Then*

$$P(|S_n - nE(n^{1/2}(\log n)^s)| > \varepsilon n^{1/2}(\log n)^s) = o(n^{-d}) \tag{26}$$

for any $\varepsilon > 0$, if and only if

$$nP(|X| > \varepsilon n^{1/2}(\log n)^s) = o(n^{-d}). \tag{27}$$

If (26) holds for some $\varepsilon > 0$, then it holds for any $\varepsilon > 0$.
If $d = 0.5, d > 0$ and (26) holds for some $\varepsilon > 0$, then (27) holds, $\sigma^2 = var X$ is finite and (26) holds for any $\varepsilon \geq (2d)^{1/2}\sigma$ and it does not hold for any $\varepsilon < (2d)^{1/2}\sigma$.

See also N.Amosova (1979). This, probably known, result follows immediately from Lemma 2 (necessity) and from the Rychlik's estimate (1983) in the central limit theorem which can be reformulated in the following way: for any sequence of independent identically distributed random variables $\{X_k\}_{k=1}^n$ with zero mean and finite positive variance σ^2, any $t > 2$, there exist constants C_1, C_2 and C_3 depending only on t and such that

$$|P(S_n < x) - \Phi(\frac{x}{n^{1/2}\sigma})|$$

$$\leq \frac{C_1}{|x|^t}\int_{-|x|}^{|x|} |y|^t\, dF(y) + nP(|X| > C_2|x|) \tag{28}$$

as $|x| \geq C_3 n^{1/2}\sigma$. Here $\Phi(z) = (2\pi)^{-1/2}\int_{-\infty}^{z} e^{-y^2/2}\, dy$.

Proof of sufficiency in Lemma 4. Clearly, (27) implies the finiteness of $\sigma^2 = varX$. Simple and routine calculation shows that the right-side part of (28) and

$$\Phi\Big(-\frac{|x|}{n^{1/2}\sigma_n}\Big) = 1 - \Phi\Big(\frac{|x|}{n^{1/2}\sigma_n}\Big) \tag{29}$$

have the order $o(n^{-d})$, if $s > 0.5, x = \pm\varepsilon n^{1/2}n^{1/t}, \varepsilon > 0, \sigma_n^2 = var(XI(|X| < |x|))$ and (27) holds. This implies (26). In the case $s = 0.5$ the same is true if $\varepsilon \geq (2d)^{1/2}\sigma_n \nearrow (2d)^{1/2}\sigma$, but the sequence from (29) tends to 0 slower than n^{-d} if $\varepsilon < (2d)^{1/2}\sigma$. This proves Lemma 3 also in the case $s = 0.5$.

Proof of Theorems 4 and 5 is the same (step by step) as of Theorem 2 with using Lemma 4 instead of the result of C.Heyde and V.Rohatgi.

THEOREM 6 *Let $s > 0.5, d > 1$. Then*

$$P(S_n - med\, S_n > \varepsilon n^{1/2}(\log\log n)^s) = o((\log n)^{-d})) \tag{30}$$

for any $\varepsilon > 0$, if and only if the conditions

$$nP(X < -n^{1/2}(\log\log n)^s) = o(1), \tag{31}$$

$$nP(X > n^{1/2}(\log\log n)^s) = o((\log n)^{-d})) \tag{32}$$

and

$$D_n^-(n/\log\log n) = o((\log\log n)^{2s}/n) \tag{33}$$

are satisfied.

The relation (30) holds for some $\varepsilon > 0$, if and only if the conditions (31) and (32) are satisfied and the condition (33) holds with O instead of o.

THEOREM 7 *Let $d > 1$. Then*

$$P(S_n - med\, S_n > \varepsilon(n\log\log n)^{1/2}) = o((\log n)^{-d})) \tag{34}$$

for some $\varepsilon > 0$, *if and only if the conditions (31) and (32) are satisfied with* $s = 0.5$. $EX^2 < \infty$ *and*

$$\int_{-n}^{-n/\log\log n} |x|\, dF(x) = o(\log\log n/n).$$

If (34) holds for any $\varepsilon > 0$, *then the distribution of* X *is degenerate.*

Proof of Theorems 6 and 7 is the same as of Theorems 4 and 5.

It is possible to prove that the least ε for which (34) holds, exists and is equal to $(2d\, var\, X)^{1/2}$. It is easy to show that (34) can not hold for a degenerate distribution if $s < 0.5$.

REFERENCES

AMOSOVA N.N. (1978) On a problem of the convergence rates in one-sided law of large numbers. *Izv. Vysshich. Uchebn. Zaved. ,N 10, 3-6* (In Russian).

AMOSOVA N.N. (1979) On probabilities of moderate deviations. *Zap. Nauch. Semin. LOMI,* **85**, *6-16.*

ESSEEN C.-G. (1968) On the concentration function of a sum of independent random variables. *Z. Wahrscheinlich. verw. Geb.,* **9**, *290-308.*

HEYDE C.C., ROHATGI V.K. (1967) A pair of complementary theorems on convergence rates in the law of large numbers. *Proc. Camb. Phil. Soc.,* **63**, *73-82.*

MARTIKAINEN A.I. (1982) One-sided variants of the law of large numbers, strong law and rates of convergence. *Proc. 16th All-Union School on Probab. Theory and Math. Statist., Bakuriani, Febr. 26 - March 5, 1982*: Metsniereba, Tbilisi, *45-61 (in Russian).*

MARTIKAINEN A.I. (1992) On convergence rates in one-sided law of large numbers. *Ann. Acad. Sci. Fenn., Ser. A. I. Mathem.* **17**, *81-84.*

MARTIKAINEN A.I. (1993) General one-sided laws of the iterated logarithm for random walks. *Preprint, Philipps-Universität Marburg, Reihe Mathematik, N 33.*

PETROV V.V. Sums of independent random variables. Berlin - Heidelberg - New-York: Springer - Verlag.

PETROV V.V., SHIROKOVA I.V. (1973) On exponential rate of convergence in the law of large numbers. *Vestnik Leningrad Univ. Mat. Mekh. Astronom., 7, 155-157.*

RYCHLIK Z. (1983) Nonuniform central limit bounds with applications to probability of deviations. *Theory Probab. Appl.,* **28**, *646-652.*

ASYMPTOTIC BEHAVIOUR OF ONE-STEP-M-ESTIMATORS IN CONTAMINATED NON-LINEAR MODELS

Christine H. Müller
Freie Universität Berlin
1. Mathematisches Institut, Arnimallee 2-6, D-14195 Berlin-Dahlem

Abstract

Extending the approach of Bickel (1975, 1981, 1984) and Rieder (1985, 1987) the asymptotic behaviour of one-step-M-estimators for θ is investigated for non-linear models $Y(t) = \mu(\theta, t) + Z(t)$ where μ is a non-linear function in θ and the errors $Z(t)$ may have different contaminated normal distributions for different experimental conditions t. These models are also called conditionally contaminated non-linear models. For these models it is shown that the one-step-M-estimators have an asymptotic bias which depends on θ as the asymptotic covariance matrix depends on θ. Using the results of Kurotschka and Müller (1992) and Müller (1992a) locally optimal robust one-step-M-estimators and corresponding optimal designs are derived by minimizing the trace of the asymptotic covariance matrix under the side condition that the asymptotic bias is bounded by some bias bound. The locally optimal solutions are appropriate to efficiency comparisons.

Key words: One-step-M-estimators, conditionally contaminated non-linear models, asymptotic normal distribution, asymptotic bias, robust estimation, locally optimal estimators, locally optimal designs, efficiency.

AMS 1980 subject classifications: 62F12, 62J02, 62F35.

1 Introduction

A general non-linear model

$$Y_{nN} = \mu(\theta, t_{nN}) + Z_{nN}, \qquad n = 1, ..., N$$

is considered, where Y_{nN} are observations, $t_{nN} \in T$ are experimental conditions, $\theta \in \Theta \subset \mathbb{R}^p$ is an unknown parameter vector, $\mu : \Theta \times T \to \mathbb{R}$ is a known non-linear response function, Z_{nN} are error variables. In classical linear models it is assumed that the error variables $Z_{1N}, ..., Z_{NN}$ are independent and identically distributed and often it is assumed that they are normally distributed with mean 0 and variances $\sigma(t_{nN})^2$, i.e.

$$(1.1) \qquad \frac{1}{\sigma(t_{nN})} Z_{nN} \sim P := n_{(0,1)}.$$

But if some outlying observations (gross errors) may appear the normal distribution is not correct. Then even for designed experiments a conditionally contaminated non-linear model is adequate. This conditionally contaminated non-linear model is a generalization of the conditionally contaminated linear model which for example is regarded in Bickel (1981, 1984), Rieder (1985, 1987), Kurotschka and Müller (1992) and Müller (1992b) for describing outlying observations in linear models. In such a model it is assumed that the error variables $Z_{1N}, ..., Z_{NN}$ are independent and distributed according to a contaminated normal distribution where the contamination may be different for different experimental conditions. I.e.

$$\frac{1}{\sigma(t_{nN})} Z_{nN} \sim Q_{nN}(dz) := (1 - R\,N^{-1/2}\,\epsilon(t_{nN}))\; n_{(0,1)}(dz) \\ + R\,N^{-1/2}\,\epsilon(t_{nN})\; q(z, t_{nN})\; n_{(0,1)}(dz)$$

where $R > 0$, $\sum_{n=1}^{N} \epsilon(t_{nN}) \leq N$ for almost all $N \in I\!N$, $\int q(z,t)\, n_{(0,1)}(dz) = 1$, $\|q\|_\infty < \infty$, $q(z,t) \geq 0$ for all $z \in I\!R$, $t \in T$. Thereby the markov kernel $q(\cdot, t)n_{(0,1)}$ models the form and $\epsilon(t) \geq 0$ the proportion of contamination. The set $\mathcal{U}_R$ of all sequences $(Q^N_{\epsilon,q} := \bigotimes_{n=1}^{N} Q_{nN})_{N \in I\!N}$ defines a conditional contamination neighbourhood around the classical model $(P^N)_{N\in I\!N}$. If θ is the true parameter then $Q^N_{\epsilon,q,\theta}$ denotes the distribution of $Y_N := (Y_{nN}, ..., Y_{NN})$ and P^N_θ denotes the distribution of Y_N for the ideal model (1.1).

To estimate θ in this model we propose in Section 2 one-step-M-estimators. These estimators are similar to that proposed in Stefanski et al. (1986) for a special score function and random carriers. While Stefanski et al. only mentioned the asymptotic normality of the estimators at the ideal model (1.1) we derive in Section 3 their asymptotic behaviour at the whole conditionally contaminated model. In particular we show that the estimators have an asymptotic bias which depends on the unknown parameter θ. In Section 4 we define robust estimators as those with finite maximum bias at the conditionally contaminated model. Because also the asymptotic covariance matrix depends on θ it is impossible to derive globally optimal robust estimators as it is the case in linear models. But the optimality results for linear models given by Hampel (1978), Krasker (1980), Rieder (1985), Kurotschka and Müller (1992) and Müller (1992a) can be used to derive locally optimal robust estimators and locally optimal designs for robust estimation. In particular it is shown that the locally optimal designs for robust estimation coincide with that for non-robust estimation based on the least squares estimator. Moreover the locally optimal robust estimators at the locally optimal designs have a very simple form so that they can be easily used for efficiency comparisons. In Section 5 the proof of the main result, the asymptotic linearity of the one-step-M-estimators, is given.

2 One-step-M-estimators

M-estimators are often used for robust estimation in linear model (see for example Huber (1981) and Hampel et al. (1986)). For non-linear models they can be generalized in a straightforward way as estimators $\widehat{\theta}_N$ which satisfy

$$\widehat{\theta}_N = \arg\min_{\theta\in\Theta} \sum_{n=1}^N \rho(y_{nN} - \mu(\theta, t_{nN}), t_{nN}),$$

where $\rho : I\!\!R \times T \to I\!\!R$ is a given function (see also Stefanski et al. (1986)). In particular if $\rho(z,t) = z^2$ then the M-estimator is the classical least squares estimator. A necessary condition for $\widehat{\theta}_N$ being a M-estimator is the condition that $\widehat{\theta}_N$ is a solution of

$$\text{(2.1)} \qquad \sum_{n=1}^N \eta(y_{nN} - \mu(\theta, t_{nN}), t_{nN})\, \dot{\mu}(\theta, t_{nN}) \;=\; 0.$$

Thereby $\eta : I\!\!R \times T \to I\!\!R$ and $\dot{\mu} : \Theta \times T \to I\!\!R^p$ are given by $\eta(z,t) := \frac{\partial}{\partial z}\rho(z,t)$ and $\dot{\mu}(\theta,t) := (\frac{\partial}{\partial\theta}\mu(\theta,t))'$, respectively, (where M' denotes the transpose of the matrix M). Solutions of (2.1) can be calculated by Newton-Raphson iterations. Starting with an initial estimator $\widehat{\theta}_N^0$ the first step of the Newton-Raphson iteration provides

$$\widehat{\theta}_N = \widehat{\theta}_N^0 \;-\; \left(\tfrac{\partial}{\partial\theta}(\textstyle\sum_{n=1}^N \eta(y_{nN} - \mu(\theta, t_{nN}), t_{nN})\, \dot{\mu}(\theta, t_{nN}))/_{\theta=\widehat{\theta}_N^0}\right)^{-1}$$
$$\cdot \textstyle\sum_{n=1}^N \eta(y_{nN} - \mu(\widehat{\theta}_N^0, t_{nN}), t_{nN})\, \dot{\mu}(\widehat{\theta}_N^0, t_{nN}).$$

Now we can also use the result of the first Newton-Raphson iteration as estimate. If we assume that

(2.2) all experimental conditions t_{nN} are lying in a finite set $T_0 \subset T$

and that the designs $d_N := (t_{1N}, ..., t_{NN})$ are converging to an asymptotic design measure δ in the following sense

$$\text{(2.3)} \qquad \lim_{N\to\infty} \frac{1}{N} \sum_{n=1}^N 1_{\{t\}}(t_{nN}) = \delta(\{t\}) \text{ for all } t \in T_0$$

then we have under some regularity conditions for the ideal model (1.1)

$$\lim_{N\to\infty} \frac{\partial}{\partial\theta}\left(\frac{1}{N}\sum_{n=1}^N \eta(y_{nN} - \mu(\theta, t_{nN}), t_{nN})\, \dot{\mu}(\theta, t_{nN})\right) \Big/_{\theta=\widehat{\theta}_N^0} \;=\; -M(\widehat{\theta}_N^0, \eta)$$

where

$$M(\theta, \eta) := \int \eta(z,t)\, z\, \dot{\mu}(\theta,t)\, \dot{\mu}(\theta,t)'\, P(dz)\, \delta(dt).$$

Taking into account that the variances $\sigma^2(t)$ may be unknown we now define one-step-M-estimators as follows:

Definition 2.1 *An estimator $\widehat{\theta}_N : I\!\!R^N \times T^N \to I\!\!R^p$ is called an one-step-M-estimator for θ with score function $\eta : I\!\!R \times T \to I\!\!R$, initial estimator $\widehat{\theta}_N^0 : I\!\!R^N \times T^N \to I\!\!R^p$ and variance estimators $\widehat{\sigma}_N^2(t) : I\!\!R^N \times T^N \to I\!\!R^+$ for $\sigma^2(t)$, $t \in T$, iff*

$$\begin{aligned}\widehat{\theta}_N \;&=\; \widehat{\theta}_N^0 \\ &+\; \frac{1}{N}\sum_{n=1}^N M(\widehat{\theta}_N^0, \eta)^{-1}\, \dot{\mu}(\widehat{\theta}_N^0, t_{nN})\, \eta\left(\tfrac{y_{nN} - \mu(\widehat{\theta}_N^0, t_{nN})}{\widehat{\sigma}_N(t_{nN})}, t_{nN}\right) \widehat{\sigma}_N(t_{nN}).\end{aligned}$$

3 Asymptotic normal distribution

To derive the asymptotic linearity and thus the asymptotic normality of the one-step-M-estimators we make the following assumptions:

(3.1) $\sqrt{N}(\widehat{\theta}_N^0 - \theta)$ is tight at the ideal model (1.1).

(3.2) $\sqrt{N}(\widehat{\sigma}_N(t) - \sigma(t))$ is tight at the ideal model (1.1) for all $t \in T_0$.

(3.3) For every $\theta_0 \in \Theta$ there exists k_0 so that
$\{\theta \in \Theta;\ |\theta - \theta_0| \leq k_0\}$ is a compact subset of $\mathbb{R}^p$
(For example Θ is an open subset of $\mathbb{R}^p$).

(3.4) $\mu(\cdot, t)$ has a continuous second order derivative on Θ for all $t \in T_0$.

(3.5) $M(\theta, \eta)$ is regular for all $\theta \in \Theta$.

(3.6) $\eta(z,t) = \eta_0(z,t) + c(t)\text{sgn}(z)$ where for all $t \in T_0$
$\eta_0(\cdot, t)$ is antisymmetric, continuous and there exists
$l_1(t), ..., l_L(t)$ so that $\eta_0(\cdot, t)$ has bounded
and continuous derivatives on $\mathbb{R} \setminus \{l_1(t), ..., l_L(t)\}$.

Condition (3.6) can be generalized (see Section 5 and Müller (1992b)) but in this form it is easy to check. Moreover it includes all relevant score functions as the sign function $\eta(z,t) = sgn(z)$ for L_1-estimation, $\eta(z,t) = z$ for least squares estimation and redescending score functions. Under some regularity conditions on Θ or μ the condition (3.1) is in particular satisfied for the least squares estimator (see for example Jennrich (1969) or Läuter (1989)). Variance estimators which satisfy condition (3.2) are for example

$$\widehat{\sigma}_N^2(t)(y_N, d_N) = \frac{1}{N_t(d_N)} \sum_{n=1}^N 1_{\{t\}}(t_{nN})(y_{nN} - \bar{y}_t(y_N, d_N))^2$$

where $N_t(d_N) = \sum_{n=1}^N 1_{\{t\}}(t_{nN})$ and $\bar{y}_t(y_N, d_N) = \frac{1}{N_t(d_N)} \sum_{n=1}^N 1_{\{t\}}(t_{nN}) y_{nN}$.

Theorem 3.1 *Under assumptions (2.2), (2.3) and (3.1)-(3.6) an one-step-M-estimator with score function η is asymptotically linear, i.e.*

$$P_\theta^N \left(|\sqrt{N}(\widehat{\theta}_N - \theta - \tfrac{1}{N} \sum_{n=1}^N M(\theta,\eta)^{-1} \dot{\mu}(\theta, t_{nN})\, \eta(\tfrac{y_{nN} - \mu(\theta, t_{nN})}{\sigma(t_{nN})}, t_{nN}) \sigma(t_{nN}))| > \epsilon \right) \overset{N\to\infty}{\longrightarrow} 0$$

for all $\epsilon > 0$ and all $\theta \in \Theta$ at the ideal model (1.1).

The proof of this theorem is given in Section 5. The asymptotic linearity provides that the one-step-M-estimators are asymptotically normally distributed at the conditionally contaminated model. For that put

$$V_\theta(\eta, \delta) := M(\theta, \eta)^{-1} \int \dot{\mu}(\theta, t)\, \dot{\mu}(\theta, t)' \eta(z, t)^2\, P(dz)\, \sigma(t)^2\, \delta(dt)\ M(\theta, \eta)^{-1}$$

and

$$b_\theta(\eta,\delta,\epsilon,q) := R\, M(\theta,\eta)^{-1} \int \dot\mu(\theta,t)\,\eta(z,t)\,\epsilon(t)\,q(z,t)\,P(dz)\,\sigma(t)\delta(dt).$$

Theorem 3.2 *Under assumptions (2.2), (2.3) and (3.1)-(3.6) an one-step-M-estimator with score function η is asymptotically normally distributed at the conditionally contaminated non-linear model, i.e.*

$$\mathcal{L}(\sqrt{N}(\widehat\theta_N-\theta)|Q^N_{\epsilon,q,\theta}) \stackrel{N\to\infty}{\longrightarrow} \mathcal{N}(b_\theta(\eta,\delta,\epsilon,q), V_\theta(\eta,\delta))$$

for all $(Q^N_{\epsilon,q})_{N\in\mathbb{N}} \in \mathcal{U}_\mathcal{R}$ and $\theta\in\Theta$
with maximum asymptotic bias

$$\sup\{b_\theta(\eta,\delta,\epsilon,q);\ (Q^N_{\epsilon,q})_{N\in\mathbb{N}}\in\mathcal{U}_\mathcal{R}\} \;=\; R\sup_{z,t}|M(\theta,\eta)^{-1}\,\dot\mu(\theta,t)\,\eta(z,t)\,\sigma(t)|.$$

The proof of Theorem 3.2 is the same as for linear models (see for example Bickel (1981), Rieder (1985, 1987) and Müller (1992b)).

4 Locally optimal robust estimators and designs

Theorem 3.2 shows that one-step-M-estimators have an asymptotic bias at the conditionally contaminated non-linear model. For robust estimators the asymptotic bias should be bounded, i.e. the maximum asymptotic bias should be finite. This is only possible if the score function η is bounded.

Now assume $\sigma(t)=1$ for all $t\in T$ and without loss of generality $R=1$. Then in conditionally contaminated linear models optimal robust estimators are derived by minimizing the trace of the asymptotic covariance matrix under the side condition that the maximum asymptotic bias is bounded by some bias bound b (see Hampel (1978), Krasker (1980), Bickel (1981, 1984), Rieder (1985, 1987), Kurotschka and Müller (1992)). But in non-linear models the maximum asymptotic bias as well as the asymptotic covariance matrix depend on the unknown parameter vector. Hence only locally optimal robust one-step-M-estimators $\widehat\theta^\theta_N$ can be defined as follows: The one-step-M-estimators $\widehat\theta^\theta_N$ at the design δ is a locally optimal robust one-step-M-estimator with bias bound b if its score function $\eta_{\theta,b,\delta}$ satisfies

$$\begin{aligned}\eta_{\theta,b,\delta} = \arg\min\{&\operatorname{tr} V_\theta(\eta,\delta);\ \eta \text{ satisfies condition (3.6) and}\\ &\sup\nolimits_{z,t}|M(\theta,\eta)^{-1}\,\dot\mu(\theta,t)\,\eta(z,t)| \le b\}.\end{aligned}$$

The locally optimal robust estimators at θ can be derived as the optimal robust estimators in linear models by regarding the response function $\overline{\mu}(\beta,t) = \dot\mu(\theta,t)'\beta$ which is linear in β.

For linear models also optimal designs for robust estimation with bias bound b were defined and characterized (see Müller (1992a)). Similarly we can define for non-linear models locally optimal designs $\delta_{\theta,b}$ for robust estimation with local bias bound b as follows:

$$\delta_{\theta,b} = \arg\min\{\mathrm{tr}\, V_\theta(\eta_{\theta,b,\delta},\delta);\ \delta \in \Delta\}$$

where Δ is some given sets of designs. Again these locally optimal designs can be derived as the optimal designs in linear models by regarding the linear function $\overline{\mu}(\beta,t) = \dot{\mu}(\theta,t)'\beta$ so that the results of Müller (1992a) concerning optimal designs for robust estimation in linear models can be used. In particular if for Δ and $\dot{\mu}(\theta,\cdot)$ the equivalence theorem for linear optimality (see Federov (1972), p. 125) holds then the locally optimal designs do not depend on the bias bound b and are given by

$$\delta_{\theta,b} = \delta_\theta = \arg\min\{\mathrm{tr}\, I_\theta(\delta)^{-1};\ \delta \in \Delta\}$$

where $I_\theta(\delta)$ is the local information matrix given by $I_\theta(\delta) := \int \dot{\mu}(\theta,t)\,\dot{\mu}(\theta,t)'\,\delta(dt)$. Hence the locally optimal designs for robust estimation coincide with the locally optimal designs for classical non-robust estimation (see for example Ford et al. (1989)).

Moreover the score functions of the locally optimal robust estimators at the locally optimal designs have a very simple form, namely (see Müller (1992a)):

$$\eta_{\theta,b}(z,t) := \eta_{\theta,b,\delta_\theta}(z,t) = \begin{cases} \mathrm{sgn}(z)\sqrt{\frac{\pi}{2}} & \text{for } b = \sqrt{\mathrm{tr}\, I_\theta(\delta_\theta)^{-1}\,\frac{\pi}{2}} \\ \mathrm{sgn}(z)\,\frac{\min\{|z|,b\,y_b\}}{y_b\sqrt{\mathrm{tr}\, I_\theta(\delta_\theta)^{-1}}} & \text{for } b > \sqrt{\mathrm{tr}\, I_\theta(\delta_\theta)^{-1}\,\frac{\pi}{2}} \end{cases}$$

where y_b is a positive solution of $y\sqrt{\mathrm{tr}\, I_\theta(\delta_\theta)^{-1}} = (2\Phi(by) - 1)$ (Φ denoting the distribution function of the standard normal distribution). Note $M(\theta,\eta_{\theta,b}) = I_\theta(\delta_\theta)$.

The locally optimal robust estimators at the locally optimal designs can be used for efficiency comparisons as Ford et al. (1989) described for designing experiments for classical non-robust estimation in non-linear models. Given an one-step-M-estimator with score function η at a design δ its efficiency is given by

$$E_\theta(\eta,\delta) := \frac{\mathrm{tr}\, V_\theta(\eta_{\theta,b(\theta)},\delta_\theta)}{\mathrm{tr}\, V_\theta(\eta,\delta)} \in [0,1],$$

where the local bias bound $b(\theta) := b(\theta,\eta,\delta) := \sup_{z,t} |M(\theta,\eta)^{-1}\,\dot{\mu}(\theta,t)\,\eta(z,t)|$ should be used to make fair comparisons.

5 Proof of the asymptotic linearity

Set

$$m(\theta,t) := M(\theta,\eta)^{-1}\,\dot{\mu}(\theta,t), \quad \lambda(b,s,t,\sigma) := \int \eta\left(\tfrac{z\sigma - b}{s},t\right) s\,P(dz),$$
$$\rho(\theta,s,t,\theta_0,\sigma) := m(\theta,t)\,\lambda(\mu(\theta,t) - \mu(\theta_0,t),s,t,\sigma).$$

Lemma 5.1 *Under assumptions (2.2) and (3.3)-(3.6) we have for all $t \in T_0$:*
(i) $m(\cdot,t)$ is continuously differentiable on Θ.
(ii) $\rho(\cdot,\cdot,t,\theta_0,\sigma)$ is continuously differentiable in a neighbourhood of $(\theta,s)=(\theta_0,\sigma)$ where

$$\frac{\partial}{\partial(\theta,s)}\rho(\theta,s,t,\theta_0,\sigma)/_{(\theta,s)=(\theta_0,\sigma)}$$
$$=(-M(\theta_0,\eta)^{-1}\dot\mu(\theta_0,t)\,\dot\mu(\theta_0,t)'\int\eta(z,t)\,z\,P(dz)\,|\,0_{p\times 1})$$

($0_{p\times1}$ denoting the zero vector in $\mathbb{R}^p$).

Proof.

(i) Because T_0 is finite and $\mu(\cdot,t)$ has continuous second order derivative, $M(\cdot,\eta)$ is continuously differentiable on Θ. Because $M(\theta,\eta)$ is regular the assertion follows.

(ii) According to Lemma 4.1 in Müller (1992b) $\lambda(\cdot,\cdot,t,\sigma)$ is continuously differentiable in a neighbourhood of $(b,s)=(0,\sigma)$ where

$$\frac{\partial}{\partial(b,\sigma)}\lambda(b,s,t,\sigma)\Big/_{(b,s)=(0,\sigma)}=\left(-\int\eta(z,t)\,z\,P(dz)\,\Big|\,0\right).$$

This implies

$$\frac{\partial}{\partial(\theta,\sigma)}\lambda(\mu(\theta,t)-\mu(\theta_0,t),s,t,\sigma)\Big/_{(\theta,s)=(\theta_0,\sigma)}=\left(-\dot\mu(\theta_0,t)'\int\eta(z,t)\,z\,P(dz)\,\Big|\,0\right).$$

Because $\eta(\cdot,t)$ is antisymmetric we get

$$\begin{aligned}\frac{\partial}{\partial(\theta,s)}\rho(\theta,s,t,\theta_0,\sigma)\Big/_{(\theta,s)=(\theta_0,\sigma)}&=\frac{\partial}{\partial(\theta,\sigma)}m(\theta,t)\Big/_{(\theta,s)=(\theta_0,\sigma)}\lambda(0,\sigma,t,\sigma)\\&\quad+M(\theta_0,\eta)^{-1}\dot\mu(\theta_0,t)\,\frac{\partial}{\partial(\theta,\sigma)}\lambda\left(\mu(\theta,t)-\mu(\theta_0,t),s,t,\sigma\right)\Big/_{(\theta,s)=(\theta_0,\sigma)}\\&=\left(-M(\theta_0,\eta)^{-1}\dot\mu(\theta_0,t)\,\dot\mu(\theta_0,t)'\int\eta(z,t)\,z\,P(dz)\,\Big|\,0_{p\times1}\right).\square\end{aligned}$$

Proof of Theorem 3.1. Set for given θ_0 and $\sigma^2: T\to\mathbb{R}^+$

$$\begin{aligned}\eta_{nNt}(\theta,s)&:=\eta\left(\frac{y_{nN}-\mu(\theta,t)}{s},t\right)s,\\\kappa_t(\theta,s)&:=\lambda(\mu(\theta,t)-\mu(\theta_0,t),s,t,\theta_0,\sigma(t)),\\\rho_t(\theta,s)&:=\rho(\theta,s,t,\theta_0,\sigma(t))\end{aligned}$$

and

$$W(t):=M(\theta_0,\eta)^{-1}\dot\mu(\theta_0,t)\,\dot\mu(\theta_0,t)'\int\eta(z,t)\,z\,P(dz).$$

Then we have

$$\left|\sqrt{N}\left(\widehat{\theta}_N-\theta_0-\frac{1}{N}\sum\nolimits_{n=1}^{N} m(\theta_0,t_{nN})\,\eta\left(\frac{y_{nN}-\mu(\theta_0,t_{nN})}{\sigma(t_{nN})},t_{nN}\right)\sigma(t_{nN})\right)\right|$$

$$\begin{aligned}
\leq\ & \sum\nolimits_{t\in T_0}\Big|\sqrt{N}\,W(t)\,\delta(\{t\})\,(\widehat{\theta}_N^0-\theta_0)\\
&\quad+\frac{1}{\sqrt{N}}\sum\nolimits_{n=1}^{N} 1_{\{t\}}(t_{nN})\left(\rho_t(\widehat{\theta}_N^0,\widehat{\sigma}_N(t))-\rho_t(\theta_0,\sigma(t))\right)\Big|\\
&+\sum\nolimits_{t\in T_0}\frac{1}{\sqrt{N}}\sum\nolimits_{n=1}^{N} 1_{\{t\}}(t_{nN})\\
&\quad\cdot|m(\widehat{\theta}_N^0,t)\eta_{nNt}(\widehat{\theta}_N^0,\widehat{\sigma}_N(t))-m(\widehat{\theta}_N^0,t)\kappa_t(\widehat{\theta}_N^0,\widehat{\sigma}_N(t))\\
&\quad+m(\widehat{\theta}_N^0,t)\eta_{nNt}(\theta_0,\sigma(t))-m(\widehat{\theta}_N^0,t)\kappa_t(\theta_0,\sigma(t))\\
&\quad-m(\widehat{\theta}_N^0,t)\eta_{nNt}(\theta_0,\sigma(t))+m(\widehat{\theta}_N^0,t)\kappa_t(\theta_0,\sigma(t))\\
&\quad-m(\theta_0,t)\eta_{nNt}(\theta_0,\sigma(t))+m(\theta_0,t)\kappa_t(\theta_0,\sigma(t))|
\end{aligned}$$

$$\leq \sum\nolimits_{t\in T_0}\left|W(t)\sqrt{N}(\widehat{\theta}_N^0-\theta_0)\left(\delta(\{t\})-\frac{1}{N}\sum\nolimits_{n=1}^{N}1_{\{t\}}(t_{nN})\right)\right| \tag{5.1}$$

$$+\sum\nolimits_{t\in T_0}\left(\frac{1}{N}\sum\nolimits_{n=1}^{N}1_{\{t\}}(t_{nN})\right)\cdot\sqrt{N}\left|W(t)(\widehat{\theta}_N^0-\theta_0)+\rho_t(\widehat{\theta}_N^0,\widehat{\sigma}_N(t))-\rho_t(\theta_0,\sigma(t))\right| \tag{5.2}$$

$$+\sum\nolimits_{t\in T_0}|m(\widehat{\theta}_N^0,t)|\frac{1}{\sqrt{N}}\sum\nolimits_{n=1}^{N}1_{\{t\}}(t_{nN})\cdot\left|\eta_{nNt}(\widehat{\theta}_N^0,\widehat{\sigma}_N(t))-\kappa_t(\widehat{\theta}_N^0,\widehat{\sigma}_N(t))-\eta_{nNt}(\theta_0,\sigma(t))+\kappa_t(\theta_0,\sigma(t))\right| \tag{5.3}$$

$$+\sum\nolimits_{t\in T_0}|m(\widehat{\theta}_N^0,t)-m(\theta_0,t)|\sqrt{N}\cdot\left|\frac{1}{N}\sum\nolimits_{n=1}^{N}1_{\{t\}}(t_{nN})\left(\eta_{nNt}(\theta_0,\sigma(t))-\kappa_t(\theta_0,\sigma(t))\right)\right|. \tag{5.4}$$

Because of Lemma 5.1 (i) as in the proof of Lemma 4.2 in Müller (1992b) term (5.1) and term (5.2) converge to zero. Because $\sqrt{N}(\widehat{\theta}_N^0-\theta_0)$ is tight there exist for every γ a $k(\gamma)\in\mathbb{R}$ with $P_{\theta_0}^N(|\widehat{\theta}_N^0-\theta_0|>\frac{k(\gamma)}{\sqrt{N}})<\gamma$ for all $N\in\mathbb{N}$. Because of condition (3.3) for sufficient large N, $N\geq N_\gamma$ say, the set $C_N=\{\theta\in\Theta;|\theta-\theta_0|\leq\frac{k(\gamma)}{\sqrt{N}}\}$ is a compact subset of $\mathbb{R}^p$ so that according to Lemma 5.1 (ii) there exists $K\in\mathbb{R}$ with

$$\sup_{\theta\in C_N}|m(\theta,t)|\leq K$$

and

$$\sup_{\theta\in C_N}\frac{|m(\theta,t)-m(\theta_0,t)|}{|\theta-\theta_0|}\leq K.$$

Hence term (5.4) converges to zero. For the convergence of term (5.3) it is sufficient to show the assertion separately for the two monotonous parts of $\eta(\cdot,t)$ and for the finite number of different experimental conditions t. I.e. we can assume without loss of generality that η is monotonously increasing and that we have only one experimental condition so that t is dropped form μ, η and σ. Then the assertion follows as in

Bickel (1975) or as in the proof of Lemma 4.2 in Müller (1992b). In particular the set C_N is discretized by $\theta_N^1, ..., \theta_N^{L(\gamma)}$ so that for every $\theta \in C_N$ there exists a θ_N^l with $|\theta_N^l - \theta| \leq \frac{\gamma}{\sqrt{N}}$. If we define

$$\nu_{nN}(b,h,s) := \eta\left(\frac{y_{nN} - \mu(\theta_0) + b + h}{\sigma + s}\right)(\sigma + s),$$

$$b_N^l := \mu(\theta_0) - \mu(\theta_N^l), \quad \text{where } |b_N^l| \leq \sup_{\theta \in C_N} |\dot{\mu}(\theta)| \frac{k(\gamma)}{\sqrt{N}}$$

$$h_N := \sup_{\theta \in C_N} |\dot{\mu}(\theta)| \frac{\gamma}{\sqrt{N}},$$

then we get for every $\theta \in C_N$ and every $N \geq N_\gamma$

$$\begin{aligned}
&|\nu_{nN}(b_N^l, 0, s) - \nu_{nN}(\mu(\theta_0) - \mu(\theta), 0, s)| \\
= & \left|\eta\left(\frac{y_{nN} - \mu(\theta_N^l)}{(\sigma + s)}\right)(\sigma + s) - \eta\left(\frac{y_{nN} - \mu(\theta_N^l) - \dot{\mu}(\overline{\theta})(\theta - \theta_N^l)}{(\sigma + s)}\right)(\sigma + s)\right| \\
\leq & \left|\eta\left(\frac{y_{nN} - \mu(\theta_N^l) + h_N}{(\sigma + s)}\right)(\sigma + s) - \eta\left(\frac{y_{nN} - \mu(\theta_N^l) - h_N}{(\sigma + s)}\right)(\sigma + s)\right| \\
= & |\nu_{nN}(b_N^l, h_N, s) - \nu_{nN}(b_N^l, -h_N, s)|.
\end{aligned}$$

With this property the rest of the proof is completely the same as in the proof of Lemma 4.2 in Müller (1992b). □

REFERENCES

[1] BICKEL, P.J. (1975). One-step Huber estimates in the linear model. *J. Amer. Statist. Assoc.* **70**, 428-434.

[2] BICKEL, P.J. (1981). Quelque aspects de la statistique robuste. In École d'Été de Probabilités de St. Flour. *Springer Lecture Notes in Math.* **876**, 1-72.

[3] BICKEL, P.J. (1984). Robust regression based on infinitesimal neighbourhoods. *Ann. Statist.* **12**, 1349-1368.

[4] FEDEROV, V.V. (1972). *Theory of Optimal Experiments.* Academic Press, New York.

[5] FORD, I., KITSOS, C.P. and TITTERINGTON, D.M. (1989). Recent advances in nonlinear experimental design. *Technometrics* **31**, 49-60.

[6] HAMPEL, F.R. (1978). Optimally bounding the gross-error-sensitivity and the influence of position in factor space. *Proceedings of the ASA Statistical Computing Section*, ASA, Washington, D.C., 59-64.

[7] HAMPEL, F.R., RONCHETTI, E.M., ROUSSEEUW, P.J. and STAHEL, W.A. (1986). *Robust Statistics - The Approach Based on Influence Functions.* John Wiley, New York.

[8] HUBER, P.J. (1981). *Robust Statistics.* John Wiley, New York.

[9] JENNRICH, R.I. (1969). Asymptotic properties of non-linear least squares estimators. *Ann. Math. Statist.* **40**, 633-643.

[10] KRASKER, W.S. (1980). Estimation in linear regression models with disparate data points. *Econometrica* **48**, 1333-1346.

[11] KUROTSCHKA, V. and MÜLLER, Ch.H. (1992). Optimum robust estimation of linear aspects in conditionally contaminated linear models. *Ann. Statist.* **20**, 331-350.

[12] LÄUTER, H. (1989). Note on the strong consistency of the least squares estimator in nonlinear regression. *Statistics* **20**, 199-210.

[13] MÜLLER, Ch.H. (1992a). Optimal designs for robust estimation in conditionally contaminated linear models. *To appear in J. Statist. Plann. Inference.*

[14] MÜLLER, Ch.H. (1992b). One-step-M-estimators in conditionally contaminated linear models. *Preprint No. A-92-11*, Freie Universität Berlin, Fachbereich Mathematik. *Submitted to Stat. Decis.*

[15] RIEDER, H. (1985). Robust estimation of functionals. *Technical Report.* Universität Bayreuth.

[16] RIEDER, H. (1987). Robust regression estimators and their least favorable contamination curves. *Stat. Decis.* **5**, 307-336.

[17] STEFANSKI, L.A., CARROLL, R.J. and RUPPERT, D. (1986). Optimally bounded score functions for generalized linear models with applications to logistic regression. *Biometrika 73*, 413-424.

CONDITIONAL RANK TESTS FOR THE TWO-SAMPLE PROBLEM WITH PARTIALLY OBSERVED DATA

Georg Neuhaus, University of Hamburg
Inst.f.Math.Stochastics, Bundesstr. 55, D-20146 Hamburg

Abstract

In the two-sample random censorship model with the additional complication that the minimum variable is observable only for the uncensored data we develop asymptotically optimal conditional rank tests for testing the null hypothesis of randomness H_0 being finite sample distribution free under H_0.

AMS 1991 subject classifications. $62G20, 62G10$

Key words. Censoring, partially observed data, conditioning, finite sample validity, asymptotic normality and optimality.

1. Introduction

In the present paper we deal with the two-sample problem for partially observed data in the following sense. Observations are made on n_k individuals from population $k (k = 1,2)$ and all $n := n_1 + n_2$ observations are independent ($a := b$ or $b := a$ means that a equals b by definition). The i'th subject has independent latent survival and censoring times X_i^0 and U_i with continuous distribution functions (df's). Let $F_1[F_2]$ denote the df of $X_i^0, 1 \leq i \leq n_1 [n_1 + 1 \leq i \leq n]$, and let all U_i have the same df G (equal censoring). The observable random variables (rv's) are $\Delta_i = 1\{X_i^0 \leq U_i\}$, and, if $\Delta_i = 1$, additionally X_i^0, where $1\{E\}$ is 1 if the event E occurs and 0 otherwise. Therefore, X_i^0 is only observed, if $X_i^0 \leq U_i$, while if $X_i^0 > U_i$, one knows just that fact, but nothing more.

As an example, see (Suzuki (1985)), let X_i^0 be the mileage of a car at the time of its first failure and U_i the mileage that the car will have run during the warranty period. If the car fails during the warranty period, its owner will ask the manufacturer to repair it under contract. In that way the manufacturer can get record of failures i.e. of X_i^0, if $\Delta_i = 1$. Otherwise, if there is no failure, the owner will not report the mileage that the car has run in the warranty period and the manufacturer will get the only information $\Delta_i := 0$.

Our hypothesis of interest is $H_0 : F_1 = F_2(=: F)$, versus the one-sided alternative $\mathcal{A} : F_1 \geq F_2, F_1 \neq F_2$. We always assume, that the probability of no censoring under H_0 denoted by τ is strictly positive, i.e., $\tau := \int (1-G)dF > 0$. In a local asymptotic setting we will construct conditional tests which are asymptotically optimal and finite sample distribution free under H_0 for arbitrary censoring nuisance df G.

2. Local asymptotic normality.

Let $t \to P_t$ be a family of probability measures on $\mathbb{R}$ where the real parameter t ranges over some neighborhood of 0. It is assumed that $(P_t)_t$ is L_2 -differentiable in 0 with derivative $h \in L_2(P_0)$, i.e.,

(2.1) $$\| 2(\tfrac{dP_t}{dP_0})^{1/2} - 2 \text{ - th } \| = o(t) \text{ as } t \to 0$$

(2.2) $$P_t(\{\tfrac{dP_0}{dP_t} = 0\}) = o(t^2) \text{ as } t \to 0 .$$

For

(2.3) $$c_{ni} = (\tfrac{n_1 n_2}{n})^{1/2} \begin{cases} \frac{1}{n_1} & \text{if } 1 \leq i \leq n_1 \\ -\frac{1}{n_2} & \text{if } n_1 + 1 \leq i \leq n \end{cases} ,$$

we assume that X_i^0 has distribution $P_{tc_{ni}}, t \in \mathbb{R}$, and U_i has df G. Put $X_i := \min(X_i^0, U_i)$. For some symbolic quantity $\xi \notin \mathbb{R}$ the observations can equivalently be described by the rv's $Z_i \in \mathbb{R} \cup \{\xi\}$ defined by $Z_i = X_i^0$ if $\Delta_i = 1$ and $Z_i = \xi$ if $\Delta_i = 0$. We denote by Q_{nt} the distribution

(2.4) $$Q_{nt} = \mathcal{L}(Z_1, \ldots, Z_n \mid t)$$

and put $Q := Q_{10}$. Moreover, let F be the df of P_0. Then the following result holds.

Theorem 2.1. (Local asymptotic normality)

Define a function $\bar{h}$ on $\mathbb{R} \cup \{\xi\}$ by $\bar{h}(z) := h(z)$ if $z \in \mathbb{R}$ and $\bar{h}(z) := \rho$ with $\rho := \rho(h) := -\int \text{GhdF} / \int \text{GdF}$ for $z = \xi$ and assume that $\sigma^2 = \sigma^2(h) :=$

$\int \bar{h}^2 dQ = \int h^2(1-G)dF + \rho^2 \cdot \int \mathrm{GdF} > 0$. For $\mathbf{Z} = (Z_1, \ldots, Z_n)$ define a remainder term R_{nt} by

(2.5) $$\log \frac{dQ_{nt}}{dQ_{n0}}(\mathbf{Z}) = tT_n(\bar{h})(\mathbf{Z}) - \frac{t^2}{2}\sigma^2 + R_{nt} \ ,$$

with

(2.6) $$T_n(\bar{h})(\mathbf{Z}) = \sum_{i=1}^n c_{ni}\bar{h}(Z_i) = \sum_{i=1}^n c_{ni}(h(X_i) - \rho)\Delta_i \ .$$

If $n_1/n \to \eta \in (0,1)$, then

(2.7) $$\mathcal{L}(T_n \mid Q_{n0}) \to_{\mathcal{D}} \mathcal{N}(0, \sigma^2)$$

where $\to_{\mathcal{D}}$ means convergence in distribution, and for every compact set $K \subset \mathbb{R}$

(2.8) $$sup_{t\in K} Q_{n0}\{\mid R_{nt} \mid > \epsilon\} \to 0 \text{ if } n \to \infty, \forall \epsilon > 0 \ .$$

Proof. It has been shown by (Janssen (1989)) that the family $\mathcal{L}((X_i, \Delta_i) \mid t)$ is L_2-differentiable with derivative $L(x,\delta) = \delta h(x) + (1-\delta)\int_{(x,\infty)} hdP_0/(1-F(x))$. Then, by (Witting (1985)), the family Q_{1t} is L_2-differentiable with derivative $E_0(L \mid Z = z)$ which is easily seen to equal $\bar{h}(z)$. Now the remainder of the proof is essentially identical to the proof of Theorem 1 of (Janssen (1989)). □

Corollary 2.2. For every $g \in L_2(P_0)$ with $\int g \, dP_0 = 0$ one has

(2.9) $$\mathcal{L}(T_n(\bar{g}) \mid Q_{nt}) \to_{\mathcal{D}} \mathcal{N}(t\int \bar{g}\bar{h}dQ, \sigma^2(\bar{g})) \text{ as } n \to \infty$$

and for any sequence S_n with $S_n - T_n \to 0$ in Q_{n0}-probability, (2.9) holds true with $T_n(\bar{g})$ replaced by S_n .

The **proof** of (2.9) is routine by an application of Lindeberg's theorem and the Cramér-Wold technique for deriving the joint asymptotic normal distribution of $(T_n(\bar{g}), T_n(\bar{h}))$ under Q_{n0} and by combining the latter result with (2.5) for an application of Le Cam's third Lemma. □

As a consequence of (2.9) the sequence of tests $\varphi_n = 1(T_n(\bar{g}) \geq u_\alpha \sigma(g)), n \geq 1$, has asymptotic power function

(2.10) $$1 - \Phi(u_\alpha - \int \bar{g}\bar{h}dQ \ /\sigma(g))$$

where Φ is the df of $\mathcal{N}(0,1)$ and u_α its α-fractile. Hence $\{\varphi_n\}$ is asymptotically optimal if $\bar{g}$ is proportional to $\bar{h}$.

Example 2.3. Let $f > 0$ be an absolutely continuous Lebesgue density on $\mathbb{R}$ with df F, derivative f', and finite Fisher information.

a) For the location family P_t corresponding to $f(\cdot - t), t \in \mathbb{R}$, it is well-known that (2.1) and (2.2) hold true with $h = -f'/f$. If $1 - G = (1-F)^\lambda$ for some $\lambda > 0$(the proportional hazard model for the censoring df) then

$$\begin{aligned} (2.11) \qquad \rho &= \int (1-(1-F)^\lambda)(-f'/f)dF / \int (1-(1-F)^\lambda)dF \\ &= \int_0^1 (1-(1-u)^\lambda)\varphi(u,f)du/(1-\int_0^1(1-u)^\lambda du) \\ &= -\int_0^1 (1-u)^\lambda \varphi(u,f)du \cdot (1+\lambda)/\lambda \end{aligned}$$

with $\varphi(u,f) := -(f'/f)\cdot F^{-1}(u)$. E.g., in the logistic case, $F(x) = 1/(1+e^{-x})$, one has $\varphi(u,f) = 2u-1$ and $\rho = -\int_0^1 (2u-1)(1-u)^\lambda du(1+\lambda)/\lambda = (\lambda+2)^{-1} = \tau/(1+\tau)$ with $\tau := Q_{10}\{\Delta_1 = 1\}$. In the latter case the optimal test statistic is

$$(2.12) \qquad \sum_{i=1}^n c_{ni}\Delta_i(\varphi(F(X_i),f) - \rho) = \sum_{i=1}^n c_{ni}\Delta_i(2F(X_i) - 1 - \tfrac{\tau}{1+\tau}) \ .$$

b) If $f \equiv 0$ on $(-\infty, 0)$(the positive case) and if P_t is the family corresponding to $e^{-t}f(e^{-t}x)$, then it is well-known that (2.1) and (2.2) hold true with $h(x) = -1$ -xf' $(x)/f(x)$. If again $(1-G) = (1-F)^\lambda, \lambda > 0$, then with id denoting the identity function

$$\begin{aligned} (2.13) \qquad \rho &= \int (1-(1-F)^\lambda)(-1 - \mathrm{id}\cdot f'/f)dF / \int (1-(1-F)^\lambda)dF \\ &= \int_0^1 (1-(1-u)^\lambda)\varphi_1(u,f)du(1+\lambda)/\lambda \text{ , with} \end{aligned}$$

$\varphi_1(u,f) = -1 - F^{-1}(u)(f'/f)(F^{-1}(u))$, see (Hájek & Sidák, p.19). E.g., in the exponential case one has $\varphi_1(u,f) = -1 - \log(1-u)$. Hence, $\rho = \int_0^1 (1-u^\lambda)(1+\log u)du(1+\lambda)/\lambda = -1/(\lambda+1) = \tau - 1$. In this case the optimal test statistic is

$$(2.14) \qquad \sum_{i=1}^n c_{ni}\Delta_i(\varphi_1(F(X_i),f) - \rho) = \sum_{i=1}^n c_{ni}\Delta_i(-\log(1-F(X_i)) - \tau) \, .$$

3. Rank- and permutation tests.

The reason for considering rank and permutation procedures is usually to produce tests being distribution free under the null hypothesis $H_0 : F_1 = F_2 = F$ with unspecified df F. But, in the present situation with only partially observed data, it is not possible to compute the usual ranks in $X_1, \ldots, X_n$, since only $X_i's$ with $\Delta_i = 1$ are known. Nevertheless, we may compute the rank R_i of X_i among the uncensored X's. (Albers and Akritas (1987)) used this kind of ranks (and a second set of ranks for the censored X's) in the usual random censoring models. While they deliberately introduced these ranks for creating statistics "which are as closely related as possible to the ordinary rank statistics in the uncensored case", for partially observed data these ranks seem to be the only senseful quantities by lack of further information.

To become precise, let $I = I_n = \{i : \Delta_i = 1\}$ be the indices of the uncensored observations, let $X_{(I)} = (X_{1:I}, \ldots, X_{|I|:I})$ denote the order statistic of $X_i, i \in I$, and define a rank vector $R_I = (R_i : i \in I)$ by $R_i = |\{j \in I : X_j \leq X_i\}|$. Moreover, define an antirank vector $D_I = (D_1, \ldots, D_{|I|})$ by $X_{D_i} = X_{i:I}, 1 \leq i \leq |I|$. The rank statistics which we will consider have the form

$$S_n = \sum_{i \in I} c_{ni} a_{|I|}\left(\frac{R_i}{|I|+1}\right) = \sum_{i=1}^{|I|} c_{nD_i} a_{|I|}\left(\frac{i}{|I|+1}\right) \tag{3.1}$$

where for some weights $a_{ki}, i = 1, \ldots, k$, a jump function $a_k(\cdot)$ on (0,1) is defined by

$$a_k(t) := \sum_{i=1}^{k} a_{ki} 1\left(\frac{i-1}{k} \leq t < \frac{i}{k}\right), 0 < t < 1 \quad . \tag{3.2}$$

For $k = |I|$, the scores a_{ki} are allowed to depend (measurably) on $X_{(I)}$.

In order to construct tests being, under H_0, finite sample distribution free, (Neuhaus (1988)) introduced conditional rank tests in the two-sample random censorship model by conditioning on $\Delta_{()} = (\Delta_{(1)}, \ldots, \Delta_{(n)})$, which is the vector of $\Delta's$ belonging to the usual order statistic $X_{1:n}, \ldots, X_{n:n}$ of $X_1 \ldots, X_n$ being not observable in the present setting. On the other hand, (Albers and Akritas (1987)), in the course of proving asymptotic normality of their combined rank statistics used conditioning on the vector $(\Delta_1, \ldots, \Delta_n)$. Yet, when constructing conditional tests the latter conditioning method would lead to tests being in general not asymptotically equivalent to their unconditional counterparts. Luckily it is possible to introduce a better conditioning device, namely conditioning on $|I|$. This is a consequence of the following result which may easily be derived by symmetry considerations.

Lemma 3.1. Under $H_0 : F_1 = F_2$, given $|I|$, the rv's $X_{(I)}$ and D_I as well as $X_{(I)}$ and R_I are independent with

$a)X_{(I)}$ being distributed as the order statistic of $|I|$ i.i.d. rv's with distribution $\mathcal{L}(X_1 \mid \Delta_1 = 1)$

$b)D_I$ being distributed as a random sample without replacement of size $|I|$ from $\{1,\dots,n\}$.

According to the lemma, fixing $|I| = k$ and $X_{(I)} = x_{()} \in \mathbb{R}^k, S_n = S_n(x_{()}, D_I) = \sum_{i \in k} c_{nD_i} a_k(\frac{i}{k+1}; x_{()})$ is a simple linear rank statistic having, under H_0, the in principle known distribution $\mathcal{L}(S_n(x_{()}, D^*))$, where $D^* = (D_1^*, \dots, D_k^*)$ is a random sample of size k taken without replacement from $\{1,\dots,n\}$. Denote the α-fractile of the last distribution by $c_n(x_{()}, k)$. Then a conditional (permutation) test $\varphi_n = \varphi_n(X_{(I)}, D_I)$ may be defined by

$$(3.3) \qquad \varphi_n = \left\{ \begin{array}{c} 1 \\ 0 \end{array} \right. \text{ if } S_n(X_{(I)}, D_I) \left\{ \begin{array}{c} \geq \\ < \end{array} \right\} c_n(X_{(I)}, |I|).$$

By construction the test φ_n is distribution free under H_0. Our aim is to study the asymptotic properties of $\varphi_n, n \geq 1$, under the local model $\{Q_{nt}\}$ from section 2 and derive conditions for their asymptotic optimality.

In a first step we relate the rank statistics $S_n = S_n(X_{(I)}, D_I)$ to statistics $T_n(\bar{h})$ from (2.6). In order to indicate for the functions a_k from (3.2) their dependence on $X_{(I)}$ write $a_k(t, x_{()})$ for $|I| = k$ and $X_{(I)} = x_{()}$. Then we have

Lemma 3.2. For $k \geq 1$ let $X_{()}^k$ denote the order statistic of k i.i.d. rv's with distribution $\mathcal{L}_0(X_1 \mid \Delta_1 = 1)$ and df denoted by F_*. Assume that

$$(3.4) \qquad E \int_0^1 (a_k(t; X_{()}^k) - a(t))^2 dt \to 0 \text{ as } k \to \infty$$

for some nonrandom function $a \in L_2(0,1)$. Then for S_n from (3.1) and

(3.5) $\qquad S_n(a) = \sum_{i=1}^{n} c_{ni} a(F_*(X_i))\Delta_i$

one has

(3.6) $\qquad S_n(a) - S_n \to 0$ in Q_{n0} -probability.

Proof. Using the independence of $X_{(I)}$ and R_I given $| I |$ under H_0, the proof of Lemma 3.5 in (Neuhaus (1988)), dealing with nonrandom a_{ki} 's may be extended to the present more general case. □

Under the assumptions of Lemma 3.2 S_n is apparently asymptotically equivalent to $T_n(\bar{h})$ from (2.6) if

(3.7) $\qquad a \circ F_* = h - \rho \qquad F_*$ -a.s.

An immediate consequence of Lemma 3.2 is that under (3.4)

(3.8) $\qquad \mathcal{L}_0(S_n) \to_{\mathcal{D}} \mathcal{N}(0, \sigma_\tau^2(a))$

with (recall $\tau = Q_{10}\{\Delta_1 = 1\} > 0$)

(3.9) $\qquad \sigma_\tau^2(a) = \tau \int_0^1 a^2(t)dt - \tau^2(\int_0^1 a(t)dt)^2$.

For $| I |= k$ and $X_{(I)} = x_{()}$ put $F_n(x; x_{()}, k) = Q_{n0}\{S_n \leq x \mid X_{(I)} = x_{()}, | I |= k\}$. Then we have the following conditional central limit theorem.

Theorem 3.3. Under the assumptions of Lemma 3.2 one has for $\sigma_\tau^2(a) > 0$ and $n_1/n \to \eta \in (0,1)$

(3.10) $$\sup_{x \in \mathbb{R}} | F_n(x; X_{(I)}, | I |) - \Phi(\tfrac{x}{\sigma_\tau(a)}) | \to 0 \text{ as } n \to \infty$$
in probability under H_0.

Consequently, for the α -fractiles $c_n(X_{(I)}, | I |)$ from (3.3)

(3.11) $\qquad c_n(X_{(I)}, | I |) \to \sigma_\tau(a) u_\alpha$
in probability under H_0.

As one of our main results (3.11) implies asymptotic equivalence of the conditional tests $\varphi_n, n \geq 1$ from (3.3) with their unconditional counterparts obtained by replacing the c_n 's by $\sigma_\tau(a)u_\alpha$.

Proof. Let $(\Omega, \mathcal{A}, \mathbb{P})$ denote the basic probability space on which all rv's are assumed to be defined with $\mathbb{P}$ representing the null hypothesis. It is an immediate consequence of (3.4) that for $I = I_n$ one has

$\int_\Omega \int_0^1 (a_{|I|}(t, X_{(I)}) - a(t))^2 dt d\mathbb{P} \to 0$ as $n \to \infty$. Hence there exists a sequence $\Omega_n \uparrow \Omega$ in $\mathcal{A}$ with $\mathbb{P}(\Omega_n) \to 1$ s.th. for any sequence $\omega_n \in \Omega_n, k_n := | I_n(\omega_n) |, x_{()}^n := X_{(I_n(\omega_n))}(\omega_n)$, one has $\int_0^1 (a_{k_n}(t, x_{()}^n) - a(t))^2 dt \to 0$ as $n \to \infty$. By the law of large numbers we may assume that $k_n/n \to \tau = \int(1-G)dF > 0$. Let $D_1^n, \ldots, D_n^n$ denote a random permutation of $(1, \ldots, n)$. Putting $k = k_n, a_{ki} = a_{ki}(x_{()}^n)$ and $a_k(t) = a_k(t, x_{()}^n)$ the assertion (3.10) will follow if

(3.12) $$S_n(a_k) := \sum_{i=1}^k c_{nD_i^n} a_k(\tfrac{i}{k+1}) \to_{\mathcal{D}} \mathcal{N}(0, \sigma_\tau^2(a)) .$$

From (Hájek and Sidák (1967), p. 61) we have with $\bar{a}_k := \sum_{i=1}^k a_{ki}/k$

$$\sigma_n^2(a_k) = \tfrac{1}{n-1} \sum_{i=1}^k a_{ki}^2 - \bar{a}_k^2 \tfrac{k^2}{n^2} \cdot \tfrac{n}{n-1} ,$$

$$= \tfrac{k}{n-1} \int_0^1 a_k^2(t)dt - \tfrac{k^2}{n^2} \tfrac{n}{n-1} (\int_0^1 a_k(t)dt)^2 ,$$

and, consequently,

(3.13) $$\sigma_n^2(a_k) \to \sigma_\tau^2(a) .$$

One has $S_n(a_k) = \sum_{i=1}^{n_1} \alpha_{nR_i}$ with $\alpha_{ni} = \sqrt{n/(n_1 n_2)}(a_{ki} - \bar{a}_k \cdot k/n)$ for $1 \le i \le k$ and $\alpha_{ni} = -\sqrt{n/(n_1 n_2)}(\bar{a}_k \cdot k/n)$ for $k < j \le n$. Apparently $\bar{\alpha}_n = \sum_{i=1}^n \alpha_{ni} = 0$ and $\sum_{i=1}^n \alpha_{ni}^2 = \frac{n}{n_1} \cdot \frac{n-1}{n_2} \sigma_n^2(a_k) \to \frac{1}{\eta(1-\eta)} \sigma_\tau^2(a) > 0$ as well as $\max \{\alpha_{ni}^2 : 1 \le i \le n\} \le \frac{n}{n_1} \frac{n}{n_2} \{\frac{k}{n}(\frac{1}{k} \cdot \max \{a_{ki}^2 : 1 \le i \le k\}) + \frac{1}{n}(\frac{k}{n})^2 \bar{a}_k^2\} \to 0$. Hence the convergence condition in (exercise 8, p. 195, in Hájek and Sidák (1967)) holds true, implying (3.12).□

Example 3.4. (Continuation of Example 2.3).

a) (Location family with proportional hazard censoring df).

We have $h(x) = \varphi(F(x), f), F_* = g \circ F$ with $g(u) = 1 - (1-u)^{\lambda+1}$. Equality (3.7) is fulfilled if

(3.14) $$a(u) = \varphi(1 - (1-u)^p, f) - \tfrac{\tau}{1+\tau} .$$

In the logistic case this results in $a(u) = 1 - 2(1-u)^\tau - \tau/(1+\tau)$. With $a_{ki} = a(\frac{i}{k+1})$(not depending on $X_{(I)}$) the rank statistic (3.1) then reads as

(3.15) $$S_n = \sum_{i=1}^k c_{nD_i} \{1 - 2(1 - \tfrac{i}{k+1})^\tau - \tau/(1+\tau)\} ,$$

with $k := | I |$.

b)(Scale family with proportional hazard censoring df).

This case is analogous to a) by replacing φ by φ_1. E.g., in the exponential case (3.1) results in $(k := | I |)$

(3.16) $$S_n = \sum_{i=1}^{k} c_{nD_i} \{ -\log(1 - (1 - \tfrac{i}{k+1})^{\tau}) - \tau \} .$$

The conditional tests (3.3) based on S_n from (3.15), resp. (3.16) are finite sample distribution free under $H_0 : F_1 = F_2$ and are asymptotically optimal for a logistic location family, resp. exponential scale family, with proportional hazard censoring df where the value of τ is given. Since $k/n \to \tau$ in probability, one may replace in (3.15) and (3.16) the value of τ by $k/$ n. It may be shown that the resulting statistics are asymptotically equivalent to those with fixed τ. Hence, the corresponding tests are asymptotically optimal for all $\tau \in (0,1]$.

REFERENCES

Albers, W. and Akritas, M.G. (1987). Combined rank tests for the two-sample problem with randomly censored data. J. American Statist. Ass., 82, 648 – 655 .

Hájek, J. and Sidák, Z. (1967). Theory of Rank Tests. Academic Press, New York.

Janssen, A. (1989). Local asymptotic normality for randomly censored models with applications to rank tests. Statistica Neerlandica 43, 109 – 125 .

Neuhaus, G. (1988). Asymptotically optimal rank tests for the two-sample problem with randomly censored data. Communications in Statistics - Theory and Methods 17, 2037 – 2058 .

Suzuki, K. (1985). Nonparametric estimation of lifetime distributions from a record of failures and follow-ups. J. American Statist. Ass. 80, 68 – 72 .

Witting, H. (1985). Mathematische Statistik I. Teubner, Stuttgart.

FINITENESS AND CONTINUITY OF DIFFERENTIAL ENTROPY

Antonín Otáhal
Institute of Information Theory and Automation
Academy of Sciences of the Czech Republic
Pod vodárenskou věží 4
182 08 Praha 8
CZECH REPUBLIC

Abstract

New sufficient conditions for finiteness of the differential entropy and for convergence of differential entropies of a given sequence of probability densities are found.

KEY WORDS: differential entropy, Rényi entropy, L_p spaces

AMS classification: 94 A 17

1 Introduction

In 1948 both Shannon (1948) and (independently) Wiener (1948) introduced the differential entropy as a natural extension of the standard (discrete) entropy — cf. the definitions below, where the Shannon differential entropy and its Rényi generalization are given for any non-negative integrable function (the extension of the classical case of a probability density is not substantial — it is just a matter of technical convenience).

Studying properties of the differential enropy is rather complicated — even a full characterization of its definition domain has not been found so far.

The present paper gives some sufficient conditions for existence and convergence of differential entropies (in terms of belonging to or convergence in L_p-spaces for positive p's near to 1). The approach is an extension of ideas of Vajda (1989), (10.20).

The results achieved seem to indicate that the (monotone increasing) union

$$\bigcup_{\alpha\nearrow 1,\beta\searrow 1} L_\alpha \cap L_\beta$$

may be rather 'near' to the definition domain of the differential entropy (though there exist examples showing that this union is *not* the definition domain).

The convergence condition of Theorem 3.3 represents, in a sense, a generalization of an old but the only known (to the author) Ikeda's (1959) result of this kind (but similar ideas appear in Koski and Persson (1992) and Peetre and Persson (1989).

2 Definitions and Inequalities

In this section basic notions are introduced and some inequalities concerning them are derived.

Let $(X, \mathcal{B}, \lambda)$ be a σ-finite measure space; for a measurable function g on $(X, \mathcal{B}, \lambda)$ we will write $\int g\, d\lambda$ instead of $\int_X g\, d\lambda$. For $p > 0$ we denote $L_p = \{g : \int |g|^p\, d\lambda < \infty\}$ and for $x > 0$ we define

$$\varphi_p(x) = \begin{cases} \dfrac{x^p - x}{1-p} & \text{for } p \neq 1, \\ -x \cdot \ln x & \text{for } p = 1. \end{cases}$$

By continuity we put $\varphi_p(0) = 0$.

Throughout this paper we will suppose, if not stated otherwise, that f is a *non-negative* integrable function on $(X, \mathcal{B}, \lambda)$. For $f \in L_p$ we define the **Rényi differential entropy**

$$I_p(f) = \int \varphi_p(f)\, d\lambda \text{ for } p \neq 1$$

(the last integral is obviously finite); for a measurable function f we define the **(Shannon) differential entropy**

$$H(f) = I_1(f) = -\int f \cdot \ln f\, d\lambda$$

whenever the last integral exists — compare these definitions with (10.19), (10.20) and (10.25) of Vajda (1989). For $p > 0$ we further denote

$$I_p^+(f) = \int_{\{f<1\}} \varphi_p(f)\, d\lambda,$$

$$I_p^-(f) = -\int_{\{f>1\}} \varphi_p(f)\, d\lambda.$$

We will study the Rényi differential entropies by means of $I_p^+(f)$ and $I_p^-(f)$; the following lemma shows the reason for that.

2.1 Lemma. *Suppose* $0 < p \leq q$. *Then the following relations hold:*

(i)

$$\int_{\{f\leq 1\}} f^p\, d\lambda \geq \int_{\{f\leq 1\}} f^q\, d\lambda,$$

(ii)

$$\int_{\{f\geq 1\}} f^p\, d\lambda \leq \int_{\{f\geq 1\}} f^q\, d\lambda,$$

(iii)

$$L_p \cap L_q = \bigcap_{r\in[p,q]} L_r.$$

Proof. Both (i) and (ii) are obvious consequences of the exponential function properties and imply

$$\int f^r\, d\lambda < \infty$$

for $p \leq r \leq q$. □

Before proving some inequalities on I_p^+ and I_p^- we will have a closer look at the function φ_p.

2.2 Lemma. *The function $\varphi_p(x)$ viewed on as a function of p is*

(i) *decreasing for a fixed $x \in (0,1)$,*
(ii) *increasing for a fixed $x > 1$.*

Proof. Consider the function

$$g(x,t) = x^t - 1 - t \cdot \ln x$$

defined for real t and $x > 0$; calculate

$$\tfrac{\partial}{\partial t} g(x,t) = (x^t - 1) \cdot \ln x.$$

This derivative is obviously positive for $t \cdot (x - 1) > 0$ and negative for $t.(x-1) < 0$. Hence

$$g(x,t) \le 0 \quad \text{for } x \in (0,1),$$

$$g(x,t) \ge 0 \quad \text{for } x > 1$$

and the proof is completed by aid of the relation

$$\tfrac{\partial}{\partial p} \varphi_p = \frac{x \cdot g(x, p-1)}{(p-1)^2}.$$

□

The following corollary is an obvious consequence of the last lemma.

2.3 Corollary. *Suppose $0 < \alpha < 1 < \beta$, $f \in L_\alpha \cap L_\beta$ and $\alpha \le p \le q \le \beta$. Then the inequalities*

(i)
$$I_p^+(f) \ge I_q^+(f),$$

(ii)
$$I_p^-(f) \le I_q^-(f)$$

hold and all these integrals are finite.

Let us notice that the last Corollary ensures finiteness of $H(f)$ for $f \in L_\alpha \cap L_\beta$.

3 Convergence Properties

Now the results of the previous section will be used in deriving some sufficient conditions for convergence of the entropies introduced above.

For a sequence (f_n) of functions from L_p we write

$$f_n \xrightarrow{p} f \;\text{ instead of }\; \lim_{n\to\infty} \int |f_n - f|^p \, d\lambda = 0.$$

Convention. *Writing a statement involving the symbol(s) $I_p^{(\pm)}(f)$ we mean that the statement is true if we substitute, for each occurence of $I_p^{(\pm)}(f)$ in this statement, any of the symbols $I_p(f), I_p^+(f), I_p^-(f)$.*

3.1 Lemma. *Suppose* $0 < \alpha < 1 < \beta$, $f_n \xrightarrow{\alpha} f, f_n \xrightarrow{\beta} f$. *Then*

(i) *both* $f_n \xrightarrow{p} f$ *and* $\lim_{n\to\infty} \int f_n^p \, d\lambda = \int f^p \, d\lambda$ *hold uniformly with respect to* $p \in [\alpha, \beta]$,

(ii) $\lim_{n\to\infty} I_p^{(\pm)}(f_n) = I_p^{(\pm)}(f)$ *is true for every* $p \in [\alpha, 1) \cup (1, \beta]$.

Proof. The assertion (i) is an easy consequence of the inequalities 2.1 (i), (ii). Hence the assertion (ii) follows from $\lim_{n\to\infty} |I_p(f_n)| = |I_p(f)|$ which also follows from (i). □

3.2 Theorem. (Continuity of Rényi entropies at $p = 1$). *For* $0 < \alpha < 1 < \beta$ *and* $f \in L_\alpha \cap L_\beta$ *the relation*

$$I_1^{(\pm)}(f) = \lim_{p\to 1} I_p^{(\pm)}(f)$$

holds.

Proof. Since $\lim_{p\to 1} \varphi_p(x) = \varphi_1(x)$ is true for every $x > 0$, from Lebesgue Dominated Convergence Theorem and from Lemma 2.2 the relations

$$I_1^+(f) = \lim_{p\to 1} I_p^+(f)$$

and

$$I_1^-(f) = \lim_{p\to 1} I_p^-(f)$$

follow. □

3.3 Theorem. *Suppose* $0 < \alpha < 1 < \beta$, $f_n \xrightarrow{\alpha} f, f_n \xrightarrow{\beta} f$. *Then*

$$\lim_{n\to\infty} H(f_n) = H(f).$$

Proof. If we write H^+ for I_1^+ and H^- for I_1^- then, according to 2.3 (i),

$$I_{\beta'}^+(f_n) - H^+(f) \leq H^+(f_n) - H^+(f) \leq I_{\alpha'}^+(f_n) - H^+(f)$$

holds for every α', β' such that $\alpha < \alpha' < 1 < \beta' < \beta$. By 3.1 (ii) we get

$$I_{\beta'}^+(f) - H^+(f) \leq H^+(f_n) - H^+(f) \leq I_{\alpha'}^+(f) - H^+(f)$$

and taking into account Theorem 3.2 we conclude

$$\lim_{n\to\infty} H^+(f_n) = H^+(f).$$

The proof will be completed if we prove

$$\lim_{n\to\infty} H^-(f_n) = H^-(f);$$

this would be done in the same way, only 2.3 (ii) would take place of 2.3 (i) and the inequalities would be reversed. □

References

[1] S. Ikeda (1959): Continuity and characterization of Shannon-Wiener information measure for continuous probability distributions. *Ann. Inst. Stat. Math.*, 11, pp. 131–144.

[2] T. Koski and L. E. Persson (1992): Some properties of generalized exponential entropies with applications to data compression.*Inf. Sciences*, 62, pp. 103–132.

[3] J. Peetre and L. E. Persson (1989): General Beckenbach's inequality with applications. In: Function Spaces, Differential Operators and Nonlinear Analysis, Pitman Research Notes in Math., Ser. 211, pp. 125–139.

[4] C. E. Shannon (1948): A mathematical theory of communication. *Bell System Tech. J.*, Vol. 27.

[5] I. ,Vajda (1989): *The Theory of Statistical Inference and Information.* Kluwer Academic Publ., Dodrecht/Boston/London.

[6] N. Wiener(1948): *Cybernetics.* J. Wiley & Sons, New York.

CHARACTERIZATIONS OF DISCRETE PROBABILITY DISTRIBUTIONS BY THE EXISTENCE OF REGULAR CONDITIONAL DISTRIBUTIONS RESPECTIVELY CONTINUITY FROM BELOW OF INNER PROBABILITY MEASURES

D. Plachky
Institute of Mathematical Statistics
Einsteinstr. 62
D-48149 Münster

Abstract

Let $(\Omega, \mathcal{B}, P)$ denote some probability space, where Ω stands for some Polish space with $\mathcal{B}$ as the corresponding Borel σ-algebra. Furthermore, $(\Omega, \mathcal{B}_P, \bar{P})$ is introduced as the completion of $(\Omega, \mathcal{B}, P)$. It is proved that P is discrete if and only if there exists a regular version of the conditional distribution $\bar{P}(A|\mathcal{B})$, $A \in \mathcal{B}_P$. It follows as a corollary that the σ-algebra consisting of the universally measurable subsets of Ω is not countably generated if and only if Ω is not countable. Furthermore it is shown that P is discrete if and only if the corresponding inner probability measure P_* is continuous from below.

AMS 1980 classification: 60 A 10
Key words: Polish space, regular conditional distribution, discrete probability measure, continuity from below, inner probability measure

One aim of this note is to prove the following

Theorem 1 *a) Let Ω denote some Polish space with $\mathcal{B}$ as the corresponding Borel σ-algebra and let $(\Omega, \mathcal{B}_P, \bar{P})$ stand for the completion of the probability space $(\Omega, \mathcal{B}, P)$. Then the existence of a regular version of the conditional distribution $P(A|\mathcal{B})$, $A \in \mathcal{B}_P$, is equivalent to the discreteness of P.*

b) Let $(\Omega, \mathcal{A}, P)$ denote some probability space with countably generated sub-σ-algebra $\mathcal{B}$ of $\mathcal{A}$ satisfying $\{\omega\} \in \mathcal{B}$, $\omega \in \Omega$. Then there exists a regular version $Q_\omega(A)$, $A \in \mathcal{A}$, $\omega \in \Omega$, for the conditional distribution $P(A|\mathcal{B})$, $A \in \mathcal{A}$, which is absolutely continuous with respect to P for any $\omega \in \Omega$ if and only if P is discrete.

Proof: For the proof of both parts of Theorem 1 one might start from the observation that there exists some P-zero set $N_o \in \mathcal{B}$ satisfying $Q_\omega(B) = I_B(\omega)$, $\omega \in N_o^c$, where $Q_\omega(A)$, $A \in \mathcal{A}$, $\omega \in \Omega$, denotes a regular version of the conditional distribution $P(A|\mathcal{B})$, $A \in \mathcal{A}$, and $(\Omega, \mathcal{A}, P)$ stands for some probability space with countably generated sub-σ- algebra $\mathcal{B}$. If in addition $\{\omega\} \in \mathcal{B}$, $\omega \in \Omega$, is fulfilled, $Q_\omega(A) = I_{A \cap N_o^c}(\omega)$, $\omega \in N_o^c$, for any $A \in \mathcal{A}$ is valid, which follows from $Q_\omega(A) \geq Q_\omega(\{\omega\}) = 1$, $\omega \in A \cap N_o^c$, and $Q_\omega(A^c) \geq Q_\omega(\{\omega\}) = 1$, $\omega \in A^c \cap N_o^c$. Therefore, for all $A \in \mathcal{A}$ there exists some $B \in \mathcal{B}$, namely $B = \{\omega \in \Omega : Q_\omega(A) = 1\}$, satisfying $A \cap N_o^c = B \cap N_o^c$, i.e. $\mathcal{A} \cap N_o^c = \mathcal{B} \cap N_o^c$ holds true. Furthermore, there exists also a regular version $\hat{Q}_\omega(A)$, $A \in \mathcal{A}$, $\omega \in \Omega$, of the conditional distribution $\hat{P}(A|\mathcal{B})$, $A \in \mathcal{A}$, where $\hat{P}$ denotes the continuous part of $\bar{P}$. A regular version $\hat{Q}_\omega(A)$, $A \in \mathcal{A}$, $\omega \in \Omega$, of $\hat{P}(A|\mathcal{B})$, $A \in \mathcal{A}$, is given by $\hat{Q}_\omega(A) = \frac{Q_\omega(AA_1)}{Q_\omega(A_1)}$, $\omega \in \{\omega \in \Omega : Q_\omega(A_1) > 0\}$, respectively $\hat{Q}_\omega(A) = \hat{P}(A)$, $\omega \in \{\omega \in \Omega : Q_\omega(A_1) = 0\}$, $A \in \mathcal{A}$, where A_1 denotes the complement of the set $\{\omega \in \Omega : P(\{\omega\}) > 0\}$. Hence, $\mathcal{A} \cap \hat{N}_o^c = \mathcal{B} \cap \hat{N}_o^c$ is valid for some $\hat{P}$-zero set $\hat{N}_o \in \mathcal{B}$, if the continuous part $\hat{P}$ of $\bar{P}$ does not vanish, i.e. if $P(A_1) > 0$ is valid. Now it will be proved that in part a) the continuous component $\hat{P}$ of $\bar{P}$ must vanish, if there exists some regular version of the conditional distribution $\bar{P}(A|\mathcal{B})$, $A \in \mathcal{A}$, where $\mathcal{A}$ stands now for $\mathcal{B}_P$. For this purpose one might observe that $\hat{N}_o^c$ is an uncountable Borel subset of the Polish space Ω, which implies the existence of some continuous and one-to-one function $f : \mathbb{N}^{\mathbb{N}} \to \Omega$ satisfying $f(\mathbb{N}^{\mathbb{N}}) \subset \hat{N}_o^c$ (cf. Cohn (1980), Proposition 8.2.13, p. 267). Furthermore, there exists some analytic subset A of $\mathbb{N}^{\mathbb{N}}$ that is not a Borel subset of $\mathbb{N}^{\mathbb{N}}$ (cf. Cohn (1980), Corollary 8.2.17, p. 269), which implies that A_o defined by $f(A)$ is an analytic subset of Ω (i.e. a continuous image of some Polish space), but not a Borel subset of Ω because of $f^{-1}(A_o) = f^{-1}(f(A)) = A$, since f is one-to-one. Finally A_o is universally measurable, since the σ-algebra of universally measurable subsets of Ω contains all analytical subsets (cf. Cohn (1980), Theorem 8.4.1, p. 278, and Corollary 8.4.2, p. 281). In particular, $A_o \in \mathcal{B}_P$, $A_o \notin \mathcal{B}$, and $A_o \subset \hat{N}_o^c$ holds true, which is a contradiction to $\mathcal{B}_P \cap \hat{N}_o^c = \mathcal{B} \cap \hat{N}_o^c$. Therefore $P(A_1) = 0$ must be valid, i.e. P is discrete, if there exists a regular version of the conditional distribution $\bar{P}(A|\mathcal{B})$, $A \in \mathcal{B}_P$. Conversely, $Q_\omega(A)$ defined by $I_{A \cap N_o^c}(\omega)$, $\omega \in N_o^c$, respectively $Q_\omega(A) = \bar{P}(A)$, $\omega \in N_o$, $A \in \mathcal{B}_P$, where N_o stands for $\{\omega \in \Omega : P(\{\omega\}) > 0\}$ is in the discrete case a regular version of $\bar{P}(A|\mathcal{B})$, which is even absolutely continuous with respect to $\bar{P}$ for any $\omega \in \Omega$. In particular, one has only to show for the proof of part b) that P is discrete, if there exist a regular version $Q_\omega(A)$, $A \in \mathcal{A}$, $\omega \in \Omega$, of $P(A|\mathcal{B})$, $A \in \mathcal{A}$, which is absolutely continuous with respect to P for any $\omega \in \Omega$. This might be seen from $Q_\omega(A) = I_{A \cap N_o^c}(\omega)$, $\omega \in \Omega$, where $N_o \in \mathcal{B}$ is a P-zero set, which implies that any P-zero set $N \in \mathcal{B}$ is already contained in N_o^c because of the absolute continuity of Q_ω with respect to P, $\omega \in N_o^c$. Therefore, N_o^c must be countable, i.e. P is discrete. □

The method of proof admits as a corollary a characterization of the property of the σ-algebra $\mathcal{B}_u$ of universally measurable subsets of a Polish space Ω to be countably generated and can be considered as a generalization of a result of B. V. Rao for the case $\Omega = \mathbb{R}$ (cf. Rao (1970) with a different proof).

Corollary. The σ-algebra of universally measurable subsets of a Polish space Ω is countably generated if and only if Ω is countable. In particular, the completion $\mathcal{B}_P$ of the Borel σ-algebra $\mathcal{B}$ of Ω with respect to any probability measure P on $\mathcal{B}$ is not countably generated if Ω is uncountable.

Proof: If Ω is countable, the Borel subsets of Ω and, therefore, also the σ-algebra $\mathcal{B}_n$ of universally measurable subsets of Ω are equal to the σ-algebra of all subsets of Ω, which is obviously countably generated. For the converse direction one might start from the observation that there exists some continuous probability measure P on the Borel subsets $\mathcal{B}$ of Ω, if Ω is not countable (cf. Parthasarathy (1967), Theorem 8.1, p. 52). Now $\bar{P}$, where $(\Omega, \mathcal{B}_P, \bar{P})$ stands again for the completion of $(\Omega, \mathcal{B}, P)$, is regular (i.e. compact approximable), since this is true for any probability measure P on $\mathcal{B}$ (cf. Cohn (1980), Proposition 8.1.10, p. 258). Hence, there would exist a regular version of the conditional distribution $\bar{P}(A|\mathcal{B})$, $A \in \mathcal{B}_u$, if $\mathcal{B}_u$ would be countably generated, since there exists regular conditional distributions, if the underlying probability measure is compact approximable and the underlying σ-algebra is countably generated. Therefore, $\mathcal{B}_u$ is not countably generated according to the method of proof of Theorem 1, if Ω is not countable. Furthermore, $\mathcal{B}_P$ is not countably generated for any probability measure P on $\mathcal{B}$, if Ω is uncountable, since $\mathcal{B}_P$ coincides for discrete probability measures P on $\mathcal{B}$ with the set $\mathcal{P}(\Omega)$ of all subsets of Ω, whose cardinality exceeds the cardinality of the continuum, since the cardinality of Ω is that of the continuum (cf. Cohn (1980), Theorem 8.3.6, p. 274), whereas the cardinality of any countably generated σ-algebra is at most equal to the cardinality of the continuum (cf. Cohn (1980), p. 271). The case, where the continuous part of some probability measure P on $\mathcal{B}$ does not vanish can be easily reduced to the continuous case, which yields a σ-algebra $\mathcal{B}_P$, which is not countably generated according to the technique of proof for Theorem 1. □

Let P_* denote the inner probability measure defined by $P_*(B) = \sup\{P(A) : A \subset B,\ A \in \mathcal{A}\}$, $B \subset \Omega$, where $(\Omega, \mathcal{A}, P)$ stands for some probability space. It is well-known that P_* is always continuous from above. It will be now shown that continuity from below of P_* might be characterized by discreteness of P under the general assumption that Ω is some Polish space and $\mathcal{A}$ stands for the corresponding Borel σ-algebra $\mathcal{B}$.

Theorem 2 *Let Ω denote some Polish space with $\mathcal{B}$ as the corresponding Borel σ-algebra and let P_* stand for the inner probability measure of the probability space $(\Omega, \mathcal{B}, P)$. Then the property of P_* to be continuous from below is equivalent to the discreteness of P.*

Proof: If P is discrete, then P_* coincides with the unique extension of P as a probability measure on the σ-algebra consisting of all subsets of Ω, which implies continuity from below of P_*. For the converse direction one might start from the observation that there exists some subset A of the unit interval of $\mathbb{R}$ such that

$A+\rho$, $\rho \in \mathbb{Q}$, are pairwise disjoint and $[0,1] \subset \bigcup_{\rho\in\mathbb{Q}}(A+\rho)$ is satisfied. This follows by means of the lemma of Zorn applied to the set of all subsets B of $[0,1]$ such that $B+\rho$, $\rho \in \mathbb{Q}$, are pairwise disjoint, which is ordered by inclusion. Therefore, $[0,1] \subset \bigcup_{\rho\in\mathbb{Q}_1}(A+\rho) \subset [-1,2]$ is fulfilled, where $\mathbb{Q}_1$ stands for $\mathbb{Q}\cap[-1,1]$. Furthermore, $\lambda_*(\bigcup_{k=1}^{n}(A+\rho_k)) = 0$ holds true for any $n \in \mathbb{N}$, where $\{\rho_1, \rho_2, \ldots\}$ denotes $\mathbb{Q}_1$, since there exists infinite many $\rho \in \mathbb{Q}_1$ satisfying $\rho + \rho_k \in \mathbb{Q}_1$, $k = 1,\ldots,n$, where λ denotes the Lebesgue measure. However, $\lambda_*(\bigcup_{n=1}^{\infty}\bigcup_{k=1}^{n}(A+\rho_k)) \geq 1$ is valid, i.e. the Lebesgue measure restricted to the Borel subsets $\mathcal{B}_{[0,1]}$ of $[0,1]$ is not continuous from below. This implies that the continuous part of P must vanish, if P_* is continuous from below, since according to [4], p. 327, $([0,1],\mathcal{B}_{[0,1]},\lambda|\mathcal{B}_{[0,1]})$ and $(\Omega,\mathcal{B},P)$ are up to zero sets of $\lambda|\mathcal{B}_{[0,1]}$ resp. P_1 pointwise isomorphic, where P_1 stands for the normed continuous part of P in the case, where P is not discrete. □

REFERENCES

Cohn, D. L. (1980): Measure Theory. Birkhäuser, Boston

Parthasarathy, K. P. (1967): Probability Measures on Metric Spaces. Academic Press, New York

Rao, B. V. (1970): Remarks on analytic sets. Fund. Math. 66, 237 - 239

Royden, H. L. (1968): Real Analysis. Macmillan, London

DIAGNOSTICS OF REGRESSION MODEL: TEST OF GOODNESS OF FIT

Asunción Maria Rubio, Jan Ámos Víšek
University of Extremadura
and
Academy of Sciences of the Czech Republic

Abstract

A test statistic for goodness of fit of the M-estimates of the linear regression model is proposed and its asymptotic behaviour is described. Numerical study on real data is included.

Key words: Diagnostic of residuals, weighted Hellinger distance, asymptotic normality, numerical study.

AMS classification: 62 J 05, 62 J 15

1 Introduction

The present robust statistic offers a large number of methods for estimating the regression model. Both for linear and nonlinear regression there are typically the whole families of procedures, parametrized by a parameter corresponding (explicitly or implicitly) to the contamination level of data (in a model of contaminacy, Víšek (1985)). The selection of family is usually based either on some general principles (like V- or B-robustness, see Hampel et al. (1986), Huber (1973) , Jurečková (1988)) or on the experiences of statisticians. Also the availability of software plays naturally its role when choosing a method. The selection of value of the "tuning" parameter(s) is still also based only on the experience. Although the selection of the value of tuning parameter is not extremely important from the efficiency point of view (see Exhibit 3 in Appendix), it has a considerable impact on the values of estimates of the regression parameters. Hence the influence of the selection of the tuning parameter(s) on the explanation or the interpretation of a structure of data as well as the influence on a quality of the prediction may be serious. For a more exhaustive discussion of the reasons for selecting a procedure see Víšek (1992c). For the effect of the selection of the family and of the tuning parameter on the values of estimates see Víšek (1992b)).

Let us give now the setup which we will use. We shall consider for all $i \in N$ a linear model

$$Y_i = X_i^{\mathrm{T}} \beta^0 + e_i$$

where $\beta^0 \in R^p$ and $\{e_i\}_{i=1}^{\infty}$ is a sequence of i. i. d. r. v. The corresponding distribution function is assumed to have a density $f(x)$ with

$$0 < v < f(x) < V < \infty \qquad \text{for} \quad x \in (A, B) \quad -\infty < A < B < \infty \tag{1}$$

and

$$f(x) = 0 \qquad \text{otherwise.} \tag{2}$$

(The assumptions (1) and (2) are in fact only technical ones and we may rid of them. However, the techniques allowing that require a triangular scheme and a long chain of approximations with an inevitable more complicated system of notations; for the details see Víšek (1992a).) In what follows we shall use a kernel estimator

$$f_n(y, Y, \beta, c_{k_n}) = \frac{1}{n\, c_{k_n}} \sum_{i=1}^{n} w\left(c_{k_n}^{-1}(y - Y_i + X_i^{\mathrm{T}}\beta)\right),$$

with $\{c_{k_n}\}_{n=1}^{\infty} \searrow 0$. We shall assume that the kernel w fulfills:

$$\lim_{|x|\to\infty} x^4\, w(x) = 0 \qquad \text{and} \qquad \int_{-\infty}^{\infty} x\, w(x)\, dx = 0.$$

Finally, we shall require

$$n^{-1} c_n^{-4} = o(1) \qquad \text{as} \quad n \to \infty.$$

Moreover, let us assume that we have $2n$ observations, say $Y^{(i)} = (Y_{i1}, Y_{i2}, \ldots, Y_{in})^{\mathrm{T}}$, $i = 1, 2$ and let us put $Y = (Y^{(1)\mathrm{T}}, Y^{(2)\mathrm{T}})^{\mathrm{T}}$. Finally, we shall consider for all $n = 1, 2, \ldots$ the statistic

$$H_n(Y, \beta) = n \int_{-\infty}^{\infty} \left[f_n^{\frac{1}{2}}(y, Y^{(1)}, \beta, c_n) - f_n^{\frac{1}{2}}(y, Y^{(2)}, \beta, c_n) \right]^2 f_{2n}^{\frac{1}{2}}(y, Y, \beta, c_n)\, dy \tag{3}$$

which represents a kind of the weighted Hellinger distance between $f_n(y, Y^{(1)}, \beta, c_n)$ and $f_n(y, Y^{(2)}, \beta, c_n)$. Since in what follows we shall integrate always over $(-\infty, \infty)$, we shall omit the bounds for integration. Moreover, the sequence $\{c_n\}_{n=1}^{\infty}$ will be fix in the rest of the paper and hence we shall write $f_n(y, Y, \beta)$ instead of $f_n(y, Y, \beta, c_{k_n})$.

In the next section we will look for the asymptotic behaviour of the statistic $H_n(Y, \beta)$ and in the last section we offer a possibility to create an idea about the behaviour of this statistic for (finite) samples of real data.

2 Asymptotics of the statistic

THEOREM 1 *Let the density f fulfill* (1) *and* (2). *Moreover, let it have a bounded second derivative and let it be Lipschitz of the first order. Then the asymptotic distribution of*

$$\mathcal{H}_n(Y, \beta^0) = \Delta_n^{-1}\{H_n(Y, \beta^0) - m_n\}$$

is $N(0, 1)$ where

$$m_n = \frac{1}{2} c_n^{-\frac{3}{2}} \int \frac{\int w^2(c_n^{-1}(t - x))\, f(x)\, dx}{\left[\int w(c_n^{-1}(t - x)\, f(x)\, dx\right]^{\frac{1}{2}}} dt \tag{4}$$

and

$$\Delta_n^2 = \frac{1}{2} c_n^{-3} \int\int \frac{\left\{\int w(c_n^{-1}(t-x))\, w(c_n^{-1}(s-x))\, f(x)\, dx\right\}^2}{\left\{\int w(c_n^{-1}(t-x))\, f(x)\, dx \int w(c_n^{-1}(s-x))\, f(x)\, dx\right\}^{\frac{1}{2}}}\, ds dt. \tag{5}$$

In order to prove Theorem 1 we prepare several lemmas, the proofs of which are simple classical approximations. Throughout the all lemmas we shall consider the assumptions of Theorem 1 to be fulfilled.

REMARK 1. Let us notice that

$$\mathsf{E}_F\, f_n(y, Y^{(1)}, \beta) = \mathsf{E}_F\, f_n(y, Y^{(2)}, \beta) = \mathsf{E}_F\, f_{2n}(y, Y, \beta) = c^{-1} \int w(c^{-1}(y-z))\, f(z)\, dz.$$

LEMMA 1 *For $i = 1, 2$ we have*

$$n \int \left[f_n^{\frac{1}{2}}(y, Y^{(i)}, \beta^0) - \mathsf{E}^{\frac{1}{2}} f_n(y, Y^{(i)}, \beta^0) \right]^4 dy = \mathcal{O}_p(n^{-1}\, c_n^{-2}).$$

PROOF. The inequality $(a-b)^2 \leq b^{-2}(a^2-b^2)^2$, valid for $a \geq 0$ and $b > 0$ yields

$$\begin{aligned}
&\mathsf{E}\left\{ n \int \left[f_n^{\frac{1}{2}}(y, Y^{(i)}, \beta^0) - \mathsf{E}^{\frac{1}{2}} f_n(y, Y^{(i)}, \beta^0) \right]^4 dy \right\} \\
\leq\ & n \int \mathsf{E} \left\{ \frac{\left[f_n(y, Y^{(i)}, \beta^0) - \mathsf{E}\, f_n(y, Y^{(i)}, \beta^0) \right]^4}{\mathsf{E}^2\, f_n(y, Y^{(i)}, \beta^0)} \right\} dy \\
=\ & n \int \mathsf{E}^{-2} f_n(y, Y^{(i)}, \beta^0) \left\{ n^{-3} c_n^{-4} \mathsf{E}[w - \mathsf{E}w]^4 + n^{-1} c_n^{-4} \mathsf{E}^2 [w - \mathsf{E}w]^2 \right\} dy
\end{aligned} \tag{6}$$

Making use of the inequalities $\mathsf{E}|w - \mathsf{E}w| \leq 2\mathsf{E}w$ and $\mathsf{E}^{-1} f_n(y, Y^{(i)}, \beta^0) \cdot \mathsf{E}|w - \mathsf{E}w| \leq c_n$, we obtain that (6) is bounded by $4c_n^{-2} \sup_{z\in R}^2 w(z)\, (B-A) \left\{ \frac{1}{n^2} + \frac{1}{n} \right\}$. An application of Chebyshev's inequality (for nonnegative random variable) concludes the proof. □

LEMMA 2

$$n \int \mathsf{E}^{-1} f_{2n}(y, Y, \beta^0) [f_{2n}(y, Y, \beta^0) - \mathsf{E} f_{2n}(y, Y, \beta^0)]^2\, dy = \mathcal{O}_p(c_n^{-1}).$$

PROOF. Similarly as above

$$\mathsf{E}\left\{ n \int \mathsf{E}^{-1} f_{2n}(y, Y, \beta^0) [f_{2n}(y, Y, \beta^0) - \mathsf{E} f_{2n}(y, Y, \beta^0)]^2\, dy \right\} \leq (B-A) c_n^{-1} \sup_{z\in R} w(z).$$

□

LEMMA 3

$$n \int \left[f_n^{\frac{1}{2}}(y, Y^{(1)}, \beta^0) - f_n^{\frac{1}{2}}(y, Y^{(2)}, \beta^0) \right]^2 \left[f_{2n}^{\frac{1}{2}}(y, Y, \beta^0) - \mathsf{E}^{\frac{1}{2}} f_{2n}(y, Y, \beta^0) \right] dy$$

$$= \mathcal{O}_p\left(n^{-\frac{1}{2}} c_n^{-\frac{3}{2}} \right).$$

PROOF. Having applied Schwarz inequality and the inequality used in the proof of Lemma 1, one finishes the proof using Lemma 1 and 2. □

LEMMA 4 *We have for* $\tau = \frac{1}{2}, 1$ *and* $i, j = 1, 2$ *with* $i \neq j$, *if* $\tau = \frac{1}{2}$

$$n \int \left[f_n^{\frac{1}{2}}(y, Y^{(1)}, \beta^0) - \mathsf{E}^{\frac{1}{2}} f_n(y, Y^{(1)}, \beta^0) \right] \left[f_n^{\tau}(y, Y^{(2)}, \beta^0) - \mathsf{E}^{\tau} f_n(y, Y^{(2)}, \beta^0) \right] dy$$
$$= \mathcal{O}_p\left(n^{-(\frac{3}{2}-\tau)} c_n^{-2} \right).$$

PROOF may be carried out in a similar way as the proofs of Lemma 1 and 2. □

LEMMA 5 *The asymptotic distribution of*

$$\tilde{\Delta}_n^{-1} \left\{ n \int \left[(f_n(y, Y^{(1)}, \beta^0) - f_n(y, Y^{(2)}, \beta^0)) \right]^2 \cdot \mathsf{E}^{-\frac{1}{2}} f_{2n}(y, Y, \beta^0)\, dy - \tilde{m}_n \right\}$$

is $N(0,1)$ *where*

$$\tilde{m}_n = 2 \cdot c_n^{-\frac{3}{2}} \int \frac{\int w^2(c_n^{-1}(t-x))\, f(x)\, dx}{\left[\int w(c_n^{-1}(t-x))\, f(x)\, dx \right]^{\frac{1}{2}}} dt \tag{7}$$

and

$$\tilde{\Delta}_n^2 = 8 c_n^{-3} \int\int \frac{\left\{ \int w(c_n^{-1}(t-x))\, w(c_n^{-1}(s-x))\, f(x)\, dx \right\}^2}{\left\{ \int w(c_n^{-1}(t-x))\, f(x)\, dx \int w(c_n^{-1}(s-x))\, f(x)\, dx \right\}^{\frac{1}{2}}} ds dt. \tag{8}$$

PROOF. For the proof of Lemma 5 we shall use Assertion 1 of Appendix. First of all, let us write $F_n^{(j)}(x)$ for the empirical d.f. of $\left\{ Y_i^{(j)} - X_i^{(j)\mathrm{T}} \beta^0 \right\}_{i=1}^n$, and for $j = 1, 2$ and $t \in [0, 1]$ put

$$B_n^{(j)}(t) = \sqrt{n} \left[F_n^{(j)}(\mathrm{inv}\, F(t)) - t \right].$$

Let us denote $\varphi_n(y, z) = c_n^{-1} w(c_n^{-1}(y - z))$. Then, keeping in mind REMARK 1, we have

$$\sqrt{n} \left\{ f_n(y, Y^{(1)}, \beta^0) - f_n(y, Y^{(2)}, \beta^0) \right\} = \sqrt{n} \left\{ \mathsf{E}_{F_n^{(1)}} \varphi_n(y, x) - \mathsf{E}_{F_n^{(2)}} \varphi_n(y, x) \right\}$$
$$= \int_0^1 B_n^{(1)}(t)\, d_t \varphi_n(y, \mathrm{inv}\, F(t)) - \int_0^1 B_n^{(2)}(t)\, d_t \varphi_n(y, \mathrm{inv}\, F(t))$$

(see also Csörgő, Révész, p. 223). Let us denote by $\mu_n(y)$ the mean value $\mathsf{E}_F f_n(y, Y^{(1)}, \beta)$ and by $G_n(y)$ the process

$$\mu_n^{-\frac{1}{4}}(y) \cdot \sqrt{n} \left\{ f_n(y, Y^{(1)}, \beta^0) - f_n(y, Y^{(2)}, \beta^0) \right\}, \quad A < y < B.$$

Then we have $\mathsf{E}_F G_n(y) \equiv 0$. Following Csörgő, Révész (1981) we obtain for $R_n(u, v)$

$$\mu_n^{-\frac{1}{4}}(u)\, \mu_n^{-\frac{1}{4}}(v)\, \mathsf{E}_F \left\{ \int_0^1 \left[B_n^{(1)}(t) - B_n^{(2)}(t) \right] d_t \varphi_n(u, \mathrm{inv}\, F(t)) \times \right.$$
$$\left. \times \int_0^1 \left[B_n^{(1)}(z) - B_n^{(2)}(z) \right] d_z \varphi_n(v, \mathrm{inv}\, F(z)) \right\}$$
$$= \mu_n^{-\frac{1}{4}}(u)\, \mu_n^{-\frac{1}{4}}(v) \int_0^1 \int_0^1 \left\{ \mathsf{E}_F \left[B_n^{(1)}(t)\, B_n^{(1)}(z) \right] + \mathsf{E}_F \left[B_n^{(2)}(t)\, B_n^{(2)}(z) \right] \right\} \times$$
$$\times d_t \varphi_n(u, \mathrm{inv}\, F(t))\, d_z \varphi_n(v, \mathrm{inv}\, F(z))$$
$$= 2 \mu_n^{-\frac{1}{4}}(u)\, \mu_n^{-\frac{1}{4}}(v) \int \varphi_n(u, x)\, \varphi_n(v, x)\, f(x)\, dx$$

Now (remember Assertion 1 of Appendix)

$$\tilde{m}_n = \int R_n(t,t)\,\mathrm{d}t = 2c_n^{-2}\int\left\{\mu^{-\frac{1}{2}}(t)\int w^2(c_n^{-1}(t-x))\,f(x)\,\mathrm{d}x\right\}\,\mathrm{d}t.$$

Similarly, for $\tilde{\Delta}_n^2$ a simple calculation yields the expression

$$\begin{aligned}
&2c_n^{-4}\int\int \mu_n^{-\frac{1}{2}}(u)\,\mu_n^{-\frac{1}{2}}(v)\left[\int w\left(\frac{u-x}{c_n}\right)w\left(\frac{v-x}{c_n}\right)f(x)\,\mathrm{d}x\right]^2\mathrm{d}u\mathrm{d}v\\
&= 2\cdot 4\cdot c_n^{-3}\int\int\frac{\left[\int w\left(\frac{u-x}{c_n}\right)w\left(\frac{v-x}{c_n}\right)f(x)\,\mathrm{d}x\right]^2}{\left\{\int w\left(\frac{u-z}{c_n}\right)f(z)\,\mathrm{d}z\int w\left(\frac{v-z}{c_n}\right)f(z)\,\mathrm{d}z\right\}^{\frac{1}{2}}}\mathrm{d}u\mathrm{d}v\\
&= 8\cdot c_n^{-1}\int\int\frac{\left\{\int w(r)\,w(v+r)\,f(t-c_nr)\,\mathrm{d}r\right\}^2}{\left\{\int w(y)\,f(t-yc_n)\,\mathrm{d}y\int w(z)\,f(t+c_n(v-z))\,\mathrm{d}z\right\}^{\frac{1}{2}}}\,\mathrm{d}v\mathrm{d}t
\end{aligned}$$

and taking into account the assumption that $f(x)$ is Lipschitz, one concludes that $\tilde{\Delta}_n^2 \to \infty$ as $n \to \infty$. In a similar way one may show that the numerator of (10) - see Appendix - is $\mathcal{O}(1)$, and hence (10) holds. □

PROOF OF THEOREM 1. Lemma 3 implies that

$$H_n(Y,\beta^0) = n\int\left[f_n^{\frac{1}{2}}(y,Y^{(1)},\beta^0) - f_n^{\frac{1}{2}}(y,Y^{(2)},\beta^0)\right]^2\mathsf{E}^{\frac{1}{2}}f_{2n}(y,Y,\beta^0)\,\mathrm{d}y + \mathcal{O}_p\left(n^{-\frac{1}{2}}c_n^{-\frac{3}{2}}\right).$$

Making use of the fact that $f_n^{\frac{1}{2}}(y,Y,\beta) - \mathsf{E}^{\frac{1}{2}}\,f_n(y,Y,\beta)$ equals to

$$\frac{1}{2}\left\{f_n(y,Y,\beta) - \mathsf{E}\,f_n(y,Y,\beta) - \left[f_n^{\frac{1}{2}}(y,Y,\beta) - \mathsf{E}^{\frac{1}{2}}\,f_n(y,Y,\beta)\right]^2\right\}\mathsf{E}^{-\frac{1}{2}}\,f_n(y,Y,\beta)$$

and Lemma 4 we obtain for the statistics $H_n(Y,\beta^0)$ a representation

$$\frac{n}{4}\int\left[f_n(y,Y^{(1)},\beta^0) - f_n(y,Y^{(2)},\beta^0)\right]^2\mathsf{E}^{-\frac{1}{2}}f_{2n}(y,Y,\beta^0)\,\mathrm{d}y + \mathcal{O}_p\left(n^{-\frac{1}{2}}c_n^{-2}d_n^{-1}\right).$$

Applying Lemma 5 we conclude the proof. □

REMARK 2. It is clear that to have a reasonable hope that the asymptotic result given in Theorem 1 already works the size of the sample of data should be sufficiently large. Taking into account the dimensionality of data we may come to the conclusion that we need hundreds or thousands of observations. Even in such a case to be able then to use Theorem 1 we need to estimate from data m_n and Δ_n^2. Again a straightforward computation gives

$$m_n = \frac{1}{2}c_n^{-1}\int\frac{\int w^2(z)\,f(t-c_nz)\,\mathrm{d}z}{\left\{\int w(z)\,f(t-c_nz)\,\mathrm{d}z\right\}^{\frac{1}{2}}}\,\mathrm{d}t.$$

Let us assume that f' exists and is bounded. Then

$$m_n = \frac{1}{2}c_n^{-1}\left[\int f^{\frac{1}{2}}(t)\,\mathrm{d}t\int w^2(z)\,\mathrm{d}z + \mathcal{O}(c_n)\right].$$

Similarly

$$\Delta_n^2 = \frac{1}{2}c_n^{-1}[1+\mathcal{O}(c_n)].$$

3 Numerical example

In 1992 Hettmansperger and Sheater studied the data originally published in Mason et al. (1989) describing the dependence of the engine knock number on the four predictors: the spark timing, the air/fuel ratio, the intake temperature and the exhaust temperature. For the convenience the data are given in the Exhibit 1.

EXHIBIT 1

Engine Knock Data

case	Spark	Air	Intake	Exhaust	Knock
1	13.3	13.9	31	697	84.4
2	13.3	14.1	30	697	84.1
3	13.4	15.2	32	700	88.4
4	12.7	13.8	31	669	84.2
5	14.4	13.6	31	631	89.8
6	14.4	13.8	30	638	84.0
7	14.5	13.9	32	643	83.7
8	14.2	13.7	31	629	84.1
9	12.2	14.8	36	724	90.5
10	12.2	15.3	35	739	90.1
11	12.2	14.9	36	722	89.4
12	12.0	15.2	37	743	90.2
13	12.9	15.4	36	723	93.8
14	12.7	16.1	35	649	93.0
15	12.9	15.1	36	721	93.3
16	12.7	15.9	37	696	93.1

When typing data into the computer they wrote falsely one coordinate of one case, namely the value of air/fuel ratio for the second case was written 15.1 rather than the correct 14.1. For these contaminated data one obtains rather different LMS and LTS estimates (of the regression model, see Exhibit 2) although both methods have (asymptotically) 50 % breakdown point. For the some practitioners this fact may be surprising although it is not difficult to give an example of data for which LMS and LTS estimates of the simple regression model are orthogonal each to other, Víšek (1993b), see also Rousseeuw, Leroy (1987). We shall use these contaminated data (i.e. the data from Exhibit 1 have to be changed in the second row of the second column from 14.1 to 15.1) to offer the reader an opportunity to create (a very first) idea about the possibility to use the weighted Hellinger statistic for the selection of the estimate of the linear regression model. The results of all computations may be found in Exhibit 2. Since the data contains only 16 cases it seems senseless to apply the asymptotic results of the paper (especially to try to create a critical region). However, the statistic $H_n(Y, \beta)$ (see (3)) may be used to select the estimate for which two subsets of residuals imply the closest densities. We have proceeded as follows:

After evaluating the estimate $\hat{\beta}$ of the model we have projected (orthogonally) the data into the regression plane. Then we have found the smallest ellipsoid containing the projections in this plane, we have found the largest axe of the ellipsoid and the plane orthogonal to this axe which divides the data into two subsets, each containing 8 points. Then we have used the residuals of these 2 sets of points (residuals in the estimated model, i. e. not the residuals of orthogonal projection) to evaluate $H_n(Y, \hat{\beta})$. The residuals were studentized by means of σ_{MAD}. We have employed the standard normal kernel and the bandwidth was selected on the base of the previous experiences with similar type of data; the Hellinger distance was used in these previous studies to optimize the bandwidth; the Romberg method was used to compute the respective integrals.

Let us explain the first column of Exhibit 2 (the others are self explaining by their heads). LS, LMS and LTS mean the Least Squares, the Least Median of Squares and the Least Trimmed Squares estimates, respectively. On the other rows the M-estimates using three-part Hampel function are presented (Hampel function:

$$\psi(x) = \begin{array}{ll} x & 0 \le x \le 1 \cdot \mathrm{Const}, \\ \mathrm{Const} & \mathrm{Const} < x \le 1.25 \cdot \mathrm{Const}, \\ (12 - 4x)\,\frac{\mathrm{Const}}{7} & 1.25 \cdot \mathrm{Const} < x \le 3 \cdot \mathrm{Const}, \\ 0 & x > 3 \cdot \mathrm{Const} \end{array}$$

and $\psi(x) = -\psi(-x)$.) The values of Const were considered as given in the first column of Exhibit 2.

EXHIBIT 2.

Const	Intercept	Spark	Air	Intake	Exhaust	$H_n(Y, \hat{\beta})$
0.60	16.08	0.589	2.189	1.002	-0.0022	0.042997
0.80	14.25	0.619	2.213	1.006	-0.0008	0.044473
1.00	14.23	0.645	2.181	1.017	-0.0011	0.042485
1.20	14.43	0.695	2.123	1.036	-0.0019	0.038375
1.25	14.49	0.709	2.106	1.041	-0.0022	0.012279
1.30	14.56	0.725	2.088	1.047	-0.0024	0.012186
1.35	14.63	0.741	2.070	1.053	-0.0027	0.012181
1.40	14.70	0.759	2.049	1.059	-0.0030	0.012275
1.45	14.78	0.778	2.027	1.066	-0.0033	0.012482
1.50	14.86	0.798	2.004	1.074	-0.0037	0.012821
1.60	15.06	0.844	1.950	1.091	-0.0045	0.014024
1.70	15.29	0.900	1.885	1.112	-0.0054	0.025393
1.80	15.58	0.969	1.805	1.137	-0.0066	0.027387
1.90	15.77	1.015	1.751	1.155	-0.0074	0.028183
2.00	15.91	1.049	1.711	1.167	-0.0080	0.029169
2.50	15.97	1.063	1.695	1.172	-0.0082	0.030281
LS	15.97	1.063	1.695	1.172	-0.0082	0.030281
LMS	48.38	0.732	3.393	0.195	-0.0109	0.039927
LTS	-88.73	4.719	1.058	1.569	0.0676	0.047861

4 Conclusions

Utilizing the residuals when forming an opinion about the goodness-of-fit of the estimate of the model to the given data seems to be quite natural. On the other hand, under assumption that the data are homogeneous the distribution of random fluctuations in any subset of the factor space has to coincide with the distribution in the complementary part. Of course, it is true for the empirical distribution of residuals only asymptotically. Nevertheless, using such decomposition of the factor space which is independent from the values of residuals we may believe that a characteristic based on the idea of comparison of the laws of residuals in disjoint parts of the factor space may properly indicate the quality of the estimate of model even for rather small samples of data. The example given above demonstrates that the statistic $H_n(Y,\hat{\beta})$ may serve for this purpose quite well. Although, it may seem that the evaluation of any statistic of Hellinger type requires an intensive computation, it is not inevitably so.

Of course, it is true that any evaluation of the statistic of this type is more complicated than the calculation of the coefficient of determination or any similar characteristic of fit. On the other hand, the coefficient of determination does not keep generally the full information brought by residuals. This is the reason for studying statistics which are easier computable than the Hellinger type ones but which may preserve the information contained in the residuals better than the coefficient of determination or similar statistics.

5 Appendix

ASSERTION 1 *(Csőrgő, Révész) Let $G_n(x)$ $(-\infty \leq A < x < B \leq \infty;\ n = 1,2\ldots)$ be a sequence of Gaussian processes with*

$$\mathsf{E}\, G_n(x) \equiv 0, \qquad R_n(u,v) = \mathsf{E}\, G_n(z)\, G_n(v)$$

and

$$\mathsf{E}\int_A^B G_n^2(x)\,\mathrm{d}x = \int_A^B \mathsf{E}\, G_n^2(x)\,\mathrm{d}x = \int_A^B R_n(x,y)\,\mathrm{d}x = m_n < \infty.$$

Assume that $R_n(u,v)$ $((u,v) \in (A,B)^2)$ is continuous at any point (u,u) $(A < u < B)$, square integrable,

$$\Delta_n^2 = \operatorname{var}\int_A^B G_n^2(x)\,\mathrm{d}x = 2\int_A^B\int_A^B R_n^2(u,v)\,\mathrm{d}u\mathrm{d}v \longrightarrow \infty \quad as \quad n \to \infty \tag{9}$$

and

$$\frac{\int_A^B \left(\int_A^B R_n(u,v)\,h(v)\,\mathrm{d}v\right)^2 \mathrm{d}u}{\int_A^B\int_A^B R_n^2(u,v)\,\mathrm{d}u\mathrm{d}v} \longrightarrow 0 \quad as \quad n \to \infty, \tag{10}$$

for any $h \in L^2(A,B)$. Then

$$\Delta_n^{-1}\left(\int_A^B g_n^2(x)\,\mathrm{d}x - m_n\right) \xrightarrow{\mathcal{D}} N(0,1). \tag{11}$$

PROOF: See Csőrgő, Révész, Theorem 6.1.2. Let us observe however that the requirement (9) is irrelevant for (10) and (11) (transform $G_n(x)$ to $\Delta_n^{-1}G_n(x)$ with "new" $\Delta_n \equiv 1$).

EXHIBIT 3

The explanation of the information given in Exhibit 3:

ψ_ε is the Huber ψ-function corresponding the contamination level ε and Φ denotes the standard normal law. $\varepsilon_1 = 0.038$ which corresponds to the "tuning" constant 1.5 (which is frequently recommended, see e.g. Hampel et al. (1986)).

$\mathrm{var}(\psi_\varepsilon, \Phi)$ - the asymptotic variance of the location estimator generated by the function ψ_ε with respect to Φ,

$\mathrm{var}(\psi_\varepsilon, F_\varepsilon)$ - the asymptotic variance of the location estimator generated by the function ψ_ε with respect to the least favorable model F_ε in the "contamination" neighbourhood

$$\mathcal{P}_\Phi(\varepsilon) = \{G \in \mathcal{F} : G = (1-\varepsilon)\Phi + \varepsilon H;\ H \in \mathcal{F}\}.$$

$\mathrm{var}_{\sup}(\psi_{\varepsilon_1})$ - the supremum of the asymptotic variances $\mathrm{var}(\psi_{\varepsilon_1}, G)$ over $\{G \in \mathcal{P}_\Phi(\varepsilon)\}$,

$\mathrm{efficiency}(\psi_\varepsilon, \psi_{\varepsilon_1})$ - the efficiency of the location estimator generated by the function ψ_{ε_1} with respect to the optimal location estimator, i.e.

$$\frac{\mathrm{var}(\psi_\varepsilon, F_\varepsilon)}{\sup_{G\in\mathcal{P}_\Phi(\varepsilon)} \mathrm{var}(\psi_{\varepsilon_1}, G)}.$$

ε	0.050	0.075	0.100	0.125	0.150	0.175
$\mathrm{var}(\psi_\varepsilon, \Phi)$	1.05	1.06	1.08	1.10	1.11	1.13
$\mathrm{var}(\psi_\varepsilon, F_\varepsilon)$	1.26	1.37	1.49	1.61	1.74	1.89
$\mathrm{var}_{\sup}(\psi_{\varepsilon_1})$	1.26	1.38	1.52	1.67	1.84	2.03
$\mathrm{efficiency}(\psi_\varepsilon, \psi_{\varepsilon_1})$	1.00	0.99	0.98	0.96	0.95	0.93

ε	0.200	0.225	0.250	0.275	0.300	0.325
$\mathrm{var}(\psi_\varepsilon, \Phi)$	1.14	1.15	1.17	1.18	1.19	1.20
$\mathrm{var}(\psi_\varepsilon, F_\varepsilon)$	2.05	2.21	2.40	2.60	2.82	3.07
$\mathrm{var}_{\sup}(\psi_{\varepsilon_1})$	2.23	2.46	2.71	3.00	3.31	3.67
$\mathrm{efficiency}(\psi_\varepsilon, \psi_{\varepsilon_1})$	0.92	0.90	0.88	0.87	0.85	0.84

ε	0.350	0.375	0.400	0.425	0.450	0.475	0.500
$\mathrm{var}(\psi_\varepsilon, \Phi)$	1.22	1.23	1.24	1.25	1.27	1.28	1.29
$\mathrm{var}(\psi_\varepsilon, F_\varepsilon)$	3.34	3.65	4.00	4.39	4.83	5.34	5.93
$\mathrm{var}_{\sup}(\psi_{\varepsilon_1})$	4.07	4.53	5.05	5.65	6.33	7.13	8.05
$\mathrm{efficiency}(\psi_\varepsilon, \psi_{\varepsilon_1})$	0.82	0.81	0.79	0.78	0.76	0.75	0.74

REFERENCES

Csörgő, M., Révész, P. (1981): *Strong Approximations in Probability and Statistics.* Budapest: Akadémiai Kiadó.

Hampel, F. R., Ronchetti, E. M., Rousseeuw, P. J., Stahel, W. A. (1986): *Robust Statistics - The Approach Based on Influence Functions.* Wiley, New York.

Hettmansperger, T. P., Sheater, S. J. (1992): A cautionary note on the method of least median of squares. *The American Statistician, May 1993, 46, 79–83.*

Hewitt, E., Stromberg, K. (1965): *Real and Abstract Analysis.* Berlin: Springer - Verlag.

Huber, P. J. (1973): Robust regression: Asymptotics, conjectures and Monte Carlo. *Ann. Statist. 1, 799–821.*

Jurečková, J. (1988): Constistency of M-estimators of vector parameters. *Proc. of the Fourth Prague Symposium on Asymptotic Statistics, Charles University, Prague, 305–312.*

Mason, R. L., Gunst, R. F. and Hess, J. L. (1989): *Statistical Design and Analysis of Experiments.* New York: John Wiley.

Rousseeuw, P.J., Leroy, A.M. (1987): *Robust Regression and Outlier Detection.* New York: J.Wiley & Sons.

Víšek, J. Á. (1985): Estimating contamination level. *Proc. of the 5th Pannonian Symposium on Mathematical Statistics, Visegrád, Hungary, 401–414.*

Víšek, J. Á. (1992a): Adaptive maximum-likelihood-like estimation in linear models. *Kybernetika 28, 357–382* and *454–471.*

Víšek, J. Á. (1992b): Stability of regression model estimates with respect to subsamples. *Computational Statistics 7, 183–203.*

Víšek, J. Á. (1992c): Problems connected with selection of robust procedure. *Proc. of PROBASTAT'91, Printing House of the Technical University of Liptovský Mikuláš, eds. A. Pázman and J. Volaufová, 189–203.*

Víšek, J. Á. (1993): A cautionary note on the method of Least Median of Squares reconsidered. Submitted to *The American Statistician.*

Authors addresses:

Asunción Maria Rubio

Departamento de Matemáticas, Facultad de Veterinaria, Universidad de Extremadura, 10071 Cáceres, Spain, Phone:34-27-25-71-00 Fax: 34-27-25-71-10

Jan Ámos Víšek

Department of Stochastic Informatics, Institute of Information Theory and Automation, Academy of Sciences the Czech Republic, Pod vodárenskou věží 4, 182 08 Prague, Czech Republic, Phone: 42-2-6605-2019 (or 2466) Fax: 42-2-0774-452, e-mail: VISEK@CSPGAS11.BITNET

ON A MULTISTAGE STOCHASTIC LINEAR PROGRAM

Karel Sladký
Institute of Information Theory and Automation
Academy of Sciences of the Czech Republic
P. O. Box 18, CZ–182 08 Prague 8
Czech Republic

In this note we consider a specific multistage linear decision problem. The randomness is defined by a Markov decision chain governing the development of the considered multistage model. We give some bounds on the growth of optimal values of the decision problem and use that bound to determine the largest value of a discount factor such that the discounted optimal values of the decision problem are finite. For a fixed value of the discount factor we find a (sub)optimal policy.

Key words: linear programming, stochastic programming, dynamic programming
AMS classification: 90C05, 90C15, 90C39

1. Motivation

Let us consider a family of linear programs

$$\widehat{F}(i,s) = \max c(i).\, u \qquad (\mathcal{P})$$

subject to

$$A(i).\, u = s, \qquad u \geq 0$$

where for $i = 1, \dots, k$

$$\begin{aligned} A(i) \in \mathcal{A} &= \{A(1), \dots, A(k)\} \\ c(i) \in \mathcal{C} &= \{c(1), \dots, c(k)\} \end{aligned}$$

$A(i)$ (resp. $c(i)$) is an $m \times n$ (resp. $1 \times m$) given real matrix ($n \geq m$), $s \in \mathbb{R}^m$ is the *vector-state*, and $i \in \mathcal{I} = \{1, \dots, k\}$ is the *index-state* of the considered linear program.
Obviously, to every $s \in \mathbb{R}^m$ and $i \in \mathcal{I}$ a solution to $(\mathcal{P})$ can be easily found.

Throughout this note we shall consider a dynamic system $\mathcal{S}$ given by a sequence $\tilde{h} \equiv \{i_t, s_t; t = 0, 1, \dots\}$ of index- and vector-states and the associated sequence of linear programs ($\mathcal{P}$). Trivially, to each pair (i_t, s_t) we can find *state-decision* $u_t = \widehat{u}_t$ such that

$$\widehat{F}(i_t, s_t) = c(i_t).\,\widehat{u}_t \geq F(i_t, s_t, u_t) = c(i_t).\,u_t$$

subject to

$$A(i_t).\,u_t = s_t, \qquad u_t \geq 0.$$

Then, if the trajectory $h \equiv \{i_t, s_t, u_t;\ t = 0, 1, \ldots\}$ of the system is followed, the performance index of the system $\mathcal{S}$ up to time T,

$$V_\beta^h(T) = \sum_{t=0}^{T} \beta^t.\,F(i_t, s_t, u_t) \leq \sum_{t=0}^{T} \beta^t.\,\widehat{F}(i_t, s_t).$$

Furthermore, if the sequence $\{s_t, t = 0, 1, \ldots\}$ is bounded, linear program ($\mathcal{P}$) is nongenerated for all $i \in \mathcal{I}$ and the discount factor $\beta < 1$, then also the performance index of the system $\mathcal{S}$ over an infinite time horizon $V_\beta^h = \liminf_{T\to\infty} V_\beta^h(T)$ is well defined and finite.

Similarly, no difficulties will arise if i_t are independent random variables taking on values in $\mathcal{I}$. In such a case it suffices to consider instead of $V_\beta^h(T)$ its expectation.

However, if the random variables i_t's are dependent and s_t is given by a linear transformation of s_{t-1}, a family of nontrivial problems will arise. According to our best knowledge, the first systematic study of above mentioned multistage linear programs can be found in Grinold (1976). In this paper infinite stage linear programs are considered and it is assumed that the "randomness" of the sequence $\{i_t, t = 0, 1, \ldots\}$ of index-states of the considered system $\mathcal{S}$ is given by a finite Markov chain; however, no decision can influence the (random) development of index-states of the considered system.

In this note we slightly adapt notations and terminology introduced in Grinold (1976) and extend some of his results to a more general model.

2. The multistage linear decision model

We consider the system $\mathcal{S}$ at discrete time points $t = 0, 1, \ldots$. At time t the system can be described by an $(m + 1)$ dimensional *state* (i_t, s_t) where
$s_t \in \mathbb{R}^m$, $s_t > 0$ is called the *vector-state* at time t, and
$i_t \in \mathcal{I} \equiv \{1, \ldots, k\}$ is the *index-state* at time t.

Supposing that $i_t = i$, an *index-decision* $d_t = \{1, \ldots, r\}$ must be selected. Given index-state $i = i_t$ at time t and selecting index-decision d_t, then the index-state at time $t + 1$ is determined by *transition probabilitites of a controlled Markov chain*, i.e.

$$\mathsf{P}\left[i_{t+1} = j | i_t = i, d_t = d\right] = p_{ij}(d) \tag{1}$$

Similarly, given the vector-state s_t at time t, a *state-decision* $u_t \in \mathbb{R}^n$ (where $n \geq m$, $u_t > 0$) must be taken. Then the vector-state at time $t + 1$ is given by a *linear transformation*

$$s_{t+1} = B(i_t, i_{t+1}; u_t).\,s_t \tag{2}$$

where $B(.,.;.)$ is an $m \times m$ nonnegative matrix. We shall assume that the state-decision u_t satisfies

$$A(i_t).\,u_t = s_t \qquad \text{and that} \quad s_{t+1} = K(i_t, i_{t+1}).\,u_t \tag{3}$$

where $A(i)$ (resp. $K(i,j)$) is a given $m \times n$ Leontief (resp. nonnegative) matrix. Recall that a matrix A is said to be Leontief if it has exactly one positive element in each column and there is some $x > 0$ for which $A.x > 0$. Notice that for matrices A, B we write $A > B$, $A \geq B$ iff $A \neq B$ and for each of its entries $A_{ij} > B_{ij}$, $A_{ij} \geq B_{ij}$, respectively.

Moreover, the selected state-decision u_t results also in a reward with the values $c(i_t).\,u_t$, where $c(.) \in \mathbb{R}^n$, $c(.) = [c_1(.), \ldots, c_n(.)] \geq 0$, discounted by β^t at time t (we only assume that $\beta \in \mathbb{R}^+$).

Throughout this note we shall consider two models:

1) The *basic model,* where we assume that the matrix $A(i)$ is an $m \times m$ identity matrix; hence $u_t \equiv s_t$ for any $t = 0, 1, \ldots$, and $B(i_t, i_{t+1}; u_t) \equiv K(i_t, i_{t+1})$.

2) The *extended model,* where equations (2), (3) hold in the general form.

A *trajectory* of the system is defined by the sequence $h \equiv \{i_t, s_t, d_t, u_t; t = 0, 1, \ldots\}$ of its index- and vector-states and the corresponding index- and state-decisions. A *policy* controlling the system $\mathcal{S}$, say ψ, is a rule how to select decisions at any t (knowing the trajectory of the system up to time t). Throughout this note we restrict our considerations on Markov policies, i.e. the decision taken at time t depends on t and the current index-state i_t and vector-state s_t of the system $\mathcal{S}$. A policy is called *stationary* if the selected index-decision and the null entries of the state-decision u_t depend only on the current index-state of the system $\mathcal{S}$, i.e. if $d_t = d$ is independent of time. A policy is called *periodic* (with *period* τ_0) if for the index-decision $d_t(i)$, taken at time t provided the index-state of the system equals i, it holds $d_t(i) = d_\tau(i)$ where $\tau \equiv t \bmod \tau_0$ and the null entries of u_t depend only on τ and the current index-state i_t.

Supposing that the initial index- and vector-state if the system $\mathcal{S}$ equals i, s respectively, for the total β-discounted rewards we have

$$V_\beta^\psi(i,s) = \liminf_{T \to \infty} \mathsf{E}_{i,s}^\psi\{\sum_{t=0}^{T} \beta^t.\,c(i_t).\,u_t\}; \qquad V_\beta(i,s) = \sup_\psi V_\beta^\psi(i,s)$$

($\mathsf{E}_{i,s}^\psi$ denotes the expectation if the policy ψ is followed and the initial index-state, vector-state equals i, s, respectively; similar notation is used for P_i^ψ.)

3. Analysis of the basic model

In this section we establish some bounds on the growth of the vector-state of the basic model. Recall that for the basic model $A(i) = I$ (throughout this note I is reserved for identity matrix of an appropriate dimension), and hence $u_t \equiv s_t$, $s_{t+1} = K(i_t, i_{t+1}).\,s_t$, $\forall t = 0, 1, \ldots$.

Let for some policy $\psi = \{d_t(i)\}$ and initial state (i_0, s_0)

$$\pi_{i_0}^{\psi}(t;i) = \mathsf{P}_{i_0}^{\psi}[i_t = i], \qquad \sigma_{i_0,s_0}^{\psi}(t;i) = \mathsf{E}_{i_0,s_0}^{\psi}[s_t | i_t = i]$$

$$\bar{\sigma}_{i_0,s_0}^{\psi}(t;i) = \pi_{i_0}^{\psi}(t;i).\,\sigma_{i_0,s_0}^{\psi}(t;i), \quad \text{and hence}$$

$$\mathsf{E}_{i_0,s_0}^{\psi}[s_t] = \sum_{i=1}^{k} \pi_{i_0}^{\psi}(t;i).\,\sigma_{i_0,s_0}^{\psi}(t;i) = \sum_{i=1}^{k} \bar{\sigma}_{i_0,s_0}^{\psi}(t;i)$$

Observe that with the above notations for the expected discounted rewards up to time T we get

$$\mathsf{E}_{i_0,s_0}^{\psi}[\sum_{t=0}^{T} \beta^t.\,c(i_t).\,s_t] = \sum_{t=0}^{T} \beta^t.\sum_{i=1}^{k} \pi_{i_0}^{\psi}(t;i).\,c(i).\,\sigma_{i_0,s_0}^{\psi}(t;i) =$$
$$= \sum_{t=0}^{T} \beta^t.\sum_{i=1}^{k} c(i).\,\bar{\sigma}_{i_0,s_0}^{\psi}(t;i) = \sum_{i=1}^{k} c(i) \sum_{t=0}^{T} \beta^t.\,\bar{\sigma}_{i_0,s_0}^{\psi}(t;i) \tag{4}$$

Taking into account (4), we can see that a recursive formula for calculating $\bar{\sigma}_{i_0,s_0}^{\psi}(t;i)$ will be very useful. To this order observe that

$$\sigma_{i_0,s_0}^{\psi}(t+1;j) = \sum_{i=1}^{k} \mathsf{P}_{i_0}^{\psi}[i_t = i | i_{t+1} = j].\,K(i,j).\,\sigma_{i_0,s_0}^{\psi}(t;i)$$

From Bayes' rule we can conclude

$$\mathsf{P}_{i_0}^{\psi}[i_t = i | i_{t+1} = j] = \frac{p_{ij}(d_t(i)).\,\pi_{i_0}^{\psi}(t;i)}{\pi_{i_0}^{\psi}(t+1;j)}$$

So for any $i, j \in \mathcal{I}$ and $t = 0, 1, \ldots$ we get

$$\bar{\sigma}_{i_0,s_0}^{\psi}(t+1;j) = \sum_{i=1}^{k} p_{ij}(d_t(i)).\,K(i,j).\,\bar{\sigma}_{i_0,s_0}^{\psi}(t;i) \tag{5}$$

where $\bar{\sigma}_{i_0,s_0}^{\psi}(0;i) = s_0 \geq 0$ if $i = i_0$ and $\bar{\sigma}_{i_0,s_0}^{\psi}(0;i) = 0$ elsewhere.

For what follows let us denote $M_{ij}(d) = p_{ij}(d).\,K(i,j)$ (considering the basic model recall that $M_{ij}(d)$ is an $m \times m$ matrix $\forall i,j = 1,\ldots,k$), and hence

$$M(\mathrm{d}) = \begin{bmatrix} M_{11}(d(1)) & \ldots & M_{k1}(d(k)) \\ \vdots & \ldots & \vdots \\ M_{1k}(d(1)) & \ldots & M_{kk}(d(k)) \end{bmatrix}; \quad \bar{\sigma}_{i_0,s_0}^{\psi}(t) = \begin{bmatrix} \bar{\sigma}_{i_0,s_0}^{\psi}(t;1) \\ \vdots \\ \bar{\sigma}_{i_0,s_0}^{\psi}(t;k) \end{bmatrix}$$

Observe that $M(\mathrm{d})$ is an $m.k \times m.k$ nonnegative matrix, $\bar{\sigma}_{i_0,s_0}^{\psi}(t)$ is an $m.k$ row vector and $\mathrm{d} = [d(1),\ldots,d(k)]$ is a k-dimensional *index-decision* vector (the symbol $\mathcal{D}$ is reserved for the set of all admissible index-decision vectors).

Having introduced the above notations equation (5) can be written in a more compact form

$$\bar{\sigma}_{i_0,s_0}^{\psi}(t+1) = M(\mathrm{d}_t).\,\bar{\sigma}_{i_0,s_0}^{\psi}(t) \qquad \text{for} \quad t = 0, 1, \ldots \tag{6}$$

where

$$\bar{\sigma}^{\psi}_{i_0,s_0}(0) = s_{i_0}(0) = \begin{bmatrix} s_0 . \delta_{i_0,1} \\ \vdots \\ s_0 . \delta_{i_0,k} \end{bmatrix}; \qquad \delta_{ij} \text{ denotes the Kronecker symbol.}$$

Iterating (6) we immediately get for $\tau = 0, 1, \ldots$

$$\bar{\sigma}^{\psi}_{i_0,s_0}(t+\tau) = M(\mathrm{d}_{t+\tau-1}). \ldots M(\mathrm{d}_t). \bar{\sigma}^{\psi}_{i_0,s_0}(t) \qquad \text{for} \quad t = 0, 1, \ldots \tag{7}$$

First we establish some bounds on the growth rate of the vector sequence $\{\bar{\sigma}^{\psi}_{i_0,s_0}(t)\}$. To simplify our analysis throughout this note we make the following assumption.

Assumption I. The matrix $M(\mathrm{d})$ is irreducible for any $\mathrm{d} \in \mathcal{D}$.

Recall that by the well-known Perron–Frobenius theorem (cf. e.g. Gantmakher (1959)) under Assumption I the spectral radius of matrix $M(.)$ is equal to its largest positive eigenvalue, denoted $\rho(M(.))$, and the corresponding eigenvector, denoted $u(M(.))$ can be selected positive, i.e. $u(M(.)) > 0$, where $u(.)$ is reserved for the left (row) eigenvector of the matrix M.

In what follows we construct the smallest real number $\rho^\star$ and a corresponding (positive row) vector $u^\star$ such that

$$\rho^\star . u^\star \geq u^\star . M(\mathrm{d}) \quad \text{for any } \mathrm{d} \in \mathcal{D}. \tag{8}$$

To this order let $m_i(\mathrm{d})$ (resp. $m_{ij}(\mathrm{d})$) be the i-th column (resp. ij-th element) of the matrix $M(\mathrm{d})$ (observe that $\mathrm{d} = [d(1), \ldots, d(k)]$ and in every column of the matrix $M(\mathrm{d})$ the decision is the same) and introduce a (finite) set of $m.k \times m.k$ nonnegative matrices

$$\mathcal{M} = \{M : m_i = m_i(\mathrm{d}) \qquad \text{for } i = 1, \ldots, m.k;\ \mathrm{d} \in \mathcal{D}\}$$

Observe that the set $\mathcal{M}$ possesses an important "column product property", i.e. if $m_i(\mathrm{d}(i))$ (resp. $m_j(\mathrm{d}(j))$) is the i-th (resp. j-th) column of some matrix $M(\mathrm{d}(i)) \in \mathcal{M}$ (resp. $M(\mathrm{d}(j)) \in \mathcal{M}$), then there exists some $M = [m_{ij}] \in \mathcal{M}$ such that $m_i = m_i(\mathrm{d}(i))$ and $m_j = m_j(\mathrm{d}(j))$.

We suggest an iterative procedure for finding $M^\star \in \mathcal{M}$ and a positive row vector $u^\star \in \mathbb{R}^{m.k}$ such that

$$\rho^\star . u^\star = u^\star . M^\star \geq u^\star . M(\mathrm{d}) \tag{9}$$

for any $\mathrm{d} \in \mathcal{D}$.

Algorithm 1.

Step 0. Choose $M = M(\mathrm{d})$ for some $\mathrm{d} \in \mathcal{D}$.

Step 1. For some $M \in \mathcal{M}$ find $\rho(M)$ along with a corresponding left eigenvector $u(M) \in \mathbb{R}^{m.k}$, $u(M) > 0$ such that $\rho(M). u(M) = u(M). M$.

Step 2. Select $N \in \mathcal{M}$ (if possible) such that $u(M).N \geq \rho(M).u(M)$, and repeat Step 1 for $M = N$.

Step 3. If there exists no $N \in \mathcal{M}$ such that $u(M).N \geq \rho(M).u(M)$, then set $\rho(M) = \rho^\star, u(M) = u^\star$ and stop.

Proposition 1. The values $\rho^\star$, $u^\star$ found by the iterative procedure of Algorithm 1 fulfil (9), and there exists no $\rho < \rho^\star$ such that (7), (9) can hold for some $u > 0$. Moreover, Algorithm 1 terminates in a finite number of steps.

The proof of Proposition 1 can be performed by standard arguments of dynamic programming (cf. e.g. Howard and Matheson (1972), Sladký (1976)).

Iterating (9) we obtain bounds on the growth rates of elements of the vector sequence $\{\bar{\sigma}^{\psi}_{i_0,s_0}(t),\ t = 0, 1, \ldots\}$.

Proposition 2. There exists (column) vector $a(s_0) \in \mathbb{R}^{m.k}$, $a(s_0) > 0$ such that

$$\bar{\sigma}^{\psi}_{i_0,s_0}(t) \leq (\rho^\star)^t \cdot a(s_0) \qquad \text{for all} \quad t = 0, 1, \ldots \tag{10}$$

From (4), (9) we immediately get the following corollary.

Corollary 1. Let the discount factor $\beta < (\rho^\star)^{-1}$. Then $V^{\psi}_{\beta}(i, s)$ is finite for any admissible policy ψ, any initial value of the vector-state s and every $i \in \mathcal{I}$.

Remark 1. It may happen that the policy maximizing the growth of $\bar{\sigma}^{\psi}_{i_0,s_0}(t)$ need not be stationary as the following example shows.

Example 1. Let $\mathcal{I} = \{1, 2\}$, $m = 2$, $r = 2$, where for some $\varepsilon > 0$

$$p_{11}(1) = 1 - \varepsilon; \quad p_{12}(1) = \varepsilon; \quad p_{11}(2) = \varepsilon; \quad p_{12}(2) = 1 - \varepsilon$$

$$p_{21}(1) = p_{21}(2) = p_{22}(1) = p_{22}(2) = p_{21} = p_{22} = \frac{1}{2}$$

$$K(1,1) = \begin{bmatrix} 1 & 10^6 \\ 0 & 1 \end{bmatrix}; K(1,2) = \begin{bmatrix} 1 & 1 \\ 1 & 1 \end{bmatrix}; K(2,1) = \begin{bmatrix} 2 & 2 \\ 2 & 2 \end{bmatrix}; K(2,2) = \begin{bmatrix} 2 & 0 \\ 0 & 2 \end{bmatrix}$$

Consider the matrices

$$C(1) = \begin{bmatrix} K(1,1) & K(2,1) \\ 0 & K(2,2) \end{bmatrix} = \begin{bmatrix} 1 & 10^6 & 1 & 1 \\ 0 & 1 & 1 & 1 \\ 0 & 0 & 1 & 0 \\ 0 & 0 & 0 & 1 \end{bmatrix}$$

$$C(2) = \begin{bmatrix} 0 & K(2,1) \\ K(1,2) & K(2,2) \end{bmatrix} = \begin{bmatrix} 0 & 0 & 1 & 1 \\ 0 & 0 & 1 & 1 \\ 1 & 1 & 1 & 0 \\ 1 & 1 & 0 & 1 \end{bmatrix}$$

Obviously, $\rho(C(1)) = 1$; $\rho(C(2)) \in (2;3)$ and after some calculation we get $\rho(C(2)) = 2,55$. However

$$C(1)\,.\,C(2) = \begin{bmatrix} 2 & 2 & 2+10^6 & 2+10^6 \\ 2 & 2 & 2 & 2 \\ 1 & 1 & 1 & 0 \\ 1 & 1 & 0 & 1 \end{bmatrix}$$

After some algebra we get that the spectral radius of the matrix product $C(1).\,C(2)$ is greater than 10^3 and the spectral radius of the matrix product $C(1)\,.\,C(1)$ equals 1, $C(2)\,.\,C(2)$ is less than 6,6 respectively.

Recalling that

$$M(1) = \begin{bmatrix} p_{11}(1).\,K(1,1) & p_{21}.\,K(2,1) \\ p_{12}(1).\,K(1,2) & p_{22}.\,K(2,2) \end{bmatrix}; \qquad M(2) = \begin{bmatrix} p_{11}(2).\,K(1,1) & p_{21}.\,K(2,1) \\ p_{12}(2).\,K(1,2) & p_{22}.\,K(2,2) \end{bmatrix}$$

we obtain

$$M(1) = \begin{bmatrix} 1-\varepsilon & (1-\varepsilon).\,10^6 & 1 & 1 \\ 0 & 1-\varepsilon & 1 & 1 \\ \varepsilon & \varepsilon & 1 & 0 \\ \varepsilon & \varepsilon & 0 & 1 \end{bmatrix}; \qquad M(2) = \begin{bmatrix} \varepsilon & \varepsilon.\,10^6 & 1 & 1 \\ 0 & \varepsilon & 1 & 1 \\ 1-\varepsilon & 1-\varepsilon & 1 & 0 \\ 1-\varepsilon & 1-\varepsilon & 0 & 1 \end{bmatrix}$$

From the above facts, for sufficiently small $\varepsilon > 0$ we can immediately conclude that on alterning in state 1 decisions 1 and 2 we obtain greater growth of elements of the resulting matrix product than for any stationary policy (recall that eigenvalues of a matrix are continuous functions of the matrix elements).

4. Optimal and suboptimal solutions in the basic model

In this section we evaluate the expected discounted rewards. Since in the basic model $u_t \equiv s_t$ we immediately get

$$V_\beta^\psi(i_0, s_0) = \liminf_{T\to\infty} \mathsf{E}_{i_0,s_0}^\psi[\sum_{t=0}^T \beta^t.\,c(i_t).\,s_t]$$

Supposing that $\beta < (\rho^\star)^{-1}$, on letting $T \to \infty$ then by Corollary 1 the expected discounted rewards are well defined and finite also over infinite time horizon. Hence by (4) we get

$$V_\beta^\psi(i_0, s_0) = \liminf_{T\to\infty} [\sum_{t=0}^T \beta^t.\,\mathrm{c}\,.\,\bar\sigma_{i_0,s_0}^\psi(t)] = \mathrm{c}\,.\sum_{t=0}^\infty \beta^t.\,\bar\sigma_{i_0,s_0}^\psi(t) \tag{11}$$

where $\mathrm{c} = [c(1), \ldots, c(k)] \geq 0$ is a row vector (observe that $\mathrm{c} \in \mathbb{R}^{m.k}$) and $\bar\sigma_{i_0,s_0}^\psi(t)$ are defined recursively by

$$\bar\sigma_{i_0,s_0}^\psi(t+1) = M(\mathrm{d}_t).\,\bar\sigma_{i_0,s_0}^\psi(t) \qquad \text{for any } t = 0, 1, \ldots \tag{12}$$

with $\bar\sigma_{i_0,s_0}^\psi(0) = s_{i_0}(0) \geq 0$ given.

Now suppose that the policy ψ is periodic with period τ_0, i.e. $d_t(.) \equiv d_\tau(.)$ where $\tau \equiv t \bmod \tau_0$ for $t = 0, 1, \ldots$. Inserting (11) into (12) and supposing that $\beta < (\rho^\star)^{-1}$ we get (in what follows we set $\prod_{\tau=i}^{j} a_\tau = 1$ if $j < i$)

$$V_\beta^\psi(i_0, s_0) = \liminf_{T\to\infty}\{\mathrm{c}.\sum_{t=0}^{T}(\sum_{\tau=1}^{\tau_0}\beta^{\tau-1}.\prod_{j=0}^{\tau-2}M(\mathrm{d}_{\tau-1-j})).(\beta^{\tau_0}.\prod_{\tau=0}^{\tau_0-1}M(\mathrm{d}_{\tau_0-1-\tau}))^t.s_{i_0}(0)\}$$

$$= \mathrm{c}.(\sum_{\tau=1}^{\tau_0}\beta^{\tau-1}.\prod_{j=0}^{\tau-2}M(\mathrm{d}_{\tau-1-j})).(I - \beta^{\tau_0}.M_\beta^\psi(\tau_0))^{-1}.s_{i_0}(0) \tag{13}$$

where $M_\beta^\psi(\tau_0) = \prod_{\tau=0}^{\tau_0-1} M(\mathrm{d}_{\tau_0-1-\tau}))$.

In the remainder of this section we shall discuss policy iteration methods for constructing upper bounds on the maximum expected discounted rewards and for finding a suboptimal policy in the class of stationary policies.

First we suggest an iterative procedure for finding $M^\star \in \mathcal{M}$ and a row vector $v_\beta^\star \in \mathbb{R}^{m.k}$ such that

$$v_\beta^\star = \mathrm{c}.(I - \beta.M_\beta^\star)^{-1} \geq v_\beta(\mathrm{d}) = \mathrm{c}.(I - \beta.M(\mathrm{d}))^{-1} \tag{14}$$

for any $\mathrm{d} \in \mathcal{D}$ (sometimes we abbreviate $v_\beta(M(\mathrm{d}))$ by $v_\beta(\mathrm{d})$).

Furthermore, from (14) also

$$v_\beta^\star = \mathrm{c} + \beta.v_\beta^\star.M_\beta^\star \geq v_\beta(\mathrm{d}) = \mathrm{c} + \beta.v_\beta(\mathrm{d}).M_\beta(\mathrm{d}) \tag{15}$$

Since $s_i(0) > 0$ from (15) we immediately get that for an arbitrary $s_i(0)$ it also holds for the scalars $\mathrm{c}.(I - \beta.M_\beta^\star)^{-1}.s_i(0)$

$$v_\beta^\star.s_i(0) = \mathrm{c}.(I - \beta.M_\beta^\star)^{-1}.s_i(0) \geq v_\beta(\mathrm{d}).s_i(0) = \mathrm{c}.(I - \beta.M(\mathrm{d}))^{-1}.s_i(0) \tag{16}$$

Observe that $v_\beta(\mathrm{d}) \in \mathbb{R}^{m.k}$ and the $(m.i + j)$-th component of $v_\beta(\mathrm{d})$ denotes (for $i = 0, \ldots, k-1$, $j = 1, \ldots, m$) the expected discounted reward if $\mathrm{d} \in \mathcal{D}$ is selected and the initial index state i_0 is equal to $i + 1$ and simultaneously, except of the j-th component of the initial vector state equal to one, all components of the initial vector-state s_0 equal zero.

The optimal matrix $M^\star$ along with $v_\beta^\star$ can be found by the following iterative procedure.

Algorithm 2.

Step 0. Choose $M = M(\mathrm{d})$ for some $\mathrm{d} \in \mathcal{D}$.

Step 1. For some $M \in \mathcal{M}$ find $v_\beta(M) \in \mathbb{R}^{m.k}$ such that such that $v_\beta(M) = \mathrm{c} + \beta.v_\beta(M).M$.

Step 2. Select $N \in \mathcal{M}$ (if possible) such that $v_\beta(M).N \geq v_\beta(M).M$, and repeat Step 1 for $M = N$.

Step 3. If there exists no $N \in \mathcal{M}$ such that $v_\beta(M).N \geq v_\beta(M).M$, then set $v_\beta(M) = v_\beta^\star$, $M = M_\beta^\star$, and stop.

Proposition 3. The values $v_\beta^\star$, $M_\beta^\star$ found by the iterative procedure of Algorithm 2 fulfil (15), (16). Moreover, Algorithm 2 terminates in a finite number of steps.

The proof of Proposition 3 can be performed by standard arguments of "transient" dynamic programming introduced in Veinott (1969). To this order observe that the matrix $M(\mathrm{d})$ is nonnegative and its spectral radius is less than β^{-1}; hence the conditions of "transient" dynamic programming are fulfilled.

Remark 2. It may happen (cf. Remark 1) that there exists no $\mathrm{d} \in \mathcal{D}$ such that $M_\beta^\star = M(\mathrm{d})$ and the first part of (15) cannot be fulfilled in the class of stationary policies. However, a "better approximation" can be obtained in the class of periodic policies. For this reason let us denote for natural number τ_0 $\mathcal{K}(\tau_0) = \{M(\mathrm{d}_{\tau_0-1}).\dots.M(\mathrm{d}_0) : \mathrm{d}_i \in \mathcal{D} \text{ for } i = 0,\dots,\tau_0 - 1\}$ and let $\mathcal{M}(\tau_0)$ be the smallest matrix set such that

(i) $\mathcal{M}(\tau_0) \supset \mathcal{K}(\tau_0)$,

(ii) the set $\mathcal{M}(\tau_0)$ possesses the "column product property".

Hence we can extend (15), (16) in the following way

$$v_\beta^\star(\tau_0) = \mathrm{c}.\,(\sum_{\tau=1}^{\tau_0} \beta^{\tau-1}.\ \prod_{j=0}^{\tau-2} M(\mathrm{d}^\star_{\tau-1-j})).\,(I - \beta^{\tau_0}.\,M_\beta^\star(\tau_0))^{-1} \geq$$

$$v_\beta(\psi,\tau_0) = \mathrm{c}.\,(\sum_{\tau=1}^{\tau_0} \beta^{\tau-1}.\ \prod_{j=0}^{\tau-2} M(\mathrm{d}_{\tau-1-j})).\,(I - \beta^{\tau_0}.\,M_\beta^\psi(\tau_0))^{-1} \tag{17}$$

$$v_\beta^\star(\tau_0).\,s_i(0) \geq v_\beta(\psi,\tau_0).\,s_i(0) \tag{18}$$

for any policy $\psi \sim (\mathrm{d}_0, \mathrm{d}_1, \dots \mathrm{d}_{\tau_0-1})$ periodic with period τ_0. Observe that a policy iteration method arising by a slight modification of Algorithm 2 can be used for finding a suboptimal matrix $M_\beta^\psi(\tau_0)$ in the class of periodic policies.

The above results will be illustrated on the following example.

Example 2. Let $\mathcal{I} = \{1,2\}$, $m = 2$, $r = 2$, where

$$p_{11}(1) = \frac{3}{4};\quad p_{12}(1) = \frac{1}{4};\quad p_{11}(2) = \frac{1}{4};\quad p_{12}(2) = \frac{3}{4};$$

$$p_{21}(1) = p_{21}(2) = p_{22}(1) = p_{22}(2) = p_{21} = p_{22} = \frac{1}{2}$$

$$K(1,1) = \begin{bmatrix} 1 & 0 \\ 0 & 1 \end{bmatrix};\quad K(1,2) = \begin{bmatrix} 1 & 1 \\ 1 & 1 \end{bmatrix};\quad K(2,1) = \begin{bmatrix} 2 & 2 \\ 2 & 2 \end{bmatrix};\quad K(2,2) = \begin{bmatrix} 2 & 0 \\ 0 & 2 \end{bmatrix}$$

and $\mathrm{c} = [10;\ 10;\ 20;\ 20]$. Since

$$M(1) = \begin{bmatrix} p_{11}(1).\,K(1,1) & p_{21}.\,K(2,1) \\ p_{12}(1).\,K(1,2) & p_{22}.\,K(2,2) \end{bmatrix};\qquad M(2) = \begin{bmatrix} p_{11}(2).\,K(1,1) & p_{21}.\,K(2,1) \\ p_{12}(2).\,K(1,2) & p_{22}.\,K(2,2) \end{bmatrix}$$

we get

$$M(1)=\begin{bmatrix} \frac{3}{4} & 0 & 1 & 1 \\ 0 & \frac{3}{4} & 1 & 1 \\ \frac{1}{4} & \frac{1}{4} & 1 & 0 \\ \frac{1}{4} & \frac{1}{4} & 0 & 1 \end{bmatrix}; \qquad M(2)=\begin{bmatrix} \frac{1}{4} & 0 & 1 & 1 \\ 0 & \frac{1}{4} & 1 & 1 \\ \frac{3}{4} & \frac{3}{4} & 1 & 0 \\ \frac{3}{4} & \frac{3}{4} & 0 & 1 \end{bmatrix}.$$

Looking at the column sums of $M(1)$, $M(2)$ we conclude that $\rho^* < 3$ and hence from Proposition 2 we immediately get that $V_\beta^\psi(i,s) < \infty$ if the discount factor $\beta < \frac{1}{3}$.

Using the above mentioned algorithm we shall find upper bounds on expected discounted reward. Employing Algorithm 2 we conclude that for

$$v_\beta(1) = \mathrm{c} + \beta.\, v_\beta(1).\, M(1) < v_\beta(2) = \mathrm{c} + \beta.\, v_\beta(2).\, M(2)$$

for the values of discount factor $\beta = 0,1;\ 0,2;\ 0,3$.

After some calculations we get

β	$M(1)$					$M(2)$				
0,1	$v_\beta(1) =$	[12,16	12,16	24,92	24,92]	$v_\beta(2) =$	[14,16	14,16	25,37	25,37]
0,2	$v_\beta(1) =$	[15,63	15,63	32,80	32,80]	$v_\beta(2) =$	[21,87	21,87	35,94	35,94]
0,3	$v_\beta(1) =$	[22,10	22,10	47,51	47,51]	$v_\beta(2) =$	[42,38	42,38	64,90	64,90]

5. Remarks on the extended model

In this section we indicate how the results obtained in Sections 3,4 can be used for investigating of the extended model. Notations introduced in Section 3 will be followed and, moreover, we denote

$$\mu^\psi_{i_0,s_0}(t;i) = \mathsf{E}^\psi_{i_0,s_0}[u_t | i_t = i]; \qquad \bar{\mu}^\psi_{i_0,s_0}(t;i) = \pi^\psi_{i_0}(t;i).\, \mu^\psi_{i_0,s_0}(t;i)$$

Obviously,

$$\mathsf{E}^\psi_{i_0,s_0}[u_t] = \sum_{i=1}^{k} \pi^\psi_{i_0}(t;i).\, \mu^\psi_{i_0,s_0}(t;i) = \sum_{i=1}^{k} \bar{\mu}^\psi_{i_0,s_0}(t;i)$$

Observe that

$$\mathsf{E}^\psi_{i_0,s_0}[\sum_{t=0}^{T} \beta^t.\, c(i_t).\, u_t] = \sum_{t=0}^{T} \beta^t.\, \sum_{i=1}^{k} \pi^\psi_{i_0}(t;i).\, c(i).\, \mu^\psi_{i_0,s_0}(t;i) =$$

$$= \sum_{t=0}^{T} \beta^t.\, \sum_{i=1}^{k} c(i).\, \bar{\mu}^\psi_{i_0,s_0}(t,i) = \sum_{i=1}^{k} c(i).\, \sum_{t=0}^{T} \beta^t.\, \bar{\mu}^\psi_{i_0,s_0}(t,i) \tag{19}$$

Similarly as for the basic model we construct a recursive formula for calculating $\bar{\mu}^\psi_{i_0,s_0}(t;i)$. Since (cf. (3))

$$A(i_{t+1}).\, u_{t+1} = K(i_t, i_{t+1}).\, u_t \qquad t = 0, 1, \ldots$$

we conclude (cf. (5)) that

$$\bar{\sigma}^{\psi}_{i_0,s_0}(t+1;j) = A(j).\,\bar{\mu}^{\psi}_{i_0,s_0}(t+1;j) = \sum_{i=1}^{k} p_{ij}(d_t).\,K(i,j).\,\bar{\mu}^{\psi}_{i_0,s_0}(t;i) \qquad (20)$$

where $A(i).\,\bar{\mu}^{\psi}_{i_0,s_0}(0;i) = s_0$ if $i = i_0$ and $\bar{\mu}^{\psi}_{i_0,s_0}(0;i) = 0$ elsewhere.

Using more compact notations (20) can be also written as

$$\bar{\sigma}^{\psi}_{i_0,s_0}(t+1) = \tilde{A}.\,\bar{\mu}^{\psi}_{i_0,s_0}(t+1) = M(\mathrm{d}_t).\,\bar{\mu}^{\psi}_{i_0,s_0}(t) \le K.\,\bar{\mu}^{\psi}_{i_0,s_0}(t) \quad \text{for } t = 0,1,\ldots \qquad (21)$$

where

$$\bar{\mu}^{\psi}_{i_0,s_0}(t) = \begin{bmatrix} \bar{\mu}^{\psi}_{i_0,s_0}(t;1) \\ \vdots \\ \bar{\mu}^{\psi}_{i_0,s_0}(t;k) \end{bmatrix};\; K = \begin{bmatrix} K(1,1) & \ldots & K(k,1) \\ \vdots & \ldots & \vdots \\ K(1,k) & \ldots & K(k,k) \end{bmatrix};\; \tilde{A} = \operatorname{diag}[A(1),\ldots,A(k)]$$

and $\tilde{A}.\,\bar{\mu}^{\psi}_{i_0,s_0}(0) = s_{i_0}(0)$. Observe that (since each of the matrices $A(i)$ is Leontief) the matrix $\tilde{A}$ is also a Leontief matrix.

Now let $\tilde{\bar{\mu}}^{\psi}_{i_0,s_0}(\beta) = \sum_{t=0}^{\infty} \beta^t.\,\bar{\mu}^{\psi}_{i_0,s_0}(t)$, $\tilde{\bar{\mu}}^{\psi}_{i_0,s_0}(\beta;i) = \sum_{t=0}^{\infty} \beta^t.\,\bar{\mu}^{\psi}_{i_0,s_0}(t;i)$ and let $\beta \in \mathbb{R}^+$ be selected so that $\tilde{\bar{\mu}}^{\psi}_{i_0,s_0}(\beta) < \infty$. Observe that on introducing $\tilde{x}(\beta) = \sum_{t=0}^{\infty} \beta^t.\,x(t)$ where $\tilde{A}.\,x(t+1) = K.\,x(t)$, $x(t) \in \mathbb{R}^{kn}$, $x(0) = s_{i_0}(0)$, then boundedness of $\tilde{\bar{\mu}}^{\psi}_{i_0,s_0}(\beta)$ follows immediately from boundedness of $\tilde{x}(\beta)$.

From (19) we get

$$V^{\psi}_{\beta}(i_0,s_0) = \liminf_{T\to\infty} \mathsf{E}^{\psi}_{i_0,s_0}\{\sum_{t=0}^{T} \beta^t.\,c(i_t).\,u_t\} = \mathrm{c}.\,\tilde{\bar{\mu}}^{\psi}_{i_0,s_0}(\beta) \qquad (22)$$

where $\mathrm{c} = [c(1),\ldots,c(k)] \ge 0$.

Now suppose that the policy ψ is stationary, i.e. $\mathrm{d}_t \equiv \mathrm{d} \quad \forall t = 0,1,\ldots$ and also the null entries of state-decision are independent of time. Then we write $\psi \sim \mathrm{d}$ and from (21) we immediately conclude

$$(\tilde{A} - \beta.\,M(\mathrm{d}))\,.\,\tilde{\bar{\mu}}^{\psi}_{i_0,s_0}(\beta) = s_{i_0}(0) \qquad (23)$$

For a fixed stationary policy $\psi \sim \mathrm{d}$, we find a (stationary) state-decision maximizing $V^{\psi}_{\beta}(i_0,s_0)$. To this order suppose that $\tilde{x}(\beta) < \infty$ and consider the problem

$$\max\ \mathrm{c}.\,x$$

subject to $\qquad (\star)$

$$(\tilde{A} - \beta K).\,x = s_{i_0}(0), \qquad x > 0$$

We make the following

Assumption II. Problem $(\star)$ has a bounded objective and the columns of the Leontief matrix $\tilde{A}$ are scaled so that the positive elements of $\tilde{A}$ are not greater than one.

Observe that since $\tilde{A}$ (resp. $M(\mathrm{d})$) is a Leontief (resp. nonnegative) matrix, for sufficiently small β the matrix $(\tilde{A} - \beta . K)$ is also Leontief and the same holds for $(\tilde{A} - \beta . M(\mathrm{d}))$ for any $\mathrm{d} \in \mathcal{D}$. Hence (cf. Veinott (1968), Thm. 6) since the RHS of $(\star)$ is positive, there exists a solution to $(\star)$ such that only $m.k$ elements of x are nonzero. For this reason we assign to every matrix $A(i)$ stationary state-decision $\delta_i \in \Delta_i$ specifying an $m \times m$ basis $A^{\delta_i}(i)$ of $A(i)$ as well as the corresponding basis of each $K(i,j)$. A^{δ} (where $\delta \in \Delta \equiv \Delta_1 \times \ldots \times \Delta_k$) is the resulting $m.k \times m.k$ matrix.

Observe that then from (23) we get

$$(I - \beta . (\tilde{A}^{\delta})^{-1} . M^{\delta}(\mathrm{d})) . \tilde{\mu}^{\psi}_{i_0,s_0}(\beta) = (\tilde{A}^{\delta})^{-1} . s_{i_0}(0) \tag{24}$$

where $\mathrm{d} = [d(1), \ldots, d(k)] \in \mathcal{D}$.

In the remainder of this section we show how to adapt algorithmic procedures suggested in the study of the basic model for finding an upper bound on maximum expected rewards.

Algorithm 3.
Step 1. Find $\delta \in \Delta$ (specifying $\tilde{A}^{\delta}$) such that x is a solution to $(\star)$.
Step 2. Set $M(\mathrm{d}) = (\tilde{A}^{\delta})^{-1} . M^{\delta}(\mathrm{d}))$ and apply Algorithm 2 for finding $v^{\star}_{\beta} = \mathrm{c} . (I - \beta . M^{\star}_{\beta})^{-1} \gtrsim \mathrm{c} . (I - \beta . (\tilde{A}^{\delta})^{-1} . M^{\delta}(\mathrm{d}))^{-1}$ for any $\mathrm{d} \in \mathcal{D}$.
Step 3. Set $F^{\star}_{\beta} = v^{\star}_{\beta} . (\tilde{A}^{\delta})^{-1} . s_{i_0}(0)$.

Proposition 4. The value $F^{\star}_{\beta}$ is an upper bound on $V^{\psi}_{\beta}(i_0, s_0)$, i.e. $F^{\star}_{\beta} \geq V^{\psi}_{\beta}(i_0, s_0)$ over all policies ψ selecting stationary state-decision δ.

Acknowledgement. This work was supported by the Grant Agency of the Czech Republic under Grant No. 402/93/0631 and by the Grant Agency of the Academy of Sciences of the Czech Republic under Grant No. 27 561.

REFERENCES

F. R. Gantmakher (1959): The Theory of Matrices. Volume I, II. Chelsea, New York.

R. C. Grinold (1976): A new approach to multistage stochastic linear programs. Mathematical Programming Study *6*, 19–29.

R. A. Howard and J. E. Matheson (1972): Risk-sensitive Markov decision processes. Manag. Science *18*, 357–369.

K. Sladký (1976): On dynamic programming recursions for multiplicative Markov decision chains. Mathematical Programming Study *6*, 216–226.

A. F. Veinott, Jr. (1968): Extreme points of Leontief substitution systems. Linear Algebra and Appl. *1*, 181–194.

A. F. Veinott, Jr. (1969): Discrete dynamic programming with sensitive discount optimality criteria. Ann. Math. Statist. *40*, 1635–1660.

CHANGE POINT AND JUMP ESTIMATES IN AN AMOC RENEWAL MODEL

Josef Steinebach, Marburg *

Abstract. A simple renewal model is studied which allows for atmost one change (AMOC) in the distribution of the underlying interoccurence times. We discuss tests and estimates concerning the change point as well as the magnitude of jump at this point. The results include limiting null distributions, consistency of tests and estimates, and asymptotic normality. Proofs make use of invariance principles for the renewal process under consideration.

Keywords: Change point analysis, jump estimate, AMOC renewal model, invariance principles, convergence rates.

AMS 1991 subject classifications. Primary 60 K 05; Secondary 62 G 05, 62 G 10, 62 G 20

1. Introduction

Consider a renewal process $\{N(t):t\geq 0\}$ based on interoccurence times $X_1,\dots,X_{[n\lambda]}$, $X_{[n\lambda]+1}+\Delta$, $X_{[n\lambda]+2}+\Delta,\dots$ $(\lambda>0,\ \Delta\in\mathbb{R})$, where $\{X_i:i\geq 1\}$ is an i.i.d. sequence of *positive* r.v.'s with $EX_1=\mu>0$, $Var(X_1)=\sigma^2>0$, and either

(H1) $EX_1^r<\infty$ $(\exists r>2)$, or

(H2) $E\exp(sX_1)<\infty$ $(\exists s>0)$.

If $\Delta<0$, we assume $P(X_1+\Delta>0)=1$ to restrict our attention to "ordinary" renewal processes. We observe $\{N(t):t\geq 0\}$ at discrete time points $t_j=j$ $(j=1,\dots,n)$, and further assume

(H3) $\lambda\mu<1$,

---

*) **Address for correspondance:** Josef Steinebach, Fachbereich Mathematik, Philipps-Universität, Hans-Meerwein-Straße, D-35032 Marburg, F.R. Germany

in order to be able to detect a possible change up to time n. Note that

$$N(t) = \max\{k : S_k \le t\} \quad (t \ge 0) ,$$

where

$$S_k = \sum_{i=1}^{k} X_i + (k-[n\lambda])^+ \Delta \quad (k=1,2,\dots) .$$

By the law of large numbers,

$$S_{[n\lambda]} \sim n\lambda\mu \qquad \text{a.s.}$$

as $n \to \infty$, which motivates assumption (H3).
We are interested in the statistical inference about λ ("change point"), $\lambda\mu$ ("time of change"), and Δ ("jump"), using the observations $N(j)$, $j=1,\dots,n$.

AMOC models as introduced above have extensively been studied in the literature, but mainly based upon direct observation of the X's (for a detailed survey we refer to Csörgö and Horváth (1988) or Krishnaiah and Miao (1988)). Chen (1988) suggests a statistical inference taking local changes of the X–sequence into account. Related problems have been discussed in regression analysis (cf. e.g. Hušková (1990 a,b), and the references mentioned therein).

Following similar ideas as in Chen (1988), we introduce the statistic

$$(1.1) \qquad M_n = (2k_n)^{-1/2} \max\{|D_{jn}| : j=0,\dots,n-2k_n\} ,$$

where

$$(1.2) \qquad D_{jn} = \{N(j+k_n) - N(j)\} - \{N(j+2k_n) - N(j+k_n)\} ,$$

and $\{k_n : n \ge 1\}$ denotes a suitable integer sequence to be specified below. Typically, $k_n \ll n$ as $n \to \infty$, so that we have a comparison of empirical intensities in all subsequent *small* intervals within the large range $[0,n]$. The statistics of type M_n as in (1.1) also play a role in the asymptotic analysis of "pontograms" (cf. e.g. Steinebach and Eastwood (1992)).

Our first result in Section 2 provides a consistent change point test based upon a Gumbel type extreme value limiting distribution. Secondly, consistent estimates of the change parameters under consideration are constructed using

(1.3) $$\hat{i}_n = \min\{i : |D_{in}| = \max\{|D_{jn}| : j=0,\dots,n-2k_n\}\}.$$

Here, we assume that the existence of a change in the model is either known in advance or evident by rejection of the null hypothesis $H_0:\Delta=0$.

In Section 3, convergence rate statements and asymptotic normality of the above estimators are established. Section 4 contains the proofs of our results, mainly based on strong approximations to renewal processes without changes (cf. Csörgõ, Horváth, Steinebach (1987) and Mason, van Zwet (1987)).

2. Consistent tests and estimates

According to the moment assumptions (H1) or (H2), we shall need the following assumptions on the integer sequence $\{k_n : n \geq 1\}$:

(K1) $k_n \uparrow,\ k_n/n \longrightarrow 0,\ n^{2/r}\log(n/k_n)/k_n \longrightarrow 0$ (as $n \longrightarrow \infty$);

(K2) $k_n \uparrow,\ k_n/n \longrightarrow 0,\ (\log^2 n)\log(n/k_n)/k_n \longrightarrow 0$ (as $n \longrightarrow \infty$).

Theorem 2.1. Assume (H1) $\wedge$ (K1) or (H2) $\wedge$ (K2). Then,

a) Under $H_0:\Delta=0$,

(2.1) $$a_n(\mu^{3/2}/\sigma)\, M_n - b_n \xrightarrow{\mathcal{D}} Y \qquad (n \to \infty),$$

where $a_n=\{2\log(n/k_n)\}^{1/2}$, $b_n=2\log(n/k_n)+\frac{1}{2}\log\log(n/k_n) - \frac{1}{2}\log(4\pi/9)$, $P(Y \leq y) = \exp(-2e^{-y})$, $y \in \mathbb{R}$;

b) Under $H_1 : \Delta \neq 0$,

(2.2) $$a_n(\mu^{3/2}/\sigma)\, M_n - b_n \xrightarrow{P} +\infty \qquad (n \to \infty),$$

so that an asymptotic test based on M_n is consistent.

Remark 2.1. In order to apply the above test in practice, one has to estimate the unknown parameter $\vartheta = \sigma/\mu^{3/2}$ not only consistently, but at a good enough rate, too, i.e.

$$(2.3) \quad \hat{\vartheta}_n - \vartheta \overset{!}{=} o_P(1/\log(n/k_n)) \quad (\text{as } n \to \infty).$$

For possible solutions to this estimation problem we refer to Steinebach (1993). A similar assumption applies in the partial sum case (cf. Chen (1988)).

In what follows, we assume the existence of a change point $\lambda > 0$ and jump $\Delta \neq 0$ in the model of Section 1.

First we provide consistent estimates of $\nu = \lambda\mu$, $\alpha = 1/\mu$ and λ. Making use of $\hat{i}_n$ as defined in (1.3) we obtain:

Theorem 2.2. Assume (H1) $\wedge$ (K1) or (H2) $\wedge$ (K2). Then, as $n \to \infty$,

a) $\hat{\nu}_n := \hat{i}_n/n \quad \xrightarrow{P} \quad \nu = \lambda\mu$,

b) $\hat{\alpha}_n := N(\hat{i}_n)/\hat{i}_n \quad \xrightarrow{P} \quad \alpha = 1/\mu$,

c) $\hat{\lambda}_n := N(\hat{i}_n)/n \quad \xrightarrow{P} \quad \lambda$.

Remark 2.2. For numerical evaluation, $\hat{i}_n$ in the above estimates can always be replaced by $\hat{i}_n + k_n$.

Next, the estimation of the jump parameter Δ will be carried out through

$$d = \frac{1}{\mu} - \frac{1}{\mu+\Delta} = \frac{\Delta}{\mu(\mu+\Delta)} ,$$

$$\hat{d}_n = \frac{N(\hat{i}_n)}{\hat{i}_n} - \frac{N(n)-N(\hat{i}_n)}{n-\hat{i}_n} .$$

Note that $\Delta = d\mu^2/(1-d\mu) = d/(\alpha^2 - d\alpha)$, where $d\mu \neq 1$, since $\Delta \neq \Delta+\mu$. Set

$$\hat{\Delta}_n = \hat{d}_n/\{\hat{\alpha}_n(\hat{\alpha}_n - \hat{d}_n)\} .$$

Theorem 2.3. Assume (H1) ∧ (K1) or (H2) ∧ (K2). Then, as $n \to \infty$,

a) $\hat{d}_n \xrightarrow{P} d$,

b) $\hat{\Delta}_n \xrightarrow{P} \Delta$.

3. Convergence rates. Asymptotic normality.

In order to establish convergence rate results, we need stronger assumptions on the integer sequence $\{k_n : n \geq 1\}$, e.g.

(K3) $k_n \uparrow$, $k_n/n^{1/2} \to 0$, $n^{2/r} \log(n/k_n)/k_n \to 0$ (as $n \to \infty$);

(K4) $k_n \uparrow$, $k_n/n^{1/2} \to 0$, $(\log^2 n) \log(n/k_n)/k_n \to 0$ (as $n \to \infty$).

Note that assumption (K3) implicitly requires a stronger moment assumption, namely

(H4) $EX_1^r < \infty$ $(\exists r > 4)$.

Theorem 3.1. Assume (K3) ∧ (H4) or (K4) ∧ (H2). Then, as $n \to \infty$,

a) $n^{1/2}(\hat{\nu}_n - \nu) \xrightarrow{\mathcal{D}} N(0,\lambda\sigma^2)$,

b) $n^{1/2}(\hat{\alpha}_n - \alpha) \xrightarrow{\mathcal{D}} N(0,\sigma^2/\lambda\mu^4)$,

c) $|\hat{\lambda}_n - \lambda| = O_P(k_n/n)$.

Remark 3.1. Asymptotic confidence intervals of size $O_P(n^{-1/2})$ for $\nu = \lambda\mu$, $\alpha = 1/\mu$ are immediate from Theorem 3.1 a,b. A corresponding confidence interval for $\lambda = \nu\alpha$ can be obtained from the latter via Bonferroni's inequality.

Analogous results are available for the parameters d and Δ.

Theorem 3.2. Assume (K3) ∧ (H4) or (K4) ∧ (H2). Then, as $n \to \infty$,

a) $n^{1/2}(\hat{d}_n - d) \xrightarrow{\mathcal{D}} N(0,\sigma_d^2)$, where $\sigma_d^2 = \sigma^2\{(1/\lambda\mu^4)+(1/(1-\lambda\mu)(\mu+\Delta)^3)\}$.

b) $$n^{1/2}(\hat{\Delta}_n-\Delta) \xrightarrow{\mathcal{D}} N(0,\sigma_\Delta^2),$$

where $\sigma_\Delta^2 = \sigma^2(\mu+\Delta-\lambda\mu\Delta)/\lambda\mu^2(1-\lambda\mu)(\mu+\Delta)$.

Remark 3.2. For consistent estimates $\hat{\vartheta}_1$ of $\vartheta_1 = \sigma/\mu^{3/2}$ and $\hat{\vartheta}_2$ of $\vartheta_2 = \sigma/(\mu+\Delta)^{3/2}$ we refer to Steinebach (1993). Using the latter, the above variances can consistently be estimated as follows:

$$\hat{\sigma}_d^2 := n\hat{\vartheta}_1^2/\hat{i}_n + n\hat{\vartheta}_2^2/(n-\hat{i}_n),$$

$$\hat{\sigma}_\Delta^2 := n\hat{\vartheta}_1^2/\hat{i}_n\hat{\alpha}_n^2 + n\hat{\vartheta}_2^2/(n-\hat{i}_n)\hat{\beta}_n^2,$$

where $\hat{\alpha}_n=N(\hat{i}_n)/\hat{i}_n$, $\hat{\beta}_n=(N(n)-N(\hat{i}_n))/(n-\hat{i}_n)$.

4. Proofs

All proofs make use of the following invariance principle for renewal processes $\{N_n(t):t\geq 0\}$ based on i.i.d. interoccurence-times $\{X_i^{(n)}:i\geq 1\}$ (i.e. under $\Delta=0$):

$$\text{(4.1)} \quad R_n := \sup_{0\leq t\leq n} |N_n(t) - t/\mu - (\sigma/\mu^{3/2})W_n(t)| = \begin{cases} o_P(n^{1/r}) & \text{under (H1)}, \\ O_P(\log n) & \text{under (H2)}, \end{cases}$$

where $\{W_n(t) : t\geq 0\}$ denotes a suitable standard Wiener process (cf. Csörgö, Horváth, Steinebach (1987), Mason and van Zwet (1987)).

Proof of Theorem 2.1. a) The limiting Gumbel type null distribution has recently been derived by Steinebach and Eastwod (1992), Theorems 2.7 and 2.8, applying the above invariance principle together with the well-known extreme value asymptotics for the approximating Gaussian process.

b) It suffices to prove that, under $H_1{:}\Delta\neq 0$,

$$M_n/\{\log(n/k_n)\}^{1/2} \xrightarrow{P} +\infty \qquad (\text{as } n\to\infty).$$

Note that for $i=0,\dots,n-2k_n$,

$$\begin{aligned} M_n &\geq |\, D_{0,n}(i) - D_{\Delta,n}(i) + k_n(\tfrac{1}{\mu} - \tfrac{1}{\mu+\Delta})\,| \,/\, (2k_n)^{1/2} \\ &\geq |d|(k_n/2)^{1/2} - (|D_{0,n}(i)| + |D_{\Delta,n}(i)|)/(2k_n)^{1/2} , \end{aligned}$$

where

$$\begin{aligned} D_{0,n}(t) &:= N(t+k_n) - N(t) - k_n/\mu \;, \\ D_{\Delta,n}(t) &:= N(t+2k_n) - N(t+k_n) - k_n/(\mu+\Delta) \;. \end{aligned}$$

Since, via (4.1) and well-known results for the increments of Wiener processes,

$$\sup_{0\leq t\leq n}\ \sup_{0\leq s\leq 1}\ |N(t+s) - N(t)|/k_n^{1/2} \xrightarrow{P} 0 \quad (n\to\infty),$$

we have with $\tau_n + k_n := S_{[n\lambda]}$,

$$M_n \geq |d|(k_n/2)^{1/2} - (|D_{0,n}(\tau_n)| + |D_{\Delta,n}(\tau_n)|)/(2k_n)^{1/2} + o_P(1).$$

Now, $D_{0,n}(\tau_n)$ depends on $X_1,\dots,X_{[n\lambda]}$, and $D_{\Delta,n}(\tau_n)$ depends on $X_{[n\lambda]+1}+\Delta$, $X_{[n\lambda]+2}+\Delta,\dots$, so both terms correspond to increments of renewal processes *without* change in distribution. In view of part a) or via the increments of Wiener processes,

$$\max\{D_{0,n}(\tau_n), D_{\Delta,n}(\tau_n)\}/k_n^{1/2} = O_P(\{\log(n/k_n)\}^{1/2}).$$

By our assumptions (K1) or (K2), $k_n/\log(n/k_n) \to \infty$, which completes the proof.

Proof of Theorem 2.2. a) Set $i_n = [S_{[n\lambda]}]-2k_n$, and $C_n = a_n^{-1}(\sigma/\mu^{3/2})(b_n+c)$, where a_n, b_n are as in Theorem 2.1, $c\in\mathbb{R}$. Now, if

$$\max_{i\notin(i_n\,,\,i_n+2k_n)} |D_{in}|/(2k_n)^{1/2} \leq C_n \,,$$

but

$$|D_{i_n+k_n,n}|/(2k_n)^{1/2} > C_n \,,$$

then $\qquad i_n \leq \hat{i}_n \leq i_n + 2k_n \,,$

D_{in} and $\hat{i}_n$ as defined in (1.2), (1.3). On the other hand, by a similar argument as in the proof of Theorem 2.1, for any $\varepsilon>0$, we can choose the constant c in C_n such that

$$\liminf_{n\to\infty} P\Big(\max_{i\notin(i_n,i_n+2k_n)} |D_{in}|/(2k_n)^{1/2} \le C_n\Big) \ge 1-\varepsilon ,$$

$$\lim_{n\to\infty} P(|D_{i+k_n,n}|/(2k_n)^{1/2} > C_n) = 1 ,$$

so that

$$\liminf_{n\to\infty} P(i_n \le \hat{i}_n \le i_n+2k_n) \ge 1-\varepsilon.$$

This shows that

$$\hat{i}_n - i_n = O_P(k_n) \quad ,$$

which completes the proof, since $i_n \overset{P}{\sim} S_{[n\lambda]} \overset{P}{\sim} n\lambda\mu$.

b) Since $P(\hat{i}_n \le S_{[n\lambda]}) \longrightarrow 1$ as $n \longrightarrow \infty$, $N(\hat{i}_n)$ essentially corresponds to the number of renewals up to time $\hat{i}_n$ of a renewal process *without* change in distribution. So, the result follows from a) and the law of large numbers for renewal processes.

c) By a) and b), as $n \longrightarrow \infty$,

$$\hat{\lambda}_n = \hat{\alpha}_n \hat{\nu}_n \xrightarrow{P} \alpha\nu = \lambda.$$

Proof of Theorem 2.3. a) Via invariance and corresponding results for the increments of Wiener processes,

$$\text{(4.2)} \qquad I_n := \sup_{0\le t\le n} \sup_{0\le s\le k_n+1} |N(t+s) - N(t)|$$
$$= R_n + O(k_n) + O_P(\{k_n \log(n/k_n)\}^{1/2}),$$

with R_n as in (4.1). By our assumptions on $\{k_n\}$, in any case $I_n/n = o_P(1)$, so that in view of $\hat{i}_n = S_{[n\lambda]} + O_P(k_n)$,

$$\hat{d}_n = \frac{N(S_{[n\lambda]})}{S_{[n\lambda]}} - \frac{N(n)-N(S_{[n\lambda]})}{n-S_{[n\lambda]}} + o_P(1) .$$

Note that $N(n) - N(S_{[n\lambda]}) = \tilde{N}_n(n-S_{[n\lambda]})$, where $\tilde{N}_n$ is the renewal process based on $X_{[n\lambda]+1}+\Delta, X_{[n\lambda]+2}+\Delta,\ldots$, which is independent of $X_1,\ldots,X_{[n\lambda]}$. So the proof can be completed by twice making use of the law of large numbers for renewal processes *without* change in distribution.

b) Combine Theorems 2.2 b) and 2.3 a).

Proof of Theorem 3.1. a) On observing that

$$\hat{i}_n - n(\lambda\mu) = S_{[n\lambda]} - [n\lambda]\mu + O_P(k_n) + O(1),$$

use of the central limit theorem completes the proof.

b) Note that

$$N(\hat{i}_n) - \hat{i}_n/\mu = N(S_{[n\lambda]}) - S_{[n\lambda]}/\mu + O_P(I_n+k_n),$$

where I_n is as in (4.2). Under both assumptions of the theorem,

$$n^{1/2}\frac{N(\hat{i}_n)-\hat{i}_n/\mu}{\hat{i}_n} = \left(\frac{\lambda\sigma^2}{\mu^2}\right)^{1/2}\frac{n}{\hat{i}_n}\frac{[n\lambda]\mu-S_{[n\lambda]}}{(n\lambda\sigma^2)^{1/2}} + o_P(1).$$

Again, the central limit theorem in combination with Theorem 2.2 a) proves the result.

c) With R_n, I_n from (4.1), (4.2) we have

$$\begin{aligned}\hat{\lambda}_n - \lambda &= (N(\hat{i}_n)-n\lambda)/n \\ &= (N(S_{[n\lambda]})-n\lambda) + O_P\{(k_n+I_n)/n\} \\ &= O_P\{(R_n+k_n)/n\} - O_P(k_n/n),\end{aligned}$$

since, by assumption, $R_n = o_P(k_n)$.

Proof of Theorem 3.2. a) With $\alpha=1/\mu$, $\beta=1/(\mu+\Delta)$, $\hat{\alpha}_n$, $\hat{\beta}_n$ as defined in Remark 3.2, one easily checks that

$$n^{1/2}(\hat{d}_n-d) = n^{1/2}(\hat{\alpha}_n-\alpha) - n^{1/2}(\hat{\beta}_n-\beta).$$

By similar arguments as in the proof of Theorem 3.1 b),

$$n^{1/2}(\hat{\alpha}_n-\alpha) - n^{1/2}(\hat{\beta}_n-\beta) = n^{1/2}(\alpha_n-\alpha) - n^{1/2}(\beta_n-\beta) + o_P(1),$$

where

$$\alpha_n = N(S_{[n\lambda]})/S_{[n\lambda]},\ \beta_n = \tilde{N}_n(n-(S_{[n\lambda]})/(n-S_{[n\lambda]}),$$

$\tilde{N}_n$ as in the proof of Theorem 2.3 a), hence independent of α_n. Via (4.1), for every small $\delta>0$,

$$\sup_{t\in nI_\delta} \left| \frac{N(t)-t\beta}{t} - \sigma\beta^{3/2}\frac{\tilde{W}_n(t)}{t} \right| = O_P(R_n/n),$$

where $I_\delta = [1-\lambda\mu-\delta,\ 1-\lambda\mu+\delta]$, and $\{\tilde{W}_n(t):t\geq 0\}$ is a Wiener process, independent of α_n. Since

$$U_n := \min_{t\in nI_\delta} \tilde{W}_n(t)/t \leq \tilde{W}_n(n-S_{[n\lambda]})/(n-S_{[n\lambda]})$$
$$\leq \max_{t\in nI_\delta} \tilde{W}_n(t)/t =: V_n$$

on $\{n-S_{[n\lambda]}\in nI_\delta\}$, and

$$n^{1/2}U_n \stackrel{\mathcal{D}}{=} \min_{s\in I_\delta} \tilde{W}_1(s)/s =: U_\delta \xrightarrow{P} \tilde{W}_1(1-\lambda\mu)/(1-\lambda\mu) \quad (\text{as } \delta\to 0),$$
$$n^{1/2}V_n \stackrel{\mathcal{D}}{=} \max_{s\in I_\delta} \tilde{W}_1(s)/s =: V_\delta \xrightarrow{P} \tilde{W}_1(1-\lambda\mu)/(1-\lambda\mu) \quad (\text{as } \delta\to 0),$$
$$P(n-S_{[n\lambda]}\in nI_\delta) \longrightarrow 1 \quad (\text{as } n\to\infty),$$

we conclude that

$$n^{1/2}(\beta_n-\beta) \xrightarrow{\mathcal{D}} N(0,\sigma^2\beta^3/(1-\lambda\mu)),$$

independently of $n^{1/2}(\alpha_n-\alpha)$. In view of Theorem 3.1 b), this completes the proof.

b) On observing that

$$\hat{\Delta}_n - \Delta = \frac{\hat{\alpha}_n-\hat{\beta}_n}{\hat{\alpha}_n\hat{\beta}_n} - \frac{\alpha-\beta}{\alpha\beta} = \frac{\hat{\alpha}_n-\alpha}{\hat{\alpha}_n\alpha} - \frac{\hat{\beta}_n-\beta}{\hat{\beta}_n\beta},$$

similar arguments as in part a) yield the desired result.

REFERENCES

[1] Chen, X. (1988). Inference in a simple change-point model. *Scientia Sinica*, Ser. A, **31**, 654-667.

[2] Csörgö, M., Horváth, L. (1988). Nonparametric methods for changepoint problems. In: *Handbook of Statistics, Vol.7* (P.R. Krishnaiah & C.R. Rao eds.), 403-425.

[3] Csörgö, M., Horváth, L., Steinebach, J. (1987). Invariance principles for renewal processes. *Ann. Probab.* **15**, 1441-1460.

[4] Hušková, M. (1990a). Asymptotics for robust MOSUM. *Comment. Math. Univ. Carolinae* **31**, 345-356.

[5] Hušková, M. (1990b). Some asymptotic results for robust procedures for testing the constancy of regression models over time. *Kybernetika* **26**, 392-403.

[6] Krishnaiah, P.R., Miao, B.Q. (1988). Review about estimation of change points. In: *Handbook of Statistics, Vol.7*, (P.R. Krishnaiah & C.R. Rao eds.), 375-402.

[7] Mason, D.M., van Zwet, W.R. (1987). A note on the strong approximation to the renewal process. *Publ. Inst. Statist. Univ. Paris* **32**, 81-93.

[8] Steinebach, J., Eastwood, V.R. (1992). On extreme value asymptotics for increments of renewal processes. Proc. "13th French-Belgian Meeting of Statisticians", Lille/France, Nov. 18-20, 1992. *J. Statist. Plann. Infer.* (submitted).

[9] Steinebach, J. (1993). Variance estimation based on invariance principles. Preprint, University of Marburg.

CONDITIONS FOR CONSISTENCY OF MLE'S [1]

Igor Vajda
Institute of Information Theory and Automation
Academy of Sciences of the Czech Republic
Pod vodárenskou věží 4
182 08 Praha 8
CZECH REPUBLIC

Abstract

It is proved that the MLE of a deterministic signal in i. i. d. noise may be inconsistent when the noise distribution has a heavy tail. A simplified version of the necessary and sufficient condition for consistency of MLE's found in Vajda (1993) is formulated for models with i. i. d. observations and with parameters from reasonable metric spaces. It is shown that this condition implies the consistency of all "approximate MLE's" and its contrary implies the inconsistency of all these estimators. Further, a statistical uncertainty of parametric sets is introduced and the mentioned condition is reformulated in terms of this uncertainty.

Key words: Maximum likelihood estimators, consistency, inconsistency, necessary and sufficient conditions, uncertainty of parametric sets.

AMS classification: 62 F 12, 62 M 09

1 Introduction

The maximum likelihood estimation is perhaps the most popular statistical method. There is an extensive statistical literature dealing rigorously with sufficient conditions for their consistency, beginning perhaps with the book of Cramér (1946) and famous paper of Wald (1949). Considerably less attention has been paid to the phenomenon of inconsistency and virtually nothing has been written about practical criteria of inconsistency. Several papers and textbooks describe rather artificial models with inconsistent MLE's. It seems that only two authors so far considered the problem of minimal complete characterization of consistency of MLE's, namely Perlman (1972) and Strasser (1981). The present paper reviews, reinterprets and completes the results of our continuation of the research initiated by them.

[1]This work was supported by GA CR grant 202/93/0232.

First we introduce a statistical model of estimation of a deterministic signal, taking on countably many values, stochastically mixed with an i. i. d. noise over the same sample space. This model is quite common e. g. in the electrical engineering. We prove that in this natural model the MLE exists and may be either consistent or inconsistent, depending on the tail behavior of the noise distribution.

Then we present a condition which is necessary and sufficient for consistency of MLE's in the above considered class of models, as well as in all similar models with discrete parameters. Further, we extend this condition to all estimation models with i. i. d. observations and with parameters from locally compact separable metric spaces, satisfying mild regularity assumptions like the upper semicontinuity of likelihood function. Then we introduce a class $\mathcal{T}$ of approximate MLE's (attaining the ML condition asymptotically) and we show that our condition is sufficient for consistency of all estimators from $\mathcal{T}$ and necessary for consistency of at least one of them. Finally, we formulate our condition equivalently in terms of uncertainties about whether certain subsets of the parameter space contain the unknown parameter. This offers a new and promising connection between the statistical estimation and information theory.

2 Notation, regularity and uncertainty

Consider a locally compact separable metric space Θ with the Borel σ-field $\mathcal{B}$. Consider a sequence of observations $\mathbf{X} = (X_1, \ldots, X_n) \sim P_{\theta_0}^n$, $n = 1, 2, \ldots$ Here P_{θ_0} is from a set $\{P_\theta : \theta \in \Theta\} \ll \mu$ of distributions on an observation space $(\mathcal{X}, \mathcal{A})$. We assume for simplicity that $P_{\theta_1} \neq P_{\theta_2}$ for $\theta_1 \neq \theta_2$ and that the mapping $\theta \mapsto P_\theta$ is continuous in the total variation distance. Further we consider the likelihood $p_\theta = dP_\theta/d\mu$ and we assume that $p_\theta(X)$ is measurable in θ and X. Finally, we suppose that

(A1) $\theta \mapsto p_\theta(X)$ is a. s. upper semicontinuous,

(A2) $\theta \rightarrow p_\theta(X)$ is a separable random process or field,

(A3) the entropy function $H(\theta) = E_{\theta_0} - \log p_\theta(X)$ exists and is finite on Θ.

Instead of the well-known log-likelihood we shall consider the normed "dislikelihood"

(1)
$$f_n(\mathbf{X}, \theta) = -\frac{1}{n} \sum_{i=1}^{n} \log p_\theta(X_i),$$

and we put

(2)
$$f_n(\mathbf{X}, B) = \inf_{\theta \in B} f_n(\mathbf{X}, \theta), \quad B \subset \Theta.$$

It follows from (A2) that, for all open and all closed sets B, the infima $f_n(\mathbf{X}, B)$ may be considered as random variables, i.e. as measurable functions of $\mathbf{X}$. Denote by C_θ the set of all compact sets $B \subset \Theta$ containing θ in their interiors. Now we can formulate our last regularity assumption:

(A4) for every $\theta \in \Theta$ there exists $B \in C_\theta$ such that for at least one n the set entropy rate

$$H_n(B) \triangleq E_{\theta_0} f_n(\mathbf{X}, B)$$

is finite.

Note that in the sequel we consider the set entropy rates $H_n(B)$ for all compact and all open sets $B \subset \Theta$. Since $H_n(B) \leq H(\theta) < \infty$ for every $\theta \in B$, the only novelty of (A4) consists in that $H_n(B)$ cannot be identically $-\infty$ for all $n = 1, 2, \ldots$ Further, for all one-point sets,

$$H_n(\{\theta\}) = H(\theta).$$

The following result follows from Lemma 5 in Vajda (1993). It extends the so-called Shannon–McMillan–Breiman theorem of information theory from one-point sets $B = \{\theta\}$ to more general open or compact sets $B \subset \Theta$.

Lemma 1. If $H_1(\theta) > -\infty$ then, for $n \to \infty$,

$$H_n(B) \uparrow H(B) \triangleq \sup_n H_n(B)$$

and $f_n(\mathbf{X}, B) \to H(B)$ a.s.

It follows from this lemma and from the identity $H_n(\{\theta\}) = H(\theta)$ that, for all one-point sets,

$$H(\{\theta\}) = H(\theta).$$

Thus $H(B)$ is nothing but an extension of the classical uncertainty $H(\theta)$ of information theory from one-element sets $\{\theta\}$ to more general open or closed sets $B \subset \Theta$. In accordance with information theory, one can interpret $H(B)$ as the *uncertainty* about whether the unknown parameter is in B.

3 Conditions for consistency

A sequence $\hat{\theta}_n(\mathbf{X})$ of Θ-valued measurable functions of the sample $\mathbf{X}$ is called an *estimator*. The estimator is said to be *consistent* if $\hat{\theta}_n(\mathbf{X}) \to \theta_0$ in probability. It is said to be *MLE* if, for all n,

$$f_n(\mathbf{X},\, \hat{\theta}_n(\mathbf{X})) = f_n(\mathbf{X}, \Theta).$$

It is said to be an *approximate* MLE (*AMLE*) if there exists a sequence $\varepsilon_n \downarrow 0$ such that

$$\lim_{n\to\infty} \mathsf{P}\left(f_n(\mathbf{X},\, \hat{\theta}_n(\mathbf{X})) \le f_n(\mathbf{X}, \Theta) + \varepsilon_n\right) = 1.$$

Note that if $f_n(\mathbf{X}, \Theta) = -\infty$ with positive probability then f_n on both sides of the last inequality should be replaced by $f_n/(1+f_n)$. Finally, denote by $\mathcal{T}$ the class of all AMLE's. In general, the MLE need not exist even under (A1)–(A4) but, as proved by Vajda (1993), under (A1), (A2) we have $\mathcal{T} \neq \emptyset$.

(a) Finite Θ

Let us consider finite $\Theta = \{\theta_1, \ldots, \theta_m\}$. By Lemma 1 we have for every $\theta \in \Theta$

$$\lim_{n\to\infty} f_n(\mathbf{X}, \theta) = H(\theta) \quad \text{a. s.}$$

Further if $\theta \neq \theta_0$ then $H(\theta) - H(\theta_0)$ is the I-divergence of P_{θ_0} and P_θ,

$$H(\theta) - H(\theta_0) = I(P_{\theta_0}, P_\theta) = \int p_{\theta_0} \log \frac{p_{\theta_0}}{p_\theta}\, \mathrm{d}\mu.$$

Thus the assumption $P_\theta \neq P_{\theta_0}$ implies that this difference is positive or, equivalently, that the uncertainty $H(\theta)$ that the unknown parameter is θ (false) strictly exceeds the uncertainty that the unknown parameter is θ_0 (true). It follows from here that for finite Θ the MLE is consistent.

(b) Countable Θ

The previous conclusion need not be true after replacing the finiteness of Θ by the infinity even if this infinity is countable. To this end let us consider a probability measure μ on $\mathcal{X} = N$ with $\mu(x)$ decreasing positive and put for some $0 < c < 1/2$

$$P_\theta = c\mu + (1-c)\,\delta_\theta, \quad \theta \in \Theta = N,$$

where δ_θ is the Dirac distribution concentrated at $\theta \in N$. Then the likelihood function is given by

$$p_\theta(x) = \begin{cases} c + \frac{1-c}{\mu(\theta)} & \text{if } x = \theta \\ c & \text{otherwise.} \end{cases} \tag{3}$$

In the following theorem and its proof, $\mathbf{X}_{(1)}, \ldots, \mathbf{X}_{(n)}$ denote the order statistics.

Theorem 1. If the probability distribution μ satisfies the condition

$$\lim_{n\to\infty} \mu(\mathbf{X}_{(n)})^{1/n} = 0 \tag{4}$$

then the MLE is inconsistent.

Proof. It follows from (3) that for every $\theta \in N$ and $\mathbf{x}_\theta = (\theta, \theta, \ldots, \theta) \in N^n$

$$f_n(\mathbf{X}, \{1,2,\ldots,\theta\}) \geq f_n(\mathbf{x}_\theta; \theta) = -\log\left(c + \frac{1-c}{\mu(\theta)}\right)$$

while if $\mathbf{X}_{(n-i)} < \mathbf{X}_{(n)}$, $\mathbf{X}_{(n-i+1)} = \mathbf{X}_{(n)}$, then

$$\begin{aligned}
f_n(\mathbf{X}, \mathbf{X}_{(n)}) &= -\frac{n-i}{n}\log c + \frac{i}{n}\log\left(c + \frac{1-c}{\mu(\mathbf{X}_{(n)})}\right) \\
&\leq -\log c - \frac{i}{n}\log\left(1 + \frac{1-c}{c\mu(\mathbf{X}_{(n)})}\right) \\
&= -\log c - \frac{i}{n}\log\left(1 + \frac{1-c}{c\mu(\mathbf{X}_{(n)})}\right) \\
&\leq -\log c - \frac{1}{n}\log\frac{1-c}{c\mu(\mathbf{X}_{(n)})} \\
&\leq -\log c - \frac{1}{n}\log\frac{1}{\mu(\mathbf{X}_{(n)})}.
\end{aligned}$$

It follows from here that for $\theta_n(\mathbf{X}) = \mathbf{X}_{(n)}$ we have

$$\lim_{n\to\infty} f_n(\mathbf{X}, \theta_n(\mathbf{X})) = -\infty \quad \text{a. s.}$$

On the other hand, since

$$\lim_{n\to\infty} \theta_n(\mathbf{X}) = \infty \quad \text{a. s.,}$$

the values $f_n(\mathbf{X}, \{\theta+1, \theta+2, \ldots\})$ are for all sufficiently large n a. s. smaller than $f_n(\mathbf{X}, \{1,2,\ldots,\theta\})$, i. e. the MLE $\hat{\theta}_n(\mathbf{X})$ takes on values from $\{\theta+1, \theta+2, \ldots\}$. Since this is true for any θ, the MLE cannot be consistent *Q. E. D.*

Using the method employed in the proof of Proposition 2 in Vajda (1993), one can prove that (4) holds if

$$\mu(x) = O\left(\frac{1}{x\sqrt{\log x}}\right) \quad \text{as } x \to \infty. \tag{5}$$

This condition is satisfied e. g. by the probability distribution

$$\mu(x) = \frac{1}{\sqrt{\ln(e+x-1)}} - \frac{1}{\sqrt{\ln(e+x)}}, \quad x \in N. \tag{6}$$

The following result follows from Theorem 3 ibid.

Theorem 2. For every model satisfying (A1) – (A4) with a discrete Θ and for the complement $\{\theta_0\}^c = \Theta - \{\theta_0\}$ it holds that the condition

$$\lim_{\varepsilon \downarrow 0} \liminf_{n\to\infty} \mathsf{P}\left(f_n(\mathbf{X}, \theta_0) < f_n(\mathbf{X}, \{\theta_0\}^c) + \varepsilon\right) = 1 \tag{7}$$

is necessary and sufficient for the consistency of MLE. Moreover, (7) is sufficient for consistency of all estimators from $\mathcal{T}$ and necessary for consistency of at least one estimator from $\mathcal{T}$.

It follows from Lemma 1 that (6) is equivalent with

$$H(\theta_0) < H(\{\theta_0\}^c), \tag{8}$$

i.e. with the condition that the uncertainty of the true θ_0 is smaller than the uncertainty that the unknown parameter differs from θ_0. (7) provides an example where, by Theorems 1, 2, (8) is not satisfied. It follows from Theorem 1 and Lemma 1 that in this example

$$H(\theta_0) > H(\{\theta_0\}^c). \tag{9}$$

(c) Arbitrary Θ

Let us now consider arbitrary Θ satisfying (A1) – (A4). The above mentioned Theorem 3 in Vajda (1993) in this case implies the following result:

Theorem 3. The assertion of Theorem 2 remains true in the general case with (7) replaced by

$$\lim_{\varepsilon \downarrow 0} \liminf_{n\to\infty} \mathsf{P}\left(f_n(\mathbf{X}, B) < f_n(\mathbf{X}, B^c) + \varepsilon\right) = 1 \quad \text{for all } B \in \mathcal{C}_{\theta_0}. \tag{10}$$

Analogically as above, it follows from Lemma 1 that (10) is equivalent with the uncertainty condition

$$H(B) < H(B^c) \quad \text{for all } B \in \mathcal{C}_{\theta_0}. \tag{11}$$

As soon as the uncertainty of at least one "true event" $B \in \mathcal{C}_{\theta_0}$ equals or exceeds the uncertainty of the "false event" B^c, MLE as well as all AMLE's become inconsistent. If this happens, then the statistical model is pathological from the information-theoretical point of view (for an example of such a model we refer to (7)). In all non-pathological statistical models the mentioned estimators are consistent.

4 Applications to statistical computations

It follows from Theorem 3 that in statistical models satisfying the regularity (A1)-(A4) there is a dichotomy in the sense that all AMLE's are either consistent or inconsistent. Thus if in these models one is using an approximate method for evaluation of consistent MLE, he or she obtains estimates which are still consistent. This may not be so in the models not satisfying (A1)-(A4). Such models are in fact very risky from the computational points of view.

As an example of the mentioned type might serve the model with the sample space $\mathcal{X} = (e, \infty)$, parameter space $\Theta = (1, 2)$ and with the likelihood function

$$p_\theta(x) = \frac{\theta}{x \ln^{1+\theta} x}.$$

Here the entropy function $H(\theta)$ is identically ∞ and the MLE is

$$\hat{\theta}_n = \left(\frac{1}{n} \sum_{i=1}^{n} \log \log x_i \right)^{-1}.$$

It is easy to verify from the strong law of large numbers that this estimator is strongly consistent. On the other hand, it follows from Theorem 2.3 of Strasser (1981) that in this example there exist inconsistent AMLE's.

References

[1] H. Cramér (1946): *Mathematical Methods of Statistics.* Princeton Univ. Press, Princeton.

[2] M. D. Perlman (1972): On the strong consistency of approximate maximum likelihood estimators. *Proc. VI-th Berkeley Symp. Prob. Math. Stat.*, 263–281.

[3] H. Strasser (1981): Consistency of maximum likelihood and Bayes estimators. *Ann. Statist.* 9, 1107–1113.

[4] I. Vajda (1993): Conditions equivalent and doubly equivalent with consistency of approximate MLE's. *J. Ital. Statist. Soc.* (in print).

[5] A. Wald (1949): Note on the consistency of maximum likelihood estimate. *Ann. Math. Statist.* 20, 595–601.

IMPROVING MAXIMUM QUASI–LIKELIHOOD ESTIMATORS

W. Wefelmeyer
University of Cologne

A quasi–likelihood model for a stochastic process is defined by parametric models for the conditional mean and variance processes given the past. The customary estimator for the parameter is the maximum quasi–likelihood estimator. We discuss some ways of improving this estimator. For simplicity we restrict attention to a Markov chain with conditional mean ϑx given the previous state x, and conditional variance a function of ϑ but not of x. Then the maximum quasi–likelihood estimator is the least squares estimator. It remains consistent if the conditional variance is misspecified. In this case, a better estimator is a weighted least squares estimator, with weights involving nonparametric predictors for the conditional variance. If the conditional variance is correctly specified, a better estimator is given by a convex combination of the least squares estimator and a function of an 'empirical' estimator for the variance.

Key words and phrases. Improved estimator, Markov chain, quasi–likelihood model.

AMS 1991 *subject classification.* 62G05, 62M05.

1. Introduction

A quasi–likelihood model for a stochastic process evolving in time is described by parametric models for the conditional mean and variance processes given the past. The name quasi–likelihood was introduced by Wedderburn (1974) in connection with independent observations. The simplest example are i.i.d. observations with known coefficient of variation. For a related regression model see Amemiya (1973). Other examples are certain generalized linear models reviewed in McCullagh and Nelder (1989), and the Markov regression models of Zeger and Qaqish (1988).

The customary estimator for the parameter in a quasi–likelihood model is the maximum quasi–likelihood estimator. We want to describe ways of improving this estimator. To keep the discussion transparent, we restrict attention to the following simple setting.

Let $X_0, \ldots, X_n$ be realizations from a real-valued, stationary and ergodic Markov chain with transition distribution $Q(x, dy)$ and invariant distribution $\pi(dx)$. Suppose that the conditional mean and variance are of the form

$$\int yQ(x,dy) = \vartheta x, \tag{1.1}$$

$$\int (y-\vartheta x)^2 Q(x,dy) = v(\vartheta), \tag{1.2}$$

but that the transition distribution remains unspecified otherwise.

Our model contains the first-order autoregressive model with a restriction on the error variance:

$$X_i = \vartheta X_{i-1} + \varepsilon_i,$$

where the ε_i are i.i.d. with density p having mean 0 and variance $v(\vartheta)$. For this submodel,

$$Q(x,dy) = p(y - \vartheta x)dy,$$

and (1.1) and (1.2) hold. Note that the additional restriction on the form of the transition distribution also involves the parameter ϑ.

We recall in Section 2 that in our setting the maximum quasi-likelihood estimator coincides with the least squares estimator

$$\hat{\vartheta} = \sum_{i=1}^{n} X_{i-1}X_i \Big/ \sum_{i=1}^{n} X_{i-1}^2.$$

In Section 3 we assume that the conditional variance is misspecified. Then the least squares estimator remains consistent. It can be improved as follows. Let v_{i-1} be a strongly consistent predictor for the conditional variance of X_i given X_{i-1}. We show that the weighted least squares estimator

$$\vartheta^* = \sum_{i=1}^{n} v_{i-1}^{-1} X_{i-1}X_i \Big/ \sum_{i=1}^{n} v_{i-1}^{-1} X_{i-1}^2$$

is better than $\hat{\vartheta}$.

In Section 4 we assume that the conditional variance is correctly specified by (1.2). An 'empirical' estimator for $v(\vartheta)$ is

$$\hat{v} = n^{-1} \sum_{i=1}^{n} (X_i - \hat{\vartheta} X_{i-1})^2.$$

If the function v has a continuous inverse, then $v^{-1}(\hat{v})$ estimates ϑ consistently. We show that a certain random convex combination of $\hat{\vartheta}$ and $v^{-1}(\hat{v})$ is better than $\hat{\vartheta}$. The arguments are heuristic throughout. We always exclude the trivial case that π is concentrated in a single point.

2. The maximum quasi-likelihood estimator

A large class of estimators for the parameter ϑ appearing in (1.1) and (1.2) is obtained from estimating equations

$$\sum_{i=1}^{n} w_\vartheta(X_{i-1})(X_i - \vartheta X_{i-1}) = 0. \tag{2.1}$$

If the weight function $w_\vartheta(x)$ is continuous in ϑ in an appropriate sense, then, by Taylor expansion, a solution of (2.1) is seen to be asymptotically normal, with variance

$$v(\vartheta) \int w_\vartheta(x)^2 \pi(dx) / \Big(\int x w_\vartheta(x) \pi(dx) \Big)^2. \tag{2.2}$$

By the Schwarz inequality, the variance is minimized for the weight function

$$w_\vartheta(x) = x.$$

The minimal variance is

$$v(\vartheta) / \int x^2 \pi(dx). \tag{2.3}$$

The optimal estimator is the *maximum quasi-likelihood estimator*. It coincides with the *least squares estimator*

$$\hat{\vartheta} = \sum_{i=1}^{n} X_{i-1} X_i / \sum_{i=1}^{n} X_{i-1}^2.$$

For general versions of the optimality result see Godambe (1985), Thavaneswaran and Thompson (1986), Hutton and Nelson (1986), Godambe and Heyde (1987). We describe improvements of the maximum quasi-likelihood estimator in Sections 3 and 4.

3. Misspecified conditional variance

Consider the larger model in which only the conditional mean (1.1) is specified, but not the conditional variance (1.2). For the unspecified conditional variance write

$$v(x) = \int (y - \vartheta x)^2 Q(x, dy).$$

We do not express the dependence of v on the unknown transition distribution Q. Note that $v(x)$ depends on ϑ both through the integrator $Q(x, dy)$ and the integrand $(y - \vartheta x)^2$. By Taylor expansion, the estimating equations (2.1) still lead to consistent and asymptotically normal estimators, now with variance

$$\int v(x) w_\vartheta(x)^2 \pi(dx) / \Big(\int x w_\vartheta(x) \pi(dx)\Big)^2. \tag{3.1}$$

In particular, the least squares estimator has asymptotic variance

$$\int v(x) x^2 \pi(dx) / \Big(\int x^2 \pi(dx)\Big)^2. \tag{3.2}$$

On the other hand, by the Schwarz inequality, the variance (3.1) is now minimized for

$$w_\vartheta(x) = x / v(x).$$

The minimal variance is

$$1 / \int v(x)^{-1} x^2 \pi(dx). \tag{3.3}$$

By the Schwarz inequality, this is strictly smaller than the asymptotic variance (3.2) of the least squares estimator unless $v(x)$ is constant π–almost surely.

However, the 'estimating equation'

$$\sum_{i=1}^n v(X_{i-1})^{-1} X_{i-1}(X_i - \vartheta X_{i-1}) = 0,$$

corresponding to the optimal weight function $w_\vartheta(x) = x/v(x)$, depends, through $v(x)$, on the unknown transition distribution $Q(x, dy)$. Hence it cannot be used for estimating ϑ. An estimator with asymptotic variance (3.3) is obtained if one replaces $v(x)$ by a suitable estimator. Specifically, let v_{i-1} be a strongly consistent predictor for $v(X_{i-1})$. For example, take $v_{i-1} = \hat{v}(X_{i-1})$ with $\hat{v}(x)$ a uniformly consistent estimator for the conditional variance $v(x)$. Recent references for such estimators are Truong and Stone (1992) and Roussas and Tran (1992). The estimating equation

$$\sum_{i=1}^n v_{i-1}^{-1} X_{i-1}(X_i - \vartheta X_{i-1}) = 0$$

has an explicit solution, the *weighted least squares estimator*

$$\vartheta^* = \sum_{i=1}^n v_{i-1}^{-1} X_{i-1} X_i \Big/ \sum_{i=1}^n v_{i-1}^{-1} X_{i-1}^2.$$

It is straightforward to check that ϑ^* is consistent and asymptotically normal, with variance (3.3). Our first main result now follows from the inequality between (3.2) and (3.3) obtained above:

The weighted least squares estimator ϑ^* is strictly better than the least squares estimator $\hat{\vartheta}$ unless $v(x)$ is constant π–almost surely.

Wefelmeyer (1993b) shows that ϑ^* is efficient. A sketch of the argument for general semimartingale models is in Wefelmeyer (1993a).

4. Correctly specified conditional variance

We return to the quasi–likelihood model, with parametric models (1.1) and (1.2) for the conditional mean and variance of a Markov chain. Since the maximum quasi–likelihood or least squares estimator remains consistent under misspecification of the conditional variance, it does not use the information about ϑ in (1.2); see Wefelmeyer (1993b). It suggests itself to improve the estimator using an estimator for the (conditional) variance $v(\vartheta)$.

Use the least squares estimator $\hat{\vartheta}$ to estimate the martingale increments $X_i - \vartheta X_{i-1}$ by $X_i - \hat{\vartheta} X_{i-1}$. Then the estimator

$$\hat{v} = n^{-1} \sum_{i=1}^{n} (X_i - \hat{\vartheta} X_{i-1})^2$$

is consistent for $v(\vartheta)$. It admits a stochastic expansion

$$n^{1/2}(\hat{v} - v(\vartheta)) = n^{-1/2} \sum_{i=1}^{n} ((x_i - \vartheta X_{i-1})^2 - v(\vartheta)) + o_P(1).$$

Note that $(X_i - \vartheta X_{i-1})^2 - v(\vartheta)$ is a martingale increment by (1.2). If v has a positive and continuous derivative v' at ϑ, the estimator $v^{-1}(\hat{v})$ is consistent for ϑ and admits a stochastic expansion

$$(4.1) \qquad n^{1/2}(v^{-1}(\hat{v}) - \vartheta) = n^{-1/2} \sum_{i=1}^{n} ((X_i - \vartheta X_{i-1})^2 - v(\vartheta)) / v'(\vartheta) + o_P(1).$$

Note that $v^{-1}(\hat{v})$ solves the estimating equation

$$\sum_{i=1}^{n} ((X_i - \hat{\vartheta} X_{i-1})^2 - v(\vartheta)) = 0.$$

On the other hand, the least squares estimator $\hat{\vartheta}$ admits a stochastic expansion

$$(4.2) \qquad n^{1/2}(\hat{\vartheta} - \vartheta) = n^{-1/2} \sum_{i=1}^{n} X_{i-1}(X_i - \vartheta X_{i-1}) / \int x^2 \pi(dx) + o_P(1).$$

We expect that an appropriate convex combination of $\hat{\vartheta}$ and $v^{-1}(\hat{v})$ is better than $\hat{\vartheta}$. Consider first an arbitrary convex combination

$$\vartheta_a = a\hat{\vartheta} + (1-a)v^{-1}(\hat{v}).$$

The stochastic expansions (4.1) and (4.2) imply that ϑ_a is asymptotically normal. To describe the asymptotic variance, we introduce some notation:

$$\begin{aligned} b &= \int x^2 \pi(dx), \\ c &= \int\int x(y-\vartheta x)^3 Q(x,dy)\pi(dx), \\ d &= \int\int (y-\vartheta x)^4 Q(x,dy)\pi(dx). \end{aligned}$$

The asymptotic variances of $\hat{\vartheta}$ and $v^{-1}(\hat{v})$, respectively, are

$$v(b)/b \quad \text{and} \quad \left(d - v(\vartheta)\right)^2 / v'(\vartheta)^2.$$

The asymptotic covariance between $\hat{\vartheta}$ and $v^{-1}(\hat{v})$ is $c/bv'(\vartheta)$. Hence the asymptotic variance of ϑ_a is

$$a^2 v(\vartheta)/b + 2a(1-a)c/bv'(\vartheta) + (1-a)^2\left(d - v(\vartheta)^2\right)/v'(\vartheta)^2.$$

It is minimized for $a = a^*$ with

$$(4.3) \qquad a^* = 1 \Big/ \Big(1 + \big(v(\vartheta)v'(\vartheta)^2 - cv'(\vartheta)\big) \big/ \big(b(d - v(\vartheta)^2) - cv'(\vartheta)\big)\Big).$$

After some calculation, the minimal variance is obtained as

$$(4.4) \quad \Big(b\big(d - v(\vartheta)^2\big)v(\vartheta) - c^2\Big) \Big/ \Big(bv(\vartheta)v'(\vartheta)^2 - 2bcv'(\vartheta) + b^2\big(d - v(\vartheta)^2\big)\Big).$$

It is easy to check that (4.4) is strictly smaller than (2.3) unless $v(\vartheta)v'(\vartheta) = c$.

The optimal coefficient a^* depends on the unknown parameter ϑ and, through b, c and d, on the unknown transition distribution Q. Hence the optimal convex combination ϑ_{a^*} cannot be used for estimating ϑ. Introduce estimators for b, c and d:

$$\begin{aligned} \hat{b} &= n^{-1}\sum_{i=1}^{n} X_{i-1}^2, \\ \hat{c} &= n^{-1}\sum_{i=1}^{n} X_{i-1}(X_i - \hat{\vartheta}X_{i-1})^3, \\ \hat{d} &= n^{-1}\sum_{i=1}^{n} (X_i - \hat{\vartheta}X_{i-1})^4. \end{aligned}$$

Then the estimator for the optimal a^* defined in (4.3) is given by

$$\hat{a}^* = 1\Big/\Big(1 + \big(v(\hat{\vartheta})v'(\hat{\vartheta})^2 - \hat{c}v'(\hat{\vartheta})\big)/\big(\hat{b}(\hat{d} - v(\hat{\vartheta})^2) - \hat{c}v'(\hat{\vartheta})\big)\Big).$$

The asymptotic variance of ϑ_{a^*} is not changed if one replaces the coefficient a^* by a consistent estimator. We arrive at our second main result:

The estimator $\vartheta^* = \hat{a}^*\hat{\vartheta}+(1-\hat{a}^*)v^{-1}(\hat{v})$ *is strictly better than the maximum quasi-likelihood estimator* $\hat{\vartheta}$ *unless* $v(\vartheta)v'(\vartheta) = c$.

The improvement of ϑ^* over the better of the two estimators $\hat{\vartheta}$ and $v^{-1}(\hat{v})$ is the larger the smaller the correlation and the closer the two variances. It is largest if the covariance c equals 0 and the two variances are equal,

$$v(\vartheta)/b = \big(d - v(\vartheta)\big)^2/v'(\vartheta)^2.$$

Then $a^* = 1/2$, and the variance of ϑ^* is half the variance of $\hat{\vartheta}$ or $v^{-1}(\hat{v})$.

The estimator ϑ^* is, however, not efficient. This is shown in Wefelmeyer (1992). The efficient estimator constructed there is considerably more complicated than ϑ^*. The estimator ϑ^* involves only the 'empirical' estimators $\hat{\vartheta}$, $\hat{b}$, $\hat{c}$ and $\hat{d}$, while the efficient estimator involves predictors for the first four centered conditional moments of Q.

REFERENCES

Amemiya, T. (1973). Regression analysis when the variance of the dependent variable is proportional to the square of its expectation. *J. Amer. Statist. Assoc.* **68** 928–934.

Godambe, V.P. (1985). The foundations of finite sample estimation in stochastic processes. *Biometrika* **72** 419–428.

Godambe, V.P. and Heyde, C.C. (1987). Quasi-likelihood and optimal estimation. *Internat. Statist. Rev.* **55** 231–244.

Hutton, J.E. and Nelson, P.I. (1986). Quasi likelihood estimation for semimartingales. *Stochastic Process. Appl.* **22** 245–257.

McCullagh, P. and Nelder, J.A. (1989). *Generalized Linear Models.* 2nd ed. Chapman and Hall, London.

Roussas, G.G. and Tran, L.T. (1992). Asymptotic normality of the recursive kernel regression estimate under dependence conditions. *Ann. Statist.* **20** 98–120.

Thavaneswaran, A. and Thompson, M.E. (1986). Optimal estimation for semimartingales. *J. Appl. Probab.* **23** 409–417.

Truong, Y.K. and Stone, C.J. (1992). Nonparametric function estimation involving time series. *Ann. Statist.* **20** 77–97.

Wedderburn, R.W.M. (1974). Quasi-likelihood functions, generalized linear models, and the Gauss–Newton method. *Biometrika* **61** 439–447.

Wefelmeyer, W. (1992). Quasi–likelihood models and optimal inference. Preprints in Statistics 138, Mathematical Institute, University of Cologne.

Wefelmeyer, W. (1993a). Estimating functions and efficiency in a filtered model. To appear in: *Proceedings of the Third Finnish–Soviet Symposium on Probability Theory and Mathematical Statistics.*

Wefelmeyer, W. (1993b). Robust and efficient estimators for parameters of the autoregression function of a Markov process. Preprints in Statistics 139, Mathematical Institute, University of Cologne.

GPSR Compliance

The European Union's (EU) General Product Safety Regulation (GPSR) is a set of rules that requires consumer products to be safe and our obligations to ensure this.

If you have any concerns about our products, you can contact us on ProductSafety@springernature.com

In case Publisher is established outside the EU, the EU authorized representative is:

Springer Nature Customer Service Center GmbH
Europaplatz 3
69115 Heidelberg, Germany

Batch number: 10725973

Printed by Printforce, the Netherlands